1983

COMPLEX VARIABLES WITH APPLICATIONS

A. DAVID WUNSCH
University of Lowell

ADDISON-WESLEY PUBLISHING COMPANY
Reading, Massachusetts ∎ Menlo Park, California
London ∎ Amsterdam ∎ Don Mills, Ontario ∎ Sydney

This book is in the
ADDISON-WESLEY SERIES IN MATHEMATICS

Lynn H. Loomis
Consulting Editor

Library of Congress Cataloging in Publication Data

Wunsch, A. David.
    Complex variables with applications.

    Includes index.
    1. Functions of complex variables. I. Title.
QA331.W86   1983   515.9   82–16288
ISBN 0–201–08885–1

ISBN 0–201–08885–1
ABCDEFGHIJ–MA–898765432

*To my family and good friends
who encouraged me with this work*

INTRODUCTION

Someone seeing this book for the first time may well wonder why I have contributed another textbook on complex variables to a world that already has quite a few of them. Some justification is required.

Complex variables and their applications have long been a staple of the undergraduate curriculum for students in engineering and physical science. Until the recent past such courses were taken, often as electives, by advanced undergraduates who already had a certain degree of mathematical sophistication. There are several excellent textbooks available for these courses.

Recently, because of the now widespread use of integral transform techniques in the engineering curriculum, it has become necessary to teach applied complex variables to undergraduates who have not had much mathematics beyond elementary calculus and differential equations. My book is directed to these students. I have tried to create a text that is clear and inviting with lots of worked examples—one that recognizes implicitly that students often fail to remember or understand large quantities of their elementary calculus. This book, therefore, includes some review of elementary topics such as limits, continuity, and series for real variables. I have dispersed physical applications of complex variables throughout the text and slightly favored those from electrical engineering since I believe that electrical engineers now comprise the largest single audience for this subject.

Almost any reader will at times be both offended and pleased by my notation since it borrows both from the pure mathematician and the engineer. Thus for the imaginary operator, I have employed the mathematician's i as opposed to the engineer's j. The i symbolism is now two hundred years old and not to be discarded lightly. However, I have at times used the symbol $\angle$ to indicate the argument of a complex number. This usage is widespread in books on circuit theory; what's more, it has the desirable feature of suggesting an angle.

My choice of topics is fairly standard, although the emphasis I have given is at times unorthodox. There is a good deal of detailed discussion of evaluation of integrals. This is partly because of my concern that students be able to use complex

v

variables together with Fourier and Laplace transforms. I have given some emphasis to branch points and branch cuts, but there is nothing on Riemann surfaces. I don't believe that the labored "cut and paste" discussions of these surfaces are very useful. Though the book is designed for a one-semester course, there is more material here than can be covered in that time. Most lecturers will want to delete, according to their tastes, some portions of the last three chapters.

ACKNOWLEDGMENTS

I should like to thank Professor Roger Baumann of the University of Lowell who suggested that I write this book. My students in Math. 315 at Lowell have given me useful suggestions for this project. I am very grateful to Frances Gedzium and Ruth Leonard for their excellent typing. Catherine Asaro has provided me with invaluable assistance by checking the problem answers, proofreading my manuscript, and preparing the index. Most of the book was written at the Cabot Science Library at Harvard and at the Boston Athenaeum. I am thankful for their fine facilities.

I am indebted to Nancy Kralowetz and Steve Quigley, my two editors at Addison-Wesley. They encouraged me during a much longer project than any of us imagined. Several consultants to Addison-Wesley offered very useful suggestions while the book was evolving. Among those whom I should very much like to thank are Steven K. Ingram, Norwich University, Northfield, Vermont; James R. King, University of Washington, Seattle, Washington; Alan C. Lazer, University of Cincinnati, Cincinnati, Ohio; Evelyn M. Silvia, University of California, Davis, California; Donald J. Wright, University of Cincinnati, Cincinnati, Ohio; and Robert C. Carson, The University of Akron, Akron, Ohio.

Any author, I am no exception, makes errors. I should be happy to have mine pointed out to me by anyone who chooses to correspond.

A. David Wunsch
Dept. of Electrical Engineering
University of Lowell
Lowell, Massachusetts

April 1982

CONTENTS

CHAPTER **1 COMPLEX NUMBERS**

1.1 INTRODUCTION

In order to prepare ourselves for a discussion of complex numbers, complex variables, and, ultimately, functions of a complex variable, let us review a little of the previous mathematical education of a hypothetical reader.

A young child learns early about those whole numbers that we with more sophistication call the positive integers. Zero, another integer, is also a concept that the child soon grasps.

Adding and multiplying two integers, the result of which is always a positive integer or zero, is learned in elementary school. Subtraction is studied, but the problems are carefully chosen; 5 minus 2 might, for instance, be asked but not 2 minus 5. The answers are always positive integers or zero.

Perhaps several years later this student is asked to ponder 2 minus 5 and similar questions. Negative integers, a seemingly logical extension of the system containing the positive integers and zero, are now required. Nevertheless, to avoid some inconsistencies one rule must be accepted that does not appeal directly to intuition, namely, $(-1)(-1) = 1$. The reader has probably forgotten how artificial this equation at first seems.

With the set of integers (the positive and negative whole numbers and zero) any feat of addition, subtraction, or multiplication can be performed by the student, and the answer will still be an integer. Some simple algebraic equations such as $m + x = n$ (m and n are any integers) can be solved for x, and the answer will be an integer. However, other algebraic equations—the solutions of which involve division—present difficulties. Given the equation $mx = n$, the student sometimes obtains an integer x as a solution. Otherwise the student must employ a kind of number called a fraction, which is specified by writing a pair of integers in a particular order; the fraction n/m is the solution of the equation just given if $m \neq 0$.

A collection of all the numbers that can be written as n/m, where n and m are any integers (excluding $m = 0$), is called the rational number system since it is based

1

on the *ratio* of whole numbers. The rationals include both fractions and whole numbers. Knowing this more sophisticated system, our hypothetical student can solve any linear algebraic equation. The result is a rational number.

Later, perhaps in our student's early teens, irrational numbers are learned. They come from two sources: algebraic equations with exponents, the quadratic $x^2 = 2$, for example; and, geometry, the ratio of the circumference to the diameter of a circle, π, for example.

For $x^2 = 2$ the unknown x is neither a whole number nor a fraction. The student learns that x can be written as a decimal expression 1.41421356... requiring an infinite number of places for its complete specification. The digits do not repeat themselves according to any pattern. The number π also requires an infinite number of nonrepeating digits when written as a decimal.

Thus for the third time the student's repertoire of numbers must be expanded. The rationals are now supplemented by the irrationals, namely, all the numbers that must be represented by infinite nonrepeating decimals.[†] The totality of these two kinds of numbers is known as the *real number system*.

The difficulties have not ended, however. Our student, given the equation $x^2 = 2$, obtains the solution $x = \pm 1.414...$, but given $x^2 = -2$ or $x^2 = -1$, the student faces a new complication since no real number times itself will yield a negative real number. To cope with this dilemma, a larger system of numbers—the *complex system*—is usually presented in high school. This system will yield solutions not only to equations like $x^2 = -1$ but also to complicated polynomial equations of the form

$$a_n z^n + a_{n-1} z^{n-1} + \cdots + a_0 = 0,$$

where $a_0, a_1, \ldots, a_n$ are complex numbers, n is a positive integer, and z is an unknown.

The following discussion, presented partly for the sake of completeness, should overlap much of what the reader probably already knows about complex numbers.

A complex number, let us call it z, is a number that is written in the form

$$z = a + ib \quad \text{or} \quad z = a + bi.$$

The letters a and b represent real numbers, and the significance of i will soon become clear.[‡]

We say that a is the real part of z and that b is the imaginary part. This is frequently written as

$$a = \text{Re}(z), \qquad b = \text{Im}(z).$$

Note that *both* the real part *and* the imaginary part of the complex number are real numbers. The complex number $-2 + 3i$ has a real part -2 and an imaginary part of 3.

[†] For a proof that $\sqrt{2}$ is irrational see G. B. Thomas, *Calculus and Analytic Geometry*, 2d ed. (Reading, Mass.: Addison-Wesley, 1958), p. 619.

[‡] Most electrical engineering texts use j instead of i since i is reserved to mean current. However, mathematics books invariably use i.

Two complex numbers are said to be equal if, and only if, the real part of one equals the real part of the other and the imaginary part of one equals the imaginary part of the other.

That is, if

$$z = a + ib, \qquad w = c + id, \tag{1.1-1}$$

and

$$z = w,$$

then

$$a = c, \qquad b = d.$$

We do not establish a hierarchy of size for complex numbers; if we did, the familiar inequalities used with real numbers would not apply. Using real numbers we can say, for example, that $5 > 3$, but it makes no sense to assert that either $(1 + i) > (2 + 3i)$ or $(2 + 3i) > (1 + i)$. An inequality like $a > b$ will always imply that both a and b are real numbers.

The words *positive* and *negative* are never applied to complex numbers, and the use of these words implies that a real number is under discussion.

We add and subtract the two complex numbers in Eq. (1.1–1) as follows:

$$z + w = (a + ib) + (c + id) = (a + c) + i(b + d), \tag{1.1-2}$$
$$z - w = (a + ib) - (c + id) = (a - c) + i(b - d). \tag{1.1-3}$$

Their product is defined by

$$zw = (a + ib)(c + id) = (ac - bd) + i(ad + bc). \tag{1.1-4}$$

The results in Eqs. (1.1–2) through (1.1–4) are obtainable through the use of the ordinary rules of algebra and one additional crucial fact: When doing the multiplication $(a + ib)(c + id)$ we must take

$$i \cdot i = i^2 = -1. \tag{1.1-5}$$

Real numbers obey the commutative, associative, and distributive laws. We readily find, with the use of the definitions shown in Eqs. (1.1–2) and (1.1–4), that complex numbers do also. Thus if w, z, and q are three complex numbers, we have

commutative law:

$$w + z = z + w \quad \text{(for addition)}, \tag{1.1-6}$$
$$wz = zw \quad \text{(for multiplication)},$$

associative law:

$$w + (z + q) = (w + z) + q \quad \text{(for addition)}, \tag{1.1-7}$$
$$w(zq) = (wz)q \quad \text{(for multiplication)},$$

distributive law:

$$w(z + q) = wz + wq. \tag{1.1-8}$$

Now, consider two complex numbers, z and w, whose imaginary parts are zero. Let $z = a + i0$ and $w = c + i0$. The sum of these numbers is

$$z + w = (a + c) + i0,$$

and for their product we find

$$(a + i0)(c + i0) = ac + i0.$$

These results show that those complex numbers whose imaginary parts are zero behave mathematically like real numbers. We can think of the complex number $a + i0$ as the real number a in different notation. The complex number system therefore *contains* the real number system.

We speak of complex numbers of the form $a + i0$ as "purely real" and, for historical reasons, those of the form $0 + ib$ as "purely imaginary." The term containing the zero is usually deleted in each case so that $0 + i$ is, for example, written i.

A multiplication (or addition) involving a real number and a complex number is treated, by definition, as if the real number were complex but with zero imaginary part. For example, if k is real,

$$(k)(a + ib) = (k + i0)(a + ib) = ka + ikb. \tag{1.1-9}$$

The complex number system has quantities equivalent to the zero and unity of the real number system. The expression $0 + i0$ plays the role of zero since it leaves unchanged any complex number to which it is added. Similarly, $1 + i0$ functions as unity since a number multiplied by it is unchanged. Thus $(a + ib)(1 + i0) = a + ib$.

Expressions such as $z^2, z^3, \ldots$, imply successive self-multiplication by z and can be calculated algebraically with the help of Eq. (1.1–5). Thus to cite some examples:

$$i^3 = i^2 \cdot i = -i, \qquad i^4 = i^3 \cdot i = -i \cdot i = 1, \qquad i^5 = i^4 \cdot i = i,$$

$$(1 + i)^3 = (1 + i)^2(1 + i) = (1 + 2i - 1)(1 + i) = 2i(1 + i) = -2 + 2i.$$

We still have not explained why the somewhat cumbersome complex numbers can yield solutions to problems unsolvable with the real numbers. Consider, however, the quadratic equation $z^2 + 1 = 0$, or $z^2 = -1$. As mentioned earlier, no real number provides a solution. Let us rewrite the problem in complex notation:

$$z^2 = -1 + i0. \tag{1.1-10}$$

We know that $(0 + i)^2 = i^2 = -1 + i0$. Thus $z = 0 + i$ (or $z = i$) is a solution of Eq. (1.1–10). Similarly, one verifies that $z = 0 - i$ (or $z = -i$) is also. We can say that $z^2 = -1$ has solutions $\pm i$. Thus we assert that in the complex system -1 has two square roots: i and $-i$.

In the case of the equation $z^2 = -N$, where N is a nonnegative real number, we can proceed in a similar fashion and find that $z = \pm i\sqrt{N}$.[†] Hence, the complex system is capable of yielding two square roots for any negative real number. Both roots are purely imaginary.

[†] The expression $\sqrt{N}$, where N is a positive real number, will mean the *positive* square root of N, and $\sqrt[n]{N}$ will mean the positive nth root of N.

For the quadratic equation

$$az^2 + bz + c = 0 \quad (a \neq 0) \quad \text{and} \quad a, b, c \text{ are real numbers,} \quad (1.1\text{--}11)$$

we are initially taught the solution

$$z = \frac{-b \pm \sqrt{b^2 - 4ac}}{2a} \tag{1.1--12}$$

provided that $b^2 \geq 4ac$. With our complex system this restriction is no longer necessary. Completing the square in Eq. (1.1–11) and using $i^2 = -1$, we have for $b^2 \leq 4ac$

$$z = \frac{-b \pm i\sqrt{(4ac - b^2)}}{2a}.$$

We will soon see that a, b, and c in Eq. (1.1–11) can themselves be complex, and we can still solve Eq. (1.1–11) in the complex system.

Having enlarged our number system so that we now use complex numbers, with real numbers treated as a special case, we will find that there is no algebraic equation whose solution requires an invention of any new numbers.

The story presented earlier of a hypothetical student's growing mathematical sophistication in some ways parallels the actual expansion of the number system by mathematicians over the ages. Complex numbers were "discovered" by people trying to solve certain algebraic equations. For example, in 1545, Girolamo Cardan, an Italian mathematician, attempted to find two numbers whose sum is 10 and whose product is 40. He concluded by writing $40 = (5 + \sqrt{-15})(5 - \sqrt{-15})$, a result he considered meaningless. Later, mathematicians began referring to numbers such as $a + \sqrt{-b}$ as "imaginary," a term still used today in lieu of "complex."

Although still uncomfortable with the concept of imaginary numbers, mathematicians had, by the end of the eighteenth century, made rather heavy use of them in both physical and abstract problems. The Swiss mathematician Leonhard Euler invented in 1779 the i notation for $\sqrt{-1}$, which we still use today, and by 1799 Karl Friedrich Gauss had used complex numbers in his proof of the fundamental theorem of algebra. Finally, in 1835, Sir William Rowan Hamilton presented the modern rigorous theory of complex numbers, which dispenses entirely with the symbols i and $\sqrt{-1}$. We will briefly look at this method in the next section.

EXERCISES

1. Consider the hierarchy of increasingly sophisticated number systems:
   integers
   rational numbers
   real numbers
   complex numbers

   For each of the following equations what is the most elementary number system, of the four listed above, in which a solution for x is obtainable?

   a) $3x + 5 = 0$ d) $3x + 6 = 0$ g) $x^2 + 3x + 2 = 0$

   b) $2x^2 - 1 = 0$ e) $3x^2 + 27 = 0$ h) $x^2 + 2x + 3 = 0$

   c) $2x^2 + 1 = 0$ f) $3x^2 - 27 = 0$

2. An infinite decimal such as $e = 2.718281...$ is an irrational number since there is no repetitive pattern in the successive digits. However, an infinite decimal such as $12.1212121...$ is a rational number. Because the digits do repeat in a specific manner, we can write this number as the ratio of two integers, as the following steps will show.

    First we rewrite the number as $12[1.01010101...]$ or $12[1 + 10^{-2} + 10^{-4} + 10^{-6} + \cdots]$.

   a) Recall from your knowledge of infinite geometric series that $1/(1 - r) = 1 + r + r^2...$, where r is a real number such that $-1 < r < 1$. Sum the series $[1 + 10^{-2} + 10^{-4} + 10^{-6} + \cdots]$.

   b) Use the result of part (a) to show that $12.1212...$ equals $1200/99$. Verify this answer by division on a pocket calculator.

   c) Using the ideas given above, express $1.23123123123...$ as the ratio of two integers. Check your result with a calculator.

3. Find the numerical value of each of the following and express it in the form $a + ib$, where a and b are real numbers.

   a) $(4 + i) + (6 - 3i)$ d) $(2 - i)^3$

   b) $(3 - i)(2 + i)$ e) $(1 + i) + (3 + 4i)^2$

   c) $(2 - i)^2$ f) $\text{Im}((1 + i) + (3 + 4i)^2)$

4. Let $z_1 = x_1 + iy_1$ and $z_2 = x_2 + iy_2$ be two complex numbers. Which of the following statements are true in general?

   a) $\text{Re}(z_1 + z_2) = \text{Re}(z_1) + \text{Re}(z_2)$ c) $\text{Im}(z_1 z_2) = \text{Im}(z_1) \, \text{Im}(z_2)$

   b) $\text{Re}(z_1 z_2) = \text{Re}(z_2 z_1)$ d) $z_1 \, \text{Re} \, z_2 = z_2 \, \text{Re} \, z_1$

5. Let z be a complex number. Show that

   a) $\text{Re}(iz) = -\text{Im} \, z$,   c) $\text{Re}(z^2) = (\text{Re} \, z)^2 - (\text{Im} \, z)^2$,

   b) $\text{Im}(iz) = \text{Re} \, z$,   d) $\text{Im}(z^2) = 2(\text{Re} \, z)(\text{Im} \, z)$.

6. a) Consider i^n, where n is a positive integer. What are the four possible numerical values of this expression?

   b) If $n > 4$, show that $i^n = i^{n-4}$.

   c) Show that if n is even, $i^n = 1$ if n is exactly divisible by 4, and $i^n = -1$ if n is not exactly divisible by 4.

   d) What are i^{961}, i^{959}, i^{12879}, i^{346673}?

7. Determine $(1 + i)^{375}$ in the form $a + ib$. *Hint:* $(1 + i)^{375} = (1 + i)^{374}(1 + i)$, and note that $(1 + i)^2 = 2i$. See also Exercise 6.

8. Consider the following equations, where x and y are real numbers. Obtain the values of x and y that provide solutions.

   a) $(1 + i)^2(x + iy) = x + i$ e) $\cos y + i \sin x = \cos x + i \sin y$

   b) $x^2 - y^2 + 2x + i(2xy + 2y) = 0$ f) $x^2 - y^2 + i2xy - 1 - i = 0$

   c) $(x + iy)^2 = i(x - iy)$ g) $(x + iy)^3 = 1$

   d) $x - iy = i(x^2 + y^2)$

1.2 MORE PROPERTIES OF COMPLEX NUMBERS

A pair of complex numbers are said to be *conjugates* of each other if they have identical real parts and imaginary parts that are identical except for their being opposite in sign.

If $z = a + ib$, then the conjugate of z, written $\bar{z}$ or z^*, is $a - ib$. Thus $\overline{(-2 + i4)}$ is $-2 - i4$. Note that $(\bar{\bar{z}}) = z$; if we take the conjugate of a complex number twice, the number emerges unaltered.

Other important identities for complex numbers $z = a + ib$ and $\bar{z} = a - ib$, are

$$z + \bar{z} = 2a + i0 = 2\,\mathrm{Re}\,z = 2\,\mathrm{Re}\,\bar{z}, \tag{1.2-1}$$

$$z - \bar{z} = 0 + 2ib = 2i\,\mathrm{Im}\,z. \tag{1.2-2}$$

Therefore the sum of a complex number and its conjugate is twice the real part of the original number. If from a complex number we subtract its conjugate, we obtain a quantity that has a real part of zero and an imaginary part twice that of the imaginary part of the original number.

The product of a complex number and its conjugate is a real number. Thus if $z = a + ib$ and $\bar{z} = a - ib$, we have

$$z\bar{z} = (a + ib)(a - ib) = a^2 + b^2 + i0 = a^2 + b^2. \tag{1.2-3}$$

This fact is particularly useful when we seek to derive the quotient of a pair of complex numbers. Suppose, for the three complex numbers α, z, and w,

$$\alpha z = w, \qquad z \neq 0, \tag{1.2-4}$$

where $z = a + ib$, $w = c + id$. Quite naturally, we call α the quotient of w and z, and write $\alpha = w/z$. To determine the value of α, we multiply both sides of Eq. (1.2–4) by $\bar{z}$. We have

$$\alpha(z\bar{z}) = w\bar{z}. \tag{1.2-5}$$

Now $z\bar{z}$ is a real number. We can remove it from the left in Eq. (1.2–5) by multiplying the entire equation by another real number $1/(z\bar{z})$. Thus $\alpha = w\bar{z}/(z\bar{z})$, or

$$\frac{w}{z} = \frac{w\bar{z}}{z\bar{z}}. \tag{1.2-6}$$

This formula says that to compute $w/z = (c + id)/(a + ib)$ we should multiply the numerator and the denominator by the conjugate of the denominator, that is,

$$\frac{c + id}{a + ib} = \frac{(c + id)(a - ib)}{(a + ib)(a - ib)} = \frac{(ac + bd) + i(ad - bc)}{a^2 + b^2},$$

or

$$\frac{c + id}{a + ib} = \frac{ac + bd}{a^2 + b^2} + i\frac{(ad - bc)}{a^2 + b^2}. \tag{1.2-7}$$

Using Eq. (1.2–7) with $c = 1$ and $d = 0$, we can obtain a useful formula for the reciprocal of $a + ib$; that is,

$$\frac{1}{a + ib} = \frac{a}{a^2 + b^2} - \frac{ib}{a^2 + b^2}. \tag{1.2-8}$$

Note in particular that with $a = 0$, and $b = 1$, we find $1/i = -i$. This result is easily checked since we know that $1 = (-i)(i)$.

Since all the preceding expressions can be derived by application of the conventional rules of algebra, and the identity $i^2 = -1$, to complex numbers, it follows that other rules of ordinary algebra, such as the following, can be applied to complex numbers.

$$\frac{z_1}{z_2} = z_1\left(\frac{1}{z_2}\right), \qquad \frac{1}{z_1 z_2} = \left(\frac{1}{z_1}\right)\left(\frac{1}{z_2}\right), \qquad \frac{z_1 z_2}{z_3 z_4} = \left(\frac{z_1}{z_3}\right)\left(\frac{z_2}{z_4}\right) \qquad (1.2\text{--}9)$$

There are a few other properties of the conjugate operation that we should know about.

The conjugate of the sum of two complex numbers is the sum of their conjugates.

Thus if $z_1 = x_1 + iy_1$ and $z_2 = x_2 + iy_2$, then

$$(\overline{z_1 + z_2}) = (x_1 + x_2) - i(y_1 + y_2) = (x_1 - iy_1) + (x_2 - iy_2) = \bar{z}_1 + \bar{z}_2.$$

A similar statement applies to the difference of two complex numbers and also to products and quotients, as will be proved in the exercises. In summary:

$$(\overline{z_1 + z_2}) = \bar{z}_1 + \bar{z}_2, \qquad (1.2\text{--}10a)$$

$$(\overline{z_1 - z_2}) = \bar{z}_1 - \bar{z}_2, \qquad (1.2\text{--}10b)$$

$$\overline{z_1 z_2} = \bar{z}_1 \bar{z}_2, \qquad (1.2\text{--}10c)$$

$$\overline{\left(\frac{z_1}{z_2}\right)} = \frac{\bar{z}_1}{\bar{z}_2}. \qquad (1.2\text{--}10d)$$

Formulas such as these can sometimes save us some labor. For example, consider

$$\frac{1 + i}{3 - 4i} + \frac{1 - i}{3 + 4i} = x + iy.$$

There can be a good deal of work involved in finding x and y. Note, however, from Eq. (1.2–10d) that the second fraction is the conjugate of the first. Thus from Eq. (1.2–1) we see that $y = 0$, whereas $x = 2\,\text{Re}((1 + i)/(3 - 4i))$. The real part of $(1 + i)/(3 - 4i)$ is found from Eq. (1.2–7) to be $(3 - 4)/25 = -1/25$. Thus the required answer is $x = -(2/25)$.

Equations (1.2–10a–d) can be extended to more than two complex numbers, for example,

$$\overline{z_1 z_2 z_3} = \overline{z_1 z_2}\,\bar{z}_3 = \bar{z}_1 \bar{z}_2 \bar{z}_3.$$

Euler's notation, $z = a + ib$, has never been entirely palatable to pure mathematicians. One reason is the presence of the plus sign. It is not clear whether this type of addition is the same as occurs in conventional algebra. Also, one might ask whether or not the implied multiplication between i and b is identical to ordinary multiplication.

A formulation of complex number theory, which dispenses with Euler's notation and the i operator, was presented in the mid-nineteenth century by the Irish mathematician William Rowan Hamilton. Any student who has done Fortran computer programming with complex numbers knows that neither the program nor the machine uses i. The approach used is identical to Hamilton's.

In his method a complex number is defined as a pair of real numbers expressed in a particular order. If this seems artificial, recall that a fraction is also expressed as a pair of numbers stated in a certain order. Hamilton's complex number z is written (a, b), where a and b are real numbers. The order is important (as it is for fractions) and such a number, in general, is not the same as (b, a). In the ordered pair (a, b), we call the first number, a, the real part of the complex number and the second, b, the imaginary part. This kind of expression is often called a *couple*. Two such complex numbers (a, b) and (c, d) are said to be equal: $(a, b) = (c, d)$ if and only if $a = c$ and $b = d$. The sum of these two complex numbers is defined by

$$(a, b) + (c, d) = (a + c, b + d), \tag{1.2-11}$$

and their product is computed as

$$(a, b) \cdot (c, d) = (ac - bd, ad + bc). \tag{1.2-12}$$

The product of a real number k and the complex number (a, b) is defined by $k(a, b) = (ka, kb)$.

Consider now all the couples whose second number in the pair is zero. Such couples handle mathematically like the ordinary real numbers. For instance, we have $(a, 0) + (c, 0) = (a + c, 0)$. Also, $(a, 0) \cdot (c, 0) = (ac, 0)$. Therefore we will say that the real numbers are really those couples, in Hamilton's notation, for which the second element is zero.

Another important identity involving ordered pairs, which is easily proved from Eq. (1.2-12), is

$$(0, 1) \cdot (0, 1) = (-1, 0). \tag{1.2-13}$$

It implies that $z^2 + 1 = 0$ when written with couples has a solution. Thus

$$z^2 + (1, 0) = (0, 0) \tag{1.2-14}$$

is satisfied by $z = (0, 1)$ since

$$(0, 1)(0, 1) + (1, 0) = (-1, 0) + (1, 0) = (0, 0).$$

The student should readily see the analogy between the $a + ib$ and the (a, b) notation. The former terminology will more often be used in these pages.

EXERCISES

1. Compute the numerical values of the following expressions. Give the answers in the form $a + ib$, where a and b are real numbers.

   a) $\dfrac{1}{1 + i}$

   b) $\dfrac{3 + 4i}{(1 + i)(2 - i)}$

   c) $\left(\dfrac{1}{1 + i}\right)\left(\dfrac{1}{1 - i}\right) + 2 + i$

   d) $\mathrm{Re}\left(\dfrac{2 + i}{1 - i}\right)$

   e) $\dfrac{\mathrm{Im}[(1 + i)(2 + i)]}{3 - 4i}$

   f) $\dfrac{1}{(1 + i)^{37}}$

g) $\operatorname{Im} \dfrac{(1 + i)(2 + i)}{\operatorname{Re}(3 - 4i)}$

h) $\left[\dfrac{3 + 4i}{3 - 4i} - \dfrac{3 - 4i}{3 + 4i} \right]^3$

2. Let $z_1 = x_1 + iy_1$, $z_2 = x_2 + iy_2$. Without using Eqs. (1.2–10a–d) show that

a) $\overline{(z_1 - z_2)} = \bar{z}_1 - \bar{z}_2$,

d) $\overline{\left(\dfrac{1}{z_1} \right)} = \dfrac{1}{\bar{z}_1}$,

b) $\overline{(z_1 z_2)} = \bar{z}_1 \bar{z}_2$,

e) $\operatorname{Re}(z_1 z_2) = \operatorname{Re}(\bar{z}_1 \bar{z}_2)$,

c) $\overline{\left(\dfrac{z_1}{z_2} \right)} = \dfrac{\bar{z}_1}{\bar{z}_2}$,

f) $\operatorname{Im}(z_1 z_2) = -\operatorname{Im}(\bar{z}_1 \bar{z}_2)$.

3. Prove that $(\bar{z})^n = \overline{(z)^n}$ when

a) n is a positive integer,

b) n is a negative integer. Use the definition $z^n = 1/(z^{-n})$ when n is negative.

4. Consider the complex numbers $z_1 = x_1 + iy_1$ and $z_2 = x_2 + iy_2$. Which of the following statements are true in general?

a) $\dfrac{\operatorname{Re}(z_1)}{\operatorname{Re}(\bar{z}_2)} = \operatorname{Re}\left(\dfrac{z_1}{\bar{z}_2} \right)$

e) $\operatorname{Im}\left(\dfrac{1}{z_1 \bar{z}_2} \right) = \operatorname{Im}\left(\dfrac{1}{\bar{z}_1 z_2} \right)$

b) $\dfrac{1}{z_1} + \dfrac{1}{\bar{z}_1} = 2 \operatorname{Re}\left(\dfrac{1}{z_1} \right)$

f) $z_1 \bar{z}_2 (\bar{z}_1 + z_2) = \overline{z_1 z_2 (z_1 + \bar{z}_2)}$

c) $\operatorname{Im}\left(\dfrac{1}{z_1 \bar{z}_1} \right) = -\operatorname{Im}\left(\dfrac{1}{z_1 \bar{z}_1} \right)$

g) $(z_1 - \bar{z}_1)(z_2 + \bar{z}_2) = 4 \operatorname{Re}(z_2) \operatorname{Im}(z_1)$

d) $\operatorname{Re}\left(\dfrac{1}{z_1 \bar{z}_2} \right) = \operatorname{Re}\left(\dfrac{1}{\bar{z}_1 z_2} \right)$

5. Let z_1, z_2, and z_3 be three arbitrary complex numbers. Which of the following statements are true in general?

a) $\overline{z_1(z_2 + z_3)} = \bar{z}_1(\bar{z}_2 + \bar{z}_3)$

b) $\dfrac{\overline{z_1 z_2}}{\bar{z}_3} = \dfrac{\bar{z}_1 z_2}{z_3}$

c) $z_3 \operatorname{Re}[z_1 - \bar{z}_1] = z_1 \operatorname{Re}[z_2 - \bar{z}_2]$

6. Suppose, following Hamilton, we regard a complex number as a pair of ordered real numbers. We want the appropriate definition for the quotient $(c, d)/(a, b)$. Let us put $(c, d)/(a, b) = (e, f)$, where e and f are real numbers to be determined.

   If our definition is to be plausible, then

$$(c, d) = (a, b) \cdot (e, f).$$

a) Perform the indicated multiplication by using the product rule for couples.

b) Equate corresponding members (real numbers and imaginary numbers) on both sides of the equation resulting from part (a).

c) In part (b) a pair of simultaneous linear equations were obtained. Solve these equations for e and f in terms of a, b, c, and d. How does the result compare with that of Eq. (1.2–7)?

1.3   COMPLEX NUMBERS AND THE ARGAND PLANE

Modulus

The *magnitude* or *modulus* of a complex number is the positive square root of the sums of the squares of its real and imaginary parts.

If the complex number is z, then its modulus is written $|z|$. If $z = x + iy$, we have, from the definition,

$$|z| = \sqrt{x^2 + y^2}. \tag{1.3-1}$$

The modulus of a complex number is a nonnegative real number. Although we cannot say one complex number is greater (or less) than another, we can say the modulus of one number exceeds that of another; for example, $|4 + i| > |2 + 3i|$ since

$$|4 + i| = \sqrt{16 + 1} = \sqrt{17} > |2 + 3i| = \sqrt{4 + 9} = \sqrt{13}.$$

A complex number has the same modulus as its conjugate because

$$|\bar{z}| = \sqrt{x^2 + (-y)^2} = \sqrt{x^2 + y^2} = |z|.$$

The product of a complex number and its conjugate is the squared modulus of the complex number. To see this note that $z\bar{z} = x^2 + y^2$. Thus, from Eq. (1.3–1),

$$z\bar{z} = |z|^2. \tag{1.3-2}$$

The square root of this expression is also useful:

$$|z| = \sqrt{z\bar{z}}. \tag{1.3-3}$$

The modulus of the product of two complex numbers is equal to the product of their moduli.

We will now prove this. Let the numbers be z_1 and z_2. Their product is $z_1 z_2$. Let $z = z_1 z_2$ in Eq. (1.3–3). We then have

$$|z_1 z_2| = \sqrt{z_1 z_2 (\overline{z_1 z_2})} = \sqrt{z_1 z_2 \bar{z}_1 \bar{z}_2} = \sqrt{z_1 \bar{z}_1} \sqrt{z_2 \bar{z}_2},$$

and finally

$$|z_1 z_2| = |z_1|\, |z_2|. \tag{1.3-4}$$

Similarly, $|z_1 z_2 z_3| = |z_1 z_2|\, |z_3| = |z_1|\, |z_2|\, |z_3|$.

The modulus of a product of numbers is the product of the moduli of each factor, regardless of how many factors are present.

It is left as an exercise to show

$$\left| \frac{z_1}{z_2} \right| = \frac{|z_1|}{|z_2|}. \tag{1.3-5}$$

The modulus of the quotient of two complex numbers is the quotient of their moduli.

The reader should not make the mistake of assuming that the modulus of the sum of two complex numbers is the same as the sum of their moduli. We will learn more about this later.

Example 1

What is the modulus of $-i + ((3 + i)/(1 - i))$?

Solution

We first simplify the fraction

$$\frac{3 + i}{1 - i} = \frac{(3 + i)(1 + i)}{(1 - i)(1 + i)} = 1 + 2i.$$

Thus

$$-i + \frac{3 + i}{1 - i} = 1 + i \quad \text{and} \quad |1 + i| = \sqrt{2}. \qquad \blacktriangleleft$$

Example 2

Find $|(3 + 4i)^5/(1 + i\sqrt{3})|$.

Solution

From Eq. (1.3–5) we have

$$\left| \frac{(3 + 4i)^5}{1 + i\sqrt{3}} \right| = \frac{|(3 + 4i)^5|}{|1 + i\sqrt{3}|}.$$

Now, $|(3 + 4i)^5| = |3 + 4i|^5 = \left(\sqrt{3^2 + 4^2}\right)^5$. Also, $|1 + i\sqrt{3}| = \sqrt{1^2 + (\sqrt{3})^2}$. Thus

$$\left| \frac{(3 + 4i)^5}{1 + i\sqrt{3}} \right| = \frac{\left(\sqrt{3^2 + 4^2}\right)^5}{\sqrt{1^2 + (\sqrt{3})^2}} = \frac{5^5}{2}. \qquad \blacktriangleleft$$

Complex or Argand Plane

If the complex number $z = x + iy$ were written as a couple $z = (x, y)$, we would perhaps be reminded of the notation for the coordinates of a point in the xy-plane. The expression $|z| = \sqrt{x^2 + y^2}$ also recalls the Pythagorean expression for the distance of that point from the origin.

It should come as no surprise to learn that the xy-plane (Cartesian plane) is frequently used to represent complex numbers. When used for this purpose, it is called the Argand plane,[†] the z-plane, or the complex plane. Under these circumstances, the x- or horizontal axis is called the axis of real numbers, whereas the y- or vertical axis is called the axis of imaginary numbers.

[†] The plane is named for Jean Argand, a Swiss mathematician who proposed this representation of complex numbers in 1806.

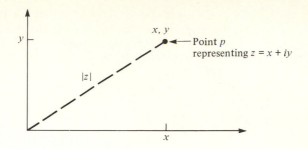

Figure 1.3-1

In Fig. 1.3–1 the point p, whose coordinates are x, y, is said to represent the complex number $z = x + iy$. The modulus of z, that is $|z|$, is the distance of x, y from the origin. Since the length of either leg of a right triangle cannot exceed the length of the hypotenuse, Fig. 1.3–1 reveals the identities

$$|\text{Re } z| = |x| \le |z|, \tag{1.3-6a}$$

$$|\text{Im } z| = |y| \le |z|. \tag{1.3-6b}$$

The | | signs have been placed around Re z and Im z since we are concerned here with physical length (which cannot be negative). Even though we have drawn Re z and Im z positive in Fig. 1.3–1, we could have easily used a figure in which one or both were negative, and Eq. (1.3–6) would still hold.

Another possible representation of z in this same plane is as a vector. We display $z = x + iy$ as a directed line that begins at the origin and terminates at the point x, y, as shown in Fig. 1.3–2.

Thus a complex number can be represented by either a point or a vector in the xy-plane. We will use both methods. Often we will refer to the point or vector as if it were the complex number itself rather than merely its representation.

When we represent a complex number as a vector, we will usually regard it as a sliding vector, that is, one whose starting point is irrelevant. Thus the line directed from the origin to $x = 3$, $y = 4$ is the complex number $3 + 4i$, and so is the directed line from $x = 1$, $y = 2$ to $x = 4$, $y = 6$ (see Fig. 1.3–3). Both vectors have length 5

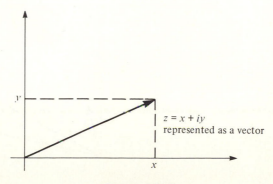

Figure 1.3-2

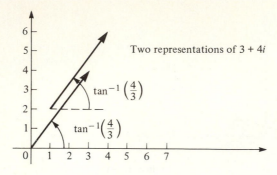

Figure 1.3–3

and point in the same direction. Each has projections of 3 and 4 on the x- and y-axes, respectively.

There are an unlimited number of directed line segments that we can draw in order to represent a complex number. All have the same magnitude and point in the same direction.

There are simple geometrical relationships between the vectors for $z = x + iy$, $-z = -x - iy$, and $\bar{z} = x - iy$, as can be seen in Fig. 1.3–4. The vector for $-z$ is the vector for z reflected through the origin, whereas $\bar{z}$ is the vector z reflected about the real axis.

The process of adding the complex number $z_1 = x_1 + iy_1$ to the number $z_2 = x_2 + iy_2$ has a simple interpretation in terms of their vectors. Their sum, $z_1 + z_2 = x_1 + x_2 + i(y_1 + y_2)$, is shown vectorially in Fig. 1.3–5. We see that the vector representing the sum of the complex numbers z_1 and z_2 is obtained by adding vectorially the vector for z_1 and the vector for z_2. The familiar parallelogram rule, which is used in adding vectors such as force, velocity, or electric field, is employed in Fig. 1.3–5 to perform the summation. We can also use a "tip-to-tail" addition, as shown in Fig. 1.3–6.

The "triangle inequalities" are derivable from this geometric picture. The length of any leg of a triangle is less than or equal to the sums of the lengths of the legs of

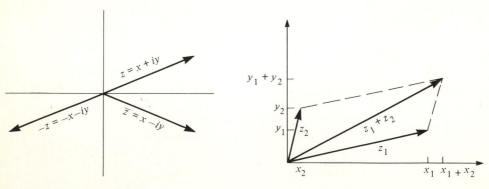

Figure 1.3–4   **Figure 1.3–5**

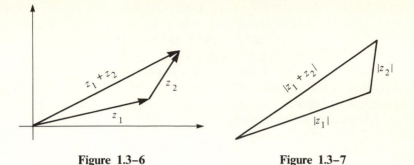

Figure 1.3–6   Figure 1.3–7

the other two sides (see Fig. 1.3–7). The length of the vector for $z_1 + z_2$ is $|z_1 + z_2|$, which must be less than or equal to the combined length $|z_1| + |z_2|$. Thus

$$|z_1 + z_2| \leq |z_1| + |z_2|. \tag{1.3-7}$$

This triangle inequality is also derivable from purely algebraic manipulations. Hence, as warned, the modulus of the sum of two complex numbers need not equal the sum of their moduli. By adding three complex numbers vectorially, as shown in Fig. 1.3–8, we see that

$$|z_1 + z_2 + z_3| \leq |z_1| + |z_2| + |z_3|.$$

This obviously can be extended to a sum having any number of elements:

$$|z_1 + z_2 + \cdots + z_n| \leq |z_1| + |z_2| + \cdots + |z_n|. \tag{1.3-8}$$

The subtraction of two complex numbers also has a counterpart in vector subtraction. Thus $z_1 - z_2$ is treated by adding together the vectors for z_1 and $-z_2$, as shown in Fig. 1.3–9. Another familiar means of vector subtraction is shown in Fig. 1.3–10.

Polar Representation

Often, points in the complex plane, which represent complex numbers, are defined by means of polar coordinates (see Fig. 1.3–11). The complex number $z = x + iy$ is

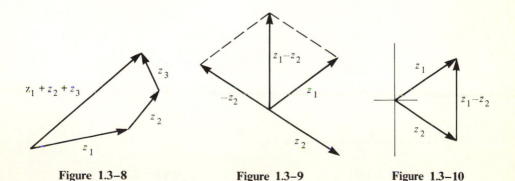

Figure 1.3–8   Figure 1.3–9   Figure 1.3–10

represented by the point p whose Cartesian coordinates are x, y, or whose polar coordinates are r, θ. We see that r is identical to the modulus of z, that is, the distance of p from the origin; and θ is the angle that the ray joining the point p to the origin makes with the positive x-axis. We call θ the *argument* of z and write $\theta = \arg z$. Occasionally, θ is referred to as the angle of z. Unless otherwise stated, θ will be expressed in radians. When the ° symbol is used, θ will be given in degrees.

The angle θ is regarded as positive when measured in the counterclockwise direction and negative when measured clockwise. The distance r is never negative. For a point at the origin, r becomes zero. Here θ is undefined since a ray like that shown in Fig. 1.3–11 cannot be constructed.

Since $r = \sqrt{x^2 + y^2}$, we have

$$r = |z|, \tag{1.3-9}$$

and a glance at Fig. 1.3–11 shows

$$\theta = \tan^{-1}\frac{y}{x}.$$

If we know (y/x), there are actually two values of θ we can draw; they differ by π. For example, if $(y/x) = 1$, θ could be drawn as $\pi/4$ or $5\pi/4$. Hence, if we specify (y/x), we must also state the quadrant of θ (or, alternatively, the sign of x or y) if θ is to be constructed.

There is one more important feature of θ. It is multivalued. Suppose that for some complex number we have found a correct value of θ in radians. Then we can add to this value any positive or negative integer multiple of 2π radians and again obtain a valid value for θ. If θ is in degrees we can add multiples of $360°$. For example, suppose $z = 1 + i$. Let us find the polar coordinates of the point that represents this complex number. Now, $r = |z| = \sqrt{1^2 + 1^2} = \sqrt{2}$, and from Fig. 1.3–12 we see that $\theta = \pi/4$ radians, or $\pi/4 + 2\pi$ radians, or $\pi/4 + 4\pi$, or $\pi/4 - 2\pi$, etc. Thus in this case $\theta = \pi/4 + k2\pi$, where $k = 0, \pm 1, \pm 2, \ldots$.

In general, all the values of θ are contained in the expression

$$\theta = \theta_0 + k2\pi, \quad k = 0, \pm 1, \pm 2, \ldots, \tag{1.3-10}$$

where θ_0 is some particular value of $\arg z$. If we work in degrees, $\theta = \theta_0 + k360°$ describes all values of θ.

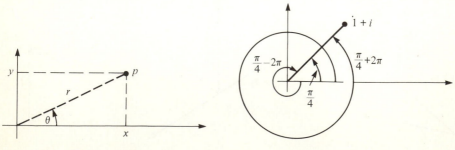

Figure 1.3–11   Figure 1.3–12

The principal value of the argument (or principal argument) of a complex number z is that value of arg z that is greater than $-\pi$ and less than or equal to π.

Thus the principal value of θ satisfies

$$-\pi < \theta \leq \pi.^\dagger \qquad (1.3\text{--}11)$$

The reader can restate this in degrees. Note that the principal value of the argument when z is a negative real number is π (or 180°).

Example 3

Using the principal argument, find the polar coordinates of the point that represents the complex number $-1 - i$.

Solution

The polar distance r for $-1 - i$ is $\sqrt{2}$, as we can see from Fig. 1.3–13. The principal value of θ is $-3\pi/4$ radians. It is *not* $5\pi/4$ since this number exceeds π. In computing principal values we should never do what was done with the broken line in Fig. 1.3–13, namely, cross the negative real axis. From Eq. (1.3–10) we see that all the values of arg $(-1 - i)$ are contained in the expression

$$\theta = \frac{-3\pi}{4} + 2k\pi, \qquad k = 0, \pm 1, \pm 2, \ldots.$$

Note that by using $k = 1$ in the above, we obtain the nonprincipal value $5\pi/4$. ◀

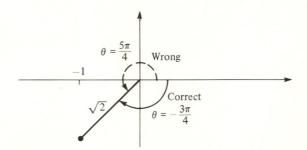

Figure 1.3–13

Let the complex number $z = x + iy$ be represented by the vector shown in Fig. 1.3–14. The point at which the vector terminates has polar coordinates r and θ. The angle θ need not be a principal value. From Fig. 1.3–14 we have $x = r \cos \theta$ and $y = r \sin \theta$. Thus $z = r \cos \theta + ir \sin \theta$, or

$$z = r(\cos \theta + i \sin \theta). \qquad (1.3\text{--}12)$$

---

†The definition presented here for the principal argument is the most common one. However, some texts use other definitions, for example, $0 \leq \theta < 2\pi$.

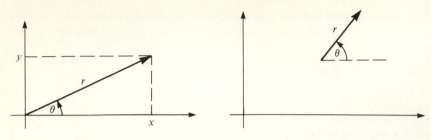

Figure 1.3–14   Figure 1.3–15

We call this the *polar form* of a complex number as opposed to the rectangular (Cartesian) form $x + iy$. The expression $\cos \theta + i \sin \theta$ is often abbreviated cis θ. We will often use $\underline{/\theta}$ to mean cis θ. Our complex number $x + iy$ becomes $r\underline{/\theta}$. This is a useful notation because it tells not only the length of the corresponding vector but also the angle made with the real axis. Note that $i = 1\underline{/\pi/2}$ and $-i = 1\underline{/-\pi/2}$.

A vector such as the one in Fig. 1.3–15 can be translated so that it emanates from the origin. It too represents a complex number $r\underline{/\theta}$.

The complex numbers $r\underline{/\theta}$ and $r\underline{/-\theta}$ are conjugates of each other, as can be seen in Fig. 1.3–16. Equivalently, we find that the conjugate of $r(\cos \theta + i \sin \theta)$ is $r(\cos(-\theta) + i \sin(-\theta)) = r(\cos \theta - i \sin \theta)$.

The polar description is particularly useful in the multiplication of complex numbers. Consider $z_1 = r_1$ cis θ_1 and $z_2 = r_2$ cis θ_2. Multiplying z_1 by z_2 we have

$$z_1 z_2 = r_1(\cos \theta_1 + i \sin \theta_1) r_2(\cos \theta_2 + i \sin \theta_2). \qquad (1.3\text{--}13)$$

With some additional multiplication we obtain

$$z_1 z_2 = r_1 r_2 \big[(\cos \theta_1 \cos \theta_2 - \sin \theta_1 \sin \theta_2) + i(\sin \theta_1 \cos \theta_2 + \cos \theta_1 \sin \theta_2) \big].$$
$$(1.3\text{--}14)$$

The reader should recall the identities

$$\cos(\theta_1 + \theta_2) = \cos \theta_1 \cos \theta_2 - \sin \theta_1 \sin \theta_2,$$
$$\sin(\theta_1 + \theta_2) = \sin \theta_1 \cos \theta_2 + \cos \theta_1 \sin \theta_2,$$

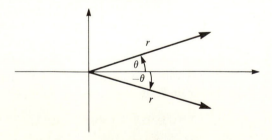

Figure 1.3–16

which we now install in Eq. (1.3–14), to get

$$z_1 z_2 = r_1 r_2 [\cos(\theta_1 + \theta_2) + i\sin(\theta_1 + \theta_2)].\qquad(1.3\text{–}15)$$

In other notation Eq. (1.3–15) becomes

$$z_1 z_2 = r_1 r_2 \,\underline{/\theta_1 + \theta_2}.\qquad(1.3\text{–}16)$$

The two preceding equations contain the following important fact.

When two complex numbers are multiplied together, the resulting product has a modulus equal to the product of the moduli of the two factors and an argument equal to the sum of the arguments of each factor.

To multiply three complex numbers we readily extend this method. Thus

$$z_1 z_2 z_3 = (z_1 z_2)(z_3) = r_1 r_2 \,\underline{/\theta_1 + \theta_2}\, r_3 \,\underline{/\theta_3} = r_1 r_2 r_3 \,\underline{/\theta_1 + \theta_2 + \theta_3}.$$

Any number of complex numbers can be multiplied in this fashion.

The modulus of the entire product is the product of the moduli of each factor, and the argument of the product is the sum of the arguments of the factors.

Example 4

Verify Eq. (1.3–16) by considering the product $(1 + i)(\sqrt{3} + i)$.

Solution

Multiplying in the usual way (see Eq. 1.1–4), we obtain

$$(1 + i)(\sqrt{3} + i) = (\sqrt{3} - 1) + i(\sqrt{3} + 1).$$

The modulus of the preceding product is

$$\sqrt{(\sqrt{3} - 1)^2 + (\sqrt{3} + 1)^2} = \sqrt{8},$$

whereas the product of the moduli of each factor is

$$\sqrt{1^2 + 1^2}\sqrt{(\sqrt{3})^2 + 1^2} = \sqrt{2}\sqrt{4} = \sqrt{8}.$$

The factor $(1 + i)$ has an argument of $\pi/4$ radians and $\sqrt{3} + i$ has an argument of $\pi/6$ (see Fig. 1.3–17).

The complex number $(\sqrt{3} - 1) + i(\sqrt{3} + 1)$ has an angle in the first quadrant equal to

$$\tan^{-1}\frac{\sqrt{3} + 1}{\sqrt{3} - 1} = \frac{5\pi}{12}.$$

But $5\pi/12 = \pi/4 + \pi/6$, that is, the sums of the arguments of the two factors. ◀

The argument of a complex number has, of course, an infinity of possible values. When we add together the arguments of two factors in order to arrive at the argument of a product, we obtain only one of the possible values for the argument of

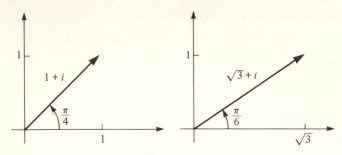

Figure 1.3–17

that product. Thus in the preceding example $(\sqrt{3} - 1) + i(\sqrt{3} + 1)$ has arguments $5\pi/12 + 2\pi$, $5\pi/12 + 4\pi$, and so forth, none of which were obtained through our procedure. However, any of these results can be derived by our adding some whole multiple of 2π on to the number $5\pi/12$ actually obtained.

Equation (1.3–16) is particularly useful when we multiply complex numbers that are given to us in polar rather than in rectangular form. For example, the product of $2/\pi/2$ and $3/3\pi/4$ is $6/5\pi/4$. We can convert the result to rectangular form:

$$6 \underline{/\frac{5\pi}{4}} = 6\cos\left(\frac{5\pi}{4}\right) + i6\sin\left(\frac{5\pi}{4}\right) = \frac{-6}{\sqrt{2}} - i\frac{6}{\sqrt{2}}.$$

The two factors in this example were written with their principal arguments $\pi/2$ and $3\pi/4$. However, when we added these angles and got $5\pi/4$, we obtained a nonprincipal argument. In fact, the principal argument of $6/5\pi/4$ is $-3\pi/4$.

When principal arguments are added together in a multiplication problem, the resulting argument need not be a principal value. Conversely, when nonprincipal arguments are combined, a principal argument may result.

When we multiply z_1 by z_2 to obtain the product z_1z_2, the operation performed with the corresponding vectors is neither scalar multiplication (dot product) nor vector multiplication (cross product), which are perhaps familiar to us from elementary vector analysis. Similarly, we can divide two complex numbers as well as, in a sense, their vectors. This too has no counterpart in any previously familiar vector operation.

It is convenient to use polar coordinates to find the reciprocal of a complex number. With $z = r\underline{/\theta}$ we have

$$\frac{1}{z} = \frac{1}{r(\cos\theta + i\sin\theta)} = \frac{\cos\theta - i\sin\theta}{r(\cos\theta + i\sin\theta)(\cos\theta - i\sin\theta)}$$

$$= \frac{\cos\theta - i\sin\theta}{r(\cos^2\theta + \sin^2\theta)} = \frac{\cos\theta - i\sin\theta}{r} = \frac{1}{r}\underline{/-\theta}.$$

Hence,

$$\frac{1}{z} = \frac{1}{r\underline{/\theta}} = \frac{1}{r}\underline{/-\theta}. \qquad (1.3–17)$$

Thus the modulus of the reciprocal of a complex number is the reciprocal of the modulus of that number, and the argument of the reciprocal of a complex number is the negative of the argument of that number.

Consider now the complex numbers $z_1 = r_1 \underline{/\theta_1}$ and $z_2 = r_2 \underline{/\theta_2}$. To divide z_1 by z_2 we multiply z_1 by $1/z_2 = 1/r_2 \underline{/-\theta_2}$. Thus

$$\frac{z_1}{z_2} = r_1 \underline{/\theta_1} \frac{1}{r_2} \underline{/-\theta_2} = \frac{r_1}{r_2} \underline{/(\theta_1 - \theta_2)} . \qquad (1.3\text{–}18)$$

The magnitude of the quotient of two complex numbers is the quotient of their magnitudes, and the argument of the quotient is the argument of the numerator less the argument of the denominator.

Example 5

Evaluate $(1 + i)/(\sqrt{3} + i)$ by using the polar form of complex numbers.

Solution

We have

$$\frac{1 + i}{\sqrt{3} + i} = \frac{\sqrt{2} \underline{/\text{arc tan}(1/1)}}{2 \underline{/\text{arc tan}(1/\sqrt{3})}} = \frac{\sqrt{2} \underline{/\pi/4}}{2 \underline{/\pi/6}} = \frac{1}{\sqrt{2}} \underline{/\pi/4 - \pi/6} = \frac{1}{\sqrt{2}} \underline{/\pi/12} .$$

The above result is convertible to rectangular form:

$$\frac{1 + i}{\sqrt{3} + i} = \frac{\cos(\pi/12)}{\sqrt{2}} + i \frac{\sin(\pi/12)}{\sqrt{2}} .$$

This problem could have been done entirely in rectangular notation with the aid of Eq. (1.2–7). There are computations, however, where switching to polar notation saves us some labor, as in the following example. ◀

Example 6

Evaluate

$$\frac{(1 + i)(3 + i)(-2 - i)}{(i)(3 + 4i)(5 + i)} = a + ib = r \underline{/\theta} ,$$

where a and b are to be determined.

Solution

Let us initially seek the polar answer $r \underline{/\theta}$. First, r is obtained from the usual properties of the moduli of products and quotients:

$$r = \frac{|(1 + i)(3 + i)(-2 - i)|}{|(i)(3 + 4i)(5 + i)|} = \frac{|1 + i||3 + i|| - 2 - i|}{|i||3 + 4i||5 + i|} = \frac{\sqrt{2}\sqrt{10}\sqrt{5}}{1\sqrt{25}\sqrt{26}} = \frac{2}{\sqrt{26}} .$$

The argument θ is the argument of the numerator, $(1 + i)(3 + i)(-2 - i)$, less that of the denominator, $(i)(3 + 4i)(5 + i)$. Thus

$$\theta = \left(\text{arc tan }\frac{1}{1} + \text{arc tan }\frac{1}{3} + \text{arc tan }\frac{-1}{-2}\right) - \left(\text{arc tan }\frac{1}{0} + \text{arc tan }\frac{4}{3} + \text{arc tan }\frac{1}{5}\right),$$

$$\theta \doteq (0.785 + 0.322 + 3.60) - (1.57 + 0.927 + 0.197) \doteq 2.01.$$

Therefore,

$$r\underline{/\theta} \doteq \frac{2}{\sqrt{26}}\underline{/2.01},$$

and

$$a + ib \doteq \frac{2}{\sqrt{26}}[\cos 2.01 + i\sin 2.01] \doteq -0.166 + i0.355. \qquad \blacktriangleleft$$

EXERCISES

1. Find the modulus of each of the following complex expressions:

   a) $(1 + i)(2 + i)(-i + 3)$,   e) $(3 + 3i)^{10}$,

   b) $\dfrac{2 + i}{3 + i}(1 + i)$,   f) $\dfrac{1}{(1 + i)^{11}}$,

   c) $(1 + i) + \dfrac{2 + i}{3 + i}$,   g) $\left(\dfrac{1 + i}{1 + i\sqrt{3}}\right)^{17}$.

   d) $\dfrac{2 + i}{-i + 2}\dfrac{-2 + i}{i - 2}$,

2. The following vectors in the complex plane represent complex numbers. State the numbers in the form $a + ib$.

   a) The vector extending from $x = 1$, $y = 2$ to $x = 4$, $y = 3$

   b) The vector extending from $x = 5$, $y = 1$ to $x = -7$, $y = 0$

   c) The vector of length 6, which begins at $x = 1$, $y = 0$ and makes a $\pi/6$ angle with the positive x-axis

3. a) What is the physical relationship among the three vectors representing the complex numbers z, iz, and z/i?

   b) What is the relationship between the vectors representing z_1 and z if $z_1 = ikz$ is satisfied, where k is some real number?

4. a) A vector begins at $x = 2$, $y = 1$ and represents the complex number $-3 + 3i$. Where does the vector terminate?

   b) A vector terminates at $x = -1$, $y = 3$. If it represents the complex number $2 + 4i$, where does it begin?

5. Under what circumstances will the equality sign hold true in Eq. (1.3–7)?

6. Let z_1 and z_2 be complex numbers.

   a) Show by means of a vector construction involving triangles that

   $$|z_1 - z_2| \le |z_1| + |z_2|. \qquad (1.3\text{–}19)$$

b) What must be the relationship between z_1 and z_2 in order to have the equality hold in part (a)?

7. a) Show by means of a diagram that the length of any leg of a triangle must equal or exceed the length of the larger of the two remaining legs less that of the smaller.

b) Let z_1 and z_2 be complex numbers. Consider the vector that represents $z_1 - z_2$. Using part (a) show that

$$|z_1 - z_2| \geq |z_1| - |z_2| \quad \text{if} \quad |z_1| \geq |z_2|,$$
$$|z_1 - z_2| \geq |z_2| - |z_1| \quad \text{if} \quad |z_2| \geq |z_1|.$$

Explain why both formulas can be reduced to the single expression

$$|z_1 - z_2| \geq ||z_1| - |z_2||. \tag{1.3–20}$$

8. Find, in the form $a + ib$, the complex numbers that are represented graphically by the points having the following polar coordinates:
a) $r = 2,\ \theta = 3$;
b) $r = 7,\ \theta = 8$;
c) $r = 2,\ \theta = 11\frac{1}{2}\pi$.

9. Convert the following complex numbers from the Cartesian form $a + ib$ to the polar form $r \operatorname{cis} \theta$. State r and give all possible values of θ (in radians). State the principal value of θ.

a) $-i + 1$ c) $-2 - i$ e) $\left(\dfrac{1 + i}{3 + 4i}\right)^2$

b) $-2i$ d) $(3 + 4i)(1 + i)$

10. What is the principal argument (in radians) of these complex numbers:
a) $2 \operatorname{cis}(1074)$,   b) $3 \operatorname{cis}(13{,}322)$,   c) $5 \operatorname{cis}(13{,}320)$.

A pocket calculator can be helpful in these calculations.

11. Convert the following expressions to polar notation $r\underline{/\theta}$. Give only the principal value of θ (radians).

a) $\dfrac{(1 + i)(2\underline{/3\pi/4})}{3\underline{/\pi}}$

c) $\dfrac{3 \operatorname{cis}(\pi/2)}{(1 + i)4 \operatorname{cis}(5\pi/6)}$

b) $\dfrac{(1 + i)(3 + 4i)(i)(2 + 3i)}{(3 + 5i)(2 - i)(-i - 1)}$

d) $\dfrac{(\operatorname{cis}(\pi/6))^4}{(\operatorname{cis}(-\pi/6))^4}$

12. a) Consider vectors representing complex numbers f and g. Show that these vectors are perpendicular if and only if $|f - g|^2 = |f|^2 + |g|^2$. *Hint:* Draw a picture.

b) Show that the preceding equation is equivalent to the requirement $\operatorname{Re}(f\bar{g}) = 0$.

13. Show that $|(z - p)/(1 - \bar{p}z)| < 1$ if $|z| < 1$ and $|p| < 1$ are both satisfied.

14. a) Consider the inequality $|z_1 + z_2|^2 \leq |z_1|^2 + |z_2|^2 + 2|z_1||z_2|$. Prove this expression by algebraic means (no triangles). *Hint:* Note that $|z_1 + z_2|^2 = (z_1 + z_2)\overline{(z_1 + z_2)} = (z_1 + z_2)(\bar{z}_1 + \bar{z}_2)$. Multiply out $(z_1 + z_2)(\bar{z}_1 + \bar{z}_2)$,

and use the facts that for a complex number, say w,

$$w + \bar{w} = 2 \operatorname{Re} w, \quad \text{and} \quad \operatorname{Re} w \leq |w|.$$

b) Observe that $|z_1|^2 + |z_2|^2 + 2|z_1||z_2| = (|z_1| + |z_2|)^2$. Show that the inequality proved in part (a) leads to the triangle inequality $|z_1 + z_2| \leq |z_1| + |z_2|$.

1.4 INTEGER AND FRACTIONAL POWERS OF A COMPLEX NUMBER

Integer Powers

In the previous section we learned to multiply any number of complex quantities together by means of polar notation. Thus with n complex numbers $z_1, z_2, \ldots, z_n$ we have

$$z_1 z_2 z_3, \ldots, z_n = r_1 r_2 r_2, \ldots, r_n \underline{/\theta_1 + \theta_2 + \theta_3 + \cdots + \theta_n}, \quad (1.4\text{--}1)$$

where $r_j = |z_j|$ and $\theta_j = \arg z_j$.

If all the values, z_1, z_2, and so on, are identical so that $z_j = z$ and $z = r\underline{/\theta}$, then Eq. (1.4–1) simplifies to

$$z^n = r^n \underline{/n\theta} = r^n \operatorname{cis}(n\theta) = r^n [\cos(n\theta) + i \sin(n\theta)] \quad (1.4\text{--}2)$$

The modulus of z^n is the modulus of z raised to the nth power, whereas the argument of z^n is n times the argument of z.

The preceding was proved valid when n is a positive integer. If we define $z^0 = 1$ (as for real numbers), Eq. (1.4–2) applies when $n = 0$ as well. With the aid of a suitable definition, we will now prove that Eq. (1.4–2) also is applicable for negative n.

Let m be a positive integer. Then, from Eq. (1.4–2) we have $z^m = r^m[\cos(m\theta) + i \sin(m\theta)]$. We define z^{-m} as being identical to $1/z^m$. Thus

$$z^{-m} = \frac{1}{r^m[\cos(m\theta) + i \sin(m\theta)]}. \quad (1.4\text{--}3)$$

If on the right in Eq. (1.4–3) we multiply numerator and denominator by the expression $\cos m\theta - i \sin m\theta$, we have

$$z^{-m} = \frac{1}{r^m} \frac{\cos m\theta - i \sin m\theta}{\cos^2 m\theta + \sin^2 m\theta} = r^{-m}[\cos m\theta - i \sin m\theta].$$

Now, since $\cos(m\theta) = \cos(-m\theta)$ and $-\sin m\theta = \sin(-m\theta)$, we obtain

$$z^{-m} = r^{-m}[\cos(-m\theta) + i \sin(-m\theta)], \quad m = 1, 2, 3, \ldots. \quad (1.4\text{--}4)$$

If we let $-m = n$ in the preceding equation, it becomes

$$z^n = r^n[\cos(n\theta) + i \sin(n\theta)], \quad n = -1, -2, -3, \ldots. \quad (1.4\text{--}5)$$

We can incorporate this result into Eq. (1.4–2) by allowing n to be any integer in that expression.

Equation (1.4–2) allows us to raise complex numbers to integer powers when the use of Cartesian coordinates and successive self-multiplication would be very tedious. For example, consider $(1 + i\sqrt{3})^{11} = a + ib$. We want a and b. We could begin with Eq. (1.1–4), square $(1 + i\sqrt{3}\,)$, multiply the result by $(1 + i\sqrt{3}\,)$, and so forth. Or, if we remember the binomial theorem, we could apply it to $(1 + i\sqrt{3}\,)^{11}$, and then combine the twelve resulting terms. Instead, we observe that $(1 + i\sqrt{3}\,) = 2\underline{/\pi/3}$, and

$$\left(2\underline{/\frac{\pi}{3}}\right)^{11} = 2^{11}\underline{/\frac{11\pi}{3}} = 2^{11}\left[\cos\left(\frac{11\pi}{3}\right) + i\sin\left(\frac{11\pi}{3}\right)\right] = 2^{10} - i2^{10}\sqrt{3}\,.$$

Equation (1.4–2) can yield an important identity. First, we put $z = r(\cos\theta + i\sin\theta)$ so that

$$[\,r(\cos\theta + i\sin\theta)\,]^{\,n} = r^n(\cos n\theta + i\sin n\theta).$$

Taking $r = 1$ in this expression, we then have

$$(\cos\theta + i\sin\theta)^n = \cos n\theta + i\sin n\theta, \qquad n = 0, \pm1, \pm2, \ldots, \qquad (1.4\text{–}6)$$

which is known as DeMoivre's theorem.

This formula can yield some familiar trigonometric identities. For example, with $n = 2$,

$$(\cos\theta + i\sin\theta)^2 = \cos 2\theta + i\sin 2\theta.$$

Expanding the left side of the preceding expression, we arrive at

$$\cos^2\theta + 2i\sin\theta\cos\theta - \sin^2\theta = \cos 2\theta + i\sin 2\theta.$$

Equating corresponding parts (real and imaginary), we obtain the pair of identities $\cos^2\theta - \sin^2\theta = \cos 2\theta$ and $2\sin\theta\cos\theta = \sin 2\theta$.

Fractional Powers

Let us try to raise z to fractional powers, that is, we want $z^{1/m}$, where m is a positive integer. We define $z^{1/m}$ so that $(z^{1/m})^m = z$. Suppose

$$z^{1/m} = \rho\underline{/\phi}\,. \qquad (1.4\text{–}7)$$

Raising both sides to the mth power we have

$$z = \left(\rho\underline{/\phi}\right)^m = z = \rho^m\underline{/m\phi} = \rho^m[\cos(m\phi) + i\sin(m\phi)]. \qquad (1.4\text{–}8)$$

Using $z = r\underline{/\theta} = r(\cos\theta + i\sin\theta)$ on the left side of the above we obtain

$$r(\cos\theta + i\sin\theta) = \rho^m[\,\cos(m\phi) + i\sin(m\phi)\,]\,. \qquad (1.4\text{–}9)$$

For this equation to hold the moduli on each side must agree. Thus

$$r = \rho^m \quad \text{or} \quad \rho = r^{1/m}.$$

Since ρ is a positive real number, we must use the positive root of $r^{1/m}$. Hence

$$\rho = \sqrt[m]{r}\,. \qquad (1.4\text{–}10)$$

The angle θ in Eq. (1.4–9) need not equal $m\phi$. The best that we can do is to conclude that these two quantities differ by an integral multiple of 2π, that is, $m\phi - \theta = 2k\pi$, which means

$$\phi = \frac{1}{m}[\theta + 2k\pi], \qquad k = 0, \pm1, \pm2, \ldots. \qquad (1.4\text{–}11)$$

Thus from Eqs. (1.4–7), (1.4–10), and (1.4–11):

$$z^{1/m} = \rho \underline{/\phi} = \sqrt[m]{r}\left[\cos\left(\frac{\theta}{m} + \frac{2k\pi}{m}\right) + i\sin\left(\frac{\theta}{m} + \frac{2k\pi}{m}\right)\right].$$

The number k on the right side of this equation can assume any integer value. Suppose we begin with $k = 0$ and allow k to increase in unit steps. With $k = 0$, we are taking the sine and cosine of θ/m; with $k = 1$, the sine and cosine $\theta/m + 2\pi/m$, and so on. Finally, with $k = m$, we take the sine and cosine of $\theta/m + 2\pi$. But $\sin(\theta/m + 2\pi)$ and $\cos(\theta/m + 2\pi)$ are numerically equal to the sine and cosine of θ/m.

If $k = m + 1$, $m + 2$, etc., we merely repeat the numerical values for the cosine and sine obtained when $k = 1, 2$, etc. Thus all the numerically distinct values of $z^{1/m}$ can be obtained by our allowing k to range from 0 to $m - 1$ in the preceding equation. Hence, $z^{1/m}$ has m different values, and they are given by the equation

$$z^{1/m} = \sqrt[m]{r}\underline{\left/\frac{\theta + 2\pi k}{m}\right.} = \sqrt[m]{r}\left[\cos\left(\frac{\theta}{m} + \frac{2k\pi}{m}\right) + i\sin\left(\frac{\theta}{m} + \frac{2k\pi}{m}\right)\right],$$

$$k = 0, 1, 2, \ldots, m - 1; \; m > 1. \tag{1.4–12}$$

Actually, we can let k range over any m successive integers (e.g., $k = 2 \rightarrow m + 1$) and still generate all the values of $z^{1/m}$.

From our previous mathematics we know that a positive real number, say 9, has two different square roots, in this case ± 3. Equation (1.4–12) tells us that any complex number also has two square roots (we put $m = 2$, $k = 0, 1$) and three cube roots ($m = 3$, $k = 0, 1, 2$) and so on.

The geometrical interpretation of Eq. (1.4–12) is important since it can quickly permit the plotting of those points in the complex plane that represent the roots of a number. The moduli of all the roots are identical and equal, $\sqrt[m]{r}$ $\left(\text{or } \sqrt[m]{|z|}\right)$. Hence, the roots are representable by points on a circle having radius $\sqrt[m]{r}$. Each of the values obtained from Eq. (1.4–12) has a different argument. While k increases as indicated in Eq. (1.4–12), the arguments grow from θ/m to $\theta/m + 2(m - 1)\pi/m$ by increasing in increments of $2\pi/m$. The points representing the various values of $z^{1/m}$, which we plot on the circle of radius $\sqrt[m]{r}$, are thus spaced uniformly at an angular separation of $2\pi/m$. One of the points ($k = 0$) makes an angle of θ/m with the positive x-axis. We thus have enough information to plot all the points (or all the corresponding vectors).

Equation (1.4–12) was derived under the assumption that m is a positive integer. If m is a negative integer the equation is still valid, except we now generate all roots by allowing k to range over $|m|$ successive values (e.g., $k = 0, 1, 2, \ldots, |m| - 1$). The $|m|$ roots are uniformly spaced around a circle of radius $\sqrt[m]{r}$, and one root makes an angle θ/m with the positive x-axis.

Example 1

Find all values of $(-1)^{1/2}$ by means of Eq. (1.4–12).

Solution

Here $r = |-1| = 1$ and $m = 2$ in Eq. (1.4–12). For θ we can use *any* valid argument of -1. We will use π. Thus

$$(-1)^{1/2} = \sqrt[2]{1}\left[\cos\left(\frac{\pi}{2} + k\pi\right) + i\sin\left(\frac{\pi}{2} + k\pi\right)\right], \qquad k = 0, 1.$$

With $k = 0$ in this formula we obtain $(-1)^{1/2} = i$, and $k = 1$ yields $(-1)^{1/2} = -i$. Points representing the two roots are plotted in Fig. 1.4–1. Their angular separation is $2\pi/n = 2\pi/2 = \pi$ radians. ◀

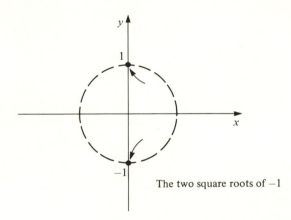

The two square roots of -1

Figure 1.4–1

Example 2

Find all values of $\left(1 + i\sqrt{3}\right)^{1/5}$.

Solution

We anticipate five roots. We use Eq. (1.4–12) with $m = 5$, $r = |1 + i\sqrt{3}| = 2$, and $\theta = \tan^{-1}\sqrt{3} = \pi/3$. Our result is

$$\left(1 + i\sqrt{3}\right)^{1/5} = \sqrt[5]{2}\left[\cos\left(\frac{\pi}{15} + \frac{2\pi}{5}k\right) + i\sin\left(\frac{\pi}{15} + \frac{2\pi k}{5}\right)\right], \qquad k = 0, 1, 2, 3, 4.$$

Expressed as decimals, these answers become approximately

$$
\begin{array}{ll}
1.12 + i0.239, & k = 0, \\
0.119 + i1.41, & k = 1, \\
-1.05 + i0.467, & k = 2, \\
-0.768 - i0.852, & k = 3, \\
0.574 - i0.994, & k = 4.
\end{array}
$$

Vectors representing the roots are plotted in Fig. 1.4–2. They are spaced $2\pi/5$ radians or 72° apart. The vector for the case $k = 0$ makes an angle of $\pi/15$ radians, or 12°, with the x-axis. Any of these results, when raised to the fifth power, must

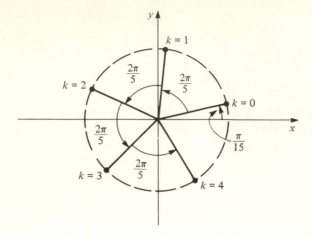

Figure 1.4–2

produce $1 + i\sqrt{3}$. For example, let us use the root for which $k = 1$. We have, with the aid of Eq. (1.4–2),

$$\left(\sqrt[5]{2}\left[\cos\left(\frac{\pi}{15} + \frac{2\pi}{5}\right) + i\sin\left(\frac{\pi}{15} + \frac{2\pi}{5}\right)\right]\right)^5 = 2\left[\cos\left(\frac{\pi}{3} + 2\pi\right) + i\sin\left(\frac{\pi}{3} + 2\pi\right)\right]$$

$$= 2\left[\frac{1}{2} + \frac{i\sqrt{3}}{2}\right] = 1 + i\sqrt{3}. \quad \blacktriangleleft$$

Our original motivation for expanding our number system from real to complex numbers was to permit our solving equations whose solution involved the square roots of negative numbers. It might have worried us that trying to find $(-1)^{1/3}$, $(-1)^{1/4}$, and so forth could lead us to successively more complicated number systems. From the work just presented we see that this is not the case. The complex system, together with Eq. (1.4–12), is sufficient to yield any such root.

From our discussion of the fractional powers of z, we can now formulate a consistent definition of z raised to any rational power, (e.gs., $z^{4/7}$, $z^{-2/3}$) and, as a result, solve such equations as $z^{4/3} + 1 = 0$. We use the definition

$$z^{n/m} = \left(z^{1/m}\right)^n.$$

With Eq. (1.4–12) we perform the inner operation and with Eq. (1.4–2) the outer one. Thus

$$z^{n/m} = \left(\sqrt[m]{r}\right)^n\left[\cos\left(\frac{n}{m}\theta + \frac{2kn\pi}{m}\right) + i\sin\left(\frac{n}{m}\theta + \frac{2kn\pi}{m}\right)\right],$$

$$k = 0, 1, 2, \ldots, m - 1, \qquad (1.4-13)$$

where, as before, $\theta = \arg z$ and $|z| = r$.

If n/m is an irreducible fraction, then our letting k range from 0 to $m - 1$ results in m numerically distinct roots. In the complex plane these roots are arranged uniformly around a circle of radius $\left(\sqrt[m]{r}\right)^n$.

However, if n/m is reducible (i.e., n and m contain common integral factors) then, when k varies from 0 to $m - 1$, some of the values obtained from Eq. (1.4–13) will be numerically identical. This is because the expression $2kn\pi/m$ will assume at least two values that differ by an integer multiple of 2π. In the extreme case where n is exactly divisible by m, all the values obtained from Eq. (1.4–13) are identical. This confirms the familiar fact that z raised to an integer power has but one value. The fraction n/m should be reduced as far as possible, *before* being used in Eq. (1.4–13), if we don't wish to waste time generating identical roots (see Exercise 8).

Assuming n/m is an irreducible fraction and z, a given complex number, let us consider the m possible values of $z^{n/m}$. Choosing any one of these values and raising it to the m/n power, we obtain n numbers. One of these is z. Thus the equation $(z^{n/m})^{m/n} = z$ must be interpreted with some care.

Example 3

Solve the equation

$$w^{4/3} + 2i = 0. \tag{1.4–14}$$

Solution

We have $w^{4/3} = -2i$, which means $w = (-2i)^{3/4}$. We now use Eq. (1.4–13) with $n = 3$, $m = 4$, $r = |z| = |-2i| = 2$, and $\theta = \arg(-2i) = -\pi/2$. Thus

$$(-2i)^{3/4} = \left(\sqrt[4]{2}\right)^3 \underline{\Big/\frac{3}{4}\left(\frac{-\pi}{2}\right) + 2k\frac{3}{4}\pi}\,, \qquad k = 0, 1, 2, 3;$$

$$\doteq 1.68 \underline{\Big/\frac{-3\pi}{8}}\,, \qquad k = 0;$$

$$\doteq 1.68 \underline{\Big/\frac{9\pi}{8}}\,, \qquad k = 1;$$

$$\doteq 1.68 \underline{\Big/\frac{21\pi}{8}}\,, \qquad k = 2;$$

$$\doteq 1.68 \underline{\Big/\frac{33\pi}{8}}\,, \qquad k = 3.$$

The four results are plotted in the complex plane shown in Fig. 1.4–3. They are

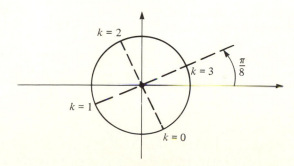

Figure 1.4–3

uniformly distributed on the circle of radius $(\sqrt[4]{2})^3$. Each of these results is a solution of Eq. (1.4–14) provided we use the appropriate 4/3 power. ◀

EXERCISES

1. Express each of the following in the form $a + ib$ and also in polar form $r\underline{/\theta}$.

   a) $(\sqrt{3} + i)^8$ c) $(-1 - i)^{-13}(-\sqrt{3} - i)^{13}$ e) $(4 + 3i)^8$

   b) $(1 - i\sqrt{3})^{-7}$ d) $[2\operatorname{cis}(5\pi/6)]^4$ f) $(2\underline{/2\pi/3})^{-4}$

2. a) Express $\cos 3\theta$ as a sum of terms involving only integer powers of $\cos\theta$ and $\sin\theta$.

   b) Express $\sin 3\theta$ as a sum of terms involving only integer powers of $\cos\theta$ and $\sin\theta$.

3. Express the following in the form $a + ib$. Give all roots. Make a polar plot of the points that represent your results.

   a) $(i)^{1/2}$ c) $(1)^{1/4}$ e) $(-1 + i)^{1/3}$ g) $(1 - i\sqrt{3})^{-1/3}$

   b) $(-i)^{1/2}$ d) $(-1)^{1/4}$ f) $(1 - i\sqrt{3})^{1/4}$ h) $(1 - i\sqrt{3})^{-1/5}$

4. Consider the quadratic equation $az^2 + bz + c = 0$, where $a \neq 0$, and a, b, and c are complex numbers. Use the method of completing the square to show that $z = (-b + (b^2 - 4ac)^{1/2})/2a$. How many solutions does this equation have in general?

5. A well-known mathematics text asks the reader to "find the effect of multiplying a complex number by $(i)^{1/2}$." The answer given is "rotation through 45°." Explain why this answer is only partially correct.

6. Find all solutions of each of the following equations. Give answers as $a + ib$.

   a) $w^3 + i = 1$ c) $w^2 + w + i = 0$ e) $w^4 + 2iw^2 + 1 = 0$

   b) $w^2 + e = -i$ d) $w^2 + 2iw + 1 = 0$ f) $4w^6 + 4iw^3 - 1 = 0$

7. We know that $z^{1/m}$ has m values and that $z^{-1/m}$ does also. For a given z and m we select at random a value of $z^{1/m}$ and one of $z^{-1/m}$.

   a) Is their product necessarily one?

   b) Is it always possible to find a value for $z^{-1/m}$ so that, for a given $z^{1/m}$, we will have $z^{1/m}z^{-1/m} = 1$?

8. Give all values of each of the following in the form $a + ib$, and draw their vector representation in the complex plane.

   a) $i^{4/3}$ d) $(1 - i)^{8/6}$ g) $i^{-1/3}i^{1/3}$ i) $\left[(-1)^4\right]^{1/6}$

   b) $(1 + i)^{2/3}$ e) $i^{2/4}$ h) $(3 + 4i)^{6/2}$ j) $\left[(-1)^{1/6}\right]^4$

   c) $(-1)^{3/8}$ f) $i^{4/6}$

9. Consider these two possible definitions of $z^{m/n}$:

$$z^{m/n} = (z^{1/n})^m,$$

$$z^{m/n} = (z^m)^{1/n}.$$

The first definition is used in this text. Suppose the second were used instead.

a) Assume m/n is an irreducible fraction. Do the two definitions always yield the same numerical results?

b) Assume m/n is reducible. Do the two definitions then yield the same results?

10. It is possible to extract the square root of the complex number $z = x + iy$ without resorting to polar coordinates. Let $a + ib = (x + iy)^{1/2}$, where x and y are known real numbers, and a and b are unknown real numbers.

a) Square both sides of this equation and show that this implies

$$1) \ x = a^2 - b^2,$$

$$2) \ y = 2ab.$$

b) Use equation (2) above to eliminate b from equation (1), and show that equation (1) now leads to a quadratic equation in a^2. Prove that

$$3) \ a^2 = \frac{x \pm \sqrt{x^2 + y^2}}{2}.$$

Explain why we must reject the minus sign in equation (3). Recall our postulate about the number a.

c) Show that

$$4) \ b^2 = \frac{-x + \sqrt{x^2 + y^2}}{2}.$$

d) From the square roots of equations (3) and (4) we obtain

$$5) \ a = \frac{\pm\sqrt{x + \sqrt{x^2 + y^2}}}{\sqrt{2}},$$

$$6) \ b = \frac{\pm\sqrt{-x + \sqrt{x^2 + y^2}}}{\sqrt{2}}.$$

Assume y is positive. What does equation (2) say about the relative signs of a and b? Show that if a, obtained from equation (5), is positive, then b, obtained from equation (6), is also. Show that if a is negative then so is b. Thus with $y > 0$ there are two possible values for $(z)^{1/2} = a + ib$.

e) Assume y is negative. Again show that $a + ib$ has two values and that a is positive and b is negative for one value and vice versa for the other.

f) Use equations (5) and (6) to obtain both values of $(i)^{1/2}$. Check your results by obtaining the same values by means of Eq. (1.4–12).

11. Recall from elementary algebra the formula for the sum of a finite geometric series:

$$1 + p + p^2 + \cdots + p^n = \left(\frac{1 - p^{n+1}}{1 - p} \right), \quad n \text{ is a positive integer, } p \neq 1.$$

This formula is also valid when p is a complex number since the same derivation applies. Show by using this formula and DeMoivre's theorem that the sum of all n roots of $(z)^{1/n}$ is zero. Give a vector interpretation of your result.

12. Use the formula for the sum of a geometric series in Exercise 11 and DeMoivre's theorem to derive the following formulas for $0 < \theta < 2\pi$.

$$1 + \cos\theta + \cos 2\theta + \cdots + \cos(n\theta) = \frac{\cos(n\theta/2)\sin\left[(n+1)\theta/2\right]}{\sin(\theta/2)},$$

$$\sin\theta + \sin 2\theta + \sin 3\theta + \cdots + \sin(n\theta) = \frac{\sin(n\theta/2)\sin\left[(n+1)\theta/2\right]}{\sin(\theta/2)}.$$

13. If n is a positive integer greater than or equal to 2, prove that

$$\cos\left(\frac{2\pi}{n}\right) + \cos\left(\frac{4\pi}{n}\right) + \cdots + \cos\left[\frac{2(n-1)\pi}{n}\right] = -1,$$

and that

$$\sin\left(\frac{2\pi}{n}\right) + \sin\left(\frac{4\pi}{n}\right) + \cdots + \sin\left[\frac{2(n-1)\pi}{n}\right] = 0.$$

Hint: Use the result of Exercise 11 and take $z = 1$.

1.5 LOCI, POINTS, SETS, AND REGIONS IN THE COMPLEX PLANE

In the previous section we saw that there is a specific point in the z-plane that represents any complex number z. Similarly, as we will see, curves and areas in the z-plane can represent equations or inequalities in the variable z.

Consider the equation $\text{Re}(z) = 1$. If this is rewritten in terms of x and y, we have $\text{Re}(x + iy) = 1$, or $x = 1$. In the complex plane the locus of all points satisfying $x = 1$ is the infinite vertical line shown in Fig. 1.5–1. Now consider the inequality $\text{Re}\, z < 1$, which is equivalent to $x < 1$. All points that satisfy this inequality must lie in the region[†] to the left of the vertical line in Fig. 1.5–1. We show this in Fig. 1.5–2.

Similarly, the double inequality $-2 \le \text{Re}\, z \le 1$, which is identical to $-2 \le x \le 1$, is satisfied by all points lying between and on the vertical lines $x = -2$ and $x = 1$. Thus $-2 \le \text{Re}\, z \le 1$ defines the infinite strip shown in Fig. 1.5–3.

More complicated regions can be likewise described. For example, consider $\text{Re}\, z \le \text{Im}\, z$. This implies $x \le y$. The equality holds when $x = y$, that is, for all the points on the infinite line shown in Fig. 1.5–4. The inequality $\text{Re}\, z < \text{Im}\, z$ describes those points that satisfy $x < y$, that is, they lie to the left of the 45° line in Fig. 1.5–4. Thus $\text{Re}\, z \le \text{Im}\, z$ represents the shaded region shown in the figure and includes the boundary line $x = y$.

The description of circles and their interiors is particularly important and easily accomplished. The locus of all points representing $|z| = 1$ is obviously the same as

---

[†] A precise definition of "region" is given further in the text. For the moment we will regard the word as meaning some portion of the total z-plane.

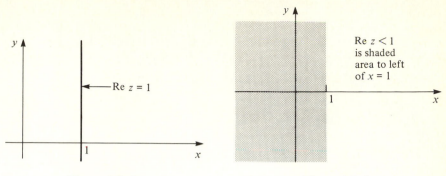

Figure 1.5–1   **Figure 1.5–2**

those for which $\sqrt{x^2 + y^2} = 1$, that is, the circumference of a circle, centered at the origin, of unit radius. The inequality $|z| < 1$ describes the points inside the circle (their modulus is less than unity), whereas $|z| \leq 1$ represents the inside and the circumference.

We need not restrict ourselves to circles centered at the origin. Let $z_0 = x_0 + iy_0$ be a complex constant. Then the points in the z-plane representing solutions of $|z - z_0| = r$, where $r > 0$, form the circumference of a circle, of radius r, centered at x_0, y_0. This statement can be proved by algebraic or geometric means. The latter course is followed in Fig. 1.5–5, where we use the vector representation of complex numbers. A vector for z_0 is drawn from the origin to the fixed point x_0, y_0 while another, for z, goes to the variable point whose coordinates are x, y. The vector difference $z - z_0$ is also shown. If this quantity is kept constant in magnitude, then z is obviously confined to the perimeter of the circle indicated. The points representing solutions of $|z - z_0| < r$ lie inside the circle, whereas those for which $|z - z_0| > r$ lie outside.

Finally, let r_1 and r_2 be a pair of nonnegative real numbers so that $r_1 < r_2$. Then the double inequality $r_1 < |z - z_0| < r_2$ is of interest. The first part, $r_1 < |z - z_0|$, specifies those points in the z-plane that lie outside a circle of radius r_1 centered at

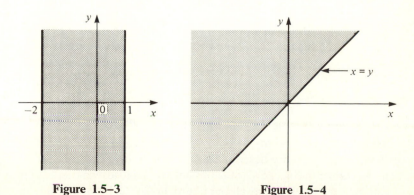

Figure 1.5–3   **Figure 1.5–4**

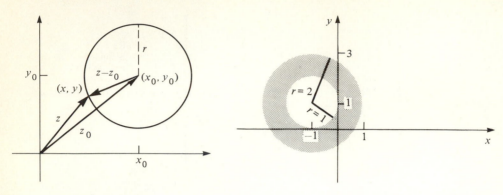

Figure 1.5–5   **Figure 1.5–6**

x_0, y_0, whereas the second part $|z - z_0| < r_2$ refers to those points inside a circle of radius r_2 centered at x_0, y_0. Points that simultaneously satisfy both inequalities must lie in the *annulus* (a disc with a hole in the center) of inner radius r_1, outer radius r_2, and center z_0.

Example 1

What region is described by the inequality $1 < |z + 1 - i| < 2$?

Solution

We can write this as $r_1 < |z - z_0| < r_2$, where $r_1 = 1$, $r_2 = 2$, $z_0 = -1 + i$. The region described is the shaded area *between*, but not including, the circles shown in Fig. 1.5–6. ◀

Points and Sets

We need to have a small vocabulary with which to describe various points and collections of points (called *sets*) in the complex plane. The following terms are worth studying and memorizing since most of the language will reappear in subsequent chapters.

A *neighborhood* of radius r of a point z_0 is the collection of all the points inside a circle, of radius r, centered at z_0. These are the points satisfying $|z - z_0| < r$. A given point can have various neighborhoods since circles of different radii can be constructed around the point.

A *deleted neighborhood* of z_0 consists of the points inside a circle centered at z_0 but excludes the point z_0 itself. These points satisfy $0 < |z - z_0| < r$.

An *open set* (of points) is one in which every member of the set has some neighborhood, all of whose points lie entirely within that set. For example, the set $|z| < 1$ is open. This inequality describes all the points inside a unit circle centered at the origin. As shown in Fig. 1.5–7, it is possible to enclose every such point with a circle C_0 (perhaps very tiny) so that all the points inside C_0 lie within the unit circle. The set $|z| \leq 1$ is not open. Points on and inside the circle $|z| = 1$ belong to this

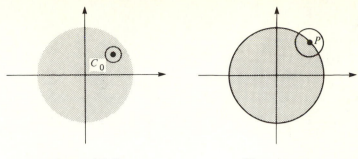

Figure 1.5–7   Figure 1.5–8

set. But every neighborhood, no matter how tiny, of a point such as P that lies on $|z| = 1$ (see Fig. 1.5–8) contains points outside the given set.

A *connected set* is one in which any two points of the set can be joined by some curve, all of whose points belong to the set. Thus the set of points shown shaded in Fig. 1.5–9(a) is not connected since we cannot join a and b by a line within the set. However, the set of points in Fig. 1.5–9(b) is connected.

A *domain* is an open connected set. For example Re $z < 3$ describes a domain. However, Re $z \le 3$ does not describe a domain since the set defined is not open.

We will often speak of *simply* and *multiply connected domains*. Loosely speaking, a simply connected domain contains no holes, but a multiply connected domain has one or more holes. An example of the former is $|z| < 2$, and an example of the latter is $1 < |z| < 2$, which contains a circular hole. More precisely, when any closed curve is constructed in a simply connected domain, every point inside the curve lies in the domain. On the other hand, it is always possible to construct some closed curve inside a multiply connected domain in such a way that one or more points inside the curve do not belong to the domain (see Fig. 1.5–10). A doubly connected domain has one hole, a triply connected domain two holes, etc.

A *boundary point* of a set is a point whose every neighborhood contains at least one point belonging to the set and one point not belonging to the set.

Consider the set described by $|z - 1| \le 1$, which consists of the points inside and on the circle shown in Fig. 1.5–11. The point $z = 1 + i$ is a boundary point of

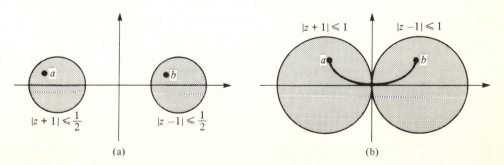

(a)   (b)

Figure 1.5–9

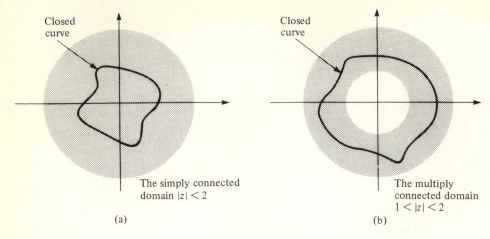

(a)

The simply connected
domain $|z| < 2$

Closed
curve

(b)

The multiply
connected domain
$1 < |z| < 2$

Closed
curve

Figure 1.5–10

the set since, inside every circle such as C_0, there are points belonging to and not belonging to the given set. Although in this case the boundary point is a member of the set in question, this need not always be so. For example, $z = 1 + i$ is a boundary point but not a member of the set $|z - 1| < 1$. One should realize that an open set cannot contain any of its boundary points.

An *interior point* of a set is a point having some neighborhood, all of whose points belong to the set. Thus $z = 1 + i/2$ is an interior point of the set $|z - 1| \leq 1$ (see Fig. 1.5–11).

A *region* is a domain plus possibly some, none, or all the boundary points of the domain. Thus every domain is a region, but not every region is a domain. The set defined by $2 < \text{Re } z \leq 3$ is a region. It contains some of its boundary points (on $\text{Re } z = 3$) but not others (on $\text{Re } z = 2$) (see Fig. 1.5–12). This particular region is not a domain.

A *closed region* consists of a domain plus all the boundary points of the domain.

A *bounded set* is one whose points can be enveloped by a circle of some finite radius. For example, the set occupying the square $0 \leq \text{Re } z \leq 1$, $0 \leq \text{Im } z \leq 1$ is bounded since we can put a circle around it (see Fig. 1.5–13). A set that cannot be

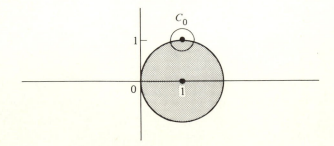

Figure 1.5–11

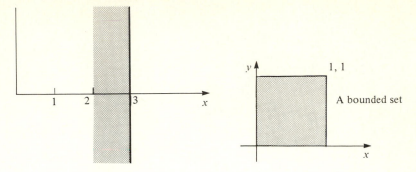

Figure 1.5–12   **Figure 1.5–13**

encompassed by a circle is called *unbounded*. An example is the infinite strip in Fig. 1.5–12.

The Complex Number Infinity and the Point at Infinity

When dealing with real numbers, we frequently use the concept of infinity and speak of "plus infinity" and "minus infinity." For example, the sequence $1, 10, 100, 1000, \ldots$ diverges to plus infinity, and the sequence $-1, -2, -4, -8, \ldots$ diverges to minus infinity.

In dealing with complex numbers we also speak of infinity, which we call "the complex number infinity." It is designated by the usual symbol ∞. We do not give a sign to the complex infinity nor do we define its argument. Its modulus, however, is an infinitely large positive real number. The complex number ∞ is defined so as to satisfy the following rules:

$$\frac{z}{\infty} = 0, \qquad z \pm \infty = \infty, \qquad (z \neq \infty), \qquad \frac{z}{0} = \infty, \qquad (z \neq 0),$$

$$z \cdot \infty = \infty, \qquad (z \neq 0), \qquad \frac{\infty}{z} = \infty, \qquad (z \neq \infty).$$

We do not define

$$\infty + \infty, \qquad \infty - \infty, \qquad \text{or} \qquad \frac{\infty}{\infty}.$$

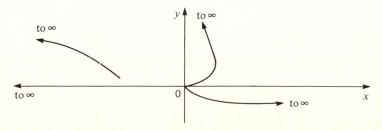

Figure 1.5–14

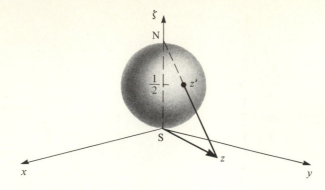

Figure 1.5–15

We can imagine that the complex number infinity is represented graphically by a point in the Argand plane—a point, unfortunately, that we can never draw in this plane. The point can be reached by proceeding along any path in which $|z|$ grows without bound, as, for instance, is shown in Fig. 1.5–14.

In order to make the notion of a point at infinity more tangible, we use an artifice called the stereographic projection illustrated in Fig. 1.5–15.

Consider the z-plane, with a third orthogonal axis, the ζ-axis,[†] added on. A sphere of radius $1/2$ is placed with center at $x = 0$, $y = 0$, $\zeta = 1/2$. The north pole, N, lies at $x = 0$, $y = 0$, $\zeta = 1$ while the south pole, S, is at $x = 0$, $y = 0$, $\zeta = 0$. This is called the Riemann number sphere.

Let us draw a straight line from N to the point in the xy-plane that represents a complex number z. This line intersects the sphere at exactly one point, which we label z'. We say that z' is the projection on the sphere of z. In this way *every* point in the complex plane can be projected on to a corresponding unique point on the sphere. Points far from the origin in the xy-plane are projected close to the top of the sphere, and, as we move farther from the origin in the plane, the corresponding projections on the sphere cluster more closely around N. Thus we conclude that N on the sphere corresponds to the point at infinity, although we are not able to draw $z = \infty$ in the complex plane.

When we regard the z-plane as containing the point at infinity, we call it the "extended z-plane." When ∞ is not included, we merely say "the z-plane" or "the finite z-plane."

EXERCISES

1. Describe with words or a sketch the portion of the complex plane corresponding to the following equations or inequalities. State which problems have no solutions.

   a) $\text{Im } z = 2$ c) $\text{Re } z = -\text{Im } z$

   b) $\text{Im } z = 2i$ d) $|z + i| \leq 2$

[†]Obviously, we don't want to call this the z-axis.

e) $|z + 1 - 2i| > 2$ i) $\mathrm{Re}\, z \leq \mathrm{Re}(z^2)$

f) $|z + 3 - 4i| = 5$ j) $z = \mathrm{Log}\,|z|$ (Do this numerically on a calculator. Log is base e.)

g) $\mathrm{Im}(z + i) = \mathrm{Re}(z - 1)$ k) $z\bar{z} + z + \bar{z} < \mathrm{Im}\, z$

h) $z\bar{z} = \mathrm{Im}(z)$

2. Represent the following regions by means of equations or inequalities in the variable z.

     a) All the points lying on and exterior to a circle of unit radius centered at $-1 - i$.

     b) All the points occupying an annular region centered at $3 + i$. The inner radius is 2, the outer is 4. Exclude points on the inner boundary but include those on the outer one.

     c) All the points, except the center, on and within a circle of radius 2, centered at $3 + 4i$.

3. Consider the open set described by $|z| < 1$. Find a neighborhood of the point $0.99i$ that lies entirely within the set. Represent the neighborhood with an inequality involving z. In the same way specify a deleted neighborhood of $0.99i$ within the set.

4. Explain why $-1 < \mathrm{Im}\, z \leq 2$ does not describe an open set.

5. a) Explain why the set in the z-plane described by $|\mathrm{Re}\, z| > |\mathrm{Im}\, z|$ is not connected.

     b) Is the set given by $|\mathrm{Re}\, z| \leq |\mathrm{Im}\, z|$ connected? Is it a domain? Explain.

6. a) Explain why the set of points in the z-plane that satisfies $|\mathrm{Re}\, z| > 1$ does not constitute a domain.

     b) Does $\mathrm{Re}\, z \geq 1$ constitute a domain? Explain. What about $\mathrm{Re}\, z > 1$?

7. What are the boundary points of the sets defined below? State which boundary points belong to the set.

     a) $0 < |z - 3i| < 1$ e) $\sin |z| < 1$

     b) $2 < |z - 3i| \leq 3$ f) $\cos(1/n) + i \sin(1/n)$, where $n < \infty$ assumes all positive integer values

     c) $\mathrm{Re}\, z > 2\,\mathrm{Im}\, z$

     d) $\sin |z| < 1/2$

8. A set of points s in the xy-plane is called *closed* if the points in the xy-plane, not in s, form an open set. Which of the following sets are closed?

     a) $|z| \leq 1$ d) $1 < |z| \leq 2$ g) $z = 1$

     b) $|z| < 1$ e) $\mathrm{Re}\, z \leq 1$ h) $0 \leq \arg(z) \leq \pi$

     c) $0 < |z| \leq 1$ f) $\mathrm{Re}\, z = 1$ i) $\mathrm{Log}\,|z| > 0$

9. a) When all the points on the unit circle $|z| = 1$ are projected stereographically on to the sphere of Fig. 1.5–15, where do they lie?

b) Where are all the points inside the unit circle projected?

c) Where are all the points outside the unit circle projected?

10. Use stereographic projection to justify the statement that the two semiinfinite lines $y = x$, $x \geq 0$ and $y = -x$, $x \leq 0$ intersect twice, once at the origin and once at infinity. What is the projection of each of these lines on the Riemann number sphere?

CHAPTER 2 THE COMPLEX FUNCTION AND ITS DERIVATIVE

2.1 INTRODUCTION

In studying elementary calculus the reader doubtless received ample exposure to the concept of a real function of a real variable. To review briefly: When y is a function of x, or $y = f(x)$, we mean that when a value is assigned to x there is at our disposal a method for determining a corresponding value of y. We term x the independent variable and y the dependent variable in the relationship. Often y will be specified only for certain values of x and left undetermined for others. In the following table, for example, y is a known function of x only when x has one of the integer values 1 through 4.

x	$y = f(x)$
1	3.141
2	1.772
3	1.331
4	1.153

Of course, there are other ways to express a functional relationship besides using a table of pairs of values of x and y. The most common method involves a mathematical formula as in the expression $y = e^x$, $-\infty < x < \infty$, which, in this case, yields a value of y for any value of x. Occasionally, we require several formulas, as in the following: $y = e^x$, $x > 0$; $y = \sin x$, $x < 0$. Taken together, these expressions determine y for any value of x except zero, that is, y is undefined at $x = 0$.

Recall that there are multivalued functions where, given x, we may find two or more corresponding values for y, as in $y = (x)^{1/2}$. In this book the word "function" applies to single-valued functions (no more than one y for each x) unless a remark is made to the contrary.

The easiest way to visualize most functional relationships is by means of a graphical plot, and the reader doubtless spent time in high school drawing y versus x, in the Cartesian plane, for various functions.

Some, but not all, of these concepts carry over directly into the study of functions of a complex variable. Here we use an independent variable, usually z, that can assume complex values. We will be concerned with functions defined in a domain or region of the complex z-plane. To each value of z in the domain there will correspond a value of a dependent variable,[†] let us say w, and we will say that w is a function of z, or $w = f(z)$ in this domain. Often the domain will be the entire z-plane. We must assume that w, like z, is capable of assuming values that are complex, real, or purely imaginary. Some examples follow.

	$w = f(z)$	Domain of definition
a)	$w = 2z$	all z
b)	$w = |z|$	all z
c)	$w = 2i|z|^2$	all z
d)	$w = 1/(z^2 + 9)$	all z except $\pm 3i$

Example (a) is quite straightforward. If z assumes a complex value, say $3 + i$, then $w = 6 + 2i$. If z happens to be purely real, w is also.

In example (b), w assumes only real values irrespective of whether z is real, complex, or purely imaginary; for example, if $z = 3 + i$, $w = \sqrt{10}$.

Conversely, in example (c), w is purely imaginary for all z; for example, if $z = 3 + i$, $w = 2i|3 + i|^2 = 20i$.

Finally, in example (d), $1/(z^2 + 9)$ cannot define a function of z whenever z equals $\pm 3i$ since the denominator of this fraction vanishes there.

The function $w(z)$ is sometimes expressed in terms of the variables x and y rather than directly in z. For example, $w(z) = 2x^2 + iy$ is a function of the variable z since, with z known, x and y are determined. Thus if $z = 3 + 4i$, then $w(3 + 4i) = 2 \cdot 3^2 + 4i = 18 + 4i$. Often an expression for w, given in terms of x and y, can be rewritten rather simply in terms of z; in other cases the z-notation is rather cumbersome. In any case, the identities

$$x = \frac{z + \bar{z}}{2}, \qquad y = \frac{1}{i}\frac{(z - \bar{z})}{2} \qquad (2.1\text{–}1)$$

are useful if we wish to convert from the xy-variables to z. An example follows.

Example 1

Express w directly in terms of z if

$$w(z) = 2x + iy + \frac{x - iy}{x^2 + y^2}.$$

Solution

Using Eq. (2.1–1), we rewrite this as

$$w(z) = (z + \bar{z}) + \frac{i(z - \bar{z})}{i2} + \frac{\bar{z}}{z\bar{z}} = \frac{3z}{2} + \frac{\bar{z}}{2} + \frac{1}{z}. \qquad ◀$$

[†]As in the real variable case, we can have single-valued and multivalued functions. A single-valued function is under discussion unless otherwise stated.

In general, $w(z)$ possesses both real and imaginary parts, and we write this function in the form $w(z) = u(z) + iv(z)$, or

$$w(z) = u(x, y) + iv(x, y), \qquad (2.1-2)$$

where u and v are real functions of the variables x and y. In Example 1 we have

$$u = 2x + \frac{x}{x^2 + y^2} \quad \text{and} \quad v = y - \frac{y}{x^2 + y^2}.$$

A difference between a function of a complex variable $u + iv = f(z)$ and a real function of a real variable $y = f(x)$ is that while we can usually plot the relationship $y = f(x)$ in the Cartesian plane, graphing is not so easily done with the complex function. Two numbers x and y are required to specify any z, and another pair of numbers is required to state the resulting values of u and v. Thus, in general, a four-dimensional space is required to plot $w = f(z)$ with two dimensions reserved for the independent variable z and the other two used for the dependent variable w.

For obvious reasons four-dimensional graphs are not a convenient means for studying a function. Instead, other techniques are employed to visualize $w = f(z)$. This matter is discussed at length in Chapter 8, and some readers may wish to skip to Sections 8.1–8.3 at this time. A small glimpse of one useful technique is in order here, however.

Two coordinate planes, the z-plane with x- and y-axes and the w-plane with u- and v-axes, are drawn side-by-side. Now consider a complex number A, which lies in the z-plane within the domain for which $f(z)$ is defined. The value of w that corresponds to A is $f(A)$. We call $A' = f(A)$. The pair of numbers A and A' are now plotted in the z- and w-planes, respectively (see Fig. 2.1–1). We say that the complex number A' is the *image* of A under the mapping $w = f(z)$.

In order to study a particular function $f(z)$ we can plot some points in the z-plane and also their corresponding images in the uv-plane. In the following table and in Fig. 2.1–2 we have investigated a few points in the case of $w = f(z) = z^2 + z$.

z	$w = z^2 + z$
$A = 0$	$0 = A'$
$B = 1$	$2 = B'$
$C = 1 + i$	$1 + 3i = C'$
$D = i$	$-1 + i = D'$

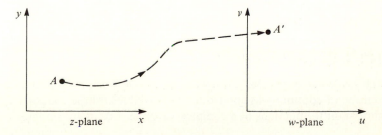

Figure 2.1–1

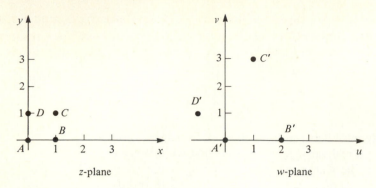

Figure 2.1–2

EXERCISES

1. Suppose $f(z) = 1/(z^2 + 1)$. Does $f(z)$ fail to be defined at any point within each of the following domains? If so, state the point or points.

   a) $|z| < 1/2$ b) $|z| < 1$ c) $|z| < 2$ d) $|z - 5i| < 6$

2. Find the numerical value of $f(1 + 2i)$ in the form $a + ib$ for each of the following functions $f(z)$.

   a) $1/(z^2 + 1)$ c) $yx^2 + i2yx$

   b) $z^3 + z$ d) $(x^2 + y^2)\sin y + i(y^2 - x^2)e^x$

3. Write the following functions of z in the form $u(x, y) + iv(x, y)$, where u and v are real explicit functions of x and y.

   a) $f(z) = z^2 + 2z + 1$ c) $f(z) = z + (\bar{z})^2$

   b) $f(z) = z^3$ d) $f(z) = z^{-1} + z^{-2}$

4. Rewrite the following complex functions entirely in terms of the complex variable z and, if necessary, its conjugate.

   a) $f(z) = xy + ixy$ b) $f(z) = (x^2 + y^2) + ixy$ c) $f(z) = y + ix$

5. If $f(\bar{z}) = \overline{f(z)}$ throughout the complex plane, show that $f(z)$ is real everywhere on the real axis.

6. For each of these functions tabulate the value of the function for these values of z: $1 + i0$, $1 + i$, $0 + i$, $-1 - i$. Indicate graphically the correspondence between values of w and values of z by means of a diagram like Fig. 2.1–2.

   a) $w = z^3$ c) $w = i/|z|$

   b) $w = 1/z$ d) $w = \arg(z)$,   principal value

2.2 LIMITS AND CONTINUITY

In elementary calculus the reader learned the notion of the limit of a function as well as the definition of continuity as applied to real variables. These concepts apply with some modification to functions of a complex variable. Let us first briefly review the real case.

The function $f(x)$ has a limit f_0 as x tends to x_0 (written $\lim_{x \to x_0} f(x) = f_0$) if the difference between $f(x)$ and f_0 can be made as small as we wish provided we choose x sufficiently close to x_0. In mathematical terms, given any positive number ε, we have

$$|f(x) - f_0| < \varepsilon \qquad (2.2\text{--}1)$$

if x satisfies

$$0 < |x - x_0| < \delta, \qquad (2.2\text{--}2)$$

where δ is a positive number typically dependent upon ε. Note that x never precisely equals x_0 in Eq. (2.2–2) and that $f(x_0)$ need not be defined for the limit to exist.

An obvious example of a limit is $\lim_{x \to 1}(1 + 2x) = 3$. To demonstrate this rigorously note that Eq. (2.2–1) requires $|1 + 2x - 3| < \varepsilon$, which is equivalent to

$$|x - 1| < \varepsilon/2. \qquad (2.2\text{--}3)$$

Since $x_0 = 1$, Eq. (2.2–2) becomes

$$0 < |x - 1| < \delta. \qquad (2.2\text{--}4)$$

Thus Eq. (2.2–3) can be satisfied if we choose $\delta = \varepsilon/2$ in Eq. (2.2–4).

A more subtle example proved in elementary calculus is

$$\lim_{x \to 0} \frac{\sin x}{x} = 1.$$

An intuitive verification can be had from a plot of $\sin x/x$ as a function of x and the use of $\sin x \approx x$ for $x \ll 1$.

Let us consider two functions that fail to possess limits at certain points. The function $f(x) = 1/(x - 1)^2$ fails to possess a limit at $x = 1$ because this function becomes unbounded as x approaches 1. The expression $|f - f_0|$ in Eq. (2.2–1) is unbounded for x satisfying Eq. (2.2–2) regardless of what value is assigned to f_0.

Now consider $f(x) = u(x)$, where $u(x)$ is the unit step function (see Fig. 2.2–1) defined by

$$u(x) = 0, \qquad x < 0, \qquad u(x) = 1, \qquad x \geq 0.$$

We investigate the limit of $f(x)$ at $x = 0$. With $x_0 = 0$, Eq. (2.2–2) becomes

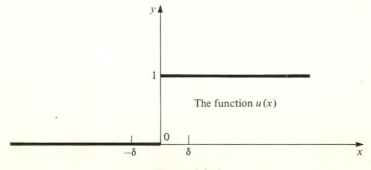

The function $u(x)$

Figure 2.2–1

$0 < |x| < \delta$. Notice x can lie to the right or to the left of 0. The left side of Eq. (2.2–1) is now either $|1 - f_0|$ or $|f_0|$ according to whether x is positive or negative. If $\varepsilon < 1/2$, it is impossible to simultaneously satisfy the inequalities $|1 - f_0| < \varepsilon$ and $|f_0| < \varepsilon$, irrespective of the value of f_0. We see that a function having a "jump" at x_0 cannot have a limit at x_0.

For $f(x)$ to be *continuous* at a point x_0, $f(x_0)$ must be defined, and $\lim_{x \to x_0} f(x)$ must exist. Furthermore, these two quantities must agree, that is,

$$\lim_{x \to x_0} f(x) = f(x_0). \tag{2.2–5}$$

A function that fails to be continuous at x_0 is said to be *discontinuous* at x_0.

The functions $1/(x - 1)^2$ and $u(x)$ fail to be continuous at $x = 1$ and $x = 0$ respectively because they do not possess limits at these points. The function

$$f(x) = \begin{cases} \dfrac{\sin x}{x}, & x \neq 0, \\[2mm] 2, & x = 0, \end{cases}$$

is discontinuous at $x = 0$. We have that $\lim_{x \to 0} f(x) = 1$, and $f(0) = 2$; thus Eq. (2.2–5) is not satisfied. However, one can show that $f(x)$ is continuous for all $x \neq 0$.

The concept of a limit can be extended to complex functions of a complex variable according to the following definition.

Definition   Limit

Let $f(z)$ be a complex function of the complex variable z, and let f_0 be a complex constant. If for every real number $\varepsilon > 0$ there exists a real number $\delta > 0$ such that

$$|f(z) - f_0| < \varepsilon \tag{2.2–6}$$

for all z satisfying

$$0 < |z - z_0| < \delta, \tag{2.2–7}$$

then we say that

$$\lim_{z \to z_0} f(z) = f_0;$$

that is, $f(z)$ has limit f_0 as z tends to z_0. ∎

The definition asserts that ε, the upper bound on the magnitude of the difference between $f(z)$ and its limit f_0, can be made arbitrarily small, provided that we confine z to a deleted neighborhood of z_0. The radius, δ, of this deleted neighborhood typically depends on ε and becomes smaller with decreasing ε.

When we compute $\lim_{x \to x_0} f(x)$ we are concerned with values of x lying to the right and left of x_0. If the limit f_0 exists, then as x approaches x_0 from either the right or left $f(x)$ must become increasingly close to f_0. In the case of the step function $u(x)$ previously considered, $\lim_{x \to 0} f(x)$ fails to exist because as x shrinks toward zero from the right (positive x), $f(x)$ remains at 1; but if x shrinks toward zero from the left (negative x), $f(x)$ remains at zero.

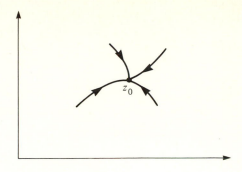

Figure 2.2–2

In the complex plane the concept of limit is more complicated because there is an infinity of *paths*, not just two directions, along which we can approach z_0. Four such paths are shown in Fig. 2.2–2. If $\lim_{z \to z_0} f(z)$ exists, $f(z)$ must tend toward the same numerical value no matter which of the infinite number of paths of approach to z_0 is selected. Let us now consider two functions that fail to possess limits at certain points.

Example 1

Let $f(z) = \arg z$ (principal value). Show that $f(z)$ fails to possess a limit on the negative real axis.

Solution

Consider a point z_0 on the negative real axis. Refer to Fig. 2.2–3. Every neighborhood of such a point contains values of $f(z)$ (in the second quadrant) that are arbitrarily near to π and values of $f(z)$ (in the third quadrant) that are arbitrarily near to $-\pi$. Approaching z_0 on two different paths such as C_1 and C_2, we see that $\arg z$ tends to two different values. Therefore $\arg z$ fails to possess a limit at z_0. ◄

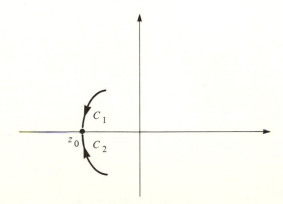

Figure 2.2–3

Example 2

Let

$$f(z) = \frac{x^2 + x}{x + y} + \frac{i(y^2 + y)}{x + y}.$$

This function is undefined at $z = 0$. Show that $\lim_{z \to 0} f(z)$ fails to exist.

Solution

Let us move toward the origin along the y-axis. With $x = 0$ in $f(z)$ we have

$$f(z) = \frac{i(y^2 + y)}{y} = i(y + 1).$$

As the origin is approached, the limit of this expression is i.

Next we move toward the origin from along the x-axis. With $y = 0$ we have $f(z) = x + 1$. As the origin is approached, the limit of this expression is 1. Because our two results disagree, $\lim_{z \to 0} f(z)$ fails to exist. ◀

Sometimes we will be concerned with the limit of a function $f(z)$ as $z \to \infty$. This limit, which is written as $\lim_{z \to \infty} f(z) = f_0$, means that, when we are given any $\varepsilon > 0$, there exists a number r such that $|f(z) - f_0| < \varepsilon$ for $|z| > r$. Thus the magnitude of the difference between $f(z)$ and f_0 can be made arbitrarily small as long as we are at any point farther than r from the origin. For example, $\lim_{z \to \infty} 1/z^2 = 0$, $\lim_{z \to \infty} 1/(1 + z^{-2}) = 1$. The proofs are left to the reader.

Formulas pertaining to limits that the reader studied in elementary calculus have counterparts for functions of a complex variable. These counterparts, stated here without proof, can be established from the definition of a limit.

Theorem 1

Let $f(z)$ have limit f_0 as $z \to z_0$ and $g(z)$ have limit g_0 as $z \to z_0$. Then

$$\lim_{z \to z_0} (f(z) + g(z)) = f_0 + g_0, \qquad (2.2\text{--}8a)$$

$$\lim_{z \to z_0} (f(z)g(z)) = f_0 g_0, \qquad (2.2\text{--}8b)$$

$$\lim_{z \to z_0} [f(z)/g(z)] = f_0/g_0, \quad \text{if } g_0 \neq 0. \quad \blacksquare \qquad (2.2\text{--}8c)$$

The definition of continuity for complex functions of a complex variable is analogous to that for real functions of a real variable.

Definition   Continuity

A function $w = f(z)$ is continuous at $z = z_0$ provided the following conditions are both satisfied:

a) $f(z_0)$ is defined,

b) $\lim_{z \to z_0} f(z)$ exists, and

$$\lim_{z \to z_0} f(z) = f(z_0). \quad \blacksquare \qquad (2.2\text{--}9)$$

Generally we will be dealing with functions that fail to be continuous only at certain points or along some locus in the z-plane. We can usually recognize points of discontinuity as places where a function becomes infinite or undefined or exhibits an abrupt change in value.

> If a function is continuous at all points in a region, we say that it is continuous in the region.

The principal value of arg z is discontinuous at all points on the negative real axis because it fails to have a limit at every such point. Moreover, arg z is undefined at $z = 0$, which means that arg z is discontinuous there as well.

Example 3

Investigate the continuity at $z = 1$ of the function

$$f(z) = \begin{cases} \left(\dfrac{z^2 - 1}{z - 1} \right), & z \neq 1, \\ 3, & z = 1. \end{cases}$$

Solution

Because $f(1)$ is defined, part (a) in our definition of continuity is satisfied.

If $z \neq 1$, we factor the numerator in the expression given above and find that $f(z)$ is identical to $z + 1$. From this we might conclude that $\lim_{z \to 1} f(z) = 2$. This is rigorously proved (and the reader should prove it) with the aid of Eqs. (2.2–6) and (2.2–7) (see Exercise 1).

Because $f(1) = 3$, which is not equal to 2, condition (b) in our definition of continuity is not satisfied. Thus $f(z)$ is discontinuous at $z = 1$. However, $f(z)$ is continuous for all $z \neq 1$. Notice too that a function identical to $f(z)$ given in the expression above, but satisfying $f(1) = 2$, is continuous for all z. ◀

There are a number of important properties of continuous functions that we will be using. Although the truth of the following theorem may seem self-evident, in certain cases the proofs are not easy, and the reader is referred to a more advanced text for them.[†]

Theorem 2

a) *Sums*, *differences*, and *products* of continuous functions are themselves continuous functions. The *quotient* of a pair of continuous functions is continuous except where the denominator equals zero.

b) A continuous function of a continuous function is a continuous function.

[†]See, for example, W. Kaplan, *Introduction to Analytic Functions* (Reading, Mass.: Addison-Wesley, 1966), p. 17.

c) Let $f(z) = u(x, y) + iv(x, y)$. The functions $u(x, y)$ and $v(x, y)$ are continuous[†] at any point where $f(z)$ is continuous. Conversely, at any point where u and v are continuous, $f(z)$ is also.

d) If $f(z)$ is continuous in some region R, then $|f(z)|$ is also continuous in R. If R is bounded and closed there exists a positive real number, say M, such that $|f(z)| \leq M$ for all z in R. The equality holds for at least one value of z in R. ∎

We can use part (a) of the theorem to investigate the continuity of the quotient $(z^2 + z + 1)/(z^2 - 2z + 1)$. Since $f(z) = z$ is obviously a continuous function of z, so is the product $z \cdot z = z^2$. Any constant is a continuous function. Thus the sum $z^2 + z + 1$ is continuous for all z and by similar reasoning so is $z^2 - 2z + 1$. The quotient of these two polynomials is therefore continuous except where $z^2 - 2z + 1 = (z - 1)^2$ is zero. This occurs only at $z = 1$.

A similar procedure applies to any rational function $P(z)/Q(z)$, where P and Q are polynomials of any degree in z. Such an expression is continuous except for values of z satisfying $Q(z) = 0$.

The usefulness of part (b) of the theorem will be more apparent in the next chapter, where we will study various transcendental functions of z. We will learn what is meant by $f(z) = e^z$, where z is complex,[‡] and we will find that this function is continuous for all z. Now, $g(z) = 1/z^2$ is continuous for all $z \neq 0$. Thus $f(g(z)) = \exp(1/z^2)$ is also continuous for $z \neq 0$.

As an illustration of parts (c) and (d), consider $f(z) = e^x \cos y + ie^x \sin y$ in the disc-shaped region R given by $|z| \leq 1$. Since $u = e^x \cos y$ and $v = e^x \sin y$ are continuous in R, $f(z)$ is also. Thus $|f(z)|$ must be continuous in R. Now, $|f(z)| = \sqrt{\exp(2x)[\cos^2 y + \sin^2 y]} = e^x$, which is indeed a continuous function. The maximum value achieved by $|f(z)|$ in R will occur when e^x is maximum, that is, at $x = 1$. Thus $|f(z)| \leq e$ in R, and the constant M in part (d) of the theorem here equals e.

EXERCISES

1. Show by using the definition of a limit that $\lim_{z \to 1}(z + 1) = 2$. Note the usefulness of this result in Example 3, p. 49.

2. Using Theorem 2 or the definition of continuity, explain why the following functions are continuous for the values of z indicated.

   a) $z^3 + z^2$, all z

   b) $z + z^{-1}$, all z except $z = 0$

   c) $|z|$, all z

[†]Continuity for $u(x,y)$, a *real* function of two real variables, is defined in a way analogous to continuity for $f(z)$. For continuity at (x_0,y_0) the difference $|u(x,y) - u(x_0,y_0)|$ can be made smaller than any positive ε for all (x,y) lying inside a circle of radius δ centered at (x_0,y_0).

[‡]e^z is also written $\exp(z)$.

d) $\bar{z}$,   all z

e) $(z^2 - |z|)^{-1}$,   all z except $z = \pm 1$ and $z = 0$

f) $\dfrac{x^2 + iy^2}{z}$,   all z except $z = 0$

g) $\dfrac{x + iy}{x - iy}$,   all z except $z = 0$

3. Consider

$$f(z) = \frac{xe^x}{\sqrt{x^2 + y^2}} + \frac{iye^y}{\sqrt{x^2 + y^2}}.$$

Show that $\lim_{z \to 0} f(z)$ fails to exist. *Hint:* Compare values of $f(z)$ as z approaches the origin first along the line $y = 0$, $x > 0$ and then along the line $x = 0$, $y > 0$.

4. Is the following function $f(z)$ continuous at $z = i$? Give an explanation like the one provided in Example 3.

$$f(z) \begin{cases} \dfrac{z^2 + 1}{z - i}, & z \neq i \\ 2i, & z = i \end{cases}$$

5. The function $f(z) = (z^2 - 4)/(z + 2)$ is defined and continuous for all $z \neq -2$. How should we define $f(-2)$ so that $f(z)$ is continuous throughout the z-plane?

6. Consider $f(z) = z^2$.

a) In the region R described by $|z| \leq 2$, we have $|f(z)| \leq M$. Find M assuming $|f(z)| = M$ for some z in R.

b) Repeat part (a) but take R as the region $|z - 1| \leq 2$.

c) Repeat part (a) but take $f(z) = 1/z$ and R as $|z - 1 - i| \leq 1$.

2.3   THE COMPLEX DERIVATIVE

Review

Before discussing the derivative of a function of a complex variable, let us briefly review some facts concerning the derivative of a function of a real variable $f(x)$. The derivative of $f(x)$ at x_0, which is written $f'(x_0)$, is given by

$$f'(x_0) = \lim_{\Delta x \to 0} \frac{f(x_0 + \Delta x) - f(x_0)}{\Delta x}. \tag{2.3–1}$$

If the limit in this expression fails to exist, $f'(x_0)$ is undefined and $f(x)$ has no derivative (is not differentiable) at x_0.

If $f(x)$ is not continuous at x_0, $f'(x_0)$ does not exist. However, a function can be continuous and *still not* have a derivative. In Eq. (2.3–1), Δx is a small increment, shrinking progressively to zero, in the argument of $f(x)$. The increment can be either a positive or a negative number. If $f'(x_0)$ is to exist, *identical* finite results must be

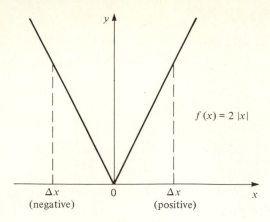

Figure 2.3–1

obtained from the right side of Eq. (2.3–1) for both the positive and negative choices. If two different numbers are obtained, $f'(x_0)$ does not exist.

As an example of how this can occur, consider $f(x) = 2|x|$ plotted in Fig. 2.3–1. Let us try to compute $f'(0)$ by means of Eq. (2.3–1). With $x_0 = 0$, $f(x_0) = 0$, and $f(x_0 + \Delta x) = 2|\Delta x|$, we have

$$f'(0) = \lim_{\Delta x \to 0} \frac{2|\Delta x|}{\Delta x}. \qquad (2.3-2)$$

Unfortunately, if Δx is positive, $2|\Delta x|/\Delta x$ has the value 2, whereas if Δx is negative, it has the value -2. The limit in Eq. (2.3–2) cannot exist, and neither does $f'(0)$. Of course, the values 2 and -2 are the slopes of the curve to the right and left of $x = 0$.

Computing the derivative of the above $f(x)$ at any point $x_0 \neq 0$, we find that the limit on the right in Eq. (2.3–1) exists. It is independent of the sign of Δx; that is, the same value is obtained irrespective of whether we approach x_0 from the right ($\Delta x > 0$) or from the left ($\Delta x < 0$). The reader should do this simple exercise.

Complex Case

Given a function of a complex variable $f(z)$, we find that its derivative at z_0, $f'(z_0)$, or $(df/dz)_{z_0}$, is defined as stated below provided the limit shown exists.

Definition Derivative

$$f'(z_0) = \lim_{\Delta z \to 0} \frac{f(z_0 + \Delta z) - f(z_0)}{\Delta z} \quad \blacksquare \qquad (2.3-3)$$

Although this definition is identical in form to the corresponding expression for real variables shown in Eq. (2.3–1), this new formula is in fact more subtle. We saw that in Eq. (2.3–1) there were two directions from which we could approach x_0. As Fig. 2.3–2 suggests, there is an infinity of different directions along which $z_0 + \Delta z$

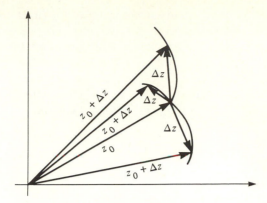

Figure 2.3–2

can approach z_0 in Eq. (2.3–3). Moreover, we need not approach z_0 along a straight line but can choose some sort of arc or spiral. If the limit in Eq. (2.3–3) exists, that is, if $f'(z_0)$ exists, then the quotient in Eq. (2.3–3) must approach the same value irrespective of the direction or locus along which Δz shrinks to zero.

In the case of the function $f(z) = z^n$ ($n = 0, 1, 2, \dots$), it is easy to verify the existence of the derivative and to obtain its value. Now $f(z_0) = z_0^n$, and $f(z_0 + \Delta z) = (z_0 + \Delta z)^n$. This last expression can be expanded with the binomial theorem:

$$(z_0 + \Delta z)^n = z_0^n + nz_0^{n-1}(\Delta z) + \frac{n(n-1)}{2}(z_0)^{n-2}(\Delta z)^2 + \text{higher powers of } \Delta z.$$

Thus

$$f'(z_0) = \lim_{\Delta z \to 0} \frac{f(z_0 + \Delta z) - f(z_0)}{\Delta z}$$

$$= \lim_{\Delta z \to 0} \left[\frac{z_0^n + nz_0^{n-1}\Delta z + \frac{(n)(n-1)}{2}(z_0)^{n-2}(\Delta z)^2 + \cdots - z_0^n}{\Delta z} \right]$$

$$= \lim_{\Delta z \to 0} \left[nz_0^{n-1}\frac{\Delta z}{\Delta z} + \frac{(n)(n-1)}{2}(z_0)^{n-2}\frac{(\Delta z)^2}{\Delta z} + \cdots + \right] = n(z_0)^{n-1}.$$

We do not need to know the path along which Δz shrinks to zero in order to obtain this result. The result is independent of the way in which $z_0 + \Delta z$ approaches z_0. Dropping the subscript zero, we have

$$\frac{d}{dz}z^n = nz^{n-1}. \tag{2.3–4}$$

Thus if n is a nonnegative integer, the derivative of z^n exists for all z. When n is a negative integer, a similar derivation can be used to show that Eq. (2.3–4) holds for all $z \neq 0$. With n negative z^n is undefined at $z = 0$, and this value of z must be avoided.

A more difficult problem occurs if we are given a function of z in the form $f(z) = u(x, y) + iv(x, y)$ and we wish to know whether its derivative exists. If the variables x and y change by incremental amounts Δx and Δy, the corresponding incremental change in z, called Δz is $\Delta x + i\Delta y$ (see Fig. 2.3–3a). Suppose, however, that Δz is constrained to lie along the horizontal line passing through z_0 depicted in Fig. 2.3–3(b). Then, y is constant and $\Delta z = \Delta x$. Now with $z_0 = x_0 + iy_0$, $f(z) = u(x, y) + iv(x, y)$, and $f(z_0) = u(x_0, y_0) + iv(x_0, y_0)$ we will assume that $f'(z_0)$ exists and apply Eq. (2.3–3):

$$f'(z_0) = \lim_{\Delta x \to 0} \frac{f(z_0 + \Delta x) - f(z_0)}{\Delta x}.$$

$$= \lim_{\Delta x \to 0} \frac{u(x_0 + \Delta x, y_0) + iv(x_0 + \Delta x, y_0) - u(x_0, y_0) - iv(x_0, y_0)}{\Delta x}.$$

We can rearrange the preceding expression to read

$$f'(z_0) = \lim_{\Delta x \to 0} \left[\frac{u(x_0 + \Delta x, y_0) - u(x_0, y_0)}{\Delta x} + i \frac{\left[v(x_0 + \Delta x, y_0) - v(x_0, y_0) \right]}{\Delta x} \right].$$

$$(2.3–5)$$

We should recognize the definition of two partial derivatives in Eq. (2.3–5). Passing to the limit we have

$$f'(z_0) = \left(\frac{\partial u}{\partial x} + i \frac{\partial v}{\partial x} \right)_{x_0, y_0}. \qquad (2.3–6)$$

Instead of having $z_0 + \Delta z$ approach z_0 from the right, as was just done, we can allow $z_0 + \Delta z$ to approach z_0 from above. If Δz is constrained to lie along the vertical line passing through z_0 in Fig. 2.3–3(b), $\Delta x = 0$ and $\Delta z = i\Delta y$. Thus

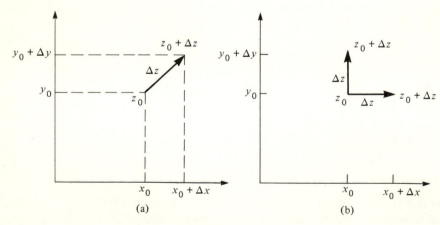

(a)   (b)

Figure 2.3–3

proceeding much as before, we have

$$f'(z_0) = \lim_{\Delta y \to 0} \frac{f(z_0 + i\Delta y) - f(z_0)}{i\Delta y}$$

$$= \lim_{\Delta y \to 0} \left[\frac{u(x_0, y_0 + \Delta y) + iv(x_0, y_0 + \Delta y) - u(x_0, y_0) - iv(x_0, y_0)}{i\Delta y} \right].$$

(2.3–7)

Passing to the limit and putting $1/i = -i$ we get

$$f'(z_0) = \left(-i\frac{\partial u}{\partial y} + \frac{\partial v}{\partial y} \right)_{x_0, y_0}.$$

(2.3–8)

Assuming $f'(z_0)$ exists, Eqs. (2.3–6) and (2.3–8) provide us with two methods for its computation. Equating these expressions, we obtain the result

$$\left(\frac{\partial u}{\partial x} + i\frac{\partial v}{\partial x} \right) = \left(-i\frac{\partial u}{\partial y} + \frac{\partial v}{\partial y} \right).$$

(2.3–9)

The real part on the left side of Eq. (2.3–9) must equal the real part on the right. A similar statement applies to the imaginaries. Thus we require at *any* point where $f(z)$ exists the set of relationships shown below.

CAUCHY–RIEMANN
EQUATIONS

$$\frac{\partial u}{\partial x} = \frac{\partial v}{\partial y},$$

(2.3–10a)

$$\frac{\partial v}{\partial x} = -\frac{\partial u}{\partial y}.$$

(2.3–10b)

This set of important relationships is known as the Cauchy–Riemann (or C–R) equations. If they fail to be satisfied for some value of z, say z_0, we know that $f'(z_0)$ cannot exist since our allowing Δz to shrink to zero along *two* different paths (Fig. 2.3–3b) leads to two contradictory limiting values for the quotient in Eq. (2.3–3). Thus we have shown that satisfaction of the C–R equations at a point is a *necessary* condition for the existence of the derivative at that point. The mere fact that a function satisfies these equations does not guarantee that *all* paths along which $z_0 + \Delta z$ approaches z_0 will yield identical limiting values for the quotient in Eq. (2.3–3). In more advanced texts[†] the following theorem is proved for $f(z) = u + iv$.

Theorem 3

If u, v and their first partial derivatives ($\partial u/\partial x, \partial v/\partial x, \partial u/\partial y, \partial v/\partial y$) are continuous throughout some neighborhood of z_0, then satisfaction of the Cauchy–Riemann equations at z_0 is both a *necessary* and *sufficient* condition for the existence of $f'(z_0)$. ∎

[†]See E. T. Copson, *An Introduction to the Theory of Functions of a Complex Variable* (London: Oxford University Press, 1960), pp. 40–42.

With the conditions of this theorem fulfilled, the limit on the right in Eq. (2.3–3) exists, that is, all paths by which $z_0 + \Delta z$ approaches z_0 yield the same finite result in this expression.

Example 1

Investigate the differentiability of $f(z) = z^2 = (x + iy)^2 = x^2 - y^2 + i2xy$.

Solution

We already know (see Eq. 2.3–4) that $f'(z)$ exists, but let us verify this result by means of the C–R equations. Here, $u = x^2 - y^2$, $v = 2xy$, $\partial u/\partial x = 2x = \partial v/\partial y$, and $\partial v/\partial x = 2y = -\partial u/\partial y$. Thus Eqs. (2.3–10) are satisfied for all z. Also, because u, v, $\partial u/\partial x$, $\partial v/\partial y$, etc. are continuous in the z-plane, $f'(z)$ exists for all z. ◄

Example 2

Investigate the differentiability of $f(z) = z\bar{z} = |z|^2$.

Solution

Here the C–R equations are helpful. We have $u + iv = |z|^2 = x^2 + y^2$. Hence, $u = x^2 + y^2$ and $v = 0$, therefore, $\partial u/\partial x = 2x$, $\partial v/\partial y = 0$, $\partial u/\partial y = 2y$, and $\partial v/\partial x = 0$. When these expressions are substituted into Eqs. (2.3–10), we obtain $2x = 0$ and $2y = 0$. These equations are simultaneously satisfied only where $x = 0$ and $y = 0$, that is, at the origin of the z-plane. Thus this function of z possesses a derivative only for $z = 0$. ◄

Let us consider why the derivative of $|z|^2$ fails to exist except at one point.

Consider the definition in Eq. (2.3–3) and refer to Fig. 2.3–4. At an arbitrary point, $z_0 = x_0 + iy_0$, we have $f(z_0) = |z_0|^2 = |x_0 + iy_0|^2 = x_0^2 + y_0^2$. With $\Delta z = \Delta x + i\Delta y$ then, $f(z_0 + \Delta z) = |z_0 + \Delta z|^2 = |(x_0 + \Delta x) + i(y_0 + \Delta y)|^2 = x_0^2 + 2x_0\Delta x + (\Delta x)^2 + y_0^2 + 2y_0\Delta y + (\Delta y)^2$. Thus

$$\lim_{\Delta z \to 0} \frac{f(z_0 + \Delta z) - f(z_0)}{\Delta z} = \lim_{\substack{\Delta x \to 0 \\ \Delta y \to 0}} \frac{2x_0\Delta x + 2y_0\Delta y + (\Delta x)^2 + (\Delta y)^2}{\Delta x + i\Delta y}.$$

$$(2.3–11)$$

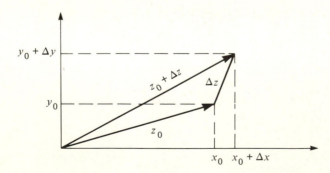

Figure 2.3–4

Now, suppose we allow Δz to shrink to zero along a straight line passing through z_0 with slope m. This means $\Delta y = m\Delta x$. With this relationship in Eq. (2.3–11) we have

$$\lim_{\substack{\Delta x \to 0 \\ \Delta y \to 0}} \frac{2x_0\Delta x + 2y_0 m\Delta x + (\Delta x)^2 + m^2(\Delta x)^2}{\Delta x(1 + im)}$$

$$= \lim_{\Delta x \to 0} \left[\frac{2x_0 + 2y_0 m}{1 + im} + \frac{\Delta x}{1 + im} + \frac{m^2\Delta x}{1 + im} \right] = \frac{2x_0 + 2y_0 m}{1 + im}.$$

Unless $x_0 = 0$ and $y_0 = 0$, this result is certainly a function of the slope m, that is, of the direction of approach to z_0. For example, if we approach z_0 along a line parallel to the x-axis, we put $m = 0$ and find that the limit is $2x_0$. However, if we approach z_0 along a line making a $45°$ angle with the horizontal, we have $\Delta y = \Delta x$, or $m = 1$. The limit is then $(2x_0 + 2y_0)/(1 + i)$.

EXERCISES

1. Consider $f(x) = \sin|x|$.

   a) Make a sketch of the function over the interval $-2\pi \le x \le 2\pi$.

   b) For what value of x does $f'(x)$ fail to exist?

   c) Is $f(x)$ continuous at the value of x found in part (b)?

2. Consider $f(x) = x^{2/3}$, where the real value of the function is used. Is $f(x)$ continuous along the x-axis at $x = 0$? Does $f'(x)$ exist at $x = 0$? Explain.

3. Assume that the second derivative, $f''(z_0)$ exists. Show that

$$f''(z_0) = \frac{\partial^2 u}{\partial x^2} + i\frac{\partial^2 v}{\partial x^2} \quad \text{and} \quad f''(z_0) = -\frac{\partial^2 u}{\partial y^2} - i\frac{\partial^2 v}{\partial y^2}.$$

   Hint: See derivation of Eqs. (2.3–6) and (2.3–8).

4. Given $f(z)$ as follows, determine where in the z-plane $f'(z)$ exists.

   a) $c + id$, c and d are constants

   b) $(x^2 + xy) - i(y + 1)$

   c) $(x + 1) + iy^2$

   d) $4ie^{(x-1)^2}$

   e) $(x^2 - y^2 - 2xy) + i(x^2 - y^2 + 2xy)$

   f) x

   g) $e^{x^2-y^2}(\cos 2xy + i\sin 2xy)$

   h) $e^{x^2+y^2}(\cos 2xy + i\sin 2xy)$

   i) $x/(x^2 + y^2) - iy/(x^2 + y^2)$

5. For what values of z do each of the following functions possess derivatives?

   a) $f(z) = \bar{z} = x - iy$

   b) $f(z) = (\bar{z})^2$

   c) $f(z) = z\,\text{Re}\,z$

   d) $f(z) = z|z|$

   e) $f(z) = z/\bar{z}$

6. Let us prove L'Hôpital's rule for functions of a complex variable. Let $f(z_0) = 0$, $g(z_0) = 0$, $g'(z_0) \ne 0$; $f(z)$ and $g(z)$ are differentiable at z_0. We must show that

$$\lim_{z \to z_0} \frac{f(z)}{g(z)} = \frac{f'(z_0)}{g'(z_0)}. \qquad (2.3\text{–}12)$$

a) Show that

$$\frac{f(z)}{g(z)} = \frac{f(z) - f(z_0)}{z - z_0} \div \frac{g(z) - g(z_0)}{z - z_0} .$$

b) Put $z = z_0 + \Delta z$ on the right in the preceding equation. Now pass to the limit $\Delta z \to 0$. Explain why Eq. (2.3–12) follows from part (c) of Theorem 1.

7. Show that if $f'(z_0)$ exists, then $f(z)$ must have a limit at z_0 and is continuous there. *Hint:* Let $z = z_0 + \Delta z$. Consider

$$\lim_{z \to z_0} \left[\frac{f(z) - f(z_0)}{z - z_0} \right] \lim_{z \to z_0} (z - z_0) = ?$$

Refer to Eq. (2.2–8b) of Theorem 1.

2.4 THE DERIVATIVE AND ANALYTICITY

Finding the Derivative

If we can establish that the derivative of $f(z) = u + iv$ exists for some z, it is a straightforward matter to find $f'(z)$. We can work directly with the definition shown in Eq. (2.3–3).

In addition, either Eq. (2.3–6), $f'(z) = \partial u/\partial x + i\partial v/\partial x$, or Eq. (2.3–8), $f'(z) = \partial v/\partial y - i\partial u/\partial y$, can be used. For example, the function $f(z) = x^2 - y^2 - y + i(2xy + x)$ is found, from the C–R equations, to have a derivative for all z. With $u = x^2 - y^2 - y$ and with $v = 2xy + x$ we have, from Eq. (2.3–6), $f'(z) = 2x + i(2y + 1)$. An identical result comes from Eq. (2.3–8).

In Section 2.3 we observed that $dz^n/dz = nz^{n-1}$, where n is any integer. This formula is identical in form to the corresponding expression in real variable calculus, $dx^n/dx = nx^{n-1}$. Thus to differentiate such expressions as z^2, $1/z^3$, etc. the usual method applies and these derivatives are respectively, $2z$ and $-3z^{-4}$.

The reason that the procedure used in differentiating x^n and z^n is identical lies in the similarity of the expressions

$$\lim_{\Delta z \to 0} \frac{f(z + \Delta z) - f(z)}{\Delta z} \quad \text{and} \quad \lim_{\Delta x \to 0} \frac{f(x + \Delta x) - f(x)}{\Delta x},$$

which define the derivatives of functions of complex and real variables.

All the identities of real differential calculus that are obtained through direct manipulation of the definition of the derivative can be carried over to functions of a complex variable.

Specifically, if $f(z)$ and $g(z)$ are differentiable for some z, then

Theorem 4

$$\frac{d}{dz}(f(z) \pm g(z)) = f'(z) \pm g'(z); \tag{2.4–1a}$$

$$\frac{d}{dz}(f(z)g(z)) = f'(z)g(z) + f(z)g'(z); \tag{2.4–1b}$$

$$\frac{d}{dz}\left(\frac{f(z)}{g(z)}\right) = \frac{f'(z)g(z) - f(z)g'(z)}{[g(z)]^2}, \quad \text{provided } g(z) \neq 0; \quad (2.4-1c)$$

$$\frac{d}{dz}f(g(z)) = f'(g(z))g'(z). \quad \blacksquare \qquad (2.4-1d)$$

Thus a function formed by the addition, subtraction, multiplication, or division of differentiable functions is itself differentiable. Equations (2.4–1a–c) provide a means for finding its derivative. Another useful formula is the "chain rule" (2.4–1d) for finding the derivative of a function of a function. It is applied in the ways familiar to us from elementary calculus, for example,

$$\frac{d}{dz}(z^3 + z^2 + 1)^{10} = 10(z^3 + z^2 + 1)^9 \frac{d}{dz}(z^3 + z^2 + 1)$$

$$= 10(z^3 + z^2 + 1)^9(3z^2 + 2z).$$

The equations contained in Eq. (2.4–1) are of no use in establishing the differentiability or in determining the derivative of any expression involving $|z|$ or $\bar{z}$. We can first rewrite such expressions in the form $u(x, y) + iv(x, y)$ and then apply the C–R equations to investigate differentiability. If the derivative exists, it can then be found from Eqs. (2.3–6) or (2.3–8). Alternatively, we might choose to investigate differentiability and determine the derivative by means of the definition given in Eq. (2.3–3).

Analytic Functions

The concept of an analytic function, although seemingly simple, is at the very core of complex variable theory, and a grasp of its meaning is essential.

Definition Analyticity

A function $f(z)$ is analytic at z_0 if $f'(z)$ exists not only at z_0 but at every point belonging to some neighborhood of z_0. $\quad \blacksquare$

Thus for a function to be analytic at a point it must not only have a derivative at that point but must have a derivative everywhere within some circle of nonzero radius centered at the point.

Definition Analyticity in a region

If a function is analytic at every point belonging to some region, we say that the function is analytic in that region. $\quad \blacksquare$

It is quite possible for a function to possess a derivative at some point yet fail to be analytic at that point. In Example 2 of Section 2.3 we considered $f(z) = |z|^2$ and found it to have a derivative only for $z = 0$. Every circle that we might draw about the point $z = 0$ will contain points at which $f'(z)$ fails to exist. Hence, $f(z)$ is not analytic at $z = 0$ (or anywhere else).

Example 1

For what values of z is the function $f(z) = x^2 + iy^2$ analytic?

Solution

From the C–R equations, with $u = x^2$, $v = y^2$, we have

$$\frac{\partial u}{\partial x} = 2x = 2y = \frac{\partial v}{\partial y} \quad \text{and} \quad \frac{\partial v}{\partial x} = 0 = -\frac{\partial u}{\partial y}.$$

Thus $f(z)$ is differentiable only for values of z that lie along the straight line $x = y$. If z_0 lies on this line, any circle centered at z_0 will contain points for which $f'(z)$ does not exist (see Fig. 2.4–1). Thus $f(z)$ is nowhere analytic. ◄

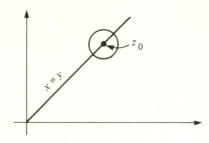

Figure 2.4–1

Equations (2.4–1a–d), which yield the derivatives of sums, products, and so forth, can be extended to give the following theorem on analyticity:

Theorem 5

If two functions are analytic in some domain, the *sum*, *difference*, and *product* of these functions are also analytic in the domain. The *quotient* of these functions is analytic in the domain except where the denominator equals zero. An analytic function of an analytic function is analytic. ∎

Definition   Entire function

A function that is analytic throughout the finite z-plane is called an *entire function*. ∎

Any constant is entire. Its derivative exists for all z and is zero. The function $f(z) = z^n$ is entire if n is a nonnegative integer (see Eq. 2.3–4). Now $a_n z^n$, where a_n is any constant, is the product of entire functions and is also entire. A polynomial expression $a_n z^n + a_{n-1} z^{n-1} + \cdots + a_0$ is entire since it is composed of a sum of entire functions.

Definition   Singularity

If a function is not analytic at z_0 but is analytic for at least one point in every neighborhood of z_0, then z_0 is called a *singularity* (or *singular point*) of that function. ∎

A rational function

$$f(z) = \frac{a_n z^n + a_{n-1} z^{n-1} + \cdots + a_0}{b_m z^m + b_{m-1} z^{m-1} + \cdots + b_0},$$

where m and n are nonnegative integers and a_n, b_m, etc. are constants, is the quotient of two polynomials and thus the quotient of two entire functions. The function $f(z)$ is analytic except for values of z satisfying $b_m z^m + b_{m-1} z^{m-1} + \cdots + b_0 = 0$. The solutions of this equation are singular points of $f(z)$.

Example 2

For what values of z does

$$f(z) = \frac{z^3 + 2}{z^2 + 1}$$

fail to be analytic?

Solution

For z satisfying $z^2 + 1 = 0$ or $z = \pm i$. Thus $f(z)$ has singularities at $+i$ and $-i$. ◀

In Chapter 3 we will be studying some transcendental functions of z. Here the portion of our theorem that deals with analytic functions of analytic functions will prove useful. For example, we will define $\sin z$ and learn that this is an entire function. Now $1/z^2$ is analytic for all $z \neq 0$. Hence, $\sin(1/z^2)$ is analytic for all $z \neq 0$.

EXERCISES

1. Consider Exercise 4, Section 2.3. Where in the z-plane, if at all, are each of these functions analytic?

2. For what values of z do the following functions fail to be analytic? In each case find an expression for the derivative, as a function of z, that applies within the domain of analyticity.

   a) $z^3 + 2z^2 + z + 1$ c) $z + \dfrac{1}{z - 1}$ e) $\displaystyle\sum_{n=0}^{2} z^{-n+1}$

   b) $\dfrac{z}{z^2 + z + 1}$ d) $\dfrac{(z + 1)}{(z - 1)^{10}}$

3. Prove that each of the following functions is entire. Find an expression for the derivative and find the numerical value of the derivative at $x = 1$, $y = 1$.

   a) $x^3 - 3xy^2 + i(3x^2 y - y^3)$
   b) $- e^{2x} \sin 2y + ie^{2x} \cos 2y$
   c) $e^{x^2 - y^2} \cos 2xy + ie^{x^2 - y^2} \sin 2xy$

4. Prove that if $f(z) = u(x, y) + iv(x, y)$ is analytic in some domain, then $g(z) = v - iu$ must also be analytic in this domain.

5. a) Prove that if an analytic function is purely real throughout some domain, then it must be constant in that domain.

   b) Repeat the question in part (a) with the word "imaginary" substituted for "real."

6. Suppose $f(z) = u + iv$ is analytic. Under what circumstances will $g(z) = u - iv$ be analytic? *Hint:* Consider the functions $f(z) + g(z)$ and $f(z) - g(z)$. Then refer to Exercise 5.

7. Consider an analytic function $f(z) = u + iv$ whose modulus $|f(z)|$ is equal to a constant k throughout some domain. Show that this can occur only if $f(z)$ is constant throughout the domain. *Hint:* The case $k = 0$ is trivial. Assuming $k \neq 0$, we have $u^2 + v^2 = k^2$ or $k^2/(u + iv) = u - iv$. Now refer to Exercise 6.

8. Polar form of the C–R equations.

   a) Suppose, for the analytic function $f(z) = u(x, y) + iv(x, y)$, that we express x and y in terms of the polar variables r and θ, where $x = r\cos\theta$ and $y = r\sin\theta$ ($r = \sqrt{x^2 + y^2}$, $\theta = \tan^{-1}(y/x)$). Then, $f(z) = u(r, \theta) + iv(r, \theta)$. We want to rewrite the C–R equations entirely in the polar variables. From the chain rule for partial differentiation we have

   $$\frac{\partial u}{\partial x} = \left(\frac{\partial u}{\partial r}\right)_\theta \left(\frac{\partial r}{\partial x}\right)_y + \left(\frac{\partial u}{\partial \theta}\right)_r \left(\frac{\partial \theta}{\partial x}\right)_y.$$

   Give the corresponding expressions for $\partial u/\partial y$, $\partial v/\partial x$, $\partial v/\partial y$.

   b) Show that

   $$\left(\frac{\partial r}{\partial x}\right)_y = \cos\theta,$$

   $$\left(\frac{\partial \theta}{\partial x}\right)_y = \frac{-\sin\theta}{r},$$

   and find corresponding expressions for $(\partial r/\partial y)_x$ and $(\partial \theta/\partial y)_x$. Use these four expressions in the equations for $\partial u/\partial x$, $\partial u/\partial y$, $\partial v/\partial x$, and $\partial v/\partial y$ found in part (a). Show that u and v satisfy the equations

   $$\frac{\partial h}{\partial x} = \frac{\partial h}{\partial r}\cos\theta - \frac{1}{r}\frac{\partial h}{\partial \theta}\sin\theta,$$

   $$\frac{\partial h}{\partial y} = \frac{\partial h}{\partial r}\sin\theta + \frac{1}{r}\frac{\partial h}{\partial \theta}\cos\theta,$$

   where h can equal u or v.

   c) Rewrite the C–R equations (2.3–10a, b) using the two equations from part (b) of this exercise. Multiply the first C–R equation by $\cos\theta$, the second by $\sin\theta$, and add to show that

   $$\frac{\partial u}{\partial r} = \frac{1}{r}\frac{\partial v}{\partial \theta}. \qquad (2.4\text{–}2a)$$

   Now multiply the first C–R equation by $-\sin\theta$, the second by $\cos\theta$, and add to show that

   $$\frac{\partial v}{\partial r} = \frac{-1}{r}\frac{\partial u}{\partial \theta}. \qquad (2.4\text{–}2b)$$

The relationships of Eqs. (2.4–2a, b) are the *polar form of the C–R equations*. If the first partial derivatives of u and v are continuous at some point whose polar coordinates are r, θ ($r \neq 0$), then Eqs. (2.4–2a, b) provide a necessary and sufficient condition for the existence of the derivative at this point.

d) Use Eq. (2.3–6) and the C–R equations in polar form to show that if the derivative of $f(r, \theta)$ exists it can be found from

$$f'(z) = \left[\frac{\partial u}{\partial r} + i\frac{\partial v}{\partial r} \right] [\cos \theta - i \sin \theta] \qquad (2.4–3)$$

or from

$$f'(z) = \left[\frac{\partial u}{\partial \theta} + i\frac{\partial v}{\partial \theta} \right] \left(\frac{-i}{r} \right) [\cos \theta - i \sin \theta]. \qquad (2.4–4)$$

9. Consider $z^{10} = [r(\cos \theta + i \sin \theta)]^{10} = [u(r, \theta) + iv(r, \theta)]$.

a) Find u and v.

b) Verify that u and v satisfy the polar C–R equations throughout the z-plane except at the origin.

2.5 HARMONIC FUNCTIONS

Given a real function of x and y, say $\phi(x, y)$, we wish to determine if there exists an analytic function $f(z)$ of either of the forms $f(z) = \phi(x, y) + iv(x, y)$ or $f(z) = u(x, y) + i\phi(x, y)$. In other words, can ϕ be regarded as either the real or the imaginary part of an analytic function? It is relatively easy to answer this question.

Consider an analytic function $f(z) = u + iv$. Then, the Cauchy–Riemann equations

$$\frac{\partial u}{\partial x} = \frac{\partial v}{\partial y}, \qquad (2.5–1a)$$

$$\frac{\partial u}{\partial y} = -\frac{\partial v}{\partial x} \qquad (2.5–1b)$$

are satisfied by u and v. Now let us assume that we can differentiate Eq. (2.5–1a) with respect to x and Eq. (2.5–1b) with respect to y. We obtain

$$\frac{\partial^2 u}{\partial x^2} = \frac{\partial}{\partial x}\frac{\partial v}{\partial y}, \qquad (2.5–2a)$$

$$\frac{\partial^2 u}{\partial y^2} = -\frac{\partial}{\partial y}\frac{\partial v}{\partial x}. \qquad (2.5–2b)$$

It can be shown[†] that if the second partial derivatives of a function are continuous, then the order of differentiation in the cross partial derivatives is immaterial. Thus $\partial^2 v/\partial x\, \partial y = \partial^2 v/\partial y\, \partial x$. With this assumption we add Eqs. (2.5–2a)

[†]See A. Taylor and W. Mann, *Advanced Calculus* (Lexington, Mass.: Xerox College Publishing, 1972), pp. 214–216.

and (2.5–2b). The right-hand sides cancel, leaving

$$\frac{\partial^2 u}{\partial x^2} + \frac{\partial^2 u}{\partial y^2} = 0. \qquad (2.5-3)$$

Alternatively, we might have differentiated Eq. (2.5–1a) with respect to y and Eq. (2.5–1b) with respect to x. Assuming that the second partial derivatives of u are continuous, we add the resulting equations and obtain

$$\frac{\partial^2 v}{\partial x^2} + \frac{\partial^2 v}{\partial y^2} = 0. \qquad (2.5-4)$$

Thus both the real and imaginary parts of an analytic function must satisfy a differential equation of the form shown below.

LAPLACE'S EQUATION $\qquad\qquad \dfrac{\partial^2 \phi}{\partial x^2} + \dfrac{\partial^2 \phi}{\partial y^2} = 0 \qquad (2.5-5)$

Laplace's equation for functions of the polar variables r and θ (Exercise 8, Section 2.4) is derived in Exercise 14 of the present section.

Definition   Harmonic function

Functions satisfying Laplace's equation in a domain are said to be *harmonic* in that domain.   ■

An example of a harmonic function is $\phi(x, y) = x^2 - y^2$ since $\partial^2\phi/\partial x^2 = 2$, $\partial^2\phi/\partial y^2 = -2$ and Laplace's equation is satisfied throughout the z-plane. A function satisfying Laplace's equation only for some set of points that does not constitute a domain is not harmonic. An example of this is presented in Exercise 1 of this section.

Equations (2.5–3) and (2.5–4) can be summarized as follows:

Theorem 6

If a function is analytic in some domain, its real and imaginary parts are harmonic in that domain.   ■

A converse to the preceding theorem can be established.[†]

Theroem 7

Given a real function $\phi(x, y)$, which is harmonic in a simply connected domain D, there exists in D an analytic function whose *real part* equals $\phi(x, y)$. There also exists in D an analytic function whose *imaginary part* is $\phi(x, y)$.   ■

Given a harmonic function $\phi(x, y)$, we may wish to find the corresponding harmonic function $v(x, y)$ such that $\phi(x, y) + iv(x, y)$ is analytic. Or, given

---

[†]R. Churchill, J. Brown, R. Verhey, *Complex Variables and Applications* (New York: McGraw-Hill, 1974), sec. 78.

$\phi(x, y)$, we might seek $u(x, y)$ such that $u(x, y) + i\phi(x, y)$ is analytic. In either case we can determine the unknown function up to an additive constant. The method is best illustrated with an example.

Example 1

Show that $\phi = x^3 - 3xy^2 + 2y$ can be the real part of an analytic function. Find the imaginary part of the analytic function.

Solution

We have

$$\frac{\partial^2 \phi}{\partial x^2} = 6x \quad \text{and} \quad \frac{\partial^2 \phi}{\partial y^2} = -6x,$$

which sums to zero throughout the z-plane. Thus ϕ is harmonic. To find $v(x, y)$ we use the C–R equations and take $u(x, y) = \phi(x, y)$:

$$\frac{\partial u}{\partial x} = 3x^2 - 3y^2 = \frac{\partial v}{\partial y}, \tag{2.5-6}$$

$$-\frac{\partial u}{\partial y} = 6xy - 2 = \frac{\partial v}{\partial x}. \tag{2.5-7}$$

Let us solve Eq. (2.5–6) for v by integrating on y:

$$v = \int (3x^2 - 3y^2)\, dy \quad \text{or} \quad v = 3x^2 y - y^3 + C(x). \tag{2.5-8}$$

It is important to recognize that the "constant" C, although independent of y, can depend on the variable x. The reader can verify this by substituting v from Eq. (2.5–8) into Eq. (2.5–6).

To evaluate $C(x)$ we substitute v from Eq. (2.5–8) into Eq. (2.5–7) and get $6xy - 2 = 6xy + dC/dx$. Obviously, $dC/dx = -2$. We integrate and obtain $C = -2x + D$, where D is a true constant, independent of x and y. Putting this value of C into Eq. (2.5–8) we finally have

$$v = 3x^2 y - y^3 - 2x + D. \tag{2.5-9}$$

Since v is a real function, D must be a real constant. Its value cannot be determined if we are given only u. However, if we are told the value of v at some point in the complex plane, the value of D can be found. For example, given $v = -2$ at $x = -1$, $y = 1$, we substitute these quantities into Eq. (2.5–9) and find that $D = -6$. ◀

Definition Harmonic conjugate

Given a harmonic function $u(x, y)$, we call $v(x, y)$ the *harmonic conjugate* of $u(x, y)$ if $u(x, y) + iv(x, y)$ is analytic. ■

This definition is not related to the notion of the conjugate of a complex number. In Example 1, just given, $3x^2 y - y^3 - 2x + D$ is the harmonic conjugate of $x^3 - 3xy^2 + 2y$ since $f(z) = x^3 - 3xy^2 + 2y + i(3x^2y - y^3 - 2x + D)$ is analytic. However, $x^3 - 3xy^2 + 2y$ *is not* the harmonic conjugate of $3x^2y - y^3 - 2x + D$ since $f(z) = 3x^2y - y^3 - 2x + D + i(x^3 - 3xy^2 + 2y)$ is not analytic. This matter is explored more fully in Exercise 7 of this section where we investigate the circumstances under which $u + iv$ and $v + iu$ can be analytic.

Conjugate functions have an interesting geometrical property. Given a harmonic function $u(x, y)$ and a real constant C_1, we find that the equation $u(x, y) = C_1$ is satisfied along some locus, typically a curve, in the xy-plane. Given a collection of such constants, $C_1, C_2, C_3, \ldots$, we can plot a family of curves by using the equations $u(x, y) = C_1$, $u(x, y) = C_2$, etc. A typical family for some $u(x, y)$ is shown in solid line in Fig. 2.5–1.

Suppose $v(x, y)$ is the harmonic conjugate of $u(x, y)$, and $k_1, k_2, k_3, \ldots$, are real constants. We can plot on the same figure the family of curves given by $v(x, y) = k_1$, $v(x, y) = k_2$, etc., which are indicated by dashed lines. We will now prove the following important theorem pertaining to the two families of curves.

Theorem 8

Let $f(z) = u(x, y) + iv(x, y)$ be an analytic function and $C_1, C_2, C_3, \ldots$ and $k_1, k_2, k_3, \ldots$ be real constants. Then the family of curves in the xy-plane along which $u = C_1$, $u = C_2$, etc. is orthogonal to the family given by $v = k_1$, $v = k_2, \ldots$; that is, the intersection of a member of one family with that of another takes place at a $90°$ angle, except possibly at any point where $f'(z) = 0$. ∎

For a proof let us consider the intersection of the curve $u = C_1$ with the curve $v = k_1$ in Fig. 2.5–1. Consider the equation $C_1 = u(x, y)$. Differentiating this implicitly with respect to x and regarding y as a function of x, we have

$$\frac{dC_1}{dx} = 0 = \frac{\partial u}{\partial x} + \frac{\partial u}{\partial y}\frac{dy}{dx}. \tag{2.5–10}$$

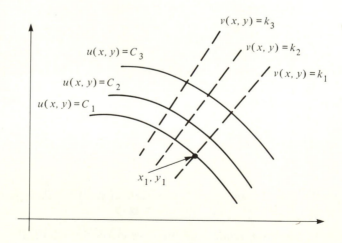

Figure 2.5–1

Let us use Eq. (2.5–10) to determine dy/dx at the point of intersection x_1, y_1. We then have

$$\frac{dy}{dx} = \left(-\frac{\partial u/\partial x}{\partial u/\partial y}\right)_{x_1, y_1}. \tag{2.5–11}$$

This is merely the slope of the curve $u = C_1$ at the point being considered. Similarly, the slope of $v = k_1$ at x_1, y_1 is

$$\frac{dy}{dx} = \left(-\frac{\partial v/\partial x}{\partial v/\partial y}\right)_{x_1, y_1}. \tag{2.5–12}$$

With the C–R equations

$$-\frac{\partial v}{\partial x} = \frac{\partial u}{\partial y} \quad \text{and} \quad \frac{\partial v}{\partial y} = \frac{\partial u}{\partial x}$$

we can rewrite Eq. (2.5–12) as

$$\frac{dy}{dx} = \left(\frac{\partial u/\partial y}{\partial u/\partial x}\right)_{x_1, y_1}. \tag{2.5–13}$$

Comparing Eqs. (2.5–11) and (2.5–13), we observe that the slopes of the curves $u = C_1$ and $v = k_1$ at the point of intersection x_1, y_1 are negative reciprocals of one another. Hence, the intersection takes place at a $90°$ angle. An identical procedure applies at any other intersection involving the families of curves. Notice that if $f'(z) = 0$ at some point, then according to Eqs. (2.3–6) and (2.3–8) the first partial derivatives of u and v vanish. The slope of the curves cannot now be found from Eqs. (2.5–11) and (2.5–12). The proof breaks down at such a point.

Example 2

With the aid of the C–R equations the function

$$f(z) = \frac{1}{2} \log\left(x^2 + y^2\right) + i \tan^{-1} \frac{y}{x},$$

where $-\pi < \tan^{-1}(y/x) \le \pi$, is readily shown to be analytic in any domain not containing $z = 0$ or points on the negative real axis. Show how this function demonstrates Theorem 8.

Solution

Letting

$$u = \frac{1}{2} \log\left(x^2 + y^2\right) \quad \text{and} \quad v = \tan^{-1}\left(\frac{y}{x}\right),$$

we have

$$\frac{\partial u}{\partial x} = \frac{x}{x^2 + y^2} = \frac{\partial v}{\partial y} \quad \text{and} \quad \frac{\partial u}{\partial y} = \frac{y}{x^2 + y^2} = \frac{-\partial v}{\partial x}.$$

Loci along which u is constant are merely the circles along which $x^2 + y^2$ assumes constant values, for example, $x^2 + y^2 = 1$, $x^2 + y^2 = 2^2$, $x^2 + y^2 = 3^2$, etc.

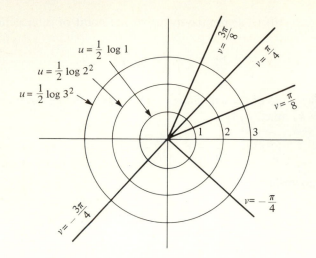

Figure 2.5–2

The function $v = \tan^{-1}(y/x)$ is identical to the argument of z. Hence, the curves along which v assumes constant values are merely rays extending outward from the origin of the z-plane. Families of curves of the form $u = C$ and $v = k$ are shown in Fig. 2.5–2. The orthogonality of the intersections should be obvious. ◄

EXERCISES

1. Find the values of z for which $x^3 - y^3$ satisfies Laplace's equation. Why is this function not harmonic?

2. Which of the following functions are harmonic?

a) $x^2 + y^2$ b) $\sin x \cos y$ c) $\dfrac{x}{x^2 + y^2}$ d) $e^x \cos y + x + y$

3. Show by direct calculation that the real and imaginary parts of the following analytic functions are harmonic.

a) $f(z) = (z - 1)^2$ b) $f(z) = \dfrac{i}{z}$ c) $f(z) = z^3$

4. Consider $\phi(x, y) = 4x^3 y - 4xy^3 + x + 1$.

a) Show that $\phi(x, y)$ could be the real or the imaginary part of an analytic function.

b) If $\phi(x, y)$ is the real part of an analytic function, find the imaginary part.

c) If $\phi(x, y)$ is the imaginary part of an analytic function, find the real part. Are your answers to parts (b) and (c) identical? If not, explain how they differ.

5. Find the harmonic conjugate of $e^{-x} \cos y + 2y + xy + 3$.

6. a) Find two values for k that make $\phi(x, y) = (\sin 2x)(e^{2y} + e^{ky})$ harmonic.

b) If $\phi(x, y)$ is the real part of an analytic function, find the imaginary part. Consider both values of k.

7. Suppose that $f(z) = u + iv$ is analytic and that $g(z) = v + iu$ is also. Show that v and u must both be constants. *Hint:* $-if(z) = v - iu$ is analytic (the product of analytic functions). Thus $g(z) \pm if(z)$ is analytic and must satisfy the C–R equations. Now refer to Exercise 5 of Section 2.4.

8. a) Show that $\phi(x, y) = \tan^{-1}(y/x)$, $-\pi < \tan^{-1}(y/x) \le \pi$, is harmonic.

   b) If $\phi(x, y)$ is the imaginary part of an analytic function, find the real part.

9. Show that $\exp(x^3 - 3xy^2)[\cos(3x^2y - y^3)]$ can be the imaginary part of an analytic function. Find the real part of the analytic function.

10. Let $u(x, y)$ and $v(x, y)$ be harmonic functions. Are the following statements necessarily true?

    a) $\phi = u + v$ is harmonic

    b) $\phi = uv$ is harmonic

    c) $u^2 - v^2$ is harmonic

    d) If v is the harmonic conjugate of u, are parts (b) and (c) necessarily true?

11. Let $v(x, y)$ be the harmonic conjugate of $u(x, y)$.

    a) Show that $\phi(x, y) = e^u \cos v$ is a harmonic function.

    b) Show that $\phi(x, y) = \log(u^2 + v^2)$ is a harmonic function.

12. Consider $f(z) = z^2 = u + iv$.

    a) Find the equation describing the curve along which $u = 1$ in the xy-plane. Repeat for $v = 1$.

    b) Find the point of intersection, in the first quadrant, of the two curves found in part (a).

    c) Find the numerical value of the slope of each curve at the point of intersection, which was found in part (b), and verify that the slopes are negative reciprocals.

13. a) Show that $f(z) = e^x \cos y + ie^x \sin y = u + iv$ is entire.

    b) Consider the curve along which $u = 1$ and the curve along which $v = 2$. Plot these loci, in the xy-plane, in the domain $0 < y < \pi/2$. Use a pocket calculator with log and trigonometric functions for the numerical data.

    c) Find mathematically the point of intersection of the curves in part (b).

    d) Take derivatives to find the slopes of the curves at their intersection, and verify that they are negative reciprocals.

14. a) Let $x = r \cos \theta$ and $y = r \sin \theta$, where r and θ are the usual polar coordinate variables. Let $f(z) = u(r, \theta) + iv(r, \theta)$ be a function that is analytic in some domain that does not include $z = 0$. Use Eqs. (2.4–2a, b) and an assumed continuity of second partial derivatives to show that in this domain u and v satisfy the differential equation

$$\frac{\partial^2 \phi}{\partial r^2} + \frac{1}{r^2}\frac{\partial^2 \phi}{\partial \theta^2} + \frac{1}{r}\frac{\partial \phi}{\partial r} = 0.$$

   This is Laplace's equation in the polar variables r and θ.

   b) Show that $u(r, \theta) = r^2 \sin 2\theta$ is a harmonic function.

2.6   SOME PHYSICAL APPLICATIONS OF HARMONIC FUNCTIONS

A number of interesting cases of natural phenomena that are described to a high degree of accuracy by harmonic functions will be discussed in this section.

Steady-state Heat Conduction[†]

Heat is said to move through a material by *conduction* when energy is transferred by collisions involving adjacent molecules and electrons. For conduction the time rate of flow of heat energy at each point within the material can be specified by means of a vector. Typically this vector will vary in both magnitude and direction throughout the material. In general, a variation with time must also be considered. However, we shall limit ourselves to steady-state problems where this will be unnecessary. Thus the intensity of heat conduction within a material is given by a vector function of spatial coordinates. Such a function is often known as a *vector field*. In the present case the vector field is called the *heat flux density* and is given the symbol Q.

Because of their close connection with complex variable theory, we will consider here only two-dimensional heat flow problems. The flow of heat takes place within a plate that we will regard as being parallel to the complex plane. The broad faces of the plate are assumed perfectly insulated. No heat can be absorbed or emitted by the insulation.

As shown in Fig. 2.6–1, some of the edge surfaces of the plate are connected to heat sources (which send out thermal energy) or heat sinks (which absorb thermal energy). The remaining edge surfaces are insulated. Heat energy cannot flow into any perfectly insulated surface. Thus the heat flux density vector Q will be assumed tangent to any insulated boundary. Since the properties of the heat sources and sinks are assumed independent of the coordinate ζ, which lies perpendicular to the xy-plane, the vector field Q within the plate depends on only the two variables x and y. The insulation on the broad faces of the plate ensures that Q has components along the x- and y-axes only; that is, Q has components $Q_x(x, y)$ and $Q_y(x, y)$. In conventional vector analysis we would write

$$Q = Q_x(x, y)\mathbf{a}_x + Q_y(x, y)\mathbf{a}_y, \tag{2.6–1}$$

where $\mathbf{a}_x$ and $\mathbf{a}_y$ are unit vectors along the x- and y-axes.

One must be careful not to confuse the vector that locates a particular point in a two-dimensional configuration with the vector representing Q at that point. For instance, if $Q_x = y + 1$, $Q_y = x$, then at the point $x = 1$, $y = 1$, we have $Q_x = 2$, $Q_y = 1$.

The direction of Q, at a particular point, is the direction in which thermal energy is being transported most rapidly.

Now consider a flat differential surface of area dS (see Fig. 2.6–2). The heat flux f through any surface is the flow of thermal energy through that surface per unit time.

[†]For a detailed discussion of this subject see L. Ingersoll, O. Zobel, and A. Ingersoll, *Heat Conduction* (Madison, Wisconsin: University of Wisconsin Press, 1954).

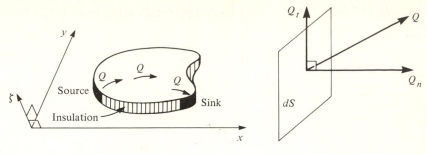

Figure 2.6–1   **Figure 2.6–2**

For dS, a differential flux of heat df passes through the surface given by

$$df = Q_n \, dS, \tag{2.6–2}$$

where Q_n is that component of Q normal to dS. The component Q_t, which is parallel to dS, carries no heat through the surface.

To obtain the heat flux f crossing a surface that is not flat and not of differential size, we must integrate the normal component of Q over the surface. The heat flux entering a volume is the total heat flux traversing inward through the surface bounding the volume.

Under steady-state conditions, the temperature in a conducting material is independent of time. The net flux of heat into any volume of the conductor is zero; otherwise, the volume would get hotter or colder depending on whether the entering flux was positive or negative. By requiring that the net flux entering a differential volume centered at x, y be zero, one can show that the components of Q satisfy

$$\frac{\partial Q_x}{\partial x} + \frac{\partial Q_y}{\partial y} = 0. \tag{2.6–3}$$

The equation is satisfied in the steady state by the two-dimensional heat flux density vector at any point where there are no heat sources or sinks. Equation (2.6–3) is the local or point form of the law of conservation of heat. The reader may recognize $\partial Q_x/\partial x + \partial Q_y/\partial y$ as the divergence of Q.

It is a familiar fact that the rate at which heat energy is conducted through a material is related to the temperature differences occurring within the material and also to the distances over which these differences occur, that is, to the rate of change of temperature with distance. Let us continue to assume two-dimensional heat flow, where the heat flow vector $Q(x, y)$ has components Q_x and Q_y. Let $\phi(x, y)$ be the temperature in the heat conducting medium. Then it can be shown that the components of the vector Q are related to $\phi(x, y)$ by

$$Q_x = -k \frac{\partial \phi}{\partial x}(x, y), \tag{2.6–4a}$$

$$Q_y = -k \frac{\partial \phi}{\partial y}(x, y). \tag{2.6–4b}$$

Here k is a constant called *thermal conductivity*. Its value depends on the material being considered. The reader may recognize Eqs. (2.6–4a, b) as being equivalent to the statement that Q is "minus k times the gradient of the temperature ϕ." The temperature ϕ serves as a "potential function" from which the heat flux density vector can be calculated by means of Eqs. (2.6–4a, b). With the aid of these equations we can rewrite Eq. (2.6–3) in terms of temperature:

$$-k\frac{\partial^2\phi}{\partial x^2} - k\frac{\partial^2\phi}{\partial y^2} = 0 \quad \text{or} \quad \frac{\partial^2\phi}{\partial x^2} + \frac{\partial^2\phi}{\partial y^2} = 0. \tag{2.6–5}$$

Thus under steady-state conditions, and where there are no sources or sinks, the temperature inside a conductor is a harmonic function.

Because the temperature $\phi(x, y)$ is a harmonic function, it can be regarded as the real part of a function that is analytic within a domain of the xy-plane corresponding to the interior of the conducting plate. This analytic function, which we call $\Phi(x, y)$ is known as the *complex temperature*. We then have

$$\Phi(x, y) = \phi(x, y) + i\psi(x, y). \tag{2.6–6}$$

Thus the real part of the complex temperature $\Phi(x, y)$ is the actual temperature $\phi(x, y)$. The imaginary part of the complex temperature, namely $\psi(x, y)$, we will call the *stream function* because of its analogy with a function describing streams along which particles flow in a fluid.

The curves along which $\phi(x, y)$ assumes constant values are called *isotherms* or *equipotentials*. These curves are just the edges of the surfaces along which the temperature is equal to a specific value. Several examples are represented by the solid line shown in Fig. 2.6–3.

From Theorem 8 we realize that the family of curves along which $\psi(x, y)$ is constant must be perpendicular to the isotherms. The $\psi =$ constant curves are called *streamlines*. Several are depicted by dashed lines in Fig. 2.6–3.

The slope of a curve along which $\psi(x, y)$ is constant is of interest. If we refer back to the derivation of Eq. (2.5–13) and replace u with ϕ and v with ψ, we find that the slope of a streamline passing through x, y is given by

$$\frac{dy}{dx} = \left(\frac{\partial\phi/\partial y}{\partial\phi/\partial x}\right). \tag{2.6–7}$$

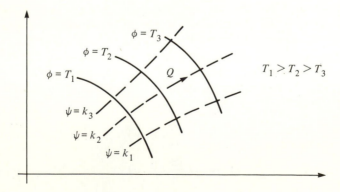

Figure 2.6–3

Now suppose we draw, at the same point, the local value of the heat flux density vector Q. From Eqs. (2.6–4a, b) we see that the slope of this vector is

$$\frac{Q_y}{Q_x} = \frac{\partial\phi/\partial y}{\partial\phi/\partial x}. \tag{2.6–8}$$

Comparing Eqs. (2.6–7) and (2.6–8) and noting the identical slopes, we can now conclude the following theorem.

Theorem 9

The heat flux density vector at a given point within a heat conducting medium is tangent to the streamline passing through that point. ■

We have illustrated this theorem by drawing Q at one point in Fig. 2.6–3. Note that the streamline establishes the slope of Q but not the actual direction of the vector. The direction is established by our realizing that the direction of heat flow is from a warmer to a colder isotherm.

A diagram of the family of curves along which $\psi(x, y)$ assumes constant values provides us with a picture of the paths along which heat is flowing. Moreover, since these streamlines are orthogonal to the isotherms, we conclude that:

The heat flux density vector, calculated at some point, is perpendicular to the isotherm passing through that point.

It is often convenient to introduce a function called the *complex heat flux density*, which is defined by

$$q(z) = Q_x(x, y) + iQ_y(x, y). \tag{2.6–9}$$

Since

$$\text{Re}(q) = Q_x \quad \text{and} \quad \text{Im}(q) = Q_y, \tag{2.6–10}$$

the vector associated with this function at any point x, y is precisely the heat flux density vector Q at that point. With the aid of Eqs. (2.6–4a, b) we rewrite q in terms of the temperature:

$$q = -k\left(\frac{\partial\phi}{\partial x} + i\frac{\partial\phi}{\partial y}\right). \tag{2.6–11}$$

The complex temperature $\Phi(z) = \phi(x, y) + i\psi(x, y)$ is an analytic function. If we want to know its derivative with respect to z, there are two convenient formulas, shown in Eqs. (2.3–6) and (2.3–8), at our disposal. Choosing the former (with $\phi = u$, $\psi = v$), we have

$$\frac{d\Phi}{dz} = \frac{\partial\phi}{\partial x} + i\frac{\partial\psi}{\partial x}. \tag{2.6–12}$$

Now ϕ and ψ satisfy the C–R equations

$$\frac{\partial\phi}{\partial x} = \frac{\partial\psi}{\partial y} \quad \text{and} \quad \frac{\partial\psi}{\partial x} = -\frac{\partial\phi}{\partial y}.$$

With the second of these equations used in the imaginary part of Eq. (2.6–12) we obtain

$$\frac{d\Phi}{dz} = \frac{\partial \phi}{\partial x} - i \frac{\partial \phi}{\partial y}.$$

Now, note that

$$\overline{\left(\frac{d\Phi}{dz}\right)} = \frac{\partial \phi}{\partial x} + i \frac{\partial \phi}{\partial y}. \tag{2.6–13}$$

Comparing Eqs. (2.6–13) and (2.6–11), we obtain the following convenient formula for the complex heat flux density:

$$q = -k \overline{\left(\frac{d\Phi}{dz}\right)}. \tag{2.6–14}$$

The real and imaginary parts of this expression then yield Q_x and Q_y. Of course, Q_x and Q_y are also obtainable if we find the temperature $\phi = \mathrm{Re}\,\Phi$ and then apply Eqs. (2.6–4a, b).

Fluid Flow

Because many of the concepts applying to heat conduction carry over directly to fluid mechanics, we can be a bit briefer about this topic.

   Let us assume we are dealing with an "ideal fluid," that is, one that's incompressible (its mass density does not alter) and nonviscous, (there are no losses due to internal friction). We assume a steady state, which means that the velocity of flow at any point of the fluid is independent of time. Like heat flow, fluid flow originates in sources and terminates in sinks.

   If a rigid impermeable obstruction is placed in the moving fluid, the fluid will move tangent to the surface of the object much as heat flows parallel to an insulated boundary.

   Earlier, we restricted ourselves to two-dimensional heat flow configurations. Heat conduction was parallel to the xy-plane and depended on only the variables x and y. Here we restrict ourselves to two-dimensional fluid flow parallel to the xy-plane. The fluid velocity V will be a vector field dependent in general on the coordinates x and y. It is analogous to the heat flux density Q. The components of V along the coordinate axes are V_x and V_y. The velocity V is the vector associated with the *complex velocity* defined by

$$v = V_x(x, y) + iV_y(x, y). \tag{2.6–15}$$

This expression is analogous to the complex heat flux density $q = Q_x(x, y) + iQ_y(x, y)$.

   Under certain conditions there is a fluid mechanical analogue of Eqs. (2.6–4a, b). There exists a real function of x and y, the *velocity potential* $\phi(x, y)$, such that

$$V_x = \frac{\partial \phi}{\partial x}, \tag{2.6–16a}$$

$$V_y = \frac{\partial \phi}{\partial y}. \tag{2.6–16b}$$

This condition is described by saying that the velocity is the *gradient* of the velocity potential. For V_x and V_y to be derivable from ϕ, as stated in Eq. (2.6–16), it is necessary that the fluid flow be what is called *irrotational*. Irrotational flow is approximated in many physical problems.[†]

Under steady-state conditions the total mass of fluid contained within any volume of space remains constant in time. For any volume not containing sources or sinks as much fluid flows in during any time interval as flows out. This should remind us of the steady-state conservation of heat. In fact, for an incompressible fluid the velocity components V_x and V_y satisfy the same conservation equation (2.6–3) as do the corresponding components Q_x and Q_y of the heat flux density vector. Using Eq. (2.6–16) to eliminate V_x and V_y from the conservation equation satisfied by the velocity vector, we have

$$\frac{\partial^2 \phi}{\partial x^2} + \frac{\partial^2 \phi}{\partial y^2} = 0. \tag{2.6–17}$$

Thus the velocity potential is a harmonic function.

An analytic function $\Phi(z)$ that has real part $\phi(x, y)$ can now be defined. Its imaginary part $\psi(x, y)$ is called the *stream function*. Thus

$$\Phi(z) = \phi(x, y) + i\psi(x, y). \tag{2.6–18}$$

The curves along which $\phi(x, y)$ assumes constant values are called equipotentials, and, as before, the curves of constant ψ are called streamlines. The two families of curves are orthogonal.

Like the heat flux density vector the fluid velocity vector at every point is tangent to the streamline passing through that point.

If we follow the progress of some specific moving droplet of fluid, we find that its path is a streamline. The fluid velocity vector at a given point is perpendicular to the equipotential passing through that point.

There is a simple relationship between the complex potential and the fluid velocity. From Eqs. (2.6–13) and (2.6–16) we have

$$\overline{\left(\frac{d\Phi}{dz} \right)} = V_x + iV_y = v. \tag{2.6–19}$$

Electrostatics[‡]

In the theory of electrostatics it is stationary (nonmoving) electric charge that plays the role of the sources and sinks we mentioned when discussing heat conduction and fluid flow.

[†] Irrotational flow presupposes the absence of vortices (whirlpools). For further discussion see M. Greenberg, *Foundations of Applied Mathematics* (Englewood Cliffs, N.J.: Prentice-Hall, 1978), sec. 9.8.

[‡] A clearly written reference on this subject is W. H. Hayt's *Engineering Electromagnetics*, 4th ed. (New York: McGraw-Hill, 1981).

According to the theory there are two kinds of charge: positive and negative. Charge is often measured in *coulombs*. Positive charge acts as a source of electric flux, negative charge acts as a sink. In other words, electric flux emanates outward from positive charge and is absorbed into negative charge.

The concentration of electric flux at a point in space is described by the electric flux density vector D. Although the notion of electric flux is something of a mathematical abstraction, we can make a physical measurement to determine D at any location. This is accomplished by our putting a point-sized test charge of q_0 coulombs at the spot in question. The test charge will experience a force because of its interaction with the source and (and sink) charges.[†] The vector force F is given by

$$F = q_0 \frac{D}{\varepsilon}. \tag{2.6--20}$$

Here, ε is a positive constant, called the *permittivity*. Its numerical value depends on the medium in which the test charge is embedded.

Often, instead of using the vector D directly, we employ the electric field vector E, which is defined by

$$\varepsilon E = D. \tag{2.6--21}$$

Thus Eq. (2.6–20) becomes

$$F = q_0 E \quad \text{or} \quad F/q_0 = E.$$

We see that the vector force on a test charge divided by the size of the charge yields the electric field E. Then Eq. (2.6–21) tells us the vector flux density D at the test charge.

We will be considering two-dimensional problems in electrostatics. This requires some explanation. All the electric charges involved in the creation of the electric flux are assumed to exist along lines, or cylinders, of infinite extent that lie perpendicular to the xy-plane. Let ζ be the coordinate perpendicular to the xy-plane. We assume that the distribution of charge along these sources or sinks of flux is independent of ζ. Any obstructions (for example, metallic conductors) placed within the electric flux must also be of infinite length, and perpendicular to the xy-plane. For this sort of configuration the electric flux density vector D is parallel to the xy-plane. Its components D_x and D_y depend, in general, on the variables x and y but are independent of ζ. Maxwell's equations show that the electric flux density vector created by static charges can be derived from a scalar potential. This electrostatic potential ϕ, usually measured in volts, bears much the same relation to the electric flux density as does the temperature to the heat flux density or the velocity potential to the fluid velocity.

The components of the electric flux density vector are obtained from $\phi(x, y)$ as follows:

$$D_x = -\varepsilon \frac{\partial \phi}{\partial x},$$

$$D_y = -\varepsilon \frac{\partial \phi}{\partial y}. \tag{2.6--22}$$

[†] This is an example of a field force that acts at a distance from its source, even through vacuum. Gravity is another example of such a force.

These equations are analogous to Eqs. (2.6–4a, b) for the heat flux density. If E_x and E_y are the components of the electric field, then, from Eq. (2.6–21), we have $D_x = \varepsilon E_x$, and $D_y = \varepsilon E_y$.

A comparison with Eq. (2.6–21) then shows that for the electric field

$$E_x = -\frac{\partial \phi}{\partial x},$$

$$E_y = -\frac{\partial \phi}{\partial y}. \tag{2.6–23}$$

One can define the electric flux crossing a surface in much the same way as one defines the heat flux crossing that surface. The amount of electric flux df that passes through a flat surface dS is obtained from Eq. (2.6–2) with D_n, the normal component of the electric flux density vector, substituted for Q_n. Thus $df = D_n\, dS$. The flux crossing a nondifferential surface is obtained by an integration of the normal component of D over that surface.

According to Maxwell's first equation the total electric flux entering any volume that contains no net electric charge is zero. This reminds us of an identical condition obeyed by the heat flux for a source-free volume. In fact, at any point in space where there is no electric charge, the electric flux density vector satisfies the same conservation equation (2.6–3) as does the heat flux density vector. If the components of D are eliminated from this conservation equation by means of Eq. (2.6–22), a familiar result is found:

$$\frac{\partial^2 \phi}{\partial x^2} + \frac{\partial^2 \phi}{\partial y^2} = 0.$$

Hence, the electrostatic potential is a harmonic function in any charge-free region.

As expected, we define an analytic function, the complex potential $\Phi = \phi + i\psi$, whose real part is the electrostatic potential. As before, the imaginary part is called the stream function.

The electric flux density vector is tangent to the streamlines generated from ψ.

The electric flux density D and electric field vector E are the vectors corresponding to the following complex functions:

$$d(z) = D_x(x, y) + iD_y(x, y),$$

$$e(z) = E_x(x, y) + iE_y(x, y).$$

These are called the *complex electric flux density* and the *complex electric field*, respectively, and they satisfy

$$d = -\varepsilon \overline{\left(\frac{d\Phi}{dz}\right)} \quad \text{and} \quad e = -\overline{\left(\frac{d\Phi}{dz}\right)}.$$

Our discussion of heat, fluids, and electrostatics is summarized in Table 1. There are other physical situations, for example, material diffusion and magnetostatics, where harmonic functions also prove useful.

TABLE 1

	Heat Conduction	Fluid Flow	Electrostatics
Flux density vector	Q = heat flux density	V = velocity	D = electric flux density
Complex flux function	$q = Q_x + iQ_y$	$v = V_x + iV_y$	$d = D_x + iD_y$
Harmonic potential function ϕ	temperature	velocity potential	electrostatic potential
Flux density components	$Q_x = -k\dfrac{\partial\phi}{\partial x}$	$V_x = \dfrac{\partial\phi}{\partial x}$	$D_x = -\varepsilon\dfrac{\partial\phi}{\partial x}$
	$Q_y = -k\dfrac{\partial\phi}{\partial y}$	$V_y = \dfrac{\partial\phi}{\partial y}$	$D_y = -\varepsilon\dfrac{\partial\phi}{\partial y}$
Complex flux density from complex potential $\Phi = \phi + i\psi$	$q = -k\overline{\left(\dfrac{d\Phi}{dz}\right)}$	$v = \overline{\left(\dfrac{d\Phi}{dz}\right)}$	$d = -\varepsilon\overline{\left(\dfrac{d\Phi}{dz}\right)}$

Example 1

A complex potential is of the form

$$\Phi(z) = Az + B, \quad \text{where } A \text{ and } B \text{ are real numbers.} \qquad (2.6\text{--}24)$$

Discuss its associated equipotentials, streamlines, and flux density in terms of electrostatics, heat conduction, and fluid flow.

Solution

The potential function is

$$\phi(x, y) = \text{Re}(Az + B) = Ax + B, \qquad (2.6\text{--}25)$$

and the stream function is

$$\psi(x, y) = \text{Im}(Az + B) = Ay. \qquad (2.6\text{--}26)$$

The equipotentials (or isotherms) are the surfaces on which $\phi(x, y)$ assumes fixed values. From Eq. (2.6–25) we see that these appear in the z-plane as lines on which x is constant. Some of these lines are drawn in Fig. 2.6–4. The streamlines on which ψ assumes fixed values are, according to Eq. (2.6–26), lines along which y is constant. These are indicated by dashes in the figure.

The reader who has studied electrostatics will recognize the potential distribution in the Fig. 2.6–4 as that existing between the plates of a parallel plate capacitor whose plates are perpendicular to the x-axis. The complex electric flux density for this configuration is, from the bottom line in Table 1,

$$d = -\varepsilon\,\overline{\frac{d}{dz}(Az + B)} = -\varepsilon A = D_x + iD_y,$$

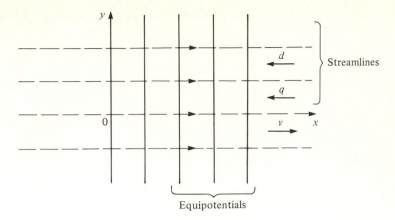

Figure 2.6–4

which implies $D_x = -\varepsilon A$, $D_y = 0$. The electric flux density vector is parallel to the x-axis. If $A > 0$, it points toward the left in Fig. 2.6–4.

If $\Phi(z)$ is the complex temperature, then the isotherms are the equipotentials in Fig. 2.6–4. The complex heat flux density is

$$q = -k\,\overline{\frac{d}{dz}\,(Az + B)} = -kA = Q_x + iQ_y$$

(see Table 1), which implies that $Q_x = -kA$, $Q_y = 0$. The heat flow is uniform and, if $A > 0$, in the negative x-direction.

Finally, if $\Phi(z)$ describes fluid flow, the fluid velocity is

$$V_x + iV_y = \overline{\frac{d}{dz}\,(Az + B)} = A,$$

so that $V_x = A$, $V_y = 0$. The fluid flow is thus uniform and in the direction of the streamlines in Fig. 2.6–4. For $A > 0$, flow is to the right. ◄

EXERCISES

1. Suppose that everywhere within a medium the components of the heat flux density vector Q are $Q_x = 3$, $Q_y = 4$ calories per square centimeter per second.

   a) Find the temperature $\phi(x, y)$ in degrees. Assume $\phi(0,0) = 0$ and that the conductivity k of the medium equals 0.1 calorie per centimeter degree second.

   b) Find the stream function $\psi(x, y)$. Assume $\psi(0,0) = 0$.

   c) Sketch the equipotentials on which ϕ equals $0, 1, -1, 2$.

   d) Sketch the streamlines $\psi = 0, 1, -1$. Verify that the lines are parallel to Q.

2. Suppose the complex potential describing a certain fluid flow is given by $\Phi(z) = 1/z$ (meter^2/sec) for $z \neq 0$.

   a) Find the complex fluid velocity at $x = 1$, $y = 1$ (meter) by differentiating the complex potential. State V_x and V_y.

b) Find the components V_x and V_y, at the same point, by finding and using the velocity potential $\phi(x, y)$.

c) Find the equation of the equipotential that passes through $x = 1$, $y = 1$. Plot this curve.

d) Find the equation of the streamline passing through $x = 1$, $y = 1$. Plot this curve.

3. Suppose that $\Phi(z) = e^x \cos y + i e^x \sin y$ represents the complex potential, in volts, for some electrostatic configuration.

a) Use the complex potential to find the complex electric field at $x = 1$, $y = 1/2$.

b) Obtain the complex electric field at the same point by first finding and using the electrostatic potential $\phi(x, y)$.

c) Assuming the configuration lies within a vacuum, find the components D_x and D_y of the electric flux density vector at $x = 1$, $y = 1/2$. In m.k.s. units, $\varepsilon = 8.85 \times 10^{-12}$ for vacuum.

4. a) Explain why $d(x, y) = y + ix$ can be the complex electric flux density in a charge-free region, but $d(x, y) = x + iy$ cannot.

b) Assume that the complex electric flux density $y + ix$ exists in a medium for which $\varepsilon = 9 \times 10^{-12}$. Find the electrostatic potential $\phi(x, y)$. Assume $\phi(0, 0) = 0$. Sketch the equipotentials $\phi(x, y) = 0$, $\phi(x, y) = 1/\varepsilon$.

c) Find the stream function $\psi(x, y)$. Assume $\psi(0, 0) = 0$.

d) Find the complex potential Φ and express it explicitly in terms of z.

e) Find the components of the electric field at $x = 1$, $y = 1$ by three different methods: from d, from $\Phi(z)$, and from $\phi(x, y)$. Show with a sketch the vector for this field and the equipotential passing through $x = 1$, $y = 1$.

5. a) Fluid flow is described by the complex potential $\Phi(z) = (\cos \alpha - i \sin \alpha)z$. $\alpha > 0$. Sketch the associated equipotentials and give their equations.

b) Sketch the streamlines and give their equations.

c) Find the components v_x and v_y of the velocity vector at (x, y). What angle does the velocity vector make with the positive x-axis?

CHAPTER 3 THE BASIC TRANSCENDENTAL FUNCTIONS

We are well acquainted with numerous functions that exist in the mathematics of real variables. We know the algebraic functions (sums, products, quotients, and powers of x^m and $x^{m/n}$), and we know transcendental functions such as e^x, $\log x$, $\sinh x$, and so forth. However, in the complex plane we have seen only algebraic functions of z. In this chapter we will enlarge our collection of functions of a complex variable. We will extend our definitions of some elementary transcendental functions of a real variable (e^x, $\sin x$, etc.) so as to yield functions of the complex variable z. These new functions reduce to our previously known real functions if z happens to be real.

3.1 THE EXPONENTIAL FUNCTION

We would like this function to have the following properties:

a) e^z reduces to our known e^x if z happens to assume real values.

b) e^z is an analytic function of z.

The function $e^x \cos y + ie^x \sin y$ will be our definition of e^z (or $\exp z$). Thus

$$e^z = e^{x+iy} = e^x[\cos y + i \sin y] \tag{3.1-1}$$

clearly satisfies condition (a) above. (Put $y = 0$.) Observe that as in the real case, $e^0 = 1$. To verify condition (b) we have

$$u + iv = e^x \cos y + ie^x \sin y, \tag{3.1-2}$$

where

$$u = \operatorname{Re} e^z = e^x \cos y, \qquad v = \operatorname{Im} e^z = e^x \sin y. \tag{3.1-3}$$

The pair of functions u and v have first partial derivatives:

$$\frac{\partial u}{\partial x} = e^x \cos y, \qquad \frac{\partial v}{\partial y} = e^x \cos y, \qquad \text{and so on,}$$

which are continuous everywhere in the xy-plane. Furthermore, u and v satisfy the Cauchy–Riemann equations,

$$\frac{\partial u}{\partial x} = \frac{\partial v}{\partial y}, \qquad \frac{\partial v}{\partial x} = \frac{-\partial u}{\partial y},$$

everywhere in this plane. Thus e^z is analytic for all z and is therefore an entire function. Condition (b) is clearly satisfied.

The derivative $d(e^z)/dz$ is easily found from Eqs. (2.3–6) and (3.1–3). Thus

$$\frac{d}{dz} e^z = \frac{\partial}{\partial x} e^x \cos y + i \frac{\partial}{\partial x} e^x \sin y = e^x \cos y + i e^x \sin y,$$

or

$$\frac{d}{dz} e^z = e^z.$$

This is a reassuring result since we already knew that e^x satisfies

$$\frac{d}{dx} e^x = e^x.$$

Note that if $g(z)$ is an analytic function, then, by the chain rule of differentiation, we have

$$\frac{d}{dz} e^{g(z)} = e^{g(z)} g'(z).$$

The function e^z shares another property with e^x. We know that if x_1 and x_2 are real, then $e^{x_1} e^{x_2} = e^{(x_1 + x_2)}$. We can show that if $z_1 = x_1 + iy_1$ and $z_2 = x_2 + iy_2$ are a pair of complex numbers, then $e^{z_1} e^{z_2} = e^{z_1 + z_2}$. Observe that

$$e^{z_1} = e^{x_1} [\cos y_1 + i \sin y_1] \quad \text{and} \quad e^{z_2} = e^{x_2} [\cos y_2 + i \sin y_2],$$

so that

$$e^{z_1} e^{z_2} = e^{x_1} e^{x_2} [\cos y_1 + i \sin y_1][\cos y_2 + i \sin y_2]$$
$$= e^{x_1 + x_2} [(\cos y_1 \cos y_2 - \sin y_1 \sin y_2) + i(\sin y_1 \cos y_2 + \cos y_1 \sin y_2)].$$

The real part of the expression in the brackets is, from elementary trigonometry, $\cos(y_1 + y_2)$. Similarly, the imaginary part is $\sin(y_1 + y_2)$. Hence,

$$e^{z_1} e^{z_2} = e^{x_1 + x_2} [\cos(y_1 + y_2) + i \sin(y_1 + y_2)]. \tag{3.1–4}$$

Now with the help of Eq. (3.1–1), we have

$$e^{z_1 + z_2} = e^{(x_1 + x_2) + i(y_1 + y_2)}$$
$$= e^{(x_1 + x_2)} [\cos(y_1 + y_2) + i \sin(y_1 + y_2)]. \tag{3.1–5}$$

Since the right sides of Eqs. (3.1–4) and (3.1–5) are identical, it is obvious that $e^{z_1} e^{z_2} = e^{z_1 + z_2}$.

Just as $e^{x_1}/e^{x_2} = e^{x_1 - x_2}$, we can show (see Exercise 4 of this section) that

$$\frac{e^{z_1}}{e^{z_2}} = e^{z_1 - z_2}.$$

If $z_1 = 0$ and $z_2 = z$ in the preceding equation, we obtain $1/e^z = e^{-z}$.

The magnitude of e^z is determined entirely by the real part of z.

We can prove this by observing that

$$|e^z| = |e^x(\cos y + i \sin y)| = |e^x||\cos y + i \sin y|.$$

But

$$|\cos y + i \sin y| = \sqrt{\cos^2 y + \sin^2 y} = 1.$$

Therefore

$$|e^z| = |e^x|$$

and, since e^x is positive for all x, we obtain

$$|e^z| = e^x. \qquad (3.1\text{–}6)$$

Equally important is the following property.

The argument of e^z is determined entirely by the imaginary part of z.

Notice that

$$e^z = e^x \cos y + ie^x \sin y$$

and that

$$\arg(e^z) = \tan^{-1}\left[\frac{\operatorname{Im} e^z}{\operatorname{Re} e^z}\right] = \tan^{-1}\left(\frac{e^x \sin y}{e^x \cos y}\right) = \tan^{-1}(\tan y) = y. \quad (3.1\text{–}7)$$

In polar form we have

$$e^z = |e^z| \,\underline{/\arg e^z} = e^x \,\underline{/y}. \qquad (3.1\text{–}8)$$

Although e^x is not a periodic function of x, e^z varies periodically as we move in the z-plane along any straight line parallel to the y-axis. Consider e^{z_0} and $e^{z_0+i2\pi}$, where $z_0 = x_0 + iy_0$. The points z_0 and $z_0 + i2\pi$ are separated a distance 2π on the line $\operatorname{Re} z = x_0$. From our definition (see Eq. 3.1–1) we have

$$e^{z_0} = e^{x_0}(\cos y_0 + i \sin y_0),$$

$$e^{z_0+i2\pi} = e^{x_0+i(y_0+2\pi)} = e^{x_0}\left[\cos(y_0 + 2\pi) + i \sin(y_0 + 2\pi)\right].$$

Since $\cos y_0 = \cos(y_0 + 2\pi)$ (and similarly for the sine), then $e^{z_0} = e^{z_0+i2\pi}$.

Thus e^z is periodic with imaginary period $2\pi i$.

Of particular interest is the behavior of $e^{i\theta}$ when θ is a real variable. In Eq. (3.1–8) we put $x = 0$, $y = \theta$ to get

$$e^{i\theta} = 1 \,\underline{/\theta} = \cos\theta + i \sin\theta. \qquad (3.1\text{–}9)$$

Thus if θ is real, $e^{i\theta}$ is a complex number of modulus 1, which lies at an angle θ with respect to the positive real axis (see Fig. 3.1–1). As θ increases, the resulting complex number progresses counterclockwise around the unit circle. Observe, in particular, that

$$e^{i0} = 1, \qquad e^{i\pi/2} = i, \qquad e^{i\pi} = -1, \qquad e^{i3\pi/2} = -i = e^{-i\pi/2}.$$

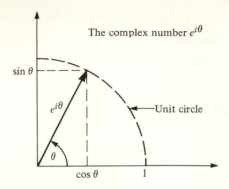

Figure 3.1–1

The relationship $e^{i\theta} = \cos\theta + i\sin\theta$ is known as Euler's identity. The complex number $1 \angle \theta$, or $\cos\theta + i\sin\theta$, can be regarded as synonymous with $e^{i\theta}$.

The exponential is often used when we perform multiplication. Thus

$$1 + i\sqrt{3} = 2\angle\frac{\pi}{3} = 2e^{i\pi/3} \quad \text{and} \quad 1 + i = \sqrt{2}\angle\frac{\pi}{4} = \sqrt{2}e^{i(\pi/4)},$$

so that

$$(1 + i\sqrt{3})(1 + i) = 2\sqrt{2}\, e^{i(\pi/3 + \pi/4)}.$$

With the aid of Eq. (3.1–9) the right side of this equation can be rewritten as

$$2\sqrt{2}\left[\cos\left(\frac{\pi}{3} + \frac{\pi}{4}\right) + i\sin\left(\frac{\pi}{3} + \frac{\pi}{4}\right)\right].$$

Some important physical applications of the complex exponential in electrical engineering and mechanics can be found in the appendix to this chapter.

EXERCISES

1. Express each of the following in the form $a + ib$, where a and b are real numbers.

   a) e^{2+i} c) e^{-2-i} e) $\mathrm{Re}(e^{-2}e^{-i})$ g) $e^{(1+i)^2}$

   b) e^{2-i} d) $e^{-2}e^{-i}$ f) $\mathrm{Im}(e^{-2}e^{-i})$ h) $e^{1/(1+i)}$

2. Express each of the following in the form $u(x, y) + iv(x, y)$, where u and v are real functions of the variables x and y.

   a) e^{z^2} b) e^{iz} c) $e^{1/z}$ d) $1/e^z$ e) e^{e^z}

3. Which of the following statements are, in general, true?

   a) $e^{|z|} = |e^z|$ c) $e^{1/z} = 1/e^z$ e) $(e^z)^m = e^{mz}$, m is any integer

   b) $e^{\bar{z}} = (\overline{e^z})$ d) $\mathrm{Re}\, e^z = e^{\mathrm{Re}\, z}$ f) $e^z + e^{\bar{z}} = 2(\mathrm{Re}\, e^z)$

4. Use the definition of e^z given in Eq. (3.1–1) to show that:

$$\frac{e^{z_1}}{e^{z_2}} = e^{z_1 - z_2}, \quad \text{where } z_1 = x_1 + iy_1, \quad z_2 = x_2 + iy_2.$$

5. If $f(z) = e^z$, show that $f(z + in\pi) = (-1)^n f(z)$, where $n = 0, \pm 1, \pm 2, \ldots$.

6. Find the numerical values of $e^{2+99\pi i}$ and $e^{2+(99.1)\pi i}$ in the Cartesian form $a + ib$.

7. Assume k is an integer. State all possible numerical values of each of the following.

    a) $e^{ik\pi}$ b) $e^{i(k\pi/2)}$ c) $e^{i(k\pi/4)}$ d) $e^{i(k\pi/3)}$

8. What are the maximum and minimum values assumed by $|e^z|$ in the region described by $|z - 2i| \le 1$?

9. Find the domain of analyticity (if any) of each of the following functions. State the derivative of the function, with respect to z, within the domain of analyticity.

    a) e^{2z} b) e^{z^2} c) $e^{|z|}$ d) $e^{\bar{z}}$ e) $e^{1/z}$

10. a) Express $e^{(e^i)}$ in the form $a + ib$, and give numerical values for a and b.

    b) Show by direct computation that $f(z) = e^{(e^z)}$ satisfies the Cauchy–Riemann equations everywhere in the z-plane. What is $f'(z)$?

11. Fluid flow is described by the complex potential $\Phi(z) = \phi(x, y) + i\psi(x, y) = e^z$.

    a) Find the velocity potential $\phi(x, y)$ and the stream function $\psi(x, y)$ explicitly in terms of x and y.

    b) In the strip $|\text{Im } z| \le \pi/2$ sketch the equipotentials $\phi = 0, +1/2, +1, +2$ and the streamlines $\psi = 0, \pm 1/2, \pm 1, \pm 2$.

    c) What is the fluid velocity vector at $x = 1$, $y = \pi/4$?

3.2 TRIGONOMETRIC FUNCTIONS

From the Euler identity (Eq. 3.1–9) we know that when θ is a real number, we have

$$e^{i\theta} = \cos \theta + i \sin \theta. \tag{3.2–1}$$

Now, if we alter the sign preceding θ in the above, we have

$$e^{-i\theta} = \cos(-\theta) + i \sin(-\theta) = \cos \theta - i \sin \theta. \tag{3.2–2}$$

The addition of Eq. (3.2–2) to Eq. (3.2–1) results in the purely real expression

$$e^{i\theta} + e^{-i\theta} = 2 \cos \theta,$$

or, finally,

$$\cos \theta = \frac{e^{i\theta} + e^{-i\theta}}{2}. \tag{3.2–3}$$

If, instead, we had subtracted Eq. (3.2–2) from Eq. (3.2–1), we would have obtained

$$e^{i\theta} - e^{-i\theta} = 2i \sin \theta,$$

or

$$\sin \theta = \frac{e^{i\theta} - e^{-i\theta}}{2i}. \tag{3.2–4}$$

Equations (3.2–3) and (3.2–4) serve to define the sine and cosine of *real* numbers in terms of complex exponentials.

It is natural to define $\sin z$ and $\cos z$, where z is complex, as follows:

$$\sin z = \frac{e^{iz} - e^{-iz}}{2i}, \tag{3.2-5}$$

$$\cos z = \frac{e^{iz} + e^{-iz}}{2}. \tag{3.2-6}$$

These definitions make sense for several reasons:

a) When z is a real number, the definitions shown in Eqs. (3.2–5) and (3.2–6) reduce to the conventional definitions shown in Eqs. (3.2–3) and (3.2–4) for the sine and cosine of real arguments.

b) e^{iz} and e^{-iz} are analytic throughout the z-plane. Therefore, $\sin z$ and $\cos z$, which are defined by the sums and differences of these functions, are also.

c) $d \sin z / dz = i[e^{iz} + e^{-iz}]/2i = \cos z$. Also, $d \cos z / dz = -\sin z$.

It is easy to show that $\sin^2 z + \cos^2 z = 1$ and that the identities satisfied by the sine and cosine of real arguments apply here too, for example,

$$\sin(z_1 \pm z_2) = \sin z_1 \cos z_2 \pm \cos z_1 \sin z_2,$$

$$\cos(z_1 \pm z_2) = \cos z_1 \cos z_2 \mp \sin z_1 \sin z_2,$$

and so on.

Using Eqs. (3.2–5) and (3.2–6), we can compute the numerical value of the sine and cosine of any complex number we please. A somewhat more convenient procedure exists, however. Recall the hyperbolic sine and cosine of real argument θ illustrated in Fig. 3.2–1 and defined by the following equations:

$$\sinh \theta = \frac{e^{\theta} - e^{-\theta}}{2}, \tag{3.2-7}$$

$$\cosh \theta = \frac{e^{\theta} + e^{-\theta}}{2}. \tag{3.2-8}$$

Consider now the expression in Eq. (3.2–5):

$$\sin z = \frac{e^{i(x+iy)} - e^{-i(x+iy)}}{2i} = \frac{e^{ix-y} - e^{-ix+y}}{2i} = \frac{e^{-y}e^{ix} - e^{y}e^{-ix}}{2i}.$$

The Euler identity (Eq. 3.2–1 or 3.2–2) can now be used to rewrite e^{ix} and e^{-ix} in the preceding equation. We then have

$$\sin z = \frac{e^{-y}(\cos x + i \sin x)}{2i} - \frac{e^{y}(\cos x - i \sin x)}{2i}$$

$$= \sin x \frac{(e^{y} + e^{-y})}{2} + i \cos x \frac{(e^{y} - e^{-y})}{2}.$$

When the expressions involving y in the last equation are compared with the hyperbolic functions in Eqs. (3.2–7) and (3.2–8), we see that

$$\sin z = \sin x \cosh y + i \cos x \sinh y. \tag{3.2-9}$$

Since many pocket calculators are equipped with these functions, the evaluation of

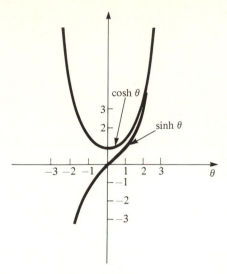

Figure 3.2–1

sin z becomes a simple matter. A similar expression can be found for cos z:

$$\cos z = \cos x \cosh y - i \sin x \sinh y. \qquad (3.2\text{--}10)$$

The other trigonometric functions of complex argument are easily defined by analogy with real argument functions, that is,

$$\tan z = \frac{\sin z}{\cos z} = \frac{1}{\cot z}, \qquad \sec z = \frac{1}{\cos z}, \qquad \operatorname{cosec} z = \frac{1}{\sin z}.$$

The derivatives of these functions are as follows:

$$\frac{d}{dz} \tan z = \sec^2 z,$$

$$\frac{d}{dz} \sec z = \tan z \sec z,$$

$$\frac{d}{dz} \operatorname{cosec} z = -\cot z \operatorname{cosec} z.$$

The sine or cosine of a real number is a real number whose magnitude is less than or equal to 1. Not only is the sine or cosine of a complex number, in general, a complex number, but the magnitude of a sine or cosine of a complex number can exceed 1 (see Exercise 1 in this section).

Example 1

Express $\sin(i\theta)$, where θ is a real number, in the form $a + ib$, and express $\cos(i\theta)$ in a similar form.

Solution

We use Eq. (3.2–9) with $x = 0$ and $y = \theta$. The result is

$$\sin(i\theta) = \sin 0 \cosh \theta + i \cos 0 \sinh \theta,$$

or

$$\sin(i\theta) = i \sinh\theta. \tag{3.2-11}$$

Similarly, with the aid of Eq. (3.2–10) we have

$$\cos(i\theta) = \cos 0 \cosh\theta - i \sin 0 \sinh\theta,$$

or

$$\cos(i\theta) = \cosh\theta. \ \blacktriangleleft \tag{3.2-12}$$

The cosine of a pure imaginary number is always a real number while the sine of a pure imaginary is always pure imaginary.

Example 2

Show that all zeros of $\cos z$ in the z-plane lie along the x-axis.

Solution

Consider the equation $\cos z = 0$, where $z = x + iy$. From Eq. (3.2–10) this becomes $\cos x \cosh y - i \sin x \sinh y = 0$. Both the real and imaginary part of the left-hand side of this equation must equal zero. We thus have

$$\cos x \cosh y = 0 \quad \text{and} \quad \sin x \sinh y = 0.$$

Consider the first equation. Since $\cosh y$ is never zero for a real number y, (see Fig. 3.2–1) evidently $\cos x = 0$. This means $x = \pm\pi/2, \pm3\pi/2,\ldots$, etc.; in other words, $x = \pm(2n + 1)\pi/2$, where $n = 0, 1, 2,\ldots$. Now consider the second equation. The first equation dictated that x be an odd multiple of $\pm\pi/2$; therefore, $\sin x$ in the second equation is ± 1. Thus $\sin x \sinh y = 0$ is only satisfied if $\sinh y = 0$. A glance at Fig. 3.2–1 shows this to be possible only when $y = 0$.

We see that $\cos z = 0$ only at those points that simultaneously satisfy $y = 0$, $x = \pm(2n + 1)\pi/2$. The first condition places all these points on the x-axis while the second spaces them at intervals of π. The values of z that solve $\cos z = 0$ are precisely the same as the solutions of $\cos x = 0$. A similar statement applies to $\sin x$ and $\sin z$ and is derived in one of the exercises below. ◀

EXERCISES

1. Use a pocket calculator or numerical tables to express each of the following in the form $a + ib$, where a, b are real numbers.

   a) $\cos(3 - 2i)$ g) $e^{\cos(i)}$

   b) $\cos(3 + 2i)$ h) $e^{\sin(im)}$, where $m = 0, \pm1, \pm2,\ldots$

   c) $\sin(3 - 2i)$ i) $\sin(\cos i)$

   d) $\sin(3 + 2i)$ j) $\cos(\sin i)$

   e) $\tan(3 - 2i)$ k) $\tan(\tan 2i)$

   f) $\cos(e^i)$

2. Consider the equation $f(\bar{z}) = \overline{f(z)}$. Do any of the following functions satisfy this equation: $\sin z$, $\cos z$, $\tan z$.

3. Use the definitions shown in Eqs. (3.2–5) and (3.2–6) to establish each of the following:

a) $\sin^2 z + \cos^2 z = 1$,

b) $\dfrac{d}{dz} \sin z = \cos z$,

c) $\sin^2 z = \dfrac{1}{2} - \dfrac{1}{2} \cos 2z$,

d) $\sin(z_1 + z_2) = \sin z_1 \cos z_2 + \cos z_1 \sin z_2$,

e) $\sin(z + 2k\pi) = \sin z,\ \cos(z + 2k\pi) = \cos z,$ where $k = 0, \pm 1, \pm 2, \dots$.

4. Show that $e^{iz} = \cos z + i \sin z$ for any complex number z. This is an extension of Euler's identity.

5. Where in the z-plane is $\tan z$ not an analytic function?

6. Show that the equation $\sin z = 0$ has solutions in the complex z-plane only where $z = n\pi$ and $n = 0, \pm 1, \pm 2, \dots$. Thus like $\cos z$, $\sin z$ has zeros only on the real axis.

7. If $f(z) = \sin(e^z)$, where in the z-plane is $f(z)$ analytic? Find $f'(z)$.

8. a) Show that $|\cos z| = \sqrt{\sinh^2 y + \cos^2 x}$. *Hint:* Recall that $\cosh^2 \theta - \sinh^2 \theta = 1$.

b) Show that $|\sin z| = \sqrt{\sinh^2 y + \sin^2 x}$.

c) Show that $|\sin z|^2 + |\cos z|^2 = \sinh^2 y + \cosh^2 y$.

9. a) Where in the z-plane are $\sec z$ and $\operatorname{cosec} z$ analytic functions of z?

b) Find $d(e^{\operatorname{cosec} z})/dz$ and give its numerical value at $z = i$.

c) Find $d(\sin(\sec z))/dz$ and give its numerical value at $z = -i$.

10. Show that
$$\tan z = \frac{\sin(2x) + i \sinh(2y)}{\cos(2x) + \cosh(2y)}.$$

3.3 HYPERBOLIC FUNCTIONS

In the previous section we used definitions of $\sin z$ and $\cos z$ that we invented by studying the definitions of sine and cosine functions of real arguments. A similar procedure will work in the case of $\sinh z$ and $\cosh z$, the hyperbolic functions of complex argument.

Equations (3.2–7) and (3.2–8), which define $\sinh \theta$ and $\cosh \theta$ for a real number θ, suggest the following definitions for complex z:

$$\sinh z = \frac{e^z - e^{-z}}{2}, \tag{3.3–1}$$

$$\cosh z = \frac{e^z + e^{-z}}{2}. \tag{3.3–2}$$

If z is a real number, these definitions reduce to those we know for the hyperbolic functions of real arguments. We see that $\sinh z$ and $\cosh z$ are composed of sums or differences of the functions e^z and e^{-z}, which are analytic in the z-plane. Thus $\sinh z$ and $\cosh z$ are analytic for all z. It is easy to verify that $d(\sinh z)/dz = \cosh z$ and that $d(\cosh z)/dz = \sinh z$.

All the identities that pertain to the hyperbolic functions of real variables carry over to these functions, for example, we may prove with Eqs. (3.3–1) and (3.3–2) that

$$\cosh^2 z - \sinh^2 z = 1, \tag{3.3-3}$$

$$\cosh(z_1 \pm z_2) = \cosh z_1 \cosh z_2 \pm \sinh z_1 \sinh z_2, \tag{3.3-4}$$

$$\sinh(z_1 \pm z_2) = \sinh z_1 \cosh z_2 \pm \cosh z_1 \sinh z_2. \tag{3.3-5}$$

Expressions for $\sinh z$ and $\cosh z$, involving real functions of real variables and analogous to Eqs. (3.2–9) and (3.2–10), are easily derived. They are

$$\sinh z = \sinh x \cos y + i \cosh x \sin y, \tag{3.3-6}$$

$$\cosh z = \cosh x \cos y + i \sinh x \sin y. \tag{3.3-7}$$

Other hyperbolic functions are readily defined in terms of the hyperbolic sine and cosine:

$$\tanh z = \frac{\sinh z}{\cosh z}, \qquad \operatorname{sech} z = \frac{1}{\cosh z}, \qquad \operatorname{cosech} z = \frac{1}{\sinh z}, \qquad \coth z = \frac{1}{\tanh z}.$$

The hyperbolic functions differ significantly from their trigonometric counterparts. Although all the roots of $\sin z = 0$ and $\cos z = 0$ lie along the real axis in the z-plane, it is shown in the exercises that all the roots of $\sinh z = 0$ and $\cosh z = 0$ lie along the imaginary axis. The trigonometric functions are periodic, with period 2π (see Exercise 3e, Section 3.2) but the hyperbolic functions have period $2\pi i$, that is, $\sinh(z + 2\pi i) = \sinh z$, $\cosh(z + 2\pi i) = \cosh z$.

EXERCISES

1. Use Eqs. (3.3–1) and (3.3–2) to show that
   a) Eqs. (3.3–6) and (3.3–7) are valid,
   b) $\cosh^2 z - \sinh^2 z = 1$,
   c) $\sinh(z_1 + z_2) = \sinh z_1 \cosh z_2 + \cosh z_1 \sinh z_2$,
   d) $\sinh(i\theta) = i \sin \theta$ and $\cosh i\theta = \cos \theta$. Thus the hyperbolic sine of a pure imaginary number is a pure imaginary number while the hyperbolic cosine of a pure imaginary number is a real number.
   e) Use parts (c) and (d) to show that $\sinh(z + 2\pi i) = \sinh z$ and also that $\cosh(z + 2\pi i) = \cosh z$.

2. Express each of the following in the form $a + ib$, where a and b are real numbers.
   a) $\sinh(2 + i)$ b) $\cosh(-1 - i)$ c) $\tanh(2 + i)$

3. Show that $d(\tanh z)/dz = \operatorname{sech}^2 z$.

4. Consider the equation $\sinh(x + iy) = 0$. Use Eq. (3.3–6) to equate the real and imaginary parts of $\sinh z$ to zero. Show that this pair of equations can be

satisfied if and only if $z = in\pi$, where $n = 0, \pm 1, \pm 2, \ldots$. Thus the zeros of $\sinh z$ all lie along the imaginary axis in the z-plane.

5. a) Carry out an argument similar to that in the previous problem to show that the zeros of $\cosh z$ must satisfy $z = \pm(2n + 1)\pi i/2$, where $n = 0, 1, 2, 3, \ldots$.

   b) Where in the z-plane is $\tanh z$ analytic?

6. Show that

   a) $|\sinh z|^2 = \sinh^2 x + \sin^2 y$,   b) $|\cosh z|^2 = \sinh^2 x + \cos^2 y$.

7. a) Where in the complex plane is $f(z) = \sinh(\sin(z^2))$ analytic?

   b) Find $f'(z)$ and give the numerical value of $f'(i)$.

8. Use L'Hopital's rule (see Eq. 2.3–12) to find the limit of the following functions as z tends to zero.

   a) $\sin(z)/z$ b) $\sinh(z)/z$ c) $(1 - \cosh(z))/z$

3.4 THE LOGARITHMIC FUNCTION†

If x is a positive real number, then, as the reader knows, $e^{\log x} = x$. The logarithm of x, which is a real number, is easily found from a calculator or numerical tables. Recall, however, that $\log 0$ is undefined.

In this section we will learn how to obtain the logarithm of a complex number z. We must anticipate that the logarithm of z may itself be a complex number. Our $\log z$ will have the property

$$e^{\log z} = z. \tag{3.4–1}$$

We shall see that $\log z$ is multivalued, that is, for a given value of z there is more than one value of $\log z$ capable of satisfying Eq. (3.4–1).

We will presently show that the following definition of $\log z$ will satisfy Eq. (3.4–1).

$$\log z = \log|z| + i \arg z, \qquad z \neq 0 \tag{3.4–2}$$

The logarithm of zero will remain undefined.

If z is expressed in polar variables $z = re^{i\theta}$, we know that $r = |z|$ and $\theta = \arg z$. Hence, Eq. (3.4–2) becomes

$$\log z = \log r + i\theta, \qquad r \neq 0. \tag{3.4–3}$$

Notice that the real part of the above expression, the logarithm of the modulus of z, is the logarithm of a positive real number. This quantity is evaluated by familiar means. The imaginary part θ is the argument (or angle) of z expressed in radians.

As promised, we see that $e^{\log z}$ is z because

$$e^{\log z} = e^{\log r + i\theta} = e^{\log r}e^{i\theta} = e^{\log r}[\cos \theta + i \sin \theta]$$
$$= r[\cos \theta + i \sin \theta] = x + iy = z. \tag{3.4–4}$$

Here we have used Euler's identity (Eq. 3.2–1) to rewrite $e^{i\theta}$ and have replaced $e^{\log r}$ by r, which is obviously valid for $r > 0$.

---

†All logarithms in this book are base e (natural) logarithms.

The chief difficulty with Eq. (3.4–2) or (3.4–3) is that $\theta = \arg z$ is not uniquely defined. We know that if θ_1 is some valid value for θ, then so is $\theta_1 + 2k\pi$, where $k = 0, \pm 1, \pm 2, \ldots$. Thus the numerical value of the imaginary part of $\log z$ is directly affected by our particular choice of the argument of z. We thus say that the logarithm of z, as defined by Eq. (3.4–2) or (3.4–3), is a multivalued function of z. Each value of $\log z$ satisfies $e^{\log z} = z$.

Even the logarithm of a positive real number, which we have been thinking is uniquely defined, has, according to Eq. (3.4–3), more than one value. However, when we consider all the possible logarithms of a positive real number, there is only one that is real; the others are complex. It is the real value that we find in numerical tables.

The *principal value* of the logarithm of z, denoted by Log z, is obtained when we use the *principal argument* of z in Eqs. (3.4–2) and (3.4–3). Recall (see Section 1.3) that the principal argument of z, which we will designate θ_p, is the argument of z satisfying $-\pi < \theta_p \leq \pi$. Thus we have

$$\text{Log } z = \text{Log } r + i\theta_p, \qquad r = |z| > 0, \quad \theta_p = \arg z, \quad -\pi < \theta_p \leq \pi. \quad (3.4\text{–}5)$$

Note that we put Log r (instead of log r) in the above equation since the natural logarithms of positive real numbers, obtained from tables or calculators, are principal values.

Any value of $\arg z$ can be obtained from the principal value θ_p by means of the formula $\arg z = \theta = \theta_p + 2k\pi$, where k has a suitable integer value. Thus all values of $\log z$ are obtainable from the expression

$$\log z = \text{Log } r + i(\theta_p + 2k\pi), \qquad k = 0, \pm 1, \pm 2, \ldots. \quad (3.4\text{–}6)$$

With $k = 0$ in this expression we obtain the principal value, Log z.

Observe that if we choose to use some nonprincipal value of $\arg z$, instead of θ_p, in Eq. (3.4–6), then Eq. (3.4–6) would still yield all possible values of $\log z$, although the principal value would not be generated by our putting $k = 0$.

Example 1

Find $\text{Log}(-1 - i)$ and find all values of $\log(-1 - i)$.

Solution

The complex number $-1 - i$ is illustrated graphically in Fig. 1.3–13. The principal argument of this number θ_p is $-3\pi/4$ while r, the absolute magnitude, is $\sqrt{2}$. From Eq. (3.4–5)

$$\text{Log}(-1 - i) = \text{Log }\sqrt{2} + i\left(\frac{-3\pi}{4}\right) \doteq 0.34657 - i\frac{3\pi}{4}.$$

All values of $\log(-1 - i)$ are easily written down with the aid of Eq. (3.4–6).

$$\log(-1 - i) = \text{Log }\sqrt{2} + i\left(2k\pi - \frac{3\pi}{4}\right)$$

$$\doteq 0.34657 + i\left(2k\pi - \frac{3\pi}{4}\right), \qquad k = 0, \pm 1, \pm 2, \ldots \quad \blacktriangleleft$$

Example 2

Find $\mathrm{Log}(-10)$ and all values of $\log(-10)$.

Solution

The principal argument of any negative real number is π. Hence, from Eq. (3.4–5) $\mathrm{Log}(-10) = \mathrm{Log}\,10 + i\pi \doteq 2.303 + i\pi$. From Eq. (3.4–6) we have

$$\log(-10) = \mathrm{Log}\,10 + i(\pi + 2k\pi).$$

We can check this result as follows:

$$e^{\log(-10)} = e^{\mathrm{Log}\,10 + i(\pi + 2k\pi)} = e^{\mathrm{Log}\,10}\big[\cos(\pi + 2k\pi) + i\sin(\pi + 2k\pi)\big] = -10. \quad \blacktriangleleft$$

In elementary calculus one learns the identity $\log(x_1 x_2) = \log x_1 + \log x_2$, where x_1 and x_2 are positive real numbers. The statement

$$\log(z_1 z_2) = \log z_1 + \log z_2, \tag{3.4–7}$$

where z_1 and z_2 are complex numbers and where we allow for the multiple values of the logarithms, requires some interpretation. The expressions $\log z_1$ and $\log z_2$ are multivalued. So is their sum, $\log z_1 + \log z_2$. If we choose particular values of each of these logarithms and add them, we will obtain *one of the possible values* of $\log(z_1 z_2)$. To establish this let $z_1 = r_1 e^{i\theta_1}$ and $z_2 = r_2 e^{i\theta_2}$. Thus $\log z_1 = \mathrm{Log}\,r_1 + i(\theta_1 + 2m\pi)$ and $\log z_2 = \mathrm{Log}\,r_2 + i(\theta_2 + 2n\pi)$. Specific integer values are assigned to m and n. By adding the logarithms we obtain

$$\log z_1 + \log z_2 = \mathrm{Log}\,r_1 + \mathrm{Log}\,r_2 + i(\theta_1 + \theta_2 + 2\pi(m + n)). \tag{3.4–8}$$

Now $\mathrm{Log}\,r_1 + \mathrm{Log}\,r_2 = \mathrm{Log}(r_1 r_2)$ since r_1 and r_2 are positive real numbers. Thus Eq. (3.4–8) becomes

$$\log z_1 + \log z_2 = \mathrm{Log}(r_1 r_2) + i(\theta_1 + \theta_2 + 2\pi(m + n)). \tag{3.4–9}$$

Notice that $r_1 r_2 = |z_1 z_2|$ while $\theta_1 + \theta_2 + 2\pi(m + n)$ is one of the values of $\arg(z_1 z_2)$. Thus Eq. (3.4–9) is *one* of the possible values of $\log(z_1 z_2)$.

Suppose $z_1 = i$ and $z_2 = -1$. Then if we take $\log(z_1) = i\pi/2$ and $\log z_2 = i\pi$ (principal values), we have $\log z_1 + \log z_2 = i3\pi/2$. Now $z_1 z_2 = -i$, and, if we use the principal value, $\log(z_1 z_2) = -i\pi/2$. Note that here $\log z_1 + \log z_2 \neq \log(z_1 z_2)$. However, $\log z_1 + \log z_2 = i3\pi/2$ is a valid value of $\log(-i)$. It just happens not to be the value we first computed. The statement $\log(z_1/z_2) = \log z_1 - \log z_2$ must also be interpreted in a manner similar to that of Eq. (3.4–7).

We are, of course, familiar with an identity from elementary calculus, $\log e^x = x$, where x is a positive real number. The corresponding complex statement

$$\log e^z = z \tag{3.4–10}$$

requires a comment. The expression e^z is, in general, a complex number. Its logarithm is multivalued. One of these values will correspond to z, and the others will not. We should not think that the principal value of the logarithm of e^z must equal z. There is nothing sacred about the principal value.

Example 3

Let $z = 1 + 3\pi i$. Find all values of $\log e^z$ and state which one is the same as z.

Solution

We have $e^z = e^{1+3\pi i} = e[\cos 3\pi + i \sin 3\pi] = -e$. Thus

$$\log e^{1+3\pi i} = \log(-e) = \text{Log}|-e| + i(\arg(-e)).$$

Now $\text{Log}|-e| = \text{Log } e = 1$ while the principal argument of $-e$ (a negative real number) is π. Thus $\arg(-e) = \pi + 2k\pi$. Therefore,

$$\log e^{1+3\pi i} = 1 + i(\pi + 2k\pi), \qquad k = 0, \pm 1, \pm 2, \dots .$$

The choice $k = 1$ will yield $\log e^{1+3\pi i} = 1 + 3\pi i$. However, the principal value of $\log e^{1+3\pi i}$ is obtained with $k = 0$ and yields $\text{Log } e^{1+3\pi i} = 1 + \pi i$. ◀

EXERCISES

1. Find all values of the logarithm of each of the following numbers and state the principal value. Put answers in the form $a + ib$.

a) 1

b) e

c) -1

d) $-e - ie$

e) $-100\left[\dfrac{1}{2} - i\dfrac{\sqrt{3}}{2}\right]$

f) $-e^2$

g) $100\left(-1 - i\sqrt{3}\right)$

h) $3 + 4i$

i) $e^{\text{Log}(-1)}$

j) $\sin i$

k) $e^{\text{Log}(\sin i)}$

l) $i\pi/2$

m) $e^{(e^i)}$

2. Find numerical values of each of the following.

a) $\sin(\log(i))$, all values

b) $\sinh(\log(-i))$, all values

c) $\text{Log}(\text{Log } i)$

d) $\text{Log}(\text{Log}(\text{Log} - i))$

3. In general, are the following equations true?

a) $\text{Log } \bar{z} = \overline{(\text{Log } z)}$ b) $\log \bar{z} = \overline{(\log z)}$

4. Use logarithms to solve for z (all values).

a) $e^z = i$ b) $e^z = ie^2$ c) $e^{e^z} = i$

5. Assume we could find the logarithm of 0. Let $w = u + iv = \log 0$. Then, since $e^{\log 0} = 0$, we have $e^{u+iv} = 0$. Use the definition of the exponential function to show why this equation cannot be satisfied for any finite choice of u and v.

6. Use logarithms to find all solutions of the following two equations.

a) $(e^z - 1)^2 = 1$ b) $(e^z - 1)^3 = -8$

7. a) Consider the identity $\log z_1 + \log z_2 = \log(z_1 z_2)$. If $z_1 = -ie$ and $z_2 = -2$, find permissible values for $\log z_1$, $\log z_2$ and $\log(z_1 z_2)$ that satisfy the identity.

b) For z_1 and z_2 given in part (a) find permissible values of $\log z_1$, $\log z_2$ and $\log(z_1/z_2)$ so that the identity $\log(z_1/z_2) = \log z_1 - \log z_2$ is satisfied.

8. We established that $\log z_1 + \log z_2 = \log(z_1 z_2)$ is valid only for certain choices of the logarithms. Let $z_1 = z_2 = z$. We then have $\log z + \log z = \log z^2$, or

$$2 \log z = \log z^2,$$

which is satisfied for specific values of $\log z$ and $\log z^2$. Similarly, $n \log z = \log z^n$, where n is an integer, is valid for appropriate choices of the logarithms on each side of the equation. Let $z = 1 + i$ and $n = 5$.

a) Find values of $\log z^n$ and $\log z$ that satisfy $n \log z = \log z^n$.

b) For the given z and n is $n \operatorname{Log} z = \operatorname{Log} z^n$ satisfied?

c) Suppose $n = 2$ and z is unchanged. Is $n \operatorname{Log} z = \operatorname{Log} z^n$ then satisfied?

9. This problem considers the relationship between $\log(8i)^{1/3}$ and $1/3 \log(8i)$.

a) Show that $(1/3)\log(8i) = \operatorname{Log} 2 + i(\pi/6 + (2/3)k\pi)$, where $k = 0, \pm 1, \pm 2, \dots$.

b) Show that $\log(8i)^{1/3} = \operatorname{Log} 2 + i(\pi/6 + (2/3)m\pi + 2n\pi)$, where $m = 0, 1, 2$ and $n = 0, \pm 1, \pm 2, \dots$. Thus there are three distinct sets (corresponding to $m = 0, 1, 2$) of values of $\log(8i)^{1/3}$. Each set has an infinity of members.

c) Show that the set of possible values of $\log(8i)^{1/3}$ is identical to the set of possible values of $(1/3)\log(8i)$. This discussion can be generalized to apply to $(1/p)\log z$ (where p is an integer) and $\log(z^{1/p})$.

3.5 ANALYTICITY OF THE LOGARITHMIC FUNCTION

To investigate the analyticity of $\log z$, let us first study the analyticity of the single-valued function $\operatorname{Log} z$, that is, the function created from the principal values. We have

$$\operatorname{Log} z = \operatorname{Log} r + i\theta, \qquad r > 0, \quad -\pi < \theta \le \pi. \qquad (3.5\text{--}1)$$

Obviously, this function is not continuous at $z = 0$ since it is not defined there; it is also not continuous along the negative real axis because θ does not possess a limit at any point along this axis (see Fig. 3.5–1).

Observe that any point on the negative real axis has an argument $\theta = \pi$. On the other hand, points in the third quadrant, which are taken arbitrarily close to the negative axis, have an argument tending toward $-\pi$. The principal argument of z goes through a "jump" of 2π as we cross the negative real axis.

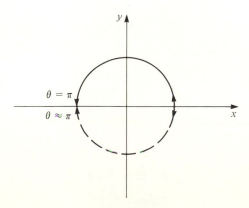

Figure 3.5–1

However, Log z is single valued and continuous in the domain D, which consists of the z-plane with the points on the negative real axis and origin "cut out." The troublesome points of discontinuity have been removed. Using the polar system with $z = re^{i\theta}$, we could describe D with the inequalities $r > 0$, $-\pi < \theta < \pi$.

Continuity is a prerequisite for analyticity. Having discovered a domain D in which Log z is continuous, we can now ask whether this function is analytic in that domain. This has already been answered affirmatively in Example 2 of Section 2.5. The function investigated there is precisely that in Eq. (3.5–1) since Log $r =$ Log $\sqrt{x^2 + y^2} = (1/2)\text{Log}(x^2 + y^2)$, and $\theta = \tan^{-1}(y/x)$. However, it is convenient in this section to repeat the same discussion in polar coordinates.

Let us write Log z as $u(r, \theta) + iv(r, \theta)$. From Eq. (3.5–1) we find

$$u = \text{Log } r, \qquad v = \theta, \qquad -\pi < \theta \leq \pi. \tag{3.5-2}$$

These functions are both defined and continuous in D. From Exercise 8 in Section 2.4 we can obtain the Cauchy–Riemann equations in polar form:

$$\frac{\partial u}{\partial r} = \frac{1}{r}\frac{\partial v}{\partial \theta}, \qquad \frac{\partial v}{\partial r} = -\frac{1}{r}\frac{\partial u}{\partial \theta}.$$

For u and v defined in Eq. (3.5–2) we have

$$\frac{\partial u}{\partial r} = \frac{1}{r}, \qquad \frac{1}{r}\frac{\partial v}{\partial \theta} = \frac{1}{r}, \qquad \frac{\partial v}{\partial r} = 0, \qquad -\frac{1}{r}\frac{\partial u}{\partial \theta} = 0.$$

Obviously, u and v do satisfy the Cauchy–Riemann equations. Moreover, the partial derivatives $\partial u/\partial r$, $\partial v/\partial \theta$, etc. are continuous in domain D. Thus the derivative of Log z must exist everywhere in this domain, and Log z is analytic there. The situation is illustrated in Fig. 3.5–2. We can readily find the derivative of Log z within this domain of analyticity.

If $f(z(r, \theta)) = u(r, \theta) + iv(r, \theta)$ is analytic, then Eq. (2.4–3) provides us with a formula for $f'(z)$. Using this equation with the substitution $e^{-i\theta} = (\cos\theta - i\sin\theta)$

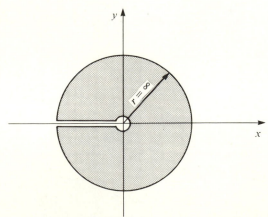

Domain of analyticity of Log z (shaded)

Figure 3.5–2

we have

$$f'(z) = e^{-i\theta}\left(\frac{\partial u}{\partial r} + i\frac{\partial v}{\partial r}\right). \tag{3.5-3}$$

Thus with u and v defined by Eq. (3.5–2) we obtain

$$\frac{d}{dz}\operatorname{Log} z = \frac{e^{-i\theta}}{r} = \frac{1}{re^{i\theta}} = \frac{1}{z} \tag{3.5-4}$$

in the domain D. An alternative derivation involving $u(x, y)$ and $v(x, y)$ is given in Exercise 1 of this section.

Equation (3.5–4) reminds us that $d(\log x)/dx = 1/x$ in real variable calculus. Note that Eq. (3.5–4) is inapplicable along the negative real axis and at the origin since $\operatorname{Log} z$ is not continuous there.

The single-valued function $w(z) = \operatorname{Log} z$ is said to be a *branch* of $\log z$.

Definition Branch

A *branch* of a multivalued function is a single-valued function *analytic* in some domain. At every point of the domain the single-valued function must assume exactly one of the various possible values that the multivalued function can assume. ■

In the case of $\operatorname{Log} z$ we found the largest such domain of analyticity and chose to call it D. Other branches of $\log z$ can be found in D. For example, $f(z) = \operatorname{Log} r + i\theta$, where $\pi < \theta \le 3\pi$. (This is the case $k = 1$ in Eq. (3.4–6).)

This function is not continuous at the origin. It is not continuous along the negative real axis where its imaginary part displays a jump discontinuity. However, its real and imaginary parts, $u = \operatorname{Log} r$, $v = \theta$, satisfy the Cauchy–Riemann equations in D. This branch is analytic in D, and it satisfies $d(\log z)/dz = 1/z$.

Obviously, an infinite number of different branches of $\log z$ are possible in D since there is no limit to our number of choices of k in Eq. (3.4–6). When $k = 0$, we obtain the *principal branch* of the logarithmic function.

The domain D was created by removing the semiinfinite line $y = 0$, $x \le 0$ from the xy-plane. This line is an example of a branch cut.

Definition Branch cut

A line used in creating a domain of analyticity is called a *branch line* or *branch cut*. ■

It is possible to create other branches of $\log z$ that are analytic in domains other than D. Consider

$$f(z) = \log z = \operatorname{Log} r + i\theta, \quad \text{where } -3\pi/2 < \theta \le \pi/2.$$

The imaginary part of this function is discontinuous on the positive imaginary axis, and both the real and imaginary parts are undefined at $z = 0$. The domain D_1 shown in Fig. 3.5–3, which is created by cutting the semiinfinite line $x = 0$, $y \ge 0$ out of the xy-plane, is a domain in which $f(z)$ is everywhere continuous. As in the discussion of

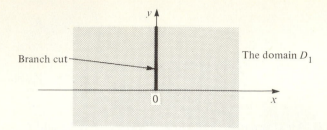

Figure 3.5–3

the principal branch, we can show that the derivative of this branch of log z exists throughout D_1 and equals $1/z$.

The reader can readily verify that branches of log z given by

$$\log z = \text{Log } r + i\theta, \qquad -\frac{3\pi}{2} + 2k\pi < \theta \le \frac{\pi}{2} + 2k\pi, \quad k = 0, \pm 1, \pm 2, \ldots$$

are all analytic throughout D_1. Note, however, that the principal branch, Log z, is not analytic at certain points in D_1. The reason is simple. The function Log z is not continuous (and therefore not differentiable) at every point along the negative real axis. Yet, the negative real axis lies within D_1.

The domains D and D_1 are just two of the infinite number of possible domains for which we can find branches of log z. Both of these domains were created by using a branch cut (or branch line) in the xy-plane. When all the possible branch lines that we can draw, whenever we are trying to establish branches of a multivalued function, share a common point, we call that point a *branch point* of the multivalued function. In the case of log z the origin is a branch point. Indeed, the two branch cuts that we investigated for this function passed through $z = 0$. Procedures for finding branch points of other functions are given in Section 3.8.

Example 1

Suppose

$$\log z = \text{Log } r + i\theta, \qquad -\frac{\pi}{2} < \theta \le \frac{3\pi}{2}.$$

a) What is the largest domain in which this function can be an analytic branch of the logarithmic function?

b) With this choice of branch what is the numerical value of $\log(-1 - i)$?

c) Find another branch of log z analytic in the domain found in part (a).

d) What is the numerical value of $\log(-1 - i)$ for the branch of part (c)?

Solution

Part (a): The function obviously fails to be continuous at the origin since log r is undefined there. In addition, θ fails to be continuous along the negative imaginary axis. For points on this axis $\theta = 3\pi/2$, while points in the fourth quadrant, which are taken arbitrarily close to this axis, have a value of θ near to $-\pi/2$, as Fig. 3.5–4 indicates. A branch cut in the xy-plane, extending from the origin outward along the negative imaginary axis, will eliminate all the singular points of the given function. Thus $\log z = \log r + i\theta$, $-\pi/2 < \theta \le 3\pi/2$ is an analytic branch of the logarithmic function in a domain consisting of the xy-plane with the origin and negative imaginary axis removed.

Part (b): An analytic function varies continuously within its domain of analyticity. Thus to reach $-1 - i$ from a point on the positive x-axis, we must use the counterclockwise path shown in Fig. 3.5–5. The argument θ at a point on the positive x-axis must be $2k\pi$ (where k is an integer). Since $-\pi/2 < \theta \le 3\pi/2$, evidently $k = 0$. Thus the argument θ begins at 0 radians and reaches the value $5\pi/4$ at $-1 - i$. It is not possible to use the broken clockwise path shown in Fig. 3.5–5 to reach the same point. In so doing we would strike the branch cut and thereby leave the domain of analyticity. Thus to answer the question,

$$\log(-1-i) = \text{Log}|-1-i| + i\frac{5\pi}{4} = \text{Log}\sqrt{2} + i\frac{5\pi}{4} \doteq 0.3466 + i\frac{5\pi}{4}.$$

Part (c): Consider $\log z = \text{Log } r + i\theta$, $-\pi/2 < \theta \le 3\pi/2$. At every point in the domain of part (a) this expression employs a value for θ lying between $-\pi/2$ and $3\pi/2$ radians. If we were to add 2π to the argument of z at each point, we would still have a function that assumes correct values of $\log z$. However, we now have

$$\log z = \text{Log } r + i\theta, \qquad -\frac{\pi}{2} + 2\pi < \theta \le \frac{3\pi}{2} + 2\pi.$$

This is a branch of $\log z$ that is analytic in the given domain. It assumes values identical to those of the original branch except for a purely imaginary constant $2\pi i$.

Part (d): The new branch of $\log z$ under consideration is one for which arg z satisfies $3\pi/2 < \theta \le 7\pi/2$. On the positive real axis, θ must be a certain integral multiple of 2π. Only $\theta = 2\pi$ satisfies the given inequality. If we move from this axis to the point $-1 - i$, we find that θ increases by an additional $5\pi/4$ radians (see Fig. 3.5–6). Thus

$$\log(-1-i) = \text{Log}\sqrt{2} + i\left(2\pi + \frac{5\pi}{4}\right) \doteq 0.3466 + i\frac{13\pi}{4}. \qquad \blacktriangleleft$$

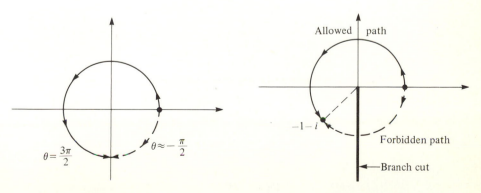

Figure 3.5–4   **Figure 3.5–5**

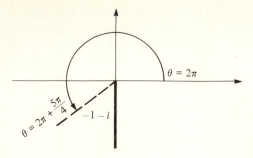

Figure 3.5–6

Example 2

a) Find the largest domain of analyticity of $f(z) = \text{Log}[z - (3 + 4i)]$.

b) Find the numerical value of $f(0)$.

Solution

Part (a): The function $\text{Log}\, w$ is analytic in the domain consisting of the entire w-plane with the semiinfinite line $\text{Im}\, w = 0$, $\text{Re}\, w \leq 0$ removed. If $w = z - (3 + 4i)$, we ensure analyticity in the z-plane by removing the points that simultaneously satisfy $\text{Im}(z - (3 + 4i)) = 0$ and $\text{Re}(z - (3 + 4i)) \leq 0$. These two conditions can be rewritten

$$\text{Im}((x + iy) - (3 + 4i)) = 0 \quad \text{or} \quad y = 4,$$
$$\text{Re}((x + iy) - (3 + 4i)) \leq 0 \quad \text{or} \quad x \leq 3.$$

The full domain of analyticity is shown in Fig. 3.5–7.

Part (b): $f(0) = \text{Log}(-3 - 4i) = \text{Log}\, 5 + i\arg(-3 - 4i)$. Since we are dealing with the principal branch, we require that $-\pi < \arg(-3 - 4i) \leq \pi$. From Fig. 3.5–8 we find this value of $\arg(-3 - 4i)$ to be approximately -2.214. Thus $f(0) \doteq \text{Log}\, 5 - i2.214$. ◀

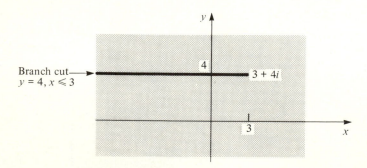

Figure 3.5–7

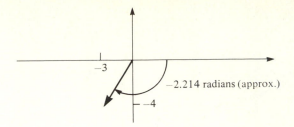

Figure 3.5–8

EXERCISES

1. Use $\text{Log } z = (1/2)\,\text{Log}\,(x^2 + y^2) + i\tan^{-1}(y/x)$, $-\pi < \tan^{-1}(y/x) \le \pi$ and Eq. (2.3–6) or Eq. (2.3–8) to show that $d(\text{Log } z)/dz = 1/z$ in the domain of Fig. 3.5–2.

2. Suppose
$$f(z) = \log z = \log r + i\theta, \qquad 0 \le \theta < 2\pi.$$

a) Find the largest domain of analyticity of this function.

b) Find the numerical value of $f(-e^2)$.

c) Explain why we cannot determine $f(e^2)$ within the domain of analyticity.

3. a) Why is the domain of analyticity of $(\text{Log } z)/(z + 1)$ the same as that of $\text{Log } z$?

b) What is the largest domain of analyticity of $(\text{Log } z)/(z - 1)$?

c) Explain why $(\text{Log } z)/(z^2 - 1)$ is analytic in the domain consisting of the z-plane with the negative real axis, the origin, and $z = 1$ removed.

4. The complex electrostatic potential $\Phi(x, y) = \phi + i\psi = \text{Log}\,(1/z)$, where $z \neq 0$, can be created by an electric line charge located at $z = 0$ and lying perpendicular to the xy-plane

a) Sketch the streamlines for this potential.

b) Sketch the equipotentials $\phi = -1, 0, 1$, and 2.

c) Find the components of the electric field at an arbitrary point x, y.

5. If the (positive) electric line charge of the previous example is located at $z = 1$ and a negative electric line charge is placed at $z = -1$, we can create the complex electrostatic potential $\Phi(x, y) = \phi + i\psi = \text{Log}\,((z + 1)/(z - 1))$, where $z \neq \pm 1$.

a) Graph $\phi(x, 0)$ for $-\infty < x < \infty$, and graph $\phi(0, y)$ for $-\infty < y < \infty$. Give equations for these functions.

b) Show that the equipotentials are circles satisfying
$$\left(x + \frac{1 + k}{1 - k}\right)^2 + y^2 = \frac{4k}{(1 - k)^2},$$

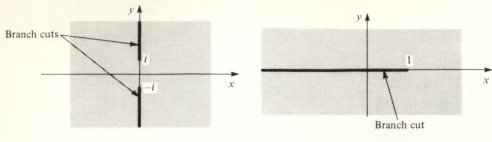

Figure 3.5–9   **Figure 3.5–10**

where k is any positive real constant. Sketch the equipotentials for which $k = 0.1$, $k = 9$, and $k \to 1$. Give the numerical value of ϕ on each of these equipotentials.

c) Show that the streamlines are circles satisfying

$$x^2 + (y - p)^2 = 1 + p^2,$$

where p is any real constant. Sketch the cases $p = 0$, $p = 1$, and $p = -1$.

6. Show that $f(z) = \text{Log}(z^2 + 1)$ is analytic in the domain shown in Fig. 3.5–9. *Hint:* Points satisfying $\text{Re}(z^2 + 1) \leq 0$ and $\text{Im}(z^2 + 1) = 0$ must not appear in the domain of analyticity. This requires a branch cut (or cuts) described by $\text{Re}((x + iy)^2 + 1) \leq 0$, $\text{Im}((x + iy)^2 + 1) = 0$. Find the locus that satisfies both these equations.

7. a) Show that $\text{Log}(\text{Log } z)$ is analytic in the domain consisting of the z-plane with a branch cut along the line $y = 0$, $-\infty < x \leq 1$ (see Fig. 3.5–10). *Hint:* Where will the inner function, $\text{Log } z$, be analytic? What restrictions must be placed on $\text{Log } z$ to render the outer logarithm an analytic function?

   b) Find $d(\text{Log}(\text{Log } z))/dz$ within the domain of analyticity found in part (a).

3.6 COMPLEX EXPONENTIALS

If we have the task of computing the numerical value of $7^{1.43}$, we can use a book of logarithms (and antilogs) and proceed as follows: We write 7 as $e^{\text{Log } 7}$ so that we have $7^{1.43} = (e^{\text{Log } 7})^{1.43}$. The latter expression is evaluated as $e^{(1.43 \, \text{Log } 7)}$. This procedure should suggest a definition for general expressions of the form z^c, where z and c are arbitrary complex numbers. We use

$$z^c = e^{c(\log z)}. \tag{3.6–1}$$

We evaluate $e^{c(\log z)}$ by means of Eq. (3.1–1). As we well know, the logarithm of z is multivalued. For this reason, depending on the value of c, z^c may have more than one numerical value. The matter is fully explored in Exercise 3 of this section. It is not hard to show that if c is a rational number n/m, then Eq. (3.6–1) yields numerical values identical to those obtained for $z^{n/m}$ in Eq. (1.4–13).

Example 1

Compute $9^{1/2}$ by means of Eq. (3.6–1).

Solution

We, of course, already know the two possible numerical values of $9^{1/2}$. Pretending otherwise, from Eq. (3.6–1) we have

$$9^{1/2} = e^{1/2\log 9} = e^{1/2[\text{Log}\,9 + i(2k\pi)]} = e^{1/2\,\text{Log}\,9 + ik\pi}$$

$$= e^{1/2\,\text{Log}\,9}[\cos(k\pi) + i\sin(k\pi)], \qquad k = 0, \pm 1, \pm 2, \dots .$$

As k ranges over all integers, the expression in the brackets yields only two possible numbers, $+1$ and -1. The term $e^{1/2\,\text{Log}\,9}$ equals $e^{\text{Log}\,3} = 3$. Thus $9^{1/2} = \pm 3$, and the familiar results are obtained. ◀

If in computing the value of z^c by means of Eq. (3.6–1) we employ the principal value of the logarithm, we obtain what is called the principal value of z^c. In the preceding example, we can put $k = 0$ and see that the principal value of $9^{1/2}$ is 3.

Example 2

Compute 9^π by means of Eq. (3.6–1). Here we are raising a number to a real, but irrational, power. Although we know there is exactly one distinct value of z^n, where n is an integer, we have not yet determined the number of values that will occur if n is irrational.

Solution

From Eq. (3.6–1) we have

$$9^\pi = e^{\pi \log 9} = e^{\pi[\text{Log}\,9 + i2k\pi]} = e^{\pi\,\text{Log}\,9 + i2k\pi^2}$$

$$= e^{\pi\,\text{Log}\,9}[\cos(2k\pi^2) + i\sin(2k\pi^2)]$$

$$\doteq e^{6.903}[\cos(2k\pi^2) + i\sin(2k\pi^2)]$$

$$\doteq 995.04[\cos(2k\pi^2) + i\sin(2k\pi^2)], \qquad k = 0, \pm 1, \pm 2, \dots .$$

With $k = 0$ we have that the principal value of 9^π is approximately 995.04. By allowing k to vary we generate other, complex, values of 9^π. All these values are numerically distinct, that is, there are no repetitions as k assumes new values. To see that this must be so assume that integers k_1 and k_2 yield identical values of 9^π. Then we would require

$$\cos(2k_1\pi^2) + i\sin(2k_1\pi^2) = \cos(2k_2\pi^2) + i\sin(2k_2\pi^2).$$

This equality can only hold if

$$2k_1\pi^2 - 2k_2\pi^2 = m\pi, \quad \text{where } m \text{ is an even integer}$$

or if

$$\pi = \frac{m}{2k_1 - 2k_2}.$$

Since π is irrational, it cannot be expressed as the ratio of integers. Thus our assumption that there are two identical roots must be false. ◄

In our previous example we generate an infinity of numerically distinct values of 9^π when we allow k to range over all the integers. This result is generalized in Exercise 3 of this section and shows that:

If c is any irrational number, then z^c possesses an infinity of different values.

Let us now consider an example in which c is complex.

Example 3

Find $(1 + i)^{3+4i}$.

Solution

Our formula in Eq. (3.6–1) yields

$$(1 + i)^{3+4i} = e^{(3+4i)(\log(1+i))}$$

$$= e^{(3+4i)[\operatorname{Log}\sqrt{2} + i(\pi/4 + 2k\pi)]}$$

$$= e^{3\operatorname{Log}\sqrt{2} - \pi - 8k\pi + i(4\operatorname{Log}\sqrt{2} + 3\pi/4 + 6k\pi)}.$$

With the aid of Eq. (3.1–1) this becomes

$$e^{3\operatorname{Log}\sqrt{2} - \pi - 8k\pi}\left[\cos\left(4\operatorname{Log}\sqrt{2} + 3\pi/4 + 6k\pi\right) + i\sin\left(4\operatorname{Log}\sqrt{2} + 3\pi/4 + 6k\pi\right)\right]$$

$$= (1 + i)^{3+4i}, \qquad k = 0, \pm 1, \pm 2, \ldots.$$

Note that $6k\pi$ can be deleted in the preceding equation. As k ranges over the integers, an infinity of complex, numerically distinct values are obtained for $(1 + i)^{3+4i}$. The principal value, with $k = 0$, is

$$e^{3\operatorname{Log}\sqrt{2} - \pi}\left[\cos\left(4\operatorname{Log}\sqrt{2} + 3\pi/4\right) + i\sin\left(4\operatorname{Log}\sqrt{2} + 3\pi/4\right)\right]. \qquad ◄$$

Example 3 is an instance of something we can prove in general:

Except when c is a rational number, z^c has an infinity of different possible values.

Numbers of the form e^c are, by definition, immune to the preceding statement. They are computed using Eq. (3.1–1), and only one value is obtained. Otherwise we would not be safe in saying, for example, that $e^{i\pi} = -1$.

If z is regarded as a variable, then z^c is a multivalued function of z. This function possesses various branches whose derivatives can be found. The principal branch, for example, is obtained with the use of the principal branch of $\log z$ in Eq. 3.6–1. This branch of z^c is analytic in the same domain as $\operatorname{Log} z$. We find the derivative as follows:

$$z^c = e^{c\operatorname{Log} z},$$

$$\frac{d}{dz}z^c = \frac{d}{dz}e^{c\operatorname{Log} z} = \frac{ce^{c\operatorname{Log} z}}{z} = \frac{ce^{c\operatorname{Log} z}}{e^{\operatorname{Log} z}} = ce^{(c-1)\operatorname{Log} z} = cz^{c-1}. \quad (3.6\text{–}2)$$

Other branches of z^c, which use other branches of $\log z$ in Eq. (3.6–1), can be differentiated in a similar manner within their domains of analyticity.

The expression c^z, where c is a constant and z a variable, is equal to $e^{z \log c}$. Having chosen a valid value for $\log c$, we find that we now have a single-valued function of z analytic in the entire z-plane. The derivative of this expression is found as follows:

$$\frac{d}{dz} c^z = \frac{d}{dz} e^{z \log c} = e^{z \log c} (\log c) = c^z \log c. \qquad (3.6\text{–}3)$$

Example 4

Find $(d/dz)i^z$.

Solution

We will use Eq. (3.6–3) taking $\log i = \operatorname{Log} i = i\pi/2$. Thus

$$\frac{d}{dz} i^z = i^z \left(\frac{i\pi}{2} \right). \qquad \blacktriangleleft$$

The multivalued function $g(z)^{h(z)}$ is defined as $e^{h(z)\log(g(z))}$. Such functions are considered in Exercises 4 and 5 of this section. Their principal value is obtained if we use the principal value of the logarithm.

EXERCISES

1. Find all values of each of the following in the form $a + ib$ and state the principal value.

   a) 1^i c) i^i e) $(\operatorname{Log}(i))^i$ g) i^{3+4i} i) $i^{(\sin i)}$

   b) $(1 + i)^{1+i}$ d) $(-i)^{-i}$ f) $(3 + 4i)^i$ h) $i^{\pi/8}$ j) $(\cos i)^i$

2. Use Eq. (3.1–1) to show that $1/e^z = e^{-z}$; then use this fact together with Eq. (3.6–1) to prove that all the values of $1/z^c$ are identical to those of z^{-c}.

3. Use Eq. (3.6–1) to show that

   a) if n is an integer, then z^n has only one value and it is the same as the one given by Eq. (1.4–2);

   b) if n and m are integers and n/m is an irreducible fraction, then $z^{n/m}$ has just m-values and they are identical to those given by Eq. (1.4–13);

   c) if c is an irrational number, then z^c has an infinity of different values;

   d) if c is complex with $\operatorname{Im} c \neq 0$, then z^c has an infinity of different values.

4. If $f(z) = z^z$ (principal value),

   a) find $f'(z)$;   b) find $f'(i)$.

5. Find $(d/dz) 10^{\sin z}$ using principal values. Where in the z-plane is $10^{\sin z}$ analytic?

6. a) Let α be a constant. Consider the principal value of z^α and a nonprincipal value of $z^{-\alpha}$. In general, is the following statement true: $z^\alpha z^{-\alpha} = 1$.

b) Suppose principal values of both z^α and $z^{-\alpha}$ are used. Then is their product always equal to 1?

3.7   INVERSE TRIGONOMETRIC AND HYPERBOLIC FUNCTIONS

If we know the logarithm of a complex number w, we can find the number itself by means of the identity $e^{\log w} = w$. We have used the fact that the exponential function is the inverse of the logarithmic function.

Suppose we know the sine of a complex number w. Let us see whether we can find w and whether w is uniquely determined.

Let $z = \sin w$. The value of w is referred to as arc sin z, or $\sin^{-1} z$ that is, the complex number whose sine is z. To find w note that

$$z = \frac{e^{iw} - e^{-iw}}{2i}. \qquad (3.7\text{–}1)$$

Now with $p = e^{iw}$ and $1/p = e^{-iw}$ in Eq. (3.7–1) we have

$$z = \frac{p - 1/p}{2i}.$$

Multiplying the above by $2ip$ and doing some rearranging, we find that

$$2izp = p^2 - 1 \quad \text{or} \quad p^2 - 2izp - 1 = 0.$$

With the quadratic formula we solve this equation for p:

$$p = zi + (1 - z^2)^{1/2} \quad \text{or} \quad e^{iw} = zi + (1 - z^2)^{1/2}.$$

We now take the logarithm of both sides of this last equation and divide the result by i to obtain

$$w = \frac{1}{i} \log\left(zi + (1 - z^2)^{1/2}\right)$$

and, since $w = \sin^{-1} z$,

$$\sin^{-1} z = -i \log\left(zi + (1 - z^2)^{1/2}\right). \qquad (3.7\text{–}2)$$

We thus, apparently, have an explicit formula for the complex number whose sine equals any given number z. The matter is not quite so simple, however, since the result is multivalued. There are two equally valid choices for the square root in Eq. (3.7–2). Having selected one such value, there are then an infinite number of possible values of the logarithm of $zi + (1 - z^2)^{1/2}$. Altogether, we see that because of the square root and logarithm, there are two different sets of values for $\sin^{-1} z$, and each set has an infinity of members. An exception occurs when $z = \pm 1$; then, the two infinite sets become identical to each other. To assure ourselves of the validity of Eq. (3.7–2) let us use it to compute a familiar result.

Example 1

Find $\sin^{-1}(1/2)$.

Solution

We see from Eq. (3.7–2) that

$$\sin^{-1}\left(\frac{1}{2}\right) = -i\log\left[\frac{i}{2} + \left(\frac{3}{4}\right)^{1/2}\right].$$

With the positive square root of 3/4 we have

$$\sin^{-1}\left(\frac{1}{2}\right) = -i\log\left[\frac{\sqrt{3}}{2} + \frac{i}{2}\right] = -i\log\left(1\angle\pi/6\right) = \frac{\pi}{6} + 2k\pi,$$

$$k = 0, \pm1, \pm2,\ldots,$$

whereas with the negative square root of 3/4 we have

$$\sin^{-1}\frac{1}{2} = -i\log\left[\frac{-\sqrt{3}}{2} + \frac{i}{2}\right] = -i\log\left(1\angle\frac{5\pi}{6}\right) = \frac{5\pi}{6} + 2k\pi,$$

$$k = 0, \pm1, \pm2,\ldots.$$

To make these answers look more familiar let us convert them to degrees. The first result says that angles with the sines of 1/2 are 30°, 390°, 750°, etc., while the second states that they are 150°, 510°, 870°, etc. We, of course, knew these results already from elementary trigonometry. ◀

In a high school trigonometry class, where one uses only real numbers, the following example would not have a solution.

Example 2

Find all the numbers whose sine is 2.

Solution

From Eq. (3.7–2) we have $\sin^{-1}2 = -i\log(2i + (-3)^{1/2})$. The two values of $(-3)^{1/2}$ are $\pm i\sqrt{3}$. With the positive sign our results are

$$\sin^{-1}2 = -i\log\left[2i + i\sqrt{3}\right] = -i\left[\operatorname{Log}\left(2 + \sqrt{3}\right) + i\left(\frac{\pi}{2} + 2k\pi\right)\right]$$

$$\doteq \left(\frac{\pi}{2} + 2k\pi\right) - i1.317, \qquad k = 0, \pm1, \pm2,$$

whereas with the negative sign we obtain

$$\sin^{-1}2 = -i\log\left(2i - i\sqrt{3}\right) = -i\left[\operatorname{Log}\left(2 - \sqrt{3}\right) + i\left(\frac{\pi}{2} + 2k\pi\right)\right]$$

$$\doteq \left(\frac{\pi}{2} + 2k\pi\right) + i1.317.$$ ◀

The equation $z = \cos w$ can be solved for w, which we call arc cos z or $\cos^{-1}z$. The procedure is similar to the one just given for $\sin^{-1}z$. Thus

$$\cos^{-1}z = -i\log\left(z + i(1 - z^2)^{1/2}\right). \qquad (3.7–3)$$

Also, $z = \tan w$ can be solved for w (or for $\tan^{-1} z$) with the following result

$$\tan^{-1} z = \frac{i}{2} \log\left(\frac{i+z}{i-z}\right). \tag{3.7-4}$$

It is not hard to show that the expressions for $\cos^{-1} z$ and $\sin^{-1} z$ just derived are purely real numbers if and only if z is a real number and $-1 \le z \le 1$. Thus $z = \sin w$ and $z = \cos w$ have real number solutions w only for z satisfying $-1 \le z \le 1$. Otherwise w is a complex number.

The functions appearing on the right in Eqs. (3.7-2), (3.7-3), and (3.7-4) are examples of inverse trigonometric functions. The inverse hyperbolic functions are similarly established. Thus

$$\sinh^{-1} z = \log\left(z + (z^2 + 1)^{1/2}\right), \tag{3.7-5}$$

$$\cosh^{-1} z = \log\left(z + (z^2 - 1)^{1/2}\right), \tag{3.7-6}$$

$$\tanh^{-1} z = \frac{1}{2} \log\left(\frac{1+z}{1-z}\right). \tag{3.7-7}$$

All the inverse functions we have derived in this section are multivalued. Analytic branches exist for all these functions. For example, a branch of $\sin^{-1} z$ can be obtained from Eq. (3.7-2) if we first specify a branch of $(1 - z^2)^{1/2}$ and then a branch of the logarithm. Having done this, we can differentiate our branch within its domain of analyticity. The subject of branches in general is given more attention in the next section. Differentiating Eq. (3.7-2), we have

$$\frac{d}{dz} \sin^{-1} z = \frac{d}{dz}\left(-i \log\left[zi + (1 - z^2)^{1/2}\right]\right) = \frac{1}{(1 - z^2)^{1/2}}. \tag{3.7-8}$$

For this identity to hold we must use the same branch of $(1 - z^2)^{1/2}$ in defining $\sin^{-1} z$ and in the expression for its derivative. Other formulas that are derived through the differentiation of branches are

$$\frac{d}{dz} \cos^{-1} z = \frac{-1}{(1 - z^2)^{1/2}}, \tag{3.7-9}$$

$$\frac{d}{dz} \tan^{-1} z = \frac{1}{(1 + z^2)}, \tag{3.7-10}$$

$$\frac{d}{dz} \sinh^{-1} z = \frac{1}{(1 + z^2)^{1/2}}, \tag{3.7-11}$$

$$\frac{d}{dz} \cosh^{-1} z = \frac{1}{(z^2 - 1)^{1/2}}, \tag{3.7-12}$$

$$\frac{d}{dz} \tanh^{-1} z = \frac{1}{(1 - z^2)}. \tag{3.7-13}$$

EXERCISES

1. a) Derive Eq. (3.7–3). b) Derive Eq. (3.7–4). c) Derive Eq. (3.7–5).

2. a) Show that if we differentiate a branch of arc cos z we obtain Eq. (3.7–9).

   b) Obtain the same result by noting that

   $$z = \cos w = (1 - \sin^2 w)^{1/2} = \left(1 - \left(\frac{dz}{dw}\right)^2\right)^{1/2}$$

   can be solved for dw/dz.

   c) Obtain Eq. (3.7–11) directly from Eq. (3.7–5); obtain it also by a procedure similar to part (b) of this exercise.

3. Show that if we use suitable branches of the arc sin and arc cos in Eqs. (3.7–2) and (3.7–3) we obtain

   $$\sin^{-1} z + \cos^{-1} z = \pi/2.$$

4. Find all possible solutions to the following equations:

   a) $\sin w = i$, d) $\cosh^2 w = i$, g) $\sin(\sin z) = 1$,

   b) $\cos w = 1 + i$, e) $\tan z = -2i$, h) $\cos(\sin z) = 0$,

   c) $\sinh w = \text{Log } i$, f) $\cos z = 4$, i) $\tan^{-1} w = -2i$.

5. Explain whether or not the following two equations are true in general.

   a) $\tan^{-1}(\tan z) = z$ b) $\tan(\tan^{-1} z) = z$

6. Explain why the values of $\sinh^{-1} x$ that are given by tables or a pocket calculator are the same as those given by $\text{Log}(x + \sqrt{x^2 + 1})$. Note the branches used for the functions in Eq. (3.7–5).

7. Show that $\tanh^{-1}(e^{i\theta}) = (1/2)\log(i \cot(\theta/2))$.

3.8 MORE ON BRANCH POINTS AND BRANCH CUTS

Let us study the branches and domains of analyticity of functions of the form $(z - z_0)^c$, where z_0 and c are complex constants. If c is an integer, such functions are single valued, have only one branch, and need not be considered here. However, if c is a noninteger, these functions are multivalued. From Eq. (3.6–1) we have

$$(z - z_0)^c = e^{c \log(z - z_0)}. \tag{3.8–1}$$

Since $\log(z - z_0)$ has a branch point at z_0, so does $(z - z_0)^c$. To study branches of $(z - z_0)^c$ and branches of algebraic combinations of such expressions, we introduce a set of polar coordinate variables measured from the branch points and examine the changes in value of our functions as their branch points are encircled.

Consider, for example, $f(z) = z^{1/2}$. Letting $r = |z|$, $\theta = \arg z$, we use Eq. (1.4–12) with $m = 2$ to find that

$$f(z) = \sqrt{r}\, e^{i(\theta/2 + k\pi)}. \tag{3.8–2}$$

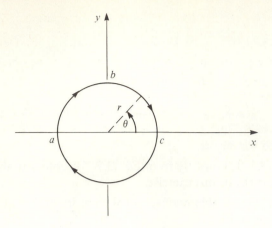

Figure 3.8–1

Suppose we set $k = 0$ in Eq. (3.8–2) and negotiate clockwise the circle of radius r shown in Fig. 3.8–1. Beginning at the point marked a and taking $\theta = \pi$, we have from Eq. (3.8–2) that $f(z) = \sqrt{r}\, e^{i\pi/2} = i\sqrt{r}$. Now, moving clockwise along the circle to b, where θ has fallen to $\pi/2$, Eq. (3.8–2) shows that $f(z) = \sqrt{r}\, e^{i\pi/4}$. Continuing on to c, where $\theta = 0$, we have $f(z) = \sqrt{r}$.

Moving clockwise to a and starting a second trip around the circle, we now have at a, $\theta = -\pi$ and $f(z) = \sqrt{r}\, e^{-i\pi/2} = -i\sqrt{r}$. Proceeding to b, where $\theta = -3\pi/2$, we find $f(z) = \sqrt{r}\, e^{-i3\pi/4}$. Advancing to c, where $\theta = -2\pi$, we have $f(z) = -\sqrt{r}$. Coming again to a and starting a third trip around the circle, we have $\theta = -3\pi$ and $f(z) = \sqrt{r}\, e^{-i3\pi/2} = i\sqrt{r}$. This was our original starting value at a.

A third trip around the circle yields values of $f(z)$ identical to those obtained on the first excursion. No new values of $f(z)$ are generated by subsequent journeys around the circle, as the reader should verify.

In encircling the branch point $z = 0$ twice we have encountered values of $(z)^{1/2}$ for two different branches of this function. In our first trip around the circle we found values of $z^{1/2} = e^{1/2\,\mathrm{Log}\,z}$. This is the principal branch and is analytic in the same domain (see Fig. 3.5–2) as $\mathrm{Log}\,z$. In our second trip we found values of the other branch of $z^{1/2}$, which is analytic throughout this domain. It is given by $z^{1/2} = e^{1/2[\mathrm{Log}\,z + i2\pi]} = -e^{1/2\,\mathrm{Log}\,z}$.

What we have seen is true in general whenever we have a branch point.

Encirclement of a branch point causes us to move from one branch of a function to another.[†]

We need not employ circular paths (as was just done) for this to happen. Any closed path surrounding the branch point will do.

[†]For functions having more than one branch point (see Example 3 in this section) encirclement of just one branch point always causes progression to another branch; encirclement of two or more branch points does not necessarily cause such as a transition, as we shall see.

To prevent our proceeding from one branch of a function to another, when we move along a path, we can construct a branch cut in the z-plane and agree never to cross this cut.

The branch cut is regarded as a barrier to encirclement of the branch point.

Alternatively, we can create a domain consisting of the z-plane minus all the points on the branch cut. One finds branches of the function that are analytic throughout this "cut" plane. We can specify a particular branch by giving its value at one point in the cut plane (see Example 1 in this section). By using this branch on paths that are confined to the domain, we cannot pass from one branch of the function to another.

Example 1

Consider a branch of $(z)^{1/2}$ that is analytic in the domain consisting of the z-plane less the points on the branch cut $y = 0$, $x \leq 0$. Suppose this function equals 2 when $z = 4$. What values does this branch assume when

$$z = 9\left[-1/2 - i\sqrt{3}/2\right].$$

Solution

With $|z| = r$ and $\theta = \arg z$ we have that

$$(z)^{1/2} = \sqrt{r}\, e^{i(\theta/2 + k\pi)}, \qquad k = 0, 1. \tag{3.8-3}$$

We will take $\theta = 0$ when $z = 4$. Then the condition $(4)^{1/2} = 2$ requires that $k = 0$ in Eq. (3.8–3). As we move along a path to $9\left[-1/2 - i\sqrt{3}/2\right]$ in Fig. 3.8–2 the argument θ in Eq. (3.8–3) changes continuously from 0 to $-2\pi/3$, and $r = |z|$ increases from 4 to 9. With $k = 0$ in Eq. (3.8–3) we have at $z = 9\left[-1/2 - i\sqrt{3}/2\right]$ that

$$(z)^{1/2} = \sqrt{9}\, e^{i(1/2)(-2\pi/3)} = 3\left[1/2 - i\sqrt{3}/2\right].$$

Notice that for our choice of branch we cannot reach $9\left[-1/2 - i\sqrt{3}/2\right]$ by way of the broken path in Fig. 3.8–2. This would take us out of the domain of analyticity of the branch. It would also involve crossing the branch cut.

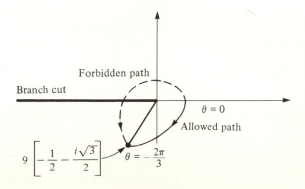

Figure 3.8–2

The derivative of our branch of $z^{1/2}$ in its domain of analyticity is

$$\frac{d}{dz} z^{1/2} = \frac{1}{(2z^{1/2})}.$$

The same branch of $z^{1/2}$ must be used on both sides of this equation. ◀

Example 2

For $(z - 1)^{1/3}$ let a branch cut be constructed along the line $y = 0$, $x \geq 1$. If we select a branch that is a negative real number when $y = 0$, $x < 1$, what value does this branch assume when $z = 1 + i$?

Solution

Introducing the variables $r_1 = |z - 1|$, $\theta_1 = \arg(z - 1)$ (see Fig. 3.8–3). We have from Eq. (1.4–12) with $m = 3$ that

$$(z - 1)^{1/3} = \sqrt[3]{r_1}\, e^{i(\theta_1/3 + 2k\pi/3)}, \qquad k = 0, 1, 2. \tag{3.8-4}$$

Taking $\theta_1 = \pi$ on the line $y = 0$, $x < 1$ we have here that

$$(z - 1)^{1/3} = \sqrt[3]{r_1}\, e^{i(\pi/3 + 2k\pi/3)}. \tag{3.8-5}$$

The left side of the equation can be a negative real number if we select $k = 1$ in Eq. (3.8–5). Proceeding to $(1 + i)$ from anywhere on the line $y = 0$, $x < 1$, we find that θ_1 has shrunk to $\pi/2$, and $r_1 = 1$. The path used in Fig. 3.8–3 for this purpose cannot cross the branch cut. With these values of r_1 and θ_1 in Eq. (3.8–4) (and $k = 1$), at $1 + i$ we have

$$(z - 1)^{1/3} = i^{1/3} = \sqrt[3]{1}\, e^{i5\pi/6} = -\sqrt{3}/2 + i/2. \qquad ◀$$

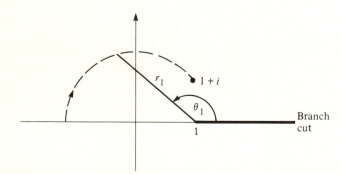

Figure 3.8–3

Example 3

Consider the multivalued function $f(z) = z^{1/2}(z - 1)^{1/2}$.

    a) Where are the branch points of the function? Verify that these are branch points by encircling them and passing from one branch of $f(z)$ to another.

b) Show that if we encircle both branch points we remain on the same branch of $f(z)$.

c) Show some possible choices of branch cut that would prevent passage from one branch of $f(z)$ to another.

Solution

Part (a): The first factor $(z)^{1/2}$ has a branch point at $z = 0$, whereas the second $(z - 1)^{1/2}$ has a branch point at $z = 1$. Thus we suspect that the product has branch points at $z = 0$ and $z = 1$. We will verify that $z = 1$ is a branch point. The proof for $z = 0$ is quite similar and will not be presented.

We have (see Fig. 3.8–4a) that

$$(z)^{1/2} = \sqrt{r}\, e^{i(\theta/2 + k\pi)}, \tag{3.8-6}$$

where $\theta = \arg z$ and $r = |z|$; and

$$(z - 1)^{1/2} = \sqrt{r_1}\, e^{i(\theta_1/2 + m\pi)}, \tag{3.8-7}$$

where $|z - 1| = r_1$ and $\theta_1 = \arg(z - 1)$. Thus

$$f(z) = z^{1/2}(z - 1)^{1/2} = \sqrt{r}\,\underline{/\tfrac{1}{2}\theta + k\pi}\ \sqrt{r_1}\,\underline{/\tfrac{1}{2}\theta_1 + m\pi}, \tag{3.8-8}$$

where k and m are assigned integer values. Let us now encircle $z = 1$ using the path $|z - 1| = \delta$, where $\delta < 1$ (see Fig. 3.8–4b). Beginning at point a, we take $\theta_1 = 0$, $\theta = 0, r_1 = \delta, r = 1 + \delta$. With these values in Eq. (3.8–8) we have

$$f(z) = \sqrt{1 + \delta}\,\underline{/k\pi}\ \sqrt{\delta}\,\underline{/m\pi}\ = \sqrt{\delta + \delta^2}\,\underline{/(k + m)\pi}. \tag{3.8-9}$$

Moving clockwise once around the circle $|z - 1| = \delta$ and returning to point a we find now that $\theta_1 = 2\pi$, while θ, after some variation, has returned to zero. With these values in Eq. (3.8–8) we have

$$f(z) = \sqrt{1 + \delta}\,\underline{/k\pi}\ \sqrt{\delta}\,\underline{/\pi + m\pi}\ = -\sqrt{\delta + \delta^2}\,\underline{/(k + m)\pi}. \tag{3.8-10}$$

Because the value obtained for $f(z)$ at a is now not the value originally obtained (see

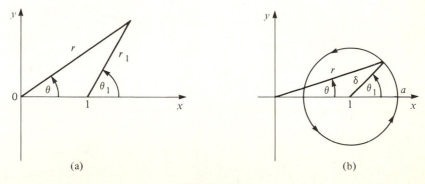

(a)   (b)

Figure 3.8–4

Eq. 3.8–9), we have progressed to another branch of $f(z)$. The preceding discussion does not require the use of a circular path. Any closed path that encloses $z = 1$ and excludes $z = 0$ will lead to the same result.

Part (b): An arbitrary closed path surrounds the branch points $z = 0$ and $z = 1$, as shown in Fig. 3.8–5. Let us evaluate $f(z)$ at the arbitrary point P lying on the path. Here $\arg z = \alpha$, $\arg(z - 1) = \beta$. Substituting these values for θ and θ_1, respectively, in Eq. (3.8–8) and combining the arguments we have

$$f(z) = \sqrt{r}\sqrt{r_1} \; \underline{/\frac{1}{2}(\alpha + \beta) + (k + m)\pi} \; . \qquad (3.8–11)$$

Moving once around the path in Fig. 3.8–5 in the indicated direction and returning to P, we now have $\arg z = \theta = \alpha + 2\pi$ and $\arg(z - 1) = \theta_1 = \beta + 2\pi$. Using these values in Eq. (3.8–8), we obtain

$$f(z) = \sqrt{r}\sqrt{r_1} \; \underline{/\frac{1}{2}(\alpha + \beta) + 2\pi + (k + m)\pi} \; . \qquad (3.8–12)$$

If Eqs. (3.8–11) and (3.8–12) are converted to Cartesian form, identical numerical values of $f(z)$ are obtained since the difference of 2π in their arguments is of no consequence. Hence, by encircling both the branch points $z = 0$ and $z = 1$ we do not pass to a new branch of $f(z)$. There are functions, however (see Exercise 4 in this section) where encirclement of two branch points does cause passage to another branch.

Part (c): If we make a circuit around either branch point of $(z)^{1/2}(z - 1)^{1/2}$, we move from one branch of this function to another. Some examples of branch cuts that prevent encirclement of one branch point are shown in Fig. 3.8–6. Notice that in case (b) we can pass around both branch points. However, as was just shown, we will remain on a given branch of $f(z)$. *Comment:* The particular choice of branch cut is dictated by the desired domain of analyticity for our branch. For example, in Fig. 3.8–6(a) we can obtain a branch of $f(z)$ analytic throughout a domain consisting of the z-plane with all points on the lines $y = 0$, $x \leq 0$ and $y = 0$, $x \geq 1$ removed. If,

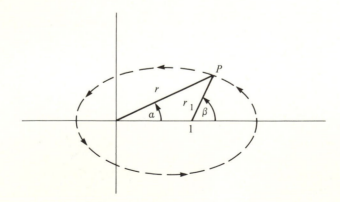

Figure 3.8–5

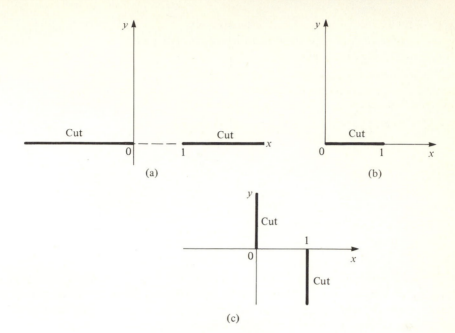

Figure 3.8–6

however, we required a branch of $f(z)$ analytic at say $y = 0$, $x = 2$, we might consider using the branch cuts shown in Fig. 3.8–6(b) or (c). ◀

Example 4

Suppose we use a branch $f(z)$ of $z^{1/2} (z - 1)^{1/2}$ analytic throughout the cut plane shown in Fig. 3.8–6(c). If $f(1/2) = i/2$, what is $f(-1)$?

Solution

Using the notation of Example 3, we have from Eq. (3.8–8) that

$$f(z) = \sqrt{r} \; \Big/ \frac{1}{2}\theta + k\pi \; \sqrt{r_1} \; \Big/ \frac{1}{2}\theta_1 + m\pi . \qquad (3.8\text{–}13)$$

At $z = 1/2$ we take $\theta = \arg z = 0$, $r = |z| = 1/2$, $\theta_1 = \arg(z - 1) = \pi$, $r_1 = |z - 1| = 1/2$ (see Fig. 3.8–7). Thus Eq. (3.8–13) becomes

$$f\left(\frac{1}{2}\right) = \frac{1}{2} \; \Big/ \frac{\pi}{2} + (k + m)\pi = \frac{i}{2} e^{i(k+m)\pi} .$$

Taking $k + m = 0$, we obtain the condition $f(1/2) = i/2$ for our branch.

Now proceeding to $z = -1$ along the path indicated in Fig. 3.8–7, we find that $\theta = -\pi$, $r = |z| = 1$, θ_1 is after some variation again π, $r_1 = |z - 1| = |-2| = 2$. Using these values in Eq. (3.8–13) together with $k + m = 0$, we obtain

$$f(-1) = \sqrt{1} \; \Big/ -\frac{\pi}{2} + k\pi \; \sqrt{2} \; \Big/ \frac{\pi}{2} + m\pi = \sqrt{2} \, e^{i(k+m)\pi} = \sqrt{2} .$$

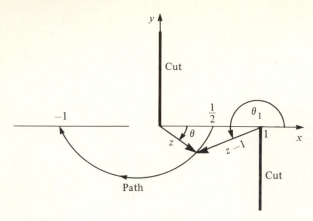

Figure 3.8–7

Note that the path taken from $1/2$ to -1 in Fig. 3.8–7 remains within the domain of analyticity. ◀

EXERCISES

1. A certain branch of $(z)^{1/2}$ is defined by means of the branch cut $x = y, y \geq 0$. If this branch equals -3 when $z = 9$, what values does this function assume for

   a) $z = -9$,   b) $z = -i$,   c) $z = i$.

   d) What is the derivative of this branch at each of these three points? Give numerical values.

2. A branch of $(z - 1)^{1/4}$ is defined by means of the branch cut $x = 1, y \geq 0$. If this branch equals 1 when $z = 2$, what value does this branch assume for

   a) $z = -2$,   b) $z = i\sqrt{3}$.

3. Consider the function $(z^2 - 1)^{1/2}$. Suppose we make a circuit around both branch points of this function and travel on the ellipse $x^2/2 + y^2/1 = 1$. We begin at an arbitrary point on the ellipse. Show that after making this circuit we remain on whatever branch we started.

4. Consider the function $z^{1/3}(z - 1)^{1/3}$.

   a) If we make a complete circuit around both branch points of this function, will we remain on a given branch? Prove your result.

   b) Describe branch cuts that can be used to define a particular branch of this function.

5. Suppose a branch of $(z - 1)^{1/3}$ $(z + 1)^{1/2}$ equals 1 when $z = 0$. There are branch cuts defined by $y = 0$, $|x| \geq 1$. What values does this branch assume at

   a) $z = i$,   b) $z = -i$.

6. If two functions each have a branch point at z_0, does their product necessarily have a branch point at z_0? Illustrate with an example.

7. If $f(z)$ has a branch point at z_0, does $1/f(z)$ necessarily have a branch point at z_0? Explain.

8. Suppose a branch of $z^{-1/3}\,(z^2 + 1)$ is a negative real number for $y = 0$, $x < 0$. There is a branch cut for $x = 0$, $y \geq 0$. What values does this function assume at

   a) $z = 1$,   b) $z = 1 - i$.

9. Consider $\sinh^{-1} z = \log(z + (z^2 + 1)^{1/2})$. Suppose we use a branch defined as follows: The principal branch of the log is employed, $(z^2 + 1)^{1/2} = 1$ when $z = 0$, and there are branch cuts satisfying $x = 0$, $|y| > 1$.

   a) What value does this branch assume at $z = 0$, $z = 1$ and $z = 1 + i$?

   b) What are the derivatives of this branch at each of these three points? Give numerical values.

APPENDIX TO CHAPTER 3
PHASORS

In the analysis of electrical circuits and many mechanical systems we must deal with functions that oscillate sinusoidally in time, grow or decay exponentially in time, or oscillate with an amplitude that grows or decays exponentially in time. Designating time as t, we find that all such functions $f(t)$ can be described by

$$f(t) = \text{Re}[Fe^{st}], \tag{A3-1}$$

where

$$s = \sigma + i\omega \tag{A3-2}$$

is called the complex frequency of oscillation of $f(t)$, and F is a complex number, independent of t, written

$$F = F_0 e^{i\theta}, \tag{A3-3}$$

where $F_0 = |F|$ and $\theta = \arg F$. We will always use real values for σ and ω.

The complex number F appearing in Eqs. (A3-1) and (A3-3) is called the *phasor* associated with $f(t)$.

Definition   Phasor

In general, the phasor associated with a given function of time, $f(t)$, is a complex number F, independent of t, such that the real part of the product of F with a complex exponential e^{st} yields $f(t)$.   ∎

We will usually use an uppercase letter to mean a phasor and the corresponding lowercase letter to stand for the associated function of t. The one exception is that the letter I means a phasor electric current but the lowercase Greek iota, ι, will be the corresponding function of t. Thus the lowercase i retains its usual meaning. As we shall see, phasors are useful in the solution of the linear differential equations with constant coefficients used to describe many electrical and mechanical configurations.

The expression Fe^{st} appearing in Eq. (A3-1) is an example of a complex function of a real variable (since t remains real). Let us consider some specific cases in Eq. (A3-1). Suppose the phasor F in Eq. (A3-3) is positive real and the complex frequency in Eq. (A3-2) is real. Then with $s = \sigma$ and $F = F_0$ in Eq. (A3-1) we obtain

$$f(t) = \text{Re}[F_0 e^{\sigma t}] = F_0 e^{\sigma t}. \tag{A3-4}$$

Here, $f(t)$ grows or decays with increasing t according to whether σ is positive or negative. If $\sigma = 0$, then $f(t)$ is constant.

Assuming that both F in Eq. (A3-3) and s in Eq. (A3-2) are complex, we have from Eq. (A3-1)

$$f(t) = \text{Re}[F_0 e^{i\theta} e^{(\sigma + i\omega)t}] = \text{Re}[F_0 e^{\sigma t} e^{i(\omega t + \theta)}].$$

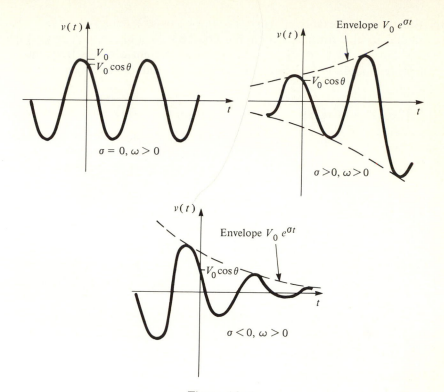

Figure A3–1

Since (see Eq. 3.1–9) Re $e^{i(\omega t + \theta)} = \cos(\omega t + \theta)$, we have

$$f(t) = F_0 e^{\sigma t} \cos(\omega t + \theta). \tag{A3-5}$$

Equation (A3–5) describes an $f(t)$ that oscillates with radian frequency ω (usually taken as positive). The amplitude of the oscillations, $F_0 e^{\sigma t}$, grows or decays with increasing t according to whether σ is positive or negative. If $\sigma = 0$, the amplitude of the oscillations remains unchanged. These three possible situations are illustrated in Fig. A3–1.

The function $f(t)$ described by Eq. (A3–5) varies as a cosinusoidal function of t. Taking $\theta = \phi - \pi/2$ in this equation, we have

$$f(t) = F_0 e^{\sigma t} \cos(\omega t + \phi - \pi/2) = F_0 e^{\sigma t} \sin(\omega t + \phi),$$

and a sinusoidal time variation is obtained. Some examples of functions of time, their complex frequencies, and their phasors, are given in the following table.

$f(t)$	F	$s = \sigma + i\omega$
$2\cos(10t + \pi/6)$	$2e^{i\pi/6}$	$10i$
$3\sin(5t + \pi/10)$	$3e^{i(-\pi/2 + \pi/10)} = -3ie^{i\pi/10}$	$5i$
$3e^{-t}\sin(5t)$	$-3i$	$-1 + 5i$
$4e^{-3t}$	4	-3

Many functions, for example, $f(t) = t \cos t$ or $\cos(t^2)$ or $\sin|t|$ or e^{t^2} are not representable in the form specified by Eq. (A3–1) and do not have phasors. Functions that are the sums of functions having different complex frequencies also are not representable in the form of Eq. (A3–1) and do not have phasors, for example,

$$f(t) = \cos(t) + \cos(2t), \qquad f(t) = e^{-t} + \sin t, \qquad f(t) = e^{-t}\cos t + \cos t.$$

The properties of phasors that make them particularly useful in the solution of linear differential equations with real constant coefficients are listed below. The proofs of these properties are, with one exception, left to the exercises.

1. a) If $f(t)$ is expressible in the form shown in Eq. (A3–1), then its phasor F is unique provided $\omega \neq 0$. There is no other phasor that can be substituted on the right in Eq. (A3–1) to yield $f(t)$. If $\omega = 0$, then only $\mathrm{Re}(F)$ is unique.[†]

    b) If $f(t)$ and $g(t)$ are identically equal for all t, then their phasors are equal provided $\omega \neq 0$. If $\omega = 0$, then the real parts of their phasors must be equal.

2. For a given complex frequency $s = \sigma + i\omega$ there is only one function of t corresponding to a given phasor.

3. The phasor for the sum of two or more time functions having identical complex frequencies is the sum of the phasors for each. The phasor for $Mf(t)$, where M is a real number, is MF, where F is the phasor for $f(t)$.

4. For a given complex frequency the function of t corresponding to the sum of two or more phasors is the sum of the time functions for each.

5. If $f(t)$ has phasor F, then df/dt has phasor sF. By extension, $d^n f/dt^n$ has phasor $s^n F$.

6. If $f(t)$ has phasor F, then $\int' f(t')\,dt'$ has phasor F/s provided the constant of integration arising from the unspecified lower limit is zero. The relationship fails if $s = 0$.

    To establish property 5, which is of major importance, we differentiate both sides of Eq. (A3–1) as follows:

$$\frac{df}{dt} = \frac{d}{dt}\,\mathrm{Re}[\,Fe^{st}\,]. \qquad\qquad (A3\text{–}6)$$

The operations d/dt and Re can be interchanged since the time derivative is a real operator. For example, if $x(t)$ and $y(t)$ are real functions of t, then $(d/dt)\,\mathrm{Re}[x(t) + iy(t)] = dx/dt$. We have also that $\mathrm{Re}[(d/dt)[x(t) + iy(t)]] = dx/dt$. Exchanging operators on the right side of Eq. (A3–6), we have $df/dt = \mathrm{Re}[sFe^{st}]$. Thus (see Eq. A3–1) df/dt possesses a phasor sF. This discussion can be extended to yield the phasor of higher-order derivatives.

[†] When $\omega = 0$ this lack of uniqueness does not matter when we use phasors in the solution of differential equations. However, by convention, when $\omega = 0$, $\mathrm{Im}(F)$ is taken as zero.

Phasors are applied to physical problems in which it is assumed that an electric circuit or mechanical configuration is excited by a real voltage, current, or mechanical force describable by Eq. (A3–1). The excitation, or forcing function, has been applied for a sufficiently long time so that all transients in the configuration have died out. Thus all voltages, currents, velocities, displacements, and so forth exhibit the same complex frequency s as the excitation.

Each quantity in the differential equation describing the physical problem is converted to its phasor. Property 5 is used to transform time derivatives in the differential equation into products of phasors and their complex frequency. The given differential equation is thus converted into an easily solved algebraic equation involving phasors. The required real function of time describing the physical problem can now be recovered from Eq. (A3–1). The uniqueness of the solution is guaranteed by the requirement that it exhibit the same complex frequency as the excitation. An example of the method follows.

Example 1

In Fig. A3–2 a series electrical circuit containing an inductor of L henries and a resistor of R ohms is driven by a voltage source $v(t) = v_0 \cos(\omega t)$. We want the unknown current $\iota(t)$. According to basic electric circuit theory[†] the voltage across the resistor is $R\iota(t)$ while that across the inductor is $L\, d\iota/dt$. According to Kirchhoff's voltage law the sum of these expressions must equal the source voltage. Thus

$$R\iota(t) + L\frac{d\iota}{dt} = v_0 \cos(\omega t). \tag{A3–7}$$

The phasor for the driving voltage $v_0 \cos(\omega t)$ is v_0, and the complex frequency is $s = i\omega$. The current also has this complex frequency.

If I is the phasor for $\iota(t)$, then by property 5 the phasor for $d\iota/dt$ must be $sI = i\omega I$. The phasor for the left side of Eq. (A3–7) is easily found from property 3 and equals $RI + i\omega LI = (R + i\omega L)I$. The phasor for the left side of this equation must equal the phasor for the right side (see property 1b). Thus

$$(R + i\omega L)I = V_0.$$

Solving for I, we have

$$I = \frac{v_0}{R + i\omega L} = \frac{v_0 e^{i\theta}}{\sqrt{R^2 + \omega^2 L^2}}, \tag{A3–8}$$

where

$$\theta = -\tan^{-1}\frac{\omega L}{R}. \tag{A3–9}$$

We can use Eq. (A3–1), taking $F = I$, $f = \iota$, and $s = i\omega$ to obtain $\iota(t)$. Thus using I

[†]See, for example, R. Kerr, *Electrical Network Science* (Englewood Cliffs, N.J.: Prentice-Hall, 1977).

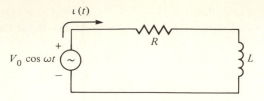

Figure A3–2

from Eq. (A3–8) and θ from Eq. (A3–9) we have

$$\iota(t) = \text{Re}\,\frac{v_0 e^{i\theta}}{\sqrt{R^2 + \omega^2 L^2}}\,e^{i\omega t}$$

$$= \frac{v_0 \cos(\omega t + \theta)}{\sqrt{R^2 + \omega^2 L^2}} = \frac{v_0 \cos\left(\omega t - \tan^{-1}\dfrac{\omega L}{R}\right)}{\sqrt{R^2 + \omega^2 L^2}}. \qquad (A3\text{–}10)$$

The reader can verify that this result satisfies the differential equation (A3–7). ◀

There are problems where the linear differential equations (with constant coefficients) describing a physical configuration are not solvable with phasors. This occurs if the complex frequency of the generator or other excitation is equal to the "natural" or resonant frequency of the physical system. Then the solution is not of the form $e^{\sigma t} \cos(\omega t + \sigma)$, $e^{\sigma t}$, etc. and does not possess a phasor. The subject is discussed in many texts.[†]

EXERCISES

1. State the time function $v(t)$ corresponding to the following phasors V and complex frequency s.

	V	s
a)	2	$-1 + i$
b)	$2i$	$-1 + i$
c)	$2ie^{i\pi/3}$	$-1 + i$
d)	$1 + i$	$1 - i$
e)	$1 + 2ie^{i\pi/3}$	$-3 + 4i$
f)	2	$3i$
g)	2	-3

[†]See, for example, C. R. Wylie, *Advanced Engineering Mathematics* (New York: McGraw-Hill, 1975), ch. 2.

2. Find the phasor corresponding to each of the following functions of time. In each case state the complex frequency. If the phasor does not exist, give the reason.

a) e^{-3t}

b) $5e^{-4t}\cos(6t + \pi/3)$

c) $\sin 3t$

d) $2\sin(3t + \pi/4)$

e) $3e^{-t}\sin(4t + \pi/3)$

f) $10e^{-t}\sin(4t + \pi/4) + 5e^{-t}\cos(4t - \pi/4)$

g) $e^{-t}\sin(4t) + t$

h) $e^{-t}\sin 4t + e^{-2t}\cos 4t$

3. Prove property 1(a) for phasors. *Hint:* Let $f(t) = \text{Re}[F_1 e^{st}]$ and let $f(t) = \text{Re}[F_2 e^{st}]$. Subtract these two equations and show that $F_1 = F_2$ when $\omega \neq 0$.

4. Prove property 2 for phasors. Let $f(t) = \text{Re}[Fe^{st}]$ and $g(t) = \text{Re}[Fe^{st}]$. Show that $f(t) = g(t)$.

5. Prove properties 3 and 4 for phasors.

6. Establish property 6 for phasors by integrating both sides of (A3–1). Justify the exchange of the order of any operations.

7. Consider an electric circuit identical to that shown in Fig. (A3–2) except that the voltage source has been changed to $v(t) = v_0 e^{\sigma t}$. The differential equation describing the current $\iota(t)$ is now

$$R\iota(t) + L\frac{d\iota}{dt} = v_0 e^{\sigma t}.$$

Assume $\sigma \neq -R/L$. Find the phasor current I and use it to find the actual current $\iota(t)$.

8. In Fig. A3–3 a series circuit containing a resistor of R ohms and a capacitor of C farads is driven by the voltage generator $v_0 \sin \omega t$. The voltage across the capacitor is given by $(1/C)\int \iota(t')\,dt'$, where ι is the current in the circuit. According to Kirchhoff's voltage law this current satisfies the integro-differential equation:

$$v_0 \sin(\omega t) = R\iota(t) + (1/C)\int^t \iota(t')\,dt'.$$

Obtain I the phasor current and use it to find $\iota(t)$. Assume $\omega > 0$.

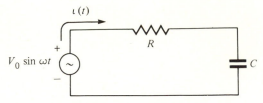

$\iota(t)$

R

$V_0 \sin \omega t$

C

Figure A3–3

9. A mass m is attached to the end of a spring and lies in a viscous fluid, as shown in Fig. A3–4. The coordinate $x(t)$ locating the mass also measures the elongation of the spring. Besides the spring force, the mass is subjected to a fluid

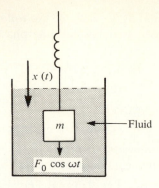

Figure A3–4

damping force proportional to the velocity of motion and also to an external mechanical force $F_0 \cos \omega t$. From Newton's law the differential equation governing $x(t)$ is

$$m\frac{d^2x}{dt^2} + \alpha\frac{dx}{dt} + kx = F_0 \cos \omega t, \qquad \omega > 0.$$

Here k is a constant determined by the stiffness of the spring and α is a damping constant determined by the fluid viscosity.

a) Find X the phasor for $x(t)$.   b) Use X to find $x(t)$.

CHAPTER 4 INTEGRATION IN THE COMPLEX PLANE

4.1 LINE INTEGRATION

When studying elementary calculus, the reader first learned to differentiate real functions of real variables and later to integrate such functions. Both indefinite and definite integrals were considered.

We have a similar agenda for complex variables. Having learned to differentiate in the complex plane and having studied the allied notion of analyticity, we turn our attention to integration. The indefinite integral, which (as for real variables) reverses the operation of differentiation, will not be considered first, however. Instead, we will initially look at a particular kind of definite integral called a line integral or contour integral.

Like the definite integral studied in elementary calculus, the line integral is a limit of a sum. However, the physical interpretation of this new integral is more elusive. Ordinarily, it cannot be considered as the area under a curve. Surprisingly, the study of line integrals will lead us to a useful theorem regarding the existence of derivatives of all orders of an analytic function and will provide us with further insight into the meaning of analyticity. Some practical physical problems solved with line integrals will be presented. In Chapter 6 we will show how evaluation of line integrals can often lead to the rapid integration of real functions; for example, an expression like $\int_{-\infty}^{+\infty} x^2/(x^4 + 1)\, dx$ is quickly evaluated if we first perform a fairly simple line integration in the complex plane.

In our discussion of integrals we require the concept of a *smooth arc* in the xy-plane. Loosely speaking, a smooth arc is a curve on which the tangent is defined everywhere and where the tangent changes its direction continuously as we move along the curve. One way to define a smooth arc is by means of a pair of equations dependent upon a real parameter, which we will call t. Thus

$$x = \psi(t), \tag{4.1-1a}$$

$$y = \phi(t), \tag{4.1-1b}$$

where $\psi(t)$ and $\phi(t)$ are real continuous functions with continuous derivatives $\psi'(t)$ and $\phi'(t)$ in the interval $t_a \leq t \leq t_b$. We also assume that there is no t in this interval for which both $\psi'(t)$ and $\phi'(t)$ are simultaneously zero. It is sometimes helpful to think that t represents time. As t advances from t_a to t_b, Eqs. (4.1–1a, b) define a locus that can be plotted in the xy-plane. This locus is a smooth arc.

An example of a smooth arc generated by such parametric equations is $x = t$, $y = 2t$, for $1 \leq t \leq 2$. Eliminating the parameter t, which connects the variables x and y, we find that the locus determined by the parametric equations lies along the line $y = 2x$. As t progresses from 1 to 2, we generate that portion of this line lying between $(1, 2)$ and $(2, 4)$ (see Fig. 4.1–1a).

Consider as another example the equations $x = \sqrt{t}$, $y = t$ for $1 \leq t \leq 4$. As t progresses from 1 to 4, the locus generated is the portion of the parabolic curve $y = x^2$ shown in Fig. 4.1–1(b). In Figs. 4.1–1(a, b) there are arrows that indicate the sense in which the arc is generated as t increases from t_a to t_b. For the right-hand arc the tangent has been constructed at some arbitrary point.

The slope of the tangent for any curve is dy/dx, which is identical to $(dy/dt)/(dx/dt) = \phi'(t)/\psi'(t)$ provided $\psi'(t) \neq 0$. If $\psi'(t) = 0$, the slope becomes infinite and the tangent is vertical. Since $\phi'(t)$ and $\psi'(t)$ are continuous, the tangent to the curve defined in Eqs. (4.1–1) alters its direction continuously as t advances through the interval $t_a \leq t \leq t_b$.

In our discussion of line integrals we must utilize the concept of a piecewise smooth curve, sometimes referred to as a contour.

Definition   Piecewise smooth curve (contour)

A *piecewise smooth curve* is a path made up of a finite number of smooth arcs connected end to end.   ■

Figure 4.1–2 shows three arcs C_1, C_2, C_3 joined to form a piecewise smooth curve.

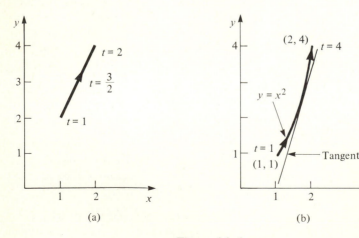

(a)   (b)

Figure 4.1–1

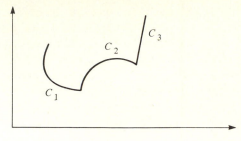

Figure 4.1–2

Where two smooth arcs join, the tangent to a piecewise smooth curve can change discontinuously. It is shown in elementary calculus that the length of a piecewise smooth curve connecting any two points is finite.

Real Line Integrals

We will begin our discussion of line integrals by using only real functions. An example of a real line integral—the integral for the length of a smooth arc[†]—is already known to the student. An approximation to the length of the arc is expressed as the sum of the lengths of chords inscribed on the arc. The actual length of the arc is obtained in the limit as the length of each chord in the sum becomes zero and the number of chords becomes infinite. In elementary calculus one learns to express this sum as an integral. The length of a piecewise smooth curve, such as is shown in Fig. 4.1–2, is obtained by adding together the lengths of the smooth arcs $C_1, C_2, \ldots$ that make up the curve.

Another type of real line integral involves not only a smooth arc C but also a function of x and y, say $F(x, y)$. It is important to realize that $F(x, y)$ is not the equation of C. Typically C is given by some equation, which, for the moment, we do not need to specify. Now, $\int_A^B F(x, y)\, ds$ integrated over C is defined as follows (refer to the arc C in Fig. 4.1–3).

We subdivide C, which goes from A to B, into n smaller arcs. The first arc goes from the point X_0, Y_0 to X_1, Y_1; the second arc goes from X_1, Y_1 to X_2, Y_2, etc.[‡] Corresponding to each of these arcs are the vector chords $\overrightarrow{\Delta s_1}, \overrightarrow{\Delta s_2}, \ldots, \overrightarrow{\Delta s_n}$. The first of these vectors is a directed line segment from X_0, Y_0 to X_1, Y_1; the second of these vectors is a directed line segment from X_1, Y_1 to X_2, Y_2, etc. These vectors, when summed, form a single vector going from A to B. The vectors form chords having lengths $\Delta s_1, \Delta s_2, \ldots$. The length of the vector $\overrightarrow{\Delta s_k}$ is thus Δs_k.

[†]See G. Thomas, *Calculus and Analytic Geometry* (Reading, Mass.: Addison-Wesley, 1972), sec. 5–7.

[‡]If the arc has been defined by a pair of parametric equations like those shown in Eqs. (4.1–1), where $t_0 \leq t \leq t_n$, we can generate the points X_0, Y_0; X_1, Y_1, etc. as follows: $X_0, Y_0 = \psi(t_0), \phi(t_0)$; $X_1, Y_1 = \psi(t_1), \phi(t_1)$; $X_n, Y_n = \psi(t_n), \phi(t_n)$, where $t_0 < t_1 < t_2, \ldots, < t_n$.

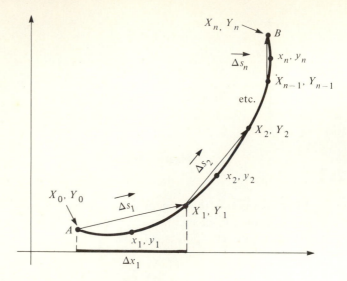

Figure 4.1–3

Let x_1, y_1 be a point at an arbitrary location on the first arc; x_2, y_2 a point somewhere on the second arc, etc. We now evaluate $F(x, y)$ at the n points (x_1, y_1), $(x_2, y_2),\ldots,(x_n, y_n)$. We define the line integral of $F(x, y)$ from A to B along C as follows:

Definition $\int_A^B F(x, y)\, ds$

$$\int_A^B F(x, y)\, ds = \lim_{n \to \infty} \sum_{k=1}^{n} F(x_k, y_k) \Delta s_k, \qquad (4.1\text{--}2)$$

where, as n, the number of subdivisions of C becomes infinite, the length Δs_k of each chord goes to zero. ■

Of course, if the limit of the sum in this definition fails to exist, we say that the integral does not exist. It can be shown that if $F(x, y)$ is continuous then the integral will exist.[†]

Evaluation of the preceding type of integral is similar to the familiar problem of evaluating integrals for arc length. A typical procedure is outlined in Exercise 1 of this section.

If $F(x, y)$ in Eq. (4.1–2) happens to be unity everywhere along C, then the summation on the right is approximately the length of the arc C connecting A with B. As $n \to \infty$, the limit of this sum is exactly the arc length. In general, however, the sum in Eq. (4.1–2) consists essentially of the sum of the lengths of the n straight line segments that approximate C, each of which is weighted by the value of the function

---

[†]W. Kaplan, *Advanced Calculus* (Reading, Mass.: Addison-Wesley, 1952), sec. 5–3 and 9–10.

$F(x, y)$ evaluated close to that segment. If the curve C is thought of as a cable and if $F(x, y)$ describes its mass density per unit length, then $F(x_k, y_k) \Delta s_k$ would be the approximate mass of the kth segment. When the summation is carried to the limit $n \to \infty$, it yields exactly the mass of the entire cable.

The line integral of a function taken over a piecewise smooth curve is obtained by adding together the line integrals over the smooth arcs in the curve. The integral of $F(x, y)$ along the contour of Fig. 4.1–2 is given by

$$\int F(x, y)\, ds = \int_{C_1} F(x, y)\, ds + \int_{C_2} F(x, y)\, ds + \int_{C_3} F(x, y)\, ds.$$

There is another type of line integral involving $F(x, y)$ and a smooth arc C that we can define. Refer to Fig. 4.1–3. Let Δx_1 be the projection of $\overrightarrow{\Delta s_1}$ on the x-axis, Δx_2 the projection of $\overrightarrow{\Delta s_2}$, etc. Note that although Δs_k is positive (because it is a length), Δx_k, which equals $X_k - X_{k-1}$, can be positive or negative depending on the direction of $\overrightarrow{\Delta s_k}$. We make the following definition:

Definition $\int_A^B F(x, y)\, dx$

$$\int_A^B F(x, y)\, dx = \lim_{n \to \infty} \sum_{k=1}^{n} F(x_k, y_k) \Delta x_k, \qquad (4.1\text{–}3)$$

where all $\Delta x_k \to 0$ as $n \to \infty$.   ■

A similar integral is definable when we instead use the projections of $\overrightarrow{\Delta s_k}$ on the y-axis. These projections are $\Delta y_1, \Delta y_2$, and so on, so that $\Delta y_k = Y_k - Y_{k-1}$. Hence:

Definition $\int_A^B F(x, y)\, dy$

$$\int_A^B F(x, y)\, dy = \lim_{n \to \infty} \sum_{k=1}^{n} F(x_k, y_k) \Delta y_k, \qquad (4.1\text{–}4)$$

where all $\Delta y_k \to 0$ as $n \to \infty$.   ■

The integrals in Eqs. (4.1–3) and (4.1–4) can be shown to exist when $F(x, y)$ is continuous along the smooth arc C. Some procedures for the evaluation of this type of integral are discussed in Example 1 of this section. Integrals along piecewise smooth curves can be defined if we add together the integrals along the arcs that make up the curves. In general, the values of the integrals defined in Eqs. (4.1–2), (4.1–3), and (4.1–4) depend not only on the function $F(x, y)$ in the integrand and the limits of integration but also on the path used to connect these limits.

What happens if we were to reverse the limits of integration in Eq. (4.1–3) or Eq. (4.1–4)? If we were to compute $\int_B^A F(x, y)\, dx$, we would go through a procedure identical to that used in computing $\int_A^B F(x, y)\, dx$, except that the vectors shown in Fig. 4.1–3 would all be reversed in direction; their sum would extend from B to A. The projections Δx_k would be reversed in sign from what they were before. Hence, along contour C

$$\int_B^A F(x, y)\, dx = -\int_A^B F(x, y)\, dx. \qquad (4.1\text{–}5)$$

A reversal in sign also occurs when we exchange A and B in the integral defined by Eq. (4.1-4). Note however that

$$\int_B^A F(x, y)\, ds = \int_A^B F(x, y)\, ds \qquad (4.1\text{-}6)$$

because the Δs_k used in the definitions of these expressions involves lengths that are positive for both directions of integration.

Integrals of the type defined in Eqs. (4.1-2), (4.1-3), and (4.1-4) can be broken up into the sum of other integrals taken along portions of the contour of integration. Let A and B be the endpoints of a piecewise smooth curve C. Let Q be a point on C. Then, one can easily show that

$$\int_A^B F(x, y)\, dx = \int_A^Q F(x, y)\, dx + \int_Q^B F(x, y)\, dx. \qquad (4.1\text{-}7)$$

Identical results hold for integrals of the form $\int_A^B F(x, y)\, dy$ and $\int_A^B F(x, y)\, ds$. Other identities that apply to all three of these kinds of integrals, but which will be written only for integration on the variable x, are

$$\int_A^B kF(x, y)\, dx = k\int_A^B F(x, y)\, dx, \quad k \text{ is any constant;} \qquad (4.1\text{-}8a)$$

$$\int_A^B [F(x, y) + G(x, y)]\, dx = \int_A^B F(x, y)\, dx + \int_A^B G(x, y)\, dx. \qquad (4.1\text{-}8b)$$

Example 1

Consider a contour consisting of that portion of the curve $y = 1 - x^2$ that goes from the point $A = (0, 1)$ to the point $B = (1, 0)$ (see Fig. 4.1-4). Let $F(x, y) = xy$. Evaluate

a) $\displaystyle\int_A^B F(x, y)\, dx,$ b) $\displaystyle\int_A^B F(x, y)\, dy.$

Solution

Part (a):

$$\int_A^B F(x, y)\, dx = \int_{0,1}^{1,0} xy\, dx.$$

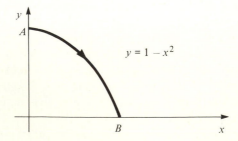

Figure 4.1-4

Along the path of integration y changes with x. The equation of the contour of integration $y = 1 - x^2$ can be used to express y as a function of x in the preceding integrand. Thus

$$\int_A^B F(x, y)\,dx = \int_{x=0,\,y=1}^{x=1,\,y=0}(x)(1 - x^2)\,dx = \left.\frac{x^2}{2} - \frac{x^4}{4}\right|_{x=0}^{x=1} = \frac{1}{4}.$$

We have converted our line integral to a conventional integral performed between constant limits; the evaluation is simple.

Part (b):

$$\int_A^B F(x, y)\,dy = \int_{0,1}^{1,0} xy\,dy.$$

To change this to a conventional integral we may regard x as a function of y along C. Since $y = 1 - x^2$ and $x \geq 0$ on the path of integration, we have $x = \sqrt{1 - y}$. Thus

$$\int_{x=0,\,y=1}^{x=1,\,y=0} xy\,dy = \int_1^0 y\sqrt{1 - y}\,dy = \frac{-4}{15}.$$

This result is negative because $F(x, y)$ is everywhere positive along the path of integration while the increments in y are everywhere negative as we proceed from A to B along the given contour.

An alternative method for performing the given integration is to notice that

$$\int F(x, y)\,dy = \int F(x, y)\frac{dy}{dx}\,dx.$$

On C, with $F(x, y) = xy$, $y = 1 - x^2$, and $dy/dx = -2x$ we obtain

$$\int_{x=0}^{x=1}(xy)(-2x)\,dx = \int_0^1 x(1 - x^2)(-2x)\,dx = \int_0^1(-2x^2 + 2x^4)\,dx = \frac{-4}{15}.$$

In part (a) we could have integrated on y instead of on x by a similar maneuver. Note that in Exercise 1 of this section the integral $\int xy\,ds$ along the same contour is evaluated. ◀

Complex Line Integrals

We now study the kind of integral encountered most often with complex functions: the complex line integral. We will find that it is closely related to the real line integrals just discussed.

We begin, as before, with a smooth arc that connects the points A and B in the xy-plane. We now regard the xy-plane as being the complex z-plane. The arc is divided into n smaller arcs and, as shown in Fig. 4.1–5, successive endpoints of the subarcs have coordinates (X_0, Y_0), $(X_1, Y_1), \ldots, (X_n, Y_n)$. Alternatively, we could say that the endpoints of these smaller arcs are at $z_0 = X_0 + iY_0$, $z_1 = X_1 + iY_1$, etc. A series of vector chords are then constructed between these points. As in our discussion of real line integrals, the vectors progress from A to B when we are integrating from A to B along the contour. Let Δz_1 be the complex number

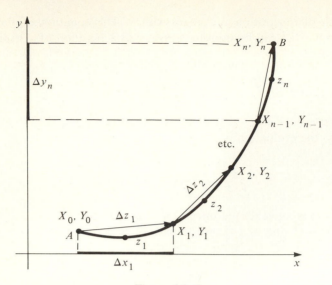

Figure 4.1–5

corresponding to the vector going from (X_0, Y_0) to (X_1, Y_1), let Δz_2 be the complex number for the vector going from (X_1, Y_1) to (X_2, Y_2), etc. There are n such complex numbers. In general,

$$\Delta z_k = \Delta x_k + i\Delta y_k, \qquad (4.1-9)$$

where Δx_k and Δy_k are the projections of the kth vector on the real and imaginary axes. Thus

$$\Delta z_k = (X_k - X_{k-1}) + i(Y_k - Y_{k-1}).$$

Let $z_k = x_k + iy_k$ be the complex number corresponding to a point lying, at an arbitrary position, on the kth arc. This arc is subtended by the vector chord Δz_k. Some study of Fig. 4.1–5 should make the notation clear.

Let us consider $f(z) = u(x, y) + iv(x, y)$, a continuous function of the complex variable z. We can evaluate this function at $z_1, z_2, \ldots, z_n$. We now define the line integral of $f(z)$ taken over the arc C.

Definition   Complex line integral

$$\int_A^B f(z)\, dz = \lim_{n \to \infty} \sum_{k=1}^n f(z_k)\, \Delta z_k, \qquad (4.1-10)$$

where all $\Delta z_k \to 0$ as $n \to \infty$.   ∎

As before, the integral only exists if the limit of the sum exists. In general, we must assume that the value of the integral depends not only on A and B, the location of the ends of the path of integration, but also on the specific path C used to connect these points. The reader is cautioned against interpreting the integral as the area under the curve in Fig. 4.1–5.

The line integral of a function over a piecewise smooth curve is computed by using Eq. (4.1–10) to determine the integral of the function over each of the arcs that make up the curve. The values of these integrals are then added together.

Let us try to develop some intuitive feeling for the sum on the right in Eq. (4.1–10). We can imagine that the arc in Fig. 4.1–5 is approximated, in shape, by the straight lines forming the n vectors. As n approaches infinity in the sum, there are more, and shorter, vectors involved in the sum. The broken line formed by these vectors more closely fits the curve C.

In the summation the complex numbers associated with each of these n vectors are added together after first having been multiplied by a weighting function $f(z)$ evaluated close to that vector. The function is evaluated on the nearby curve. If the weighting function were identically equal to 1, the sum in Eq. (4.1–10) would be represented, for all n, by a single vector extending from A to B.

A summation up to a finite value of n in Eq. (4.1–10) can be used to approximate the integral on the left in this equation. Such a procedure is often used when one performs a complex line integration on the computer. Let us consider an example.

Example 2

The function $f(z) = z + 1$ is to be integrated from $0 + i0$ to $1 + i$ along the arc $y = x^2$, as shown in Fig. 4.1–6. We will consider one-term series and two-term series approximations to the result.

a) One-term series: A single vector, associated with the complex number $\Delta z_1 = 1 + i$, goes from $(0, 0)$ to $(1, 1)$. The point z_1 can be chosen anywhere on the arc shown. We arbitrarily choose it to lie so that its x-coordinate is in the middle of the projection of Δz_1 on the x-axis. Since z_1 is on the curve $y = x^2$, we have $\operatorname{Re} z_1 = 1/2$, $\operatorname{Im} z_1 = 1/4$. Now, because $f(z) = z + 1$, we find that $f(z_1) = (1/2 + i/4 + 1)$. Thus

$$\int_{0+i0}^{1+i}(z + 1)\, dz \doteq f(z_1)\,\Delta z_1 = \left(\frac{1}{2} + \frac{i}{4} + 1\right)(1 + i) = 1.25 + 1.75i.$$

b) Two-term series: Referring to Fig. 4.1–7, we see that

$$\Delta z_1 = \frac{1}{2} + \frac{i}{4}, \qquad \Delta z_2 = \frac{1}{2} + \frac{3i}{4}, \qquad z_1 = \frac{1}{4} + \frac{i}{16},$$

$$f(z_1) = \frac{1}{4} + \frac{i}{16} + 1, \qquad z_2 = \frac{3}{4} + i\frac{9}{16}, \qquad f(z_2) = \frac{3}{4} + i\frac{9}{16} + 1.$$

We have used the sum of two vectors to connect $(0, 0)$ with $(1, 1)$ and have chosen z_1 and z_2 according to the same criterion used in part (a). Thus

$$\int_{0+i0}^{1+i}(z + 1)\, dz \doteq f(z_1)\,\Delta z_1 + f(z_2)\,\Delta z_2$$

$$= \left(\frac{5}{4} + \frac{i}{16}\right)\left(\frac{1}{2} + \frac{i}{4}\right) + \left(\frac{7}{4} + \frac{i9}{16}\right)\left(\frac{1}{2} + \frac{3i}{4}\right) = 1.0625 + i1.9375.$$

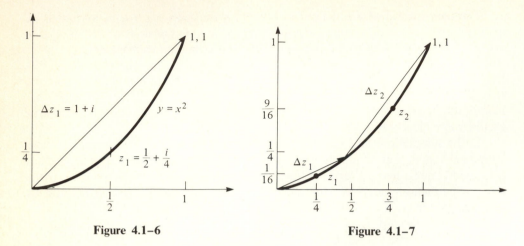

Figure 4.1–6   Figure 4.1–7

We will see in Exercise 2 at the end of this section that the exact value of the given integral is $1 + 2i$, a result that is surprisingly well approximated by the two-term series. Note that this result is unrelated to the area under the curve $y = x^2$. ◀

When the function $f(z)$ is written in the form $u(x, y) + iv(x, y)$, line integrals involving $f(z)$ can be expressed in terms of real line integrals. Thus referring back to Eq. (4.1–10) and noting that $z_k = x_k + iy_k$ and that $\Delta z_k = \Delta x_k + i\Delta y_k$, we have

$$\int_A^B f(z)\, dz = \lim_{n \to \infty} \sum_{k=1}^{n} \left(u(x_k, y_k) + iv(x_k, y_k) \right)(\Delta x_k + i\Delta y_k). \quad (4.1\text{–}11)$$

We now multiply the terms under the summation sign in Eq. (4.1–11) and separate the real and imaginary parts. Thus

$$\int_A^B f(z)\, dz = \lim_{n \to \infty} \left[\sum_{k=1}^{n} u(x_k, y_k)\, \Delta x_k - \sum_{k=1}^{n} v(x_k, y_k)\, \Delta y_k \right.$$
$$\left. + i \sum_{k=1}^{n} v(x_k, y_k)\, \Delta x_k + i \sum_{k=1}^{n} u(x_k, y_k)\, \Delta y_k \right]. \quad (4.1\text{–}12)$$

Upon comparing the four summations in Eq. (4.1–12) with the definitions of the real line integrals $\int F(x, y)\, dx$, $\int F(x, y)\, dy$ (see Eqs. 4.1–3 and 4.1–4) we find that

$$\int_C f(z)\, dz = \int_C u(x, y)\, dx - \int_C v(x, y)\, dy + i\int_C v(x, y)\, dx + i\int_C u(x, y)\, dy.$$
$$(4.1\text{–}13)$$

The letter C signifies that all these integrals are to be taken, in a specific direction, along contour C. The continuity of u and v (or the continuity of $f(z)$) is sufficient to guarantee the existence of all the integrals in Eq. (4.1–13). The four real line integrals on the right are of a type that we have already studied; thus Eq. (4.1–13) provides us with a method for computing complex line integrals. Note that, as a useful mnemonic, Eq. (4.1–13) can be obtained from the following manipulation:

$$\int f(z)\, dz = \int (u + iv)(dx + i\, dy) = \int u\, dx - v\, dy + iv\, dx + iu\, dy.$$

We merely multiplied out the integrand $(u + iv)(dx + i\, dy)$.

When the path of integration for a complex line integral lies parallel to the real axis, we have $dy = 0$. There then remains

$$\int_C f(z)\, dz = \int_C [u(x, y) + iv(x, y)]\, dx, \quad y \text{ is constant.}$$

This is the conventional type of integral encountered in elementary calculus, except that the integrand is complex if $v \neq 0$.

Example 3

a) Compute $\int_{0+i}^{1+2i} (\bar{z})^2\, dz$ taken along the contour $y = x^2 + 1$ (see Fig. 4.1–8a).

b) Perform an integration like that in part (a) using the same integrand and limits, but take as a contour the piecewise smooth curve C shown in Fig. 4.1–8b.

Solution

Part (a): To apply Eq. (4.1–13) we put $f(z) = (\bar{z})^2 = (x - iy)^2 = x^2 - y^2 - 2ixy = u + iv$. Thus with $u = x^2 - y^2$, $v = -2xy$, we have

$$\int_{0+i}^{1+2i} (\bar{z})^2\, dz = \int_{0,1}^{1,2} (x^2 - y^2)\, dx + \int_{0,1}^{1,2} 2xy\, dy + i \int_{0,1}^{1,2} (-2xy)\, dx$$

$$+ i \int_{0,1}^{1,2} (x^2 - y^2)\, dy. \tag{4.1–14}$$

In the first and third integrals on the right we substitute the relationship $y = x^2 + 1$ that holds along the contour. These line integrals become ordinary integrals whose limits are $x = 0$ and $x = 1$. After integration their values are found to be $-23/15$ and $-3/2$, respectively. The equation $y = x^2 + 1$ yields $x = \sqrt{y - 1}$ on the contour. This can be used to convert the second and fourth integrals on the right to ordinary integrals with limits $y = 1$ and $y = 2$. The integrals are found to have the

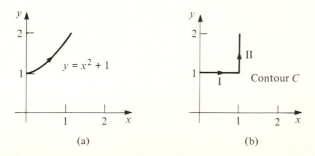

(a)   (b)

Figure 4.1–8

numerical values $32/15$ and $-11/6$, respectively. Having evaluated the four line integrals on the right side of Eq. (4.1–14) we finally obtain

$$\int_{0+i}^{1+2i} (\bar{z})^2 \, dz = \frac{3}{5} - i\frac{10}{3}.$$

Part (b): Referring to Fig. 4.1–8(b), we break the path of integration into a part taken along path I and a part taken along path II.

Along I we have $y = 1$ so that $f(z) = (\bar{z})^2 = (x + i)^2 = x^2 - 1 - 2xi = u + iv$. Thus $u = x^2 - 1$, $v = -2x$. Since $y = 1$, $dy = 0$. The limits of integration along path I are $(0, 1)$ and $(1, 1)$. Using this information in Eq. (4.1–13), we obtain

$$\int_I f(z) \, dz = \int_0^1 (x^2 - 1) \, dx + i\int_0^1 (-2x) \, dx = -\frac{2}{3} - i.$$

Along path II, $x = 1$, $dx = 0$, $f(z) = (\overline{1 + iy})^2 = 1 - y^2 - 2iy = u + iv$. The limits of integration are $(1, 1)$ and $(1, 2)$. Referring to Eq. (4.1–13), we have

$$\int_{II} f(z) \, dx = \int_1^2 (2y) \, dy + i\int_1^2 (1 - y^2) \, dy = 3 - \frac{4}{3}i.$$

The value of the integral along C is obtained by summing the contributions from I and II. Thus

$$\int_C (\bar{z})^2 \, dz = -\frac{2}{3} - i + 3 - \frac{4}{3}i = \frac{7}{3} - \frac{7}{3}i.$$

This result is different from that of part (a) and illustrates how the value of a line integral between two points can depend on the contour used to connect them. ◀

Since Eq. (4.1–13) allows us to express a complex line integral in terms of real line integrals, the properties of real line integrals contained in Eqs. (4.1–5), (4.1–7), and (4.1–8) also apply to complex line integrals. Thus the following relationships are satisfied by integrals taken along a piecewise smooth curve C that connects points A and B.

$$\int_A^B f(z) \, dz = -\int_B^A f(z) \, dz \tag{4.1–15a}$$

$$\int_A^B \Gamma f(z) \, dz = \Gamma \int_A^B f(z) \, dz, \quad \text{where } \Gamma \text{ is any constant} \tag{4.1–15b}$$

$$\int_A^B [f(z) + g(z)] \, dz = \int_A^B f(z) \, dz + \int_A^B g(z) \, dz \tag{4.1–15c}$$

$$\int_A^B f(z) \, dz = \int_A^Q f(z) \, dz + \int_Q^B g(z) \, dz, \quad \text{where } Q \text{ lies on } C \tag{4.1–15d}$$

Sometimes it is easier to perform a line integration without using the variables x and y in Eq. (4.1–13). Instead, we integrate on a single real variable that is the parameter used in generating the contour of integration. Let a smooth arc C be generated by the pair of parametric equations in (4.1–1). Then

$$z(t) = x(t) + iy(t) = \psi(t) + i\phi(t) \tag{4.1–16}$$

is a complex function of the real variable t with derivative

$$\frac{dz}{dt} = \frac{d\psi}{dt} + i\frac{d\phi}{dt}. \tag{4.1–17}$$

As t advances from t_a to t_b in the interval $t_a \le t \le t_b$, the locus of $z(t)$ in the complex plane is the arc C connecting $z_a = \psi(t_a), \phi(t_a)$ with $z_b = \psi(t_b), \phi(t_b)$. To evaluate $\int_C f(z)\, dz$ we can make a change of variables as follows:

$$\int_C f(z)\, dz = \int_{t_a}^{t_b} f(z(t)) \frac{dz}{dt}\, dt, \tag{4.1-18}$$

where the left-hand integration is performed along C from z_a to z_b. A rigorous justification for this equation is given in several texts.[†] Note that the integral on the right involves complex functions integrated on a real variable. This integration is performed with the familiar methods of elementary calculus. An application of Eq. (4.1–18) is given in the following example.

Example 4

Evaluate $\int_C z\, dz$, where C is the parabolic arc $y = x^2$, $1 \le x \le 2$ shown in Fig. 4.1–1(b). The direction of integration is from $(1, 1)$ to $(2, 4)$.

Solution

We showed earlier that this arc can be generated by the parametric equations $x = \sqrt{t}$, $y = t$, where $1 \le t \le 4$. Thus the arc can be described as the locus of $z(t) = \sqrt{t} + it$ for $1 \le t \le 4$; notice that $dz/dt = 1/(2\sqrt{t}) + i$. The integrand is $f(z) = z = \sqrt{t} + it$. Using Eq. (4.1–18) with $t_a = 1$ and $t_b = 4$, we have

$$\int_C z\, dz = \int_1^4 (\sqrt{t} + it)\left(\frac{1}{2\sqrt{t}} + i\right) dt$$

$$= \int_1^4 \left(\frac{1}{2} + \frac{3i\sqrt{t}}{2} - t\right) dt = -6 + 7i. \qquad \blacktriangleleft$$

Bounds on Line Integrals; the "ML Inequality"

Given a line integral to evaluate $\int_C f(z)\, dz$, we can often, without going through the labor of performing the integration, obtain an upper bound on the absolute value of the answer. That is, we can find a positive number that is known to be greater than or equal to the magnitude of the still unknown integral.

In Eq. (4.1–10) we defined, with the aid of Fig. 4.1–5, $\int_C f(z)\, dz$. A related integral will now be defined with the use of the smooth arc C of the same figure.

Definition $\int_C |f(z)|\,|dz|$

$$\int_C |f(z)|\,|dz| = \lim_{n \to \infty} \sum_{k=1}^n |f(z_k)|\,|\Delta z_k|, \quad \text{all } |\Delta z_k| \to 0 \text{ as } n \to \infty. \quad \blacksquare \tag{4.1-19}$$

This integration results in a positive real number. Since $|\Delta z_k| = \Delta s_k$ (refer to Figs.

[†]See, for example, E. T. Copson, *An Introduction to the Theory of Functions of a Complex Variable* (London: Oxford University Press, 1960), sec. 4.13.

4.1–3 and 4.1–5), we see from Eq. (4.1–2) that the preceding integral is identical to $\int_C |f(z)| \, ds$. Note that if $|f(z)| = 1$, then Eq. (4.1–19) simplifies to

$$\int_C |dz| = \lim_{n \to \infty} \sum_{k=1}^{n} |\Delta z_k| = \lim_{n \to \infty} \sum_{k=1}^{n} \Delta s_k = L, \qquad (4.1\text{–}20)$$

where L, the length of C, is the sum of the chord lengths of Fig. 4.1–5 in the limit indicated. Let us compare the magnitude of the sum appearing on the right side of Eq. (4.1–10) with the sum on the right side of Eq. (4.1–19).

Recall that the magnitude of a sum of complex numbers is less than or equal to the sum of their magnitudes, and the magnitude of the product of two complex numbers equals the product of the magnitude of the numbers.

Using these two facts, it follows that

$$\left| \sum_{k=1}^{n} f(z_k) \, \Delta z_k \right| \le \sum_{k=1}^{n} |f(z_k)| \, |\Delta z_k|. \qquad (4.1\text{–}21)$$

The preceding inequality remains valid as $n \to \infty$ and $|\Delta z_k| \to 0$. Thus, combining Eqs. (4.1–10), (4.1–19), and (4.1–21), we have

$$\left| \int_C f(z) \, dz \right| \le \int_C |f(z)| \, |dz|, \qquad (4.1\text{–}22)$$

which will occasionally prove useful.

Now assume that M, a positive real number, is an upper bound for $|f(z)|$ on C. Thus $|f(z)| \le M$ for z on C. In particular, $|f(z_1)|$, $|f(z_2)|$, etc. on the right in Eq. (4.1–21) satisfy this inequality. Using this fact in Eq. (4.1–21), we obtain

$$\left| \sum_{k=1}^{n} f(z_k) \, \Delta z_k \right| \le \sum_{k=1}^{n} |f(z_k)| \, |\Delta z_k| \le \sum_{k=1}^{n} M|\Delta z_k| = M \sum_{k=1}^{n} |\Delta z_k|. \quad (4.1\text{–}23)$$

Now observe that $\sum_{k=1}^{n} |\Delta z_k| \le L$ since the sum of the chord lengths, as in Fig. 4.1–5, cannot exceed the length L of the arc C. Combining this inequality with Eq. (4.1–23) we have

$$\left| \sum_{k=1}^{n} f(z_k) \, \Delta z_k \right| \le ML.$$

As $n \to \infty$, the preceding inequality still holds. Passing to this limit, with $\Delta z_k \to 0$, and referring to the definition of the line integral in Eq. (4.1–10), we have

ML INEQUALITY $$\left| \int_A^B f(z) \, dz \right| \le ML. \qquad (4.1\text{–}24)$$

The above equation is known as the *ML inequality;* stated in words, it reads:

If $|f(z)|$ does not exceed a constant M everywhere along a smooth arc C and if L is the length of C, then the magnitude of the integral of $f(z)$ along C cannot exceed ML.

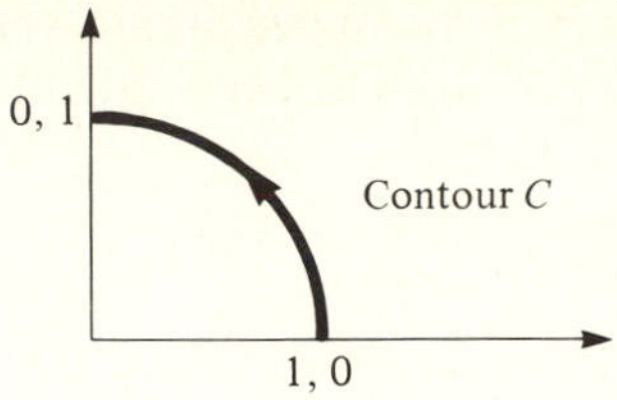

0, 1

Contour C

1, 0

Figure 4.1–9

Example 5

Find an upper bound on the absolute value of $\int_{1+i0}^{0+i1} e^{1/z}\, dz$, where the integral is taken along the contour C, which is the quarter circle $|z| = 1, 0 \le \arg z \le \pi/2$ (see Fig. 4.1–9).

Solution

Let us first find M, an upper bound on $|e^{1/z}|$. We require that on C

$$|e^{1/z}| \le M. \tag{4.1--25}$$

Now, notice that

$$e^{1/z} = e^{1/(x+iy)} = e^{x/(x^2+y^2) - iy/(x^2+y^2)} = e^{x/(x^2+y^2)} e^{i(-y)/(x^2+y^2)}.$$

Hence

$$\left|e^{1/z}\right| = \left|e^{x/(x^2+y^2)}\right|\left|e^{-i(y)/(x^2+y^2)}\right| = \left|e^{x/(x^2+y^2)}\right|.$$

Since $e^{x/(x^2+y^2)}$ is always positive, we can drop the magnitude signs on the right side of the preceding equation. On contour C, $x^2 + y^2 = 1$. Thus

$$|e^{1/z}| = e^x \quad \text{on } C.$$

The maximum value achieved by e^x on the given quarter circle occurs when x is maximum, that is, at $x = 1, y = 0$. On C, therefore, $e^x \le e$. Thus

$$|e^{1/z}| \le e$$

on the given contour. A glance at Eq. (4.1–25) now shows that we can take M as equal to e.

The length L of the path of integration is simply the circumference of the given quarter circle, namely, $\pi/2$. Thus, applying the ML inequality,

$$\left|\int_{1+i0}^{0+i1} e^{1/z}\, dz\right| \le e\frac{\pi}{2}. \qquad \blacktriangleleft$$

EXERCISES

1. Using the contour of Example 1, show that

$$\int_{0,1}^{1,0} xy\, ds = \int_0^1 x(1 - x^2)\sqrt{1 + 4x^2}\, dx = \sqrt{5}\,(5/24) - 11/120.$$

Hint: Recall from elementary calculus that $ds = \sqrt{1 + (dy/dx)^2}\, dx$.

2. In Example 2 we determined the approximate value of $\int_{0+i0}^{1+i}(z+1)\,dz$ taken along the contour $y = x^2$. Find the exact value of the integral and compare it with the approximate result.

3. Consider $\int_i^1 z^2\,dz$ performed along the contour $y = 1 - x^2$. Find the approximate value of this integral by means of the two-term series approximation $f(z_1)\Delta z_1 + f(z_2)\Delta z_2$. Take $z_1, z_2, \Delta z_1, \Delta z_2$, as shown in Fig. 4.1–10. Now find the exact value of this integral and compare it with the approximate result.

4. Evaluate $\int_{0+i}^{1+ie}\bar{z}\,dz$ along the contour C, where

   a) C is the curve $y = e^x$,

   b) C is composed of the horizontal line $y = 1$, $0 \le x \le 1$, and the vertical line $x = 1$, $1 \le y \le e$. Compare the results of parts (a) and (b).

5. Evaluate $\int e^z\,dz$

   a) from $z = 0$ to $z = 1$ along the line $y = 0$,

   b) from $z = 1$ to $z = 1 + i$ along the line $x = 1$,

   c) from $z = 1 + i$ to $z = 0$ along the line $y = x$. Verify that the sum of your three answers is zero. The reason is given in the next section.

6. Evaluate $\int (x + i\,\mathrm{Log}|z|)\,dz$ along the arc $|z| = 1$, $\pi/4 \le \arg z \le \pi/2$ from $\exp(i\pi/4)$ to i.

7. Consider $I = \int_{0+i0}^{1+i}|z|\bar{z}\,dz$ taken along the contour $y = x^2$. Without actually doing the integration show that $|I| \le 2.96$.

8. Consider $I = \int_{1+i0}^{0+i}\mathrm{Log}|z|\,dz$ taken along the contour $x + y = 1$. Without actually doing the integration show that $|I| \le \sqrt{2}\,\mathrm{Log}\sqrt{2}$.

9. a) Sketch the contour in the z-plane whose parametric representation is $x = \sin t$, $y = 2\cos t$; $0 \le t \le \pi/2$.

   b) By integrating directly on the parameter t integrate $z + 1$ along this contour from $z = 2i$ to $z = 1$.

10. Perform the integration $\int_1^{-1}1/z\,dz$ around the semicircular arc: $|z| = 1$, $0 \le \arg z \le \pi$ by using a suitable parametric representation of the arc and integrating on this real parameter. *Hint:* Consider $z = e^{it}$, where t varies between certain limits.

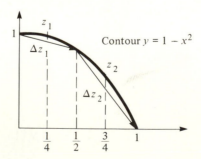

Figure 4.1–10

11. a) Let $g(t)$ be a complex function of the real variable t.

Express $\int_a^b g(t)\, dt$ as the limit of a sum. Using an argument similar to the one used in deriving Eq. (4.1–22), show that for $b > a$ we have

$$\left| \int_a^b g(t)\, dt \right| \le \int_a^b |g(t)|\, dt. \qquad (4.1\text{–}26)$$

b) Use Eq. (4.1–26) to prove that

$$\left| \int_0^1 \sqrt{t}\, e^{it}\, dt \right| \le \frac{2}{3}.$$

4.2   CONTOUR INTEGRATION AND GREEN'S THEOREM

In the preceding section we discussed piecewise smooth curves, called contours, that connect two points A and B. If these two points happen to coincide, the resulting curve is called a *closed contour*.

Definition   Simple closed contour

A *simple closed contour* is a contour that creates two domains, a bounded one and an unbounded one; each domain has the contour for its boundary. The bounded domain is said to be the *interior* of the contour. ■

Examples of two different closed contours, one of which is simple, are shown in Fig. 4.2–1.

We will often be concerned with line integrals taken around a simple closed contour.

The integration is said to be performed in the positive sense around the contour if the interior of the contour is on our left as we do the integration.

For the curve in Fig. 4.2–1(a) the positive direction of integration is indicated by the arrow.

When an integration around a simple closed contour is done in the positive direction, it will be indicated by the operator $\oint$ while an integration in the negative

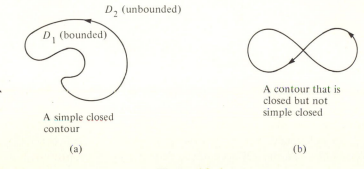

D_2 (unbounded)

D_1 (bounded)

A simple closed contour

(a)

A contour that is closed but not simple closed

(b)

Figure 4.2–1

sense is denoted by $\oint$. Note that

$$\oint f(z)\, dz = -\oint f(z)\, dz,$$

$$\oint f(x, y)\, dx = -\oint f(x, y)\, dx,$$

$$\oint f(x, y)\, dy = -\oint f(x, y)\, dy.$$

The following important theorem, known as Green's theorem in the plane, applies to real functions integrated around closed contours.

Theorem 1   Green's theorem in the plane

Let $P(x, y)$ and $Q(x, y)$ and their first partial derivatives be continuous functions throughout a region R consisting of the interior of a simple closed contour C plus the points on C. Then

$$\oint_C P\, dx + Q\, dy = \iint_R \left(\frac{\partial Q}{\partial x} - \frac{\partial P}{\partial y} \right) dx\, dy. \quad \blacksquare \qquad (4.2\text{-}1)$$

Thus Green's theorem converts a line integral around C into an integral over the area enclosed by C. A brief proof of the theorem is presented in the appendix to this chapter.

Complex line integrals can be expressed in terms of real line integrals (see Eq. 4.1–13) and it is here that Green's theorem proves useful. Consider a function $f(z) = u(x, y) + iv(x, y)$ that is not only analytic in the region R (of the preceding theorem) but whose first derivative is continuous in R. Since $f'(z) = \partial u / \partial x + i \partial v / \partial x = \partial v / \partial y - i \partial u / \partial y$, it follows that the first partial derivatives $\partial u / \partial x$, $\partial v / \partial x$, etc. are continuous in R also. Now refer to Eq. (4.1–13). We restate this equation and perform the integrations around the simple closed contour C.

$$\oint_C f(z)\, dz = \oint_C u\, dx - v\, dy + i \oint_C v\, dx + u\, dy. \qquad (4.2\text{-}2)$$

We can rewrite the pair of integrals appearing on the right by means of Green's theorem. For the first integral we apply Eq. (4.2–1) with $P = u$ and $Q = -v$. For the second integral we use Eq. (4.2–1) with $P = v$ and $Q = u$. Hence

$$\oint_C u\, dx - v\, dy = \iint_R \left(-\frac{\partial v}{\partial x} - \frac{\partial u}{\partial y} \right) dx\, dy, \qquad (4.2\text{-}3a)$$

$$\oint_C v\, dx + u\, dy = \iint_R \left(\frac{\partial u}{\partial x} - \frac{\partial v}{\partial y} \right) dx\, dy. \qquad (4.2\text{-}3b)$$

Recalling the C–R equations $\partial u / \partial x = \partial v / \partial y$, $\partial v / \partial x = -\partial u / \partial y$, we see that both integrands on the right in Eq. (4.2–3) vanish. Thus both line integrals on the left in these equations are zero. Referring back to Eq. (4.2–2), we find that $\oint f(z)\, dz = 0$.

Our proof, which relied on Green's theorem, was presented first by Cauchy in the early nineteenth century. We required that $f'(z)$ be continuous in R since otherwise Green's theorem is inapplicable. There is a less restrictive proof, formulated by Goursat,[†] that eliminates this requirement on $f'(z)$. The result contained in the previous equation, together with the less restrictive conditions of Goursat's derivation, are known as the Cauchy–Goursat theorem or sometimes just the Cauchy integral theorem.

Theorem 2   Cauchy–Goursat

Let C be a simple closed contour and let $f(z)$ be a function that is analytic in the interior of C as well as on C itself. Then

$$\oint_C f(z)\,dz = 0. \quad \blacksquare \tag{4.2–4}$$

An alternative statement of the theorem is this:

Let $f(z)$ be analytic in a simply connected domain D. Then, for any simple closed contour C in D, we have $\oint_C f(z)\,dz = 0$.

The Cauchy–Goursat theorem is capable of saving us a great deal of labor when we seek to perform certain integrations. For example, such integrals as $\oint_C \sin z\,dz$, $\oint_C e^z\,dz$, $\oint_C \cosh z\,dz$ must be zero when C is any simple closed contour. The integrands in each case are entire functions.

Note that the direction of integration in Eq. (4.2–4) is immaterial since $-\oint_C f(z)\,dz = \oint_C f(z)\,dz$.

We can verify the truth of the Cauchy–Goursat theorem in some simple cases. Consider $f(z) = z^n$, where n is a nonnegative integer. Now, since z^n is an entire function, we have, according to the theorem

$$\oint_C z^n\,dz = 0, \qquad n = 0, 1, 2, \ldots, \tag{4.2–5}$$

where C is any simple closed contour.

If n is a negative integer, then z^n fails to be analytic at $z = 0$. The theorem cannot be applied when the origin is inside C. However, if the origin lies outside C, the theorem can again be used and we have that $\oint_C z^n\,dz$ is again zero.

Let us verify Eq. (4.2–5) in a specific case. We will take C as a circle of radius r centered at the origin. Let us switch to polar notation and express C parametrically by using the polar angle θ. At any point on C, $z = re^{i\theta}$ (see Fig. 4.2–2). As θ advances from 0 to 2π or through any interval of 2π radians, the locus of z is the circle C generated in the counterclockwise sense. Note that $dz/d\theta = ire^{i\theta}$. Employing Eq. (4.1–18), with θ used instead of t, we have

$$\oint_{|z|=r} z^n\,dz = \int_0^{2\pi} (re^{i\theta})^n ire^{i\theta}\,d\theta = ir^{n+1}\int_0^{2\pi} e^{i(n+1)\theta}\,d\theta. \tag{4.2–6}$$

[†]See, for example, N. Levinson and R. Redheffer, *Complex Analysis* (San Francisco, Calif.: Holden-Day, 1970), p. 124.

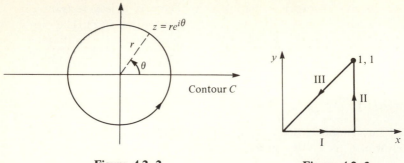

Figure 4.2–2   **Figure 4.2–3**

Proceeding on the assumption $n \geq 0$, we integrate Eq. (4.2–6) as follows:

$$ir^{n+1} \int_0^{2\pi} e^{i(n+1)\theta} \, d\theta = ir^{n+1} \int_0^{2\pi} \text{cis} \, (n + 1)\theta \, d\theta$$

$$= \frac{r^{n+1}}{n + 1} \left[\cos(n + 1)\theta + i \sin(n + 1)\theta \right]_0^{2\pi} = 0. \quad (4.2-7)$$

This is precisely the result predicted by the Cauchy–Goursat theorem.

Suppose n is a negative integer and C, the contour of integration, is still the same circle. Because z^n is not analytic at $z = 0$ and $z = 0$ is enclosed by C, we *cannot* use the Cauchy–Goursat theorem. To find $\oint_C z^n \, dz$, we must evaluate the integral directly. Fortunately Eq. (4.2–6) is still valid if n is a negative integer. Moreover, Eq. (4.2–7) is still usable except, because of a vanishing denominator, when $n = -1$. Thus

$$\oint z^n \, dz = 0, \qquad n = -2, -3, \ldots . \quad (4.2-8)$$

Finally, to evaluate $\oint z^{-1} \, dz$ we employ Eq. (4.2–6) with $n = -1$ and obtain

$$\oint_{|z|=r} z^{-1} \, dz = i \int_0^{2\pi} e^0 \, d\theta = 2\pi i.$$

In summary, if n is any integer

$$\oint_{|z|=r} z^n \, dz = \begin{cases} 0, & n \neq -1, \\ 2\pi i, & n = -1. \end{cases} \quad (4.2-9)$$

An important generalization of this result is contained in Exercise 6 at the end of this section. With z_0 an arbitrary complex constant it is shown that

$$\oint_{|z-z_0|=r} (z - z_0)^n \, dz = \begin{cases} 0, & n \neq -1, \\ 2\pi i, & n = -1, \end{cases} \quad (4.2-10)$$

where the contour of integration is a circle centered at z_0.

Example 1

The Cauchy–Goursat theorem asserts that $\oint_C e^z \, dz$ equals zero when C is the triangular contour shown in Fig. 4.2–3. Verify this result by direct computation.

Solution

Refer to Exercise 5 of the previous section. Here the separate values of the integrals along paths I, II, and III in Fig. 4.2–3 are determined and found to be $e - 1$, $e[\cos(1) - 1] + ie\sin(1)$, and $1 - e\cos(1) - ie\sin(1)$, respectively. The sum of these results is zero. ◄

There are situations in which an extension of the Cauchy–Goursat theorem establishes that two contour integrals are equal without necessarily telling us the value of either integral. The extension is as follows:

Theorem 3 Deformation of contours

Consider two simple closed contours C_1 and C_2 such that all the points of C_2 lie interior to C_1. If a function $f(z)$ is analytic not only on C_1 and C_2 but at all points of the doubly connected domain D whose boundaries are C_1 and C_2, then

$$\oint_{C_1} f(z)\,dz = \oint_{C_2} f(z)\,dz.$$ ■

This theorem is easily proved. The contours C_1 and C_2 are displayed in solid line in Fig. 4.2–4(a). The domain D is shown shaded. We introduce, using broken lines, a pair of straight line cuts, which connect C_1 and C_2. By means of these cuts we have created a pair of simple closed contours, C_U and C_L, which are drawn, slightly separated, in Fig. 4.2–4(b). The integral of $f(z)$ is now taken around C_U and also around C_L. In each case the Cauchy–Goursat theorem is applicable since $f(z)$ is analytic on and interior to both C_U and C_L. Thus

$$\oint_{C_U} f(z)\,dz = 0 \quad \text{and} \quad \oint_{C_L} f(z)\,dz = 0.$$

Adding these results yields

$$\oint_{C_U} f(z)\,dz + \oint_{C_L} f(z)\,dz = 0. \tag{4.2–11}$$

Now refer to Fig. 4.2–4(b) and observe that the integral along the straight line segment from a to b on C_U is the negative of the integral on the line from b to a on

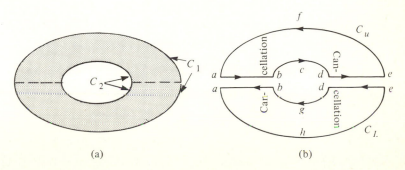

(a)   (b)

Figure 4.2–4

C_L. A similar statement applies to the integral from d to e on C_U and the integral from e to d on C_L.

If we write out the integrals in Eq. (4.2–11) in detail and combine those portions along the straight line segments that cancel, we are left only with integrations performed around C_1 and C_2 in Fig. 4.2–4(a). Thus

$$\oint_{C_2} f(z)\, dz + \oint_{C_1} f(z)\, dz = 0,$$

or

$$\oint_{C_1} f(z)\, dz = -\oint_{C_2} f(z)\, dz = \oint_{C_2} f(z)\, dz.$$

We have eliminated the minus sign in the middle expression and obtained the right-hand expression by reversing the direction of integration. The preceding equation is the desired result.

Another way of stating the theorem just proved is this:

The line integrals of an analytic function $f(z)$ around each of two simple closed contours will be identical in value if one contour can be continuously deformed into the other without passing through any singularity of $f(z)$.

In Fig. 4.2–4 we can regard C_2 as a deformed version of C_1 or vice versa. We call this approach *the principle of deformation of contours*.

Example 2

What is the value of $\oint_C dz/z$, where the contour C is the square shown in Fig. 4.2–5?

Solution

If the integration is performed instead around the broken circle, we obtain, using Eq. (4.2–9), the value $2\pi i$. Since $1/z$ is analytic on this circle, on the given square, and at all points lying between these contours, the principle of deformation of contours applies. Thus

$$\oint_C \frac{dz}{z} = 2\pi i. \qquad \blacktriangleleft$$

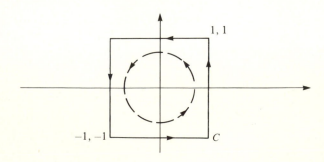

Figure 4.2–5

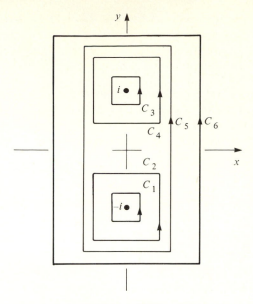

Figure 4.2-6

Example 3

Let $f(z) = \cos z/(z^2 + 1)$. The contours $C_1, C_2, \ldots, C_6$ are illustrated in Fig. 4.2–6. Explain why the following equations are valid.

a)
$$\oint_{C_1} f(z)\, dz = \oint_{C_2} f(z)\, dz$$

b)
$$\oint_{C_3} f(z)\, dz = \oint_{C_4} f(z)\, dz$$

c)
$$\oint_{C_5} f(z)\, dz = \oint_{C_6} f(z)\, dz$$

Solution

Except at points satisfying $z^2 + 1 = 0$, $f(z)$ is analytic. Hence, $f(z)$ is analytic in any domain not containing $z = \pm i$. If the contour C_1 is continuously deformed into the contour C_2, no singularity of $f(z)$ is crossed. Thus we establish equation (a). Similarly, C_3 can be deformed into C_4 to establish (b) and C_5 into C_6 to establish (c). Note that we *cannot* conclude that $\oint_{C_2} f(z)\, dz$ equals $\oint_{C_6} f(z)\, dz$ since the domain bounded by C_2 and C_6 contains the singular point of $f(z)$ at $z = i$. ◀

EXERCISES

1. Let C be an arbitrary simple closed contour. Use Green's theorem to find a simple interpretation of the line integral $(1/2) \oint_C (-y\, dx + x\, dy)$.

2. Let C be the unit square with corners at $(0,0)$, $(0, 1)$, $(1,0)$, $(1, 1)$ in the xy-plane.

   a) Do a line integration to obtain $\oint_C e^y\, dx + e^x\, dy$.

   b) Use Green's theorem to solve the problem in part (a) by means of an area integration. Do your answers agree?

3. In the discussion of Green's theorem in the appendix to this chapter it is shown that if $P(x, y)$ and $Q(x, y)$ are a pair of functions with continuous partial derivatives $\partial P/\partial y$ and $\partial Q/\partial x$ inside some simply connected domain D and if $\oint_C P\, dx + Q\, dy = 0$ for every simple closed contour in D, then $\partial Q/\partial x = \partial P/\partial y$ in D.

   Let $f(z) = u(x, y) + iv(x, y)$ be a function with a continuous derivative $f'(z)$ in a simply connected domain D. Given that $\oint f(z)\, dz = 0$ for every simple closed contour in D, use the preceding result to show that $f(z)$ must be analytic in D.

   This is a converse of the Cauchy–Goursat theorem. There is another derivation that eliminates the requirement that $f'(z)$ be continuous in D. The resulting converse of the Cauchy–Goursat theorem is known as Morera's theorem.

4. To which of the following integrals is the Cauchy–Goursat theorem directly applicable?

   a) $\oint_{|z|=1} \dfrac{\cos z}{z}\, dz$

   b) $\oint_{|z-i|=1} \dfrac{\cos z}{z}\, dz$

   c) $\oint_{|z-2i|=1} \dfrac{\cos z}{z}\, dz$

   d) $\oint_{|z-(1+i)|=1} \operatorname{Log} z\, dz$

   e) $\oint_C \operatorname{Log} z\, dz$, where C is the square having corners at $1 \pm i$ and $-1 \pm i$.

   f) $\oint_C (az^2 + bz + c)\, dz$, where C is the contour $x^2/a^2 + y^2/b^2 = 1$, and a and b are real.

   g) $\oint_C \dfrac{dz}{z^2 + z + 1}$, where C is the contour $4x^2 + 9y^2 = 1$.

   h) $\oint_{|z|=1} \dfrac{dz}{z^2 + z + 1}$

5. Prove the following:

   a) $\int_0^{2\pi} e^{\cos\theta} [\cos(\sin\theta + \theta)]\, d\theta = 0$,   b) $\int_0^{2\pi} e^{\cos\theta} [\sin(\sin\theta + \theta)]\, d\theta = 0$.
   Hint: Consider $\oint e^z\, dz$ performed around $|z| = 1$. Make a change of variables to polar coordinates and apply the Cauchy–Goursat theorem. Equate real parts on each side of the equation. Repeat with imaginary parts.

   Let $n \geq 0$ be an integer. Prove the following:

   c) $\int_0^{2\pi} e^{\sin n\theta} \cos(\theta - \cos n\theta)\, d\theta = 0$,   d) $\int_0^{2\pi} e^{\sin n\theta} \sin(\theta - \cos n\theta)\, d\theta = 0$.

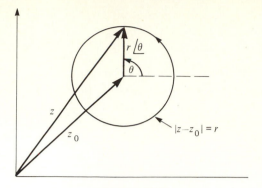

Figure 4.2–7

6. Let n be any integer, r a positive real number, and z_0 a complex constant. Show that

$$\oint_{|z-z_0|=r} (z - z_0)^n \, dz = \begin{cases} 0, & n \neq -1, \\ 2\pi i, & n = -1. \end{cases}$$

Hint: Refer to the derivation of Eq. (4.2–9) and follow a similar procedure. Consider the change of variable $z = z_0 + re^{i\theta}$ indicated in Fig. 4.2–7.

7. Consider the rectangular contour C formed from the four straight lines $x = 0$, $x = 2$, $y = -1$, $y = 1$. Find the value of the following integrals taken around C. Refer, if necessary, to the results of the previous problem.

a) $\displaystyle\oint \frac{dz}{(z - 1)}$

b) $\displaystyle\oint \frac{dz}{(z - 1)^2}$

c) $\displaystyle\oint (z - 1) \, dz$

d) $\displaystyle\oint (1 - z)^{-1} + 2(1 - z)^{-2} \, dz$

e) $\displaystyle\oint \frac{dz}{(z - 1 + 0.5i)^{50}}$

f) $\displaystyle\oint \frac{dz}{(z - 3)}$

8. Evaluate

$$\oint_{|z-1|=3} \left[(z - 2) + (z - 2)^{-1} \right]^3 \, dz.$$

9. Show that

$$\oint_{|z|=1/2} \frac{\cosh z}{(z)(z - 1)} \, dz = \oint_{|z|=1/2} -\frac{\cosh z}{z} \, dz.$$

Hint: Try writing $1/[(z)(z - 1)]$ as a sum of partial fractions.

10. Consider the n-tuply connected domain D whose nonoverlapping boundaries are the simple closed contours $C_0, C_1, \ldots, C_{n-1}$ as shown in Fig. 4.2–8.[†] Let $f(z)$ be a function that is analytic in D and on its boundaries. Show that

$$\oint_{C_0} f(z) \, dz = \oint_{C_1} f(z) \, dz + \oint_{C_2} f(z) \, dz + \cdots + \oint_{C_{n-1}} f(z) \, dz.$$

[†]An n-tuply connected domain has $n - 1$ holes. See Section 1.5.

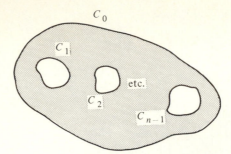

Figure 4.2–8

Hint: Consider the derivation of the principle of deformation of contours. Make a set of cuts similar to those made in Fig. 4.2–4 in order to link up the boundaries.

11. Use the result derived in Exercise 10 to show

$$\oint_{|z|=2} \frac{\sin z}{(z^2 - 1)}\, dz = \oint_{|z-1|=1/2} \frac{\sin z}{(z^2 - 1)}\, dz + \oint_{|z+1|=1/2} \frac{\sin z}{(z^2 - 1)}\, dz.$$

4.3 PATH INDEPENDENCE AND INDEFINITE INTEGRALS

The Cauchy–Goursat theorem is a useful tool when we must integrate an analytic function around a closed contour. However, there also exist some techniques that we can use when the given contour is not closed. We can, for example, prove the following theorem:

Theorem 4   Principle of path independence

Let $f(z)$ be a function that is analytic throughout a simply connected domain D, and let z_1 and z_2 lie in D. Then, provided we consider only those contours lying in D, the value of $\int_{z_1}^{z_2} f(z)\, dz$ will not depend on the particular contour used to connect z_1 and z_2. ∎

The preceding theorem is sometimes known as the *principle of path independence*. It is really just a restatement of the Cauchy–Goursat theorem. To establish this principle we will consider two nonintersecting contours C_1 and C_2, each of which connects z_1 and z_2. Each contour is assumed to lie within the simply connected domain D in which $f(z)$ is analytic. We will show that

$$\int_{z_1}^{z_2} f(z)\, dz = \int_{z_1}^{z_2} f(z)\, dz. \tag{4.3–1}$$

along C_1 \qquad along C_2

We begin by reversing the sense of integration along C_1 in Eq. (4.3–1) and placing a minus sign in front of the integral to compensate. Thus

$$-\int_{z_2}^{z_1} f(z)\, dz = \int_{z_1}^{z_2} f(z)\, dz,$$

along C_1 \qquad along C_2

or, with an obvious rearrangement,

$$0 = \int_{z_1}^{z_2} f(z)\, dz + \int_{z_2}^{z_1} f(z)\, dz.$$

along C_2 along C_1

The preceding merely states that the line integral of $f(z)$ taken around the closed loop formed by C_1 and C_2 (see Fig. 4.3–1) is zero. The correctness of this result follows directly from the Cauchy–Goursat theorem.

Although we have assumed that C_1 and C_2 in Eq. (4.3–1) do not intersect, a slightly different derivation dispenses with this restriction. Exercise 1 in this section treats the case of a single intersection.

Since *any* two piecewise smooth curves (in domain D) that connect z_1 and z_2 can be used in deriving Eq. (4.3–1), it follows that all such paths lying in D must yield the same result. Thus the value of $\int_{z_1}^{z_2} f(z)\, dz$ is independent of the path connecting z_1 and z_2 as long as that path lies within a simply connected domain in which $f(z)$ is analytic.

Example 1

Compute $\int_0^2 e^z\, dz$, where the integration is along the arc C_1 shown in Fig. 4.3–2.

Solution

Since e^z is an entire function, we can use any other contour that connects $z = 0 + i0$ with $z = 2 + i0$. An obvious choice is the straight line C_2. Along C_2, we have $z = x$, $dz = dx$. Thus

$$\int_{C_1} e^z\, dz = \int_{C_2} e^z\, dz = \int_0^2 e^x\, dx = e^2 - 1. \qquad \blacktriangleleft$$

In elementary calculus integration is generally performed by our recognizing that the integrand $f(x)$ is the derivative of some particular function $F(x)$. Then $F(x)$ is evaluated at the limits of integration. Thus

$$\int_{x_1}^{x_2} f(x)\, dx = \int_{x_1}^{x_2} \left(\frac{dF}{dx}\right) dx = \int_{x_1}^{x_2} dF = F(x_2) - F(x_1).$$

To cite a specific example:

$$\int_1^4 x^2\, dx = \int_1^4 \frac{d}{dx}\left(\frac{x^3}{3}\right) dx = \int_1^4 d\left(\frac{x^3}{3}\right) = \left.\frac{x^3}{3}\right|_1^4 = \frac{63}{3}.$$

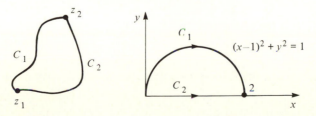

Figure 4.3–1   **Figure 4.3–2**

Does a similar procedure work for complex line integrals? We shall see that with certain restrictions it does.

Let $F(z)$ be analytic in a domain D. Assume $dF/dz = f(z)$ in D. Consider $\int_{z_1}^{z_2} f(z)\,dz$ integrated along a smooth arc C lying in D and connecting z_1 with z_2. We will assume that C has a parametric representation of the form $z(t) = x(t) + iy(t)$, where $t_1 \leq t \leq t_2$. Observe that $z(t_1) = z_1$, that $z(t_2) = z_2$, and that dz/dt exists in the interval $t_1 \leq t \leq t_2$. Now from Eq. (4.1–18) we have

$$\int_{z_1}^{z_2} f(z)\,dz = \int_{t_1}^{t_2} f(z(t))\frac{dz}{dt}\,dt.$$

We can replace $f(z(t))$ on the right by dF/dz. Hence,

$$\int_{z_1}^{z_2} f(z)\,dz = \int_{t_1}^{t_2}\frac{dF}{dz}\frac{dz}{dt}\,dt.$$

The expression on the above right, $(dF/dz)(dz/dt)$ is, from the chain rule of differentiation, merely dF/dt. Thus

$$\int_{z_1}^{z_2} f(z)\,dz = \int_{t_1}^{t_2}\frac{dF}{dt}\,dt = \int_{t_1}^{t_2}dF = F(z(t_2)) - F(z(t_1)) = F(z_2) - F(z_1).$$
$$(4.3–2)$$

If the contour C is not restricted to being a smooth arc but is permitted to be a piecewise smooth curve, the result

$$\int_{z_1}^{z_2} f(z)\,dz = F(z_2) - F(z_1), \qquad (4.3–3)$$

which we obtained in Eq. (4.3–2) is still valid. However, a slightly more elaborate proof, not presented here, is required since dz/dt may fail to exist at those points along C where smooth arcs are joined together.

The preceding discussion is summarized in the following theorem.

Theorem 5   Integration of analytic functions

Let $F(z)$ be analytic in a domain D. Let $dF/dz = f(z)$ in D. Then, if z_1 and z_2 are in D,

$$\int_{z_1}^{z_2} f(z)\,dz = F(z_2) - F(z_1), \qquad (4.3–4)$$

where the integration can be performed along any contour in D that connects z_1 and z_2. ∎

Thus within the constraints of the theorem the conventional rules of integration apply, and ordinary tables of integrals (which are based on such rules) can be employed. For example, the theorem justifies the following evaluation.

$$\int_{1+i}^{2+2i} z^2\,dz = \int_{1+i}^{2+2i}\frac{d}{dz}\frac{z^3}{3}\,dz = \left.\frac{z^3}{3}\right|_{1+i}^{2+2i} = \frac{1}{3}\left[(2 + 2i)^3 - (1 + i)^3\right]$$

Since $z^3/3$ is an entire function, the path of integration was not, and need not, be specified.

Example 2

Evaluate $\int_{-i}^{+i} 1/z\, dz$ along the contour C shown in Fig. 4.3–3.

Solution

Recall from Chapter 3 that $(d/dz)\log z = 1/z$. The logarithm is a multivalued function. In order to specify a particular analytic branch of the log, let us call it $F(z)$, we must employ a branch cut. Any branch of the logarithm whose branch cut does not intersect C can be used to perform the given integration. A possible cut is shown by the broken line in the figure. Since the branch cut contains all the singular points of $F(z)$, the contour C lies in a domain of analyticity of $F(z)$.

Using our analytic branch of $\log z$, we have

$$\int_{-i}^{+i} \frac{1}{z}\, dz = \log z \Big|_{-i}^{+i} = \mathrm{Log}|z| + i\arg z \Big|_{-i}^{+i}.$$

Note that $\mathrm{Log}|i| = \mathrm{Log}|-i| = 0$. Thus $\int_{-i}^{+i} 1/z\, dz = i(\arg i - \arg(-i))$. Along contour C the argument of z varies continuously. At $-i$ the argument of z is $-\pi/2 + 2k\pi$, where k is an integer, while at $+i$ the argument becomes $\pi/2 + 2k\pi$, where k must have the *same value* in both cases. Hence,

$$\int_{-i}^{+i} \frac{1}{z}\, dz = i\left(\frac{\pi}{2} + 2k\pi\right) - i\left(-\frac{\pi}{2} + 2k\pi\right) = \pi i. \qquad \blacktriangleleft$$

EXERCISES

1. In Fig. 4.3–4 contour C_1 (solid line) and contour C_2 (broken line) each connect points z_1 and z_2. The contours also intersect at one other point, designated z_3. Let $f(z)$ be analytic in a simply connected domain containing C_1 and C_2. Show that

$$\int_{z_1}^{z_2} f(z)\, dz = \int_{z_1}^{z_2} f(z)\, dz.$$

$$\text{along } C_1 \qquad\qquad \text{along } C_2$$

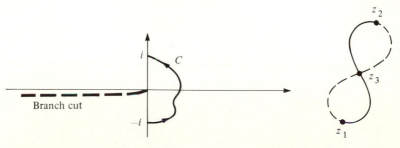

Figure 4.3–3　　　　　　　　　　　　　　Figure 4.3–4

2. Evaluate the following integrals:

a) $\int_{0+i}^{1+ie} \sin z \, dz$, along the contour $y = e^x$;

b) $\int_{0+i}^{1+ie} e^z \cos z \, dz$, along the contour of part (a);

c) $\int_{0+i0}^{3\pi/2-i} (z^2 + z + 1) \, dz$, along the contour $y = \sin x$.

3. a) What, if anything, is incorrect about the following two integrations? The integrals are both along the line $y = x$.

$$\int_{0+i0}^{1+i1} z \, dz = \frac{z^2}{2} \Big|_{0+i0}^{1+i} = \frac{(1+i)^2}{2} = i$$

$$\int_{0+i0}^{1+i} \bar{z} \, dz = \frac{(\bar{z})^2}{2} \Big|_{0+i0}^{1+i} = \frac{(1-i)^2}{2} = -i$$

b) What is the correct numerical value of each of the above integrals?

4. a) Find the value of $\int_{1+i0}^{0+i1} \text{Log } z \, dz$ taken along the contour $x + y = 1$. Why is it necessary to specify the contour?

b) What would the answer be for part (a) if we instead used the contour $y = \cos(\pi x/2)$? The limits of integration are kept the same.

5. Consider contours C_1 and C_2 shown in Fig. 4.3–5. We can use the result derived in Example 2 of this section to show that along C_1, $\int_{-i}^{+i} 1/z \, dz = \pi i$. Can we use the principle of path independence to show that along C_2, $\int_{-i}^{+i} 1/z \, dz = \pi i$? If not, what is the correct value of this integral?

6. a) Find $\int_0^{2i} dz/(z - i)$ taken along the arc satisfying $|z - i| = 1$, Re $z \geq 0$.

b) Repeat part (b) with the same limits, but use the arc $|z - i| = 1$, Re $z \leq 0$.

7. Let z_1 and z_2 be a pair of arbitrary points in the complex plane. Contours C_1 and C_2 each connect points z_1 and z_2. The contours do not otherwise intersect, and neither passes through $z = 0$. Explain why

$$\int_{z_1}^{z_2} \frac{1}{z^2} dz = \int_{z_1}^{z_2} \frac{1}{z^2} dz.$$

$$\text{along } C_1 \qquad \text{along } C_2$$

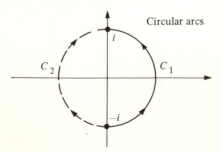

Figure 4.3–5

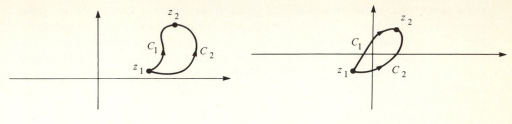

Figure 4.3–6   Figure 4.3–7

Consider two cases:

a) $z = 0$ does not belong to the domain whose boundaries are C_1 and C_2 (see Fig. 4.3–6).

b) $z = 0$ does belong to the domain whose boundaries are C_1 and C_2 (see Fig. 4.3–7).

8. Find $\int_{\sqrt{3}/2+i/2}^{1+i0}(z)^{1/2}\,dz$ taken along an arbitrary contour lying in the first quadrant. Let $(z)^{1/2}$ be defined by a branch cut along the line $\operatorname{Im} z = 0$, $\operatorname{Re} z \le 0$. Take $(z)^{1/2} = 1$ when $z = 1$.

4.4   THE CAUCHY INTEGRAL FORMULA AND ITS CONSEQUENCES

Consider a simple closed contour C (see, for example, Fig. 4.4–1). Starting with the Cauchy integral theorem, we will derive a remarkable fact. Suppose $f(z)$ is a function that is analytic on C and at all points belonging to the interior of C. Then, if we know the values of $f(z)$ only on C, we have sufficient information to find $f(z)$ at any point z_0 in the interior of C. The means for finding $f(z_0)$ is the Cauchy integral formula, which we will now obtain. The impatient reader may wish to skip directly to the simple and nonrigorous formal derivation outlined in Exercise 1 at the end of this section.

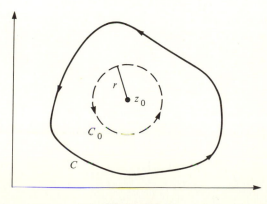

Figure 4.4–1

To derive the result rigorously we proceed as follows: Let C_0 be a circle, centered at z_0, of radius r. The value of r is sufficiently small so that C_0 lies entirely within C. The function $f(z)/(z - z_0)$ is analytic at all points for which $f(z)$ is analytic except $z = z_0$. Thus $f(z)/(z - z_0)$ is analytic on C_0, C, and at all points lying outside C_0 but inside C. Using the principle of deformation of contours, we can assert that

$$\oint_C \frac{f(z)}{z - z_0}\, dz = \oint_{C_0} \frac{f(z)}{z - z_0}\, dz. \tag{4.4-1}$$

Let us show that the expression on the right equals $2\pi i f(z_0)$. Recall from Exercise 6, Section 4.2 that $\oint_{C_0} dz/(z - z_0) = 2\pi i$. Thus

$$2\pi i f(z_0) = \oint_{C_0} \frac{dz}{(z - z_0)} f(z_0) = \oint_{C_0} \frac{f(z_0)}{z - z_0}\, dz. \tag{4.4-2}$$

Since the quantity $f(z_0)$ is a constant, it was taken under the integral sign on the right in Eq. (4.4–2). If we subtract both the left side and the right side of Eq. (4.4–2) from the right-hand integral in Eq. (4.4–1), we obtain

$$\oint_{C_0} \frac{f(z)}{z - z_0}\, dz - 2\pi i f(z_0) = \oint_{C_0} \frac{f(z) - f(z_0)}{z - z_0}\, dz. \tag{4.4-3}$$

Our goal is to show that the integral on the right is zero. The value of this integral, although still unknown, must, by the principle of deformation of contours, be independent of r, the radius of C_0.

The ML inequality shown in Eq. (4.1–24) can be applied to obtain a bound on the magnitude of the right-hand side of Eq. (4.4–3). The length of path, L, is here merely the circumference of the circle C_0, that is, $2\pi r$. The quantity M must have the property that

$$\frac{|f(z) - f(z_0)|}{|z - z_0|} \le M, \quad \text{for } z \text{ on } C_0. \tag{4.4-4}$$

On the contour of integration we have $|z - z_0| = r$.

Since $f(z)$ is continuous at z_0, we can apply the definition of continuity (see Section 2.2) and assert that, given a positive number ε, there exists a positive number δ such that $|f(z) - f(z_0)| < \varepsilon$ for $|z - z_0| < \delta$. If the radius r of C_0 is chosen to be less than δ, it follows that on C_0 we have $|f(z) - f(z_0)| < \varepsilon$. Hence, we have on C_0 that

$$\left| \frac{f(z) - f(z_0)}{z - z_0} \right| = \left| \frac{f(z) - f(z_0)}{r} \right| < \frac{\varepsilon}{r},$$

and we can take M in Eq. (4.4–4) as ε/r.

Now, knowing M and L, we apply the ML inequality to the right side of Eq. (4.4–3) and obtain

$$\left| \oint_{C_0} \frac{f(z) - f(z_0)}{r}\, dz \right| \le \frac{\varepsilon}{r} 2\pi r = 2\pi\varepsilon. \tag{4.4-5}$$

Since ε on the right in Eq. (4.4–5) can be made arbitrarily small, the absolute value of the integral on the left can likewise be made arbitrarily small. Reducing ε merely implies that we must shrink the radius r of C_0.

We observed earlier that the value of the integral within the absolute magnitude signs in Eq. (4.4–5) must be independent of r. Since the absolute value of this integral can be made as small as we please, we conclude that the actual value of the integral is zero.

Because we have shown that the right side of Eq. (4.4–3) is zero, we can rearrange the left side to yield

$$2\pi i f(z_0) = \oint_{C_0} \frac{f(z)}{(z - z_0)}\, dz. \tag{4.4-6}$$

Now Eq. (4.4–6) shows that the right side of Eq. (4.4–1) is $2\pi i f(z_0)$. Dividing both sides of Eq. (4.4–1) by $2\pi i$, we obtain the Cauchy integral formula.

Theorem 6   Cauchy integral formula

Let $f(z)$ be analytic on and in the interior of a simple closed contour C. Let z_0 be a point in the interior of C. Then

$$f(z_0) = \frac{1}{2\pi i} \oint_C \frac{f(z)\, dz}{(z - z_0)}. \quad \blacksquare \tag{4.4-7}$$

Example 1

    a) Find $\oint (\cos z)/(z - 1)\, dz$, where C is the triangular contour shown in Fig. 4.4–2

    b) Find $\oint_C (\cos z)/(z + 1)\, dz$, where C is the same as in part (a).

Solution

Part (a): Since $\cos z$ is an entire function and the point $z = 1$ lies within C, we can apply Eq. (4.4–7). Thus

$$\frac{1}{2\pi i} \oint_C \frac{\cos z}{z - 1}\, dz = \cos 1, \quad \text{or} \quad \oint_C \frac{\cos z}{(z - 1)}\, dz = 2\pi i \cos 1 \doteq 3.39i.$$

Part (b): The integrand can be written as $(\cos z)/(z - (-1))$. Employing Eq. (4.4–7), we find $z_0 = -1$ lies outside the contour of integration. The Cauchy

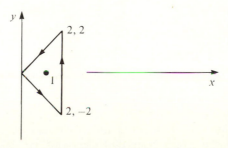

Figure 4.4–2

integral formula *does not* apply here. However, because $(\cos z)/(z + 1)$ is analytic both on C and at all points in the interior of C, the Cauchy–Goursat theorem does apply. Hence, the value of the given integral is zero. ◀

Example 2

Find $(1/2\pi i)\oint_C(\cos z)/(z^2 + 1)\, dz$, where C is the circle $|z - 2i| = 2$.

Solution

It is not immediately apparent whether the Cauchy integral formula or the Cauchy–Goursat theorem is applicable here. Factoring the denominator, we have

$$\frac{1}{2\pi i} \oint \frac{\cos z}{(z - i)(z + i)}\, dz.$$

Notice that the factor $(z - i)$ goes to zero within the contour of integration (at $z = i$), and $z + i$ remains nonzero both on and inside the contour (see Fig. 4.4–3). Writing the given integral as

$$\frac{1}{2\pi i} \oint \frac{\dfrac{\cos z}{z + i}}{(z - i)}\, dz,$$

we see that because $\cos z/(z + i)$ is analytic both on and inside $|z - 2i| = 2$, the Cauchy integral formula is applicable. In Eq. (4.4–7) we take $f(z) = \cos z/(z + i)$ and $z_0 = i$. Hence, the value of the given integral is

$$\left(\frac{\cos z}{z + i}\right)_{z=i} = \frac{\cos i}{2i} = \frac{-i}{2} \cosh 1. \quad ◀$$

An integration such as

$$\oint_{|z|=2} \frac{\cos z}{(z^2 + 1)}\, dz = \oint_{|z|=2} \frac{\cos z}{(z - i)(z + i)}\, dz,$$

where $z^2 + 1$ goes to zero *at two* points inside the contour of integration cannot be directly evaluated by means of the Cauchy integral formula. However, the formula is readily extended (see Exercise 7 in this section) to problems of this type.

One of the interesting consequences of the Cauchy integral formula is stated below.

Theorem 7   Gauss's mean value theorem

Let $f(z)$ be analytic in a simply connected domain. Consider any circle lying in this domain. The value assumed by $f(z)$ at the center of the circle equals the average of the values assumed by $f(z)$ on its circumference. ∎

To establish this fact let the circle be centered at z_0 and have radius r. A glance at Fig. 4.4–4 shows that any point z on the circle can be expressed in the form $z = z_0 + re^{i\theta}$, where $0 \le \theta \le 2\pi$. Note that $dz = re^{i\theta}i\, d\theta$. With these substitutions

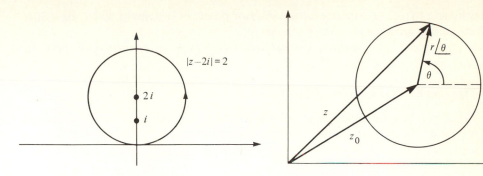

Figure 4.4–3   **Figure 4.4–4**

for z and dz made in Eq. (4.4–7), we obtain

$$f(z_0) = \frac{1}{2\pi i} \int_0^{2\pi} \frac{f(z_0 + re^{i\theta})}{re^{i\theta}} ire^{i\theta} \, d\theta.$$

With some obvious cancellations this becomes

$$f(z_0) = \frac{1}{2\pi} \int_0^{2\pi} f(z_0 + re^{i\theta}) \, d\theta. \tag{4.4–8}$$

The expression on the right in Eq. (4.4–8) is the arithmetic mean (average value) of $f(z)$ on the circumference of the circle. Equation (4.4–8) is the mathematical statement of Gauss's mean value theorem.

If the function $f(z)$ is expressed in terms of its real and imaginary parts, $f(z) = u(x, y) + iv(x, y)$, we can recast Eq. (4.4–8) as follows:

$$u(z_0) + iv(z_0) = \frac{1}{2\pi} \int_0^{2\pi} u(z_0 + re^{i\theta}) \, d\theta + \frac{i}{2\pi} \int_0^{2\pi} v(z_0 + re^{i\theta}) \, d\theta. \tag{4.4–9}$$

Taking $z_0 = x_0 + iy_0$ and equating corresponding parts of each side of Eq. (4.4–9), we obtain

$$u(x_0, y_0) = \frac{1}{2\pi} \int_0^{2\pi} u(z_0 + re^{i\theta}) \, d\theta, \tag{4.4–10a}$$

$$v(x_0, y_0) = \frac{1}{2\pi} \int_0^{2\pi} v(z_0 + re^{i\theta}) \, d\theta. \tag{4.4–10b}$$

We see from Eq. (4.4–10a) that the real part u of the analytic function evaluated at the center of the circle is equal to u averaged over the circumference of the circle. A similar statement contained in Eq. (4.4–10b) applies to the imaginary part v.

If the value of u is known at certain discrete points along the circumference of a circle, these values can be used to determine, approximately, the value of the integral on the right in Eq. (4.4–10a). In this way an approximation can be obtained for the value of u at the center of the circle (see Exercise 11 in this section). Typically, four uniformly spaced points on the circumference might be used. This technique forms the basis of a numerical procedure called the *finite difference method*, which is used to evaluate a harmonic function at the interior points of a domain when the values of the function are known on the boundaries.[†]

Gauss's mean value theorem can be used to establish an important property of analytic functions:

Theorem 8   Maximum modulus theorem

Let a nonconstant function $f(z)$ be continuous throughout a closed bounded region R. Let $f(z)$ be analytic at every interior point of R. Then the maximum value of $|f(z)|$ in R must occur at a point on the boundary of R.   ∎

Loosely stated, the theorem asserts that the maximum value of the modulus of $f(z)$ occurs on the boundary of a region.

To prove the theorem we must employ without proof the following property of analytic functions: If an analytic function fails to assume a constant value over all the interior points of a region, then it is not constant throughout any neighborhood of any interior point of that region.[‡] From Exercise 7 Section 2.4 we see, in addition, that $|f(z)|$ will not be constant in any such neighborhood.

Returning to the maximum modulus theorem, let us assume that the maximum value of $|f(z)|$ in R occurs at z_0, an interior point of R. At z_0 we have $|f(z_0)| = m$ (the maximum value). We assume R to be the set of points on and inside the contour C of Fig. 4.4–5.

Consider a circle C_0 of radius r centered at z_0 and lying entirely within C. Since $|f(z)|$ is not constant in any neighborhood of z_0, we can choose r so that C_0 passes through at least one point where $|f(z)| < m$. If we describe C_0 by the equation $z = z_0 + re^{i\theta}$, $0 \le \theta \le 2\pi$, we have at this point $|f(z_0 + re^{i\theta})| < m$.

Because $f(z)$ is a continuous function, there must be a finite segment of arc on C_0 along which $|f(z_0 + re^{i\theta})| \le m - b$. Here b is a positive constant such that $b < m$. For simplicity let the arc in question extend from $\theta = 0$ to $\theta = \beta$. Along the remainder of the arc, $\beta \le \theta \le 2\pi$, we have $|f(z_0 + re^{i\theta})| \le m$ since $|f(z_0)| = m$ is the maximum value of $f(z)$.

Now we refer to Eq. (4.4–8) and write the integral around C_0 in two parts:

$$f(z_0) = \frac{1}{2\pi} \int_0^\beta f(z_0 + re^{i\theta})\, d\theta + \frac{1}{2\pi} \int_\beta^{2\pi} f(z_0 + re^{i\theta})\, d\theta.$$

Let us take the absolute magnitude of both sides of this equation and also apply a triangle inequality. We then have

$$|f(z_0)| \le \frac{1}{2\pi} \left| \int_0^\beta f(z_0 + re^{i\theta})\, d\theta \right| + \frac{1}{2\pi} \left| \int_\beta^{2\pi} f(z_0 + re^{i\theta})\, d\theta \right|.$$

We can apply the ML inequality to each of these integrals. For the first integral on

---

[†] For an application to electrostatics see W. H. Hayt, *Engineering Electromagnetics* (New York: McGraw-Hill, 1967), ch. 6.

[‡] See, for example, R. V. Churchill, J. Brown, R. Verhey, *Complex Variables and Applications* (New York: McGraw-Hill, 1974), pp. 135, 284.

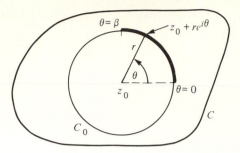

Figure 4.4–5

the right we know that $|f(z_0 + re^{i\theta})| \leq m - b$, and for the second we can assert that $|f(z_0 + re^{i\theta})| \leq m$. The quantity L in each case is just the interval of integration: β and $2\pi - \beta$, respectively. Thus

$$|f(z_0)| \leq \frac{1}{2\pi}(m - b)\beta + \frac{m}{2\pi}(2\pi - \beta).$$

Adding the terms on the right side of this equation we obtain

$$|f(z_0)| \leq m - \frac{b\beta}{2\pi}.$$

The quantity on the left, $|f(z_0)|$, is m, the maximum value of $|f(z)|$. But m cannot be less than $m - b\beta/2\pi$. We have obtained a contradiction. Our assumption that z_0 is an interior point of R must be false. Since $|f(z)|$ must be maximum *somewhere* in R (see Theorem 2, Chapter 2) this maximum must be at a boundary point. For the region R of Fig. 4.4–5 such a point would be on the boundary C.

There is a similar theorem, which is proved in Exercise 13 of this section, pertaining to the minimum value of $|f(z)|$ achieved in R:

Theorem 9 Minimum modulus theorem

Let a nonconstant function $f(z)$ be continuous and nowhere zero throughout a closed bounded region R. Let $f(z)$ be analytic at every interior point of R. Then the minimum value of $|f(z)|$ in R must occur on the boundary of R. ■

Note the additional requirement $f(z) \neq 0$.[†]

We will see in Exercises 14, 15, and 16 of this section that the maximum and minimum modulus theorems can tell us some useful properties about the behavior of harmonic functions in bounded regions. These properties have direct physical

---

[†] There are other versions of the maximum and minimum modulus theorems. They are called the "local" versions and are stated as follows: a) Let $f(z)$ be analytic and not constant in a neighborhood N of z_0. Then there are points in N lying arbitrarily close to z_0 where $|f(z)| > |f(z_0)|$, that is, $|f(z)|$ cannot have a local maximum at z_0. b) If, in addition, $f(z_0) \neq 0$, there are points in N lying arbitrarily close to z_0, where $|f(z)| < |f(z_0)|$, that is, $|f(z)|$ cannot have a local minimum at z_0.

application to problems involving heat conduction (see, for example, Exercise 17) and electrostatics.

Example 3

Consider $f(z) = e^z$ in the region $|z| \leq 1$. Find the points in this region where $|f(z)|$ achieves its maximum and minimum values.

Solution

Because e^z is an entire function and e^z is never 0 in the given region, both the maximum and minimum modulus theorems should be confirmed by our result. We have $|f(z)| = |e^z| = |e^{x+iy}| = |e^x||e^{iy}| = |e^x| = e^x$. Because e^x is nonnegative, we were able to drop the absolute magnitude signs. Now $|f(z)|$ is maximum at the point in the region where e^x achieves its largest value, that is, at $x = 1$, $y = 0$, and $|f(z)|$ is minimum where e^x is smallest, that is, at $x = -1$, $y = 0$. Both points are on the boundary of R (see Fig. 4.4–6). ◄

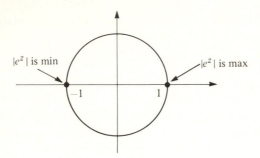

Figure 4.4–6

As we know, the Cauchy integral formula yields the value of an analytic function at a point when we know the values assumed by that function on a surrounding simple closed contour. This formula can be extended. With the "extended formula" the derivatives of any order of the function at this same point are obtainable provided we again know the function everywhere along a surrounding curve.

The extended formula is obtainable by the following series of manipulations, which do not constitute a proof. With $f(z)$ analytic on and interior to a simple closed contour C and with z_0 a point inside C, we have, from Eq. (4.4–7),

$$f(z_0) = \frac{1}{2\pi i} \oint_C \frac{f(z)}{z - z_0} dz. \tag{4.4–11}$$

We now regard $f(z)$ as a function of the variable z_0, and we differentiate both sides of Eq. (4.4–11) with respect to z_0. We will assume that it is permissible to take the

d/dz_0 operator under the integral sign. Thus

$$\frac{d}{dz_0} f(z_0) = \frac{d}{dz_0} \frac{1}{2\pi i} \oint_C \frac{f(z)}{z - z_0} dz = \frac{1}{2\pi i} \oint_C f(z) \frac{d}{dz_0} \left(\frac{1}{z - z_0} \right) dz$$

$$= \frac{1}{2\pi i} \oint_C \frac{f(z)}{(z - z_0)^2} dz. \tag{4.4-12}$$

In effect, we have assumed that Leibnitz's rule[†] applies not only to real integrals but also to contour integrals. When the second and nth derivatives are found in this way, we have

$$f^{(2)}(z_0) = \frac{1}{2\pi i} \oint_C f(z) \frac{d^2}{dz_0^2} \left(\frac{1}{z - z_0} \right) dz = \frac{2}{2\pi i} \oint_C \frac{f(z)}{(z - z_0)^3} dz, \tag{4.4-13}$$

$$f^{(n)}(z_0) = \frac{1}{2\pi i} \oint_C f(z) \frac{d^n}{dz_0^n} \left(\frac{1}{z - z_0} \right) dz = \frac{n!}{2\pi i} \oint_C \frac{f(z)}{(z - z_0)^{n+1}} dz. \tag{4.4-14}$$

The formulas obtained in Eqs. (4.4–12) through (4.4–14) can, in fact, be rigorously justified. To properly obtain $f'(z_0)$ we use the definition of the first derivative and the Cauchy integral formula Eq. (4.4–11):

$$f'(z_0) = \lim_{\Delta z_0 \to 0} \frac{f(z_0 + \Delta z_0) - f(z_0)}{\Delta z_0}$$

$$= \lim_{\Delta z_0 \to 0} \frac{1}{\Delta z_0} \left[\frac{1}{2\pi i} \oint_C \frac{f(z)\, dz}{z - (z_0 + \Delta z_0)} - \frac{1}{2\pi i} \oint_C \frac{f(z)}{(z - z_0)} dz \right]$$

$$= \lim_{\Delta z_0 \to 0} \frac{1}{2\pi i\, \Delta z_0} \oint_C \left[\frac{1}{z - (z_0 + \Delta z_0)} - \frac{1}{z - z_0} \right] f(z)\, dz$$

$$= \lim_{\Delta z_0 \to 0} \frac{1}{2\pi i} \oint_C \frac{f(z)\, dz}{[z - (z_0 + \Delta z_0)](z - z_0)}. \tag{4.4-15}$$

If we could interchange the order of the $\lim_{\Delta z_0 \to 0}$ operation and the integration in the last term in Eq. (4.4–15), we would obtain the expression for $f'(z_0)$ presented in Eq. (4.4–12). However, we have no obvious means of justifying this step. Instead, what can be done is to show that the absolute value of the difference between

$$\frac{1}{2\pi i} \oint_C \frac{f(z)\, dz}{[z - (z_0 + \Delta z_0)](z - z_0)} \quad \text{and} \quad \frac{1}{2\pi i} \oint_C \frac{f(z)\, dz}{(z - z_0)^2}$$

goes to zero as $\Delta z_0 \to 0$. This would establish the validity of Eq. (4.4–12) for $f'(z_0)$. The procedure is quite straightforward and involves the ML inequality in a manner similar to that used in deriving the Cauchy integral formula. The reader can find the details outlined in Exercise 4 at the end of this section.

[†]See W. Kaplan, *Advanced Calculus* (Reading, Mass.: Addison-Wesley, 1957), p. 219.

Once the validity of $f'(z_0) = (1/2\pi i) \oint_C f(z)/(z - z_0)^2 \, dz$ is established, one can, by a similar, rigorous procedure justify the formula for $f^{(2)}(z_0)$ given in Eq. (4.4–13). Since the derivative of $f'(z_0)$ with respect to z_0 not only exists but exists in any domain in which $f(z_0)$ is an analytic function, we can assert that $f'(z_0)$ is itself an analytic function of z_0. The preceding procedure can be carried out any number of times so as to yield any derivative of $f(z_0)$. A formula for the nth derivative is obtained and is given in Eq. (4.4–14). Summarizing these results, we have the following theorem.

Theorem 10   Extension of Cauchy integral formula

If a function is analytic within a domain, then it possesses derivatives of all orders in that domain. These derivatives are themselves analytic functions in the domain. If $f(z)$ is analytic on and in the interior of a simple closed contour C and if z_0 is inside C, then

$$f^{(n)}(z_0) = \frac{n!}{2\pi i} \oint_C \frac{f(z)}{(z - z_0)^{n+1}} \, dz. \quad \blacksquare \qquad (4.4\text{–}16)$$

Note that if we interpret $f^{(0)}(z_0)$ as $f(z_0)$, and $0! = 1$, then Eq. (4.4–16) contains the Cauchy integral formula for $f(z_0)$.

If $f(z)$ is expressed in the form $u(x, y) + iv(x, y)$, then the various derivatives of $f(z)$ can be written in terms of the partial derivatives of u and v (see Section 2.3 and Exercise 3 at the end of that section). For example,

$$f'(z) = \frac{\partial u}{\partial x} + i\frac{\partial v}{\partial x} = \frac{\partial v}{\partial y} - i\frac{\partial u}{\partial y}, \qquad (4.4\text{–}17)$$

$$f''(z) = \frac{\partial^2 u}{\partial x^2} + i\frac{\partial^2 v}{\partial x^2} = -\frac{\partial^2 u}{\partial y^2} - i\frac{\partial^2 v}{\partial y^2}. \qquad (4.4\text{–}18)$$

The extension of the Cauchy integral formula tells us that if $f(z)$ is analytic it possesses derivatives of all orders. Since these derivatives are defined by Eqs. (4.4–17) and (4.4–18) as well as similar equations involving higher-order partial derivatives, we see that the partial derivatives of u and v of all orders must exist. Since a harmonic function can be regarded as the real (or imaginary) part of an analytic function we can assert Theorem 11.

Theorem 11

A function that is harmonic in a domain will possess partial derivatives of *all* orders in that domain. $\blacksquare$

Example 4

Determine the value of $\oint_C [(z^3 + 2z + 1)/(z - 1)^3] \, dz$, where C is the contour $|z| = 2$.

Solution

Considering the form of the denominator in the integrand, we will use Eq. (4.4–16) with $n = 2$. We then have

$$f^{(2)}(z_0) = \frac{2!}{2\pi i} \oint_C \frac{f(z)}{(z - z_0)^3} \, dz.$$

With $z_0 = 1$ and a simple multiplication the preceding equation becomes

$$\frac{2\pi i}{2} f^{(2)}(1) = \oint_C \frac{f(z)}{(z - 1)^3} \, dz.$$

This formula, with $f(z) = z^3 + 2z + 1$, yields the value of the given integral. Thus

$$\oint_C \frac{z^3 + 2z + 1}{(z - 1)^3} \, dz = \pi i \frac{d^2}{dz^2} (z^3 + 2z + 1)_{z=1} = \pi i (6z)_{z=1} = 6\pi i.$$

Notice that if the contour C were $|z| = 1/2$, we would apply the Cauchy integral theorem. This is because $(z^3 + 2z + 1)/(z - 1)^3$ is analytic on and inside this circle. ◄

Example 5

Find $\oint_C [\cos z / ((z - 1)^3 (z - 5)^2)] \, dz$, where C is the circle $|z - 4| = 2$.

Solution

Let us examine the two factors in the denominator. The term $(z - 1)^3$ is nonzero both inside and on the contour of integration. However, $(z - 5)^2$ does become zero at the point $z = 5$ inside C. We therefore rewrite the integral as

$$\oint_C \frac{\left(\dfrac{\cos z}{(z - 1)^3} \right)}{(z - 5)^2} \, dz$$

and apply Eq. (4.4–16) with $n = 1$, $z_0 = 5$, $f(z) = \cos z / (z - 1)^3$. Thus

$$\frac{1}{2\pi i} \oint_C \frac{\left(\dfrac{\cos z}{(z - 1)^3} \right)}{(z - 5)^2} \, dz = \frac{d}{dz} \frac{\cos z}{(z - 1)^3} \bigg|_{z=5} = \frac{-64 \sin 5 - 48 \cos 5}{4^6}.$$

The value of the *given* integral is $2\pi i$ times the preceding result. ◄

An interesting theorem is easily proved from the extension of the Cauchy integral formula.

Theorem 12   Liouville's theorem

An entire function whose absolute value is bounded (that is, does not exceed some constant) throughout the z-plane is a constant. ∎

To prove this theorem consider a circle C, of radius r, centered at z_0. Since $f(z)$ is everywhere analytic, we can use Eq. (4.4–16) with $n = 1$ to integrate $f(z)/(z - z_0)^2$ around C. Thus

$$f'(z_0) = \frac{1}{2\pi i} \oint_C \frac{f(z)}{(z - z_0)^2} dz.$$

Taking magnitudes we have

$$|f'(z_0)| = \frac{1}{2\pi} \left| \oint_C \frac{f(z)}{(z - z_0)^2} dz \right|.$$

We now apply the ML inequality to the integral and take L as the circumference of C, that is, $2\pi r$. Thus

$$|f'(z_0)| = \frac{1}{2\pi} \left| \oint_C \frac{f(z)}{(z - z_0)^2} dz \right| \le \frac{1}{2\pi} M 2\pi r, \tag{4.4–19}$$

where M is a constant satisfying

$$\left| \frac{f(z)}{(z - z_0)^2} \right| \le M$$

along C. Because $|z - z_0| = r$ on C, the preceding can be rewritten

$$\frac{|f(z)|}{r^2} \le M. \tag{4.4–20}$$

We have assumed that $|f(z)|$ is bounded throughout the z-plane. Thus there is a constant m such that $|f(z)| \le m$ for all z. Dividing both sides of this inequality by r^2 results in

$$\frac{|f(z)|}{r^2} \le \frac{m}{r^2}. \tag{4.4–21}$$

A comparison of Eqs. (4.4–20) and (4.4–21) shows that we can take M as m/r^2. Rewriting Eq. (4.4–19) with this choice of M, we have $|f'(z_0)| \le m/r$.

The preceding inequality can be applied to any circle centered at z_0. Because we can consider circles of arbitrary large radius r, the right-hand side of this inequality can be made arbitrarily small. Thus the derivative of $f(z)$ at z_0, which is some specific number, has a magnitude of zero. Hence $f'(z_0) = 0$. Since the preceding argument can be applied at any point z_0, the expression $f'(z)$ must be zero throughout the z-plane. This is only possible if $f(z)$ is constant.

EXERCISES

1. To arrive at a formal (nonrigorous) derivation of the Cauchy integral formula let $f(z)$ be analytic on and inside a simple closed contour C, let z_0 lie inside C, and let C_0 be a circle centered at z_0 and lying completely inside C. From the

principle of deformation of contours we then have

$$\oint_C \frac{f(z)}{(z - z_0)} \, dz = \oint_{C_0} \frac{f(z)}{(z - z_0)} \, dz.$$

a) Rewrite the integral on the right by means of the change of variables $z = z_0 + re^{i\theta}$, where r is the radius of C_0 and θ increases from 0 to 2π (see Fig. 4.2–7). Note that $dz/d\theta = ire^{i\theta}$.

b) For the integral obtained in part (a) let $r \to 0$ in the integrand. Now perform the integration and use your result to show that

$$\oint_C \frac{f(z)}{(z - z_0)} \, dz = 2\pi i f(z_0).$$

c) What makes this derivation nonrigorous?

2. a) Find $\oint_C (\cos z \sin z)/(z + 2) \, dz$, where C is the ellipse $x^2/9 + y^2/16 = 1$.

b) Repeat part (a), but use the ellipse $x^2 + y^2/16 = 1$.

c) Find $\oint_C (\cos z \sin z)/[(z + 2)(z - 5)] \, dz$, where C is the same as in part (a).

d) Perform the integration of part (c) around the ellipse of part (b).

3. a) Find $\oint_C (\cosh z)/(z^2 + z + 1) \, dz$, where C is the circle $|z + (1 + i)| = 1$.

b) Repeat part (a), but take C as the circle $|z - (1 + i)| = 1$.

4. The rigorous proof of the extended Cauchy integral formula for the first derivative, which was begun on page 162, requires for its completion our showing that

$$\lim_{\Delta z_0 \to 0} \frac{1}{2\pi} \left| \oint_C \frac{f(z) \, dz}{(z - (z_0 + \Delta z_0))(z - z_0)} - \oint_C \frac{f(z) \, dz}{(z - z_0)^2} \right| = 0.$$

Complete the proof. *Hint:* Let b equal the shortest distance from z_0 to any point on the contour C, let m be the maximum value of $|f(z)|$ on C, let L be the length of C, and assume $|\Delta z_0| < b$. Show that you can rewrite the preceding limit with a single integral:

$$\lim_{\Delta z_0 \to 0} \frac{1}{2\pi} \left| \oint_C \frac{f(z)}{(z - z_0)^2} \left(\frac{\Delta z_0}{z - (z_0 + \Delta z_0)} \right) dz \right|.$$

Apply the ML inequality to this integral using m, b, L, etc., and then pass to the limit indicated.

5. a) Find $\oint_C \text{Log } z/[z(z - 2)^2] \, dz$, where C is the circle $|z - 2| = 1$.

b) Find $\oint_C e^{2z}/z^6 \, dz$, where C is the closed contour formed by the line $x = 1$ and the parabola $y^2 = x + 1$.

6. Find $\oint_C \cos z/(z - \pi)^2 \, dz$, where C is the contour $|z - \pi| = 1$.

7. Let $f(z)$ be analytic on and inside a simple closed contour C. Let z_1 and z_2 lie inside C (see Fig. 4.4–7).

a) Show that

$$\frac{1}{2\pi i} \oint_C \frac{f(z)}{(z - z_1)(z - z_2)}\, dz = \frac{f(z_1)}{z_1 - z_2} + \frac{f(z_2)}{z_2 - z_1}.$$

Hint: Integrate around the two contours shown by the broken line in Fig. 4.4–7 and combine the results.

b) Let $f(z)$ have the same properties as in part (a), and let $z_1, z_2, \ldots, z_n$ lie inside C. Extend the method used in part (a) to show that

$$\frac{1}{2\pi i} \oint \frac{f(z)\, dz}{(z - z_1)(z - z_2) \cdots (z - z_n)}$$

$$= \frac{f(z_1)}{(z_1 - z_2)(z_1 - z_3) \cdots (z_1 - z_n)}$$

$$+ \frac{f(z_2)}{(z_2 - z_1)(z_2 - z_3) \cdots (z_2 - z_n)}$$

$$+ \cdots + \frac{f(z_{n'})}{(z_n - z_1)(z_n - z_2) \cdots (z_n - z_{n-1})}.$$

8. The following problems require the result or technique of Exercise 7. Evaluate these integrals:

a) $\dfrac{1}{2\pi i} \oint \dfrac{e^z \cos z\, dz}{(z - 1)(z + 2)}$,   around $|z| = 3$;

b) $\oint \dfrac{dz}{z^2 - 3z + 2}$,   around the square formed by $x = \pm 4, y = \pm 4$;

c) $\oint \dfrac{\sin(z - 2)}{(z - 2)(z + 1)}\, dz$,   around the contour in part (b);

d) $\oint \dfrac{e^z}{(z - 1)^2(z + 1)}\, dz$,   around the contour in part (b).

9. a) Let D be a doubly connected domain bounded by the simple closed contours C_0 and C_1 as shown in Fig. 4.4–8. Let $f(z)$ be analytic in D and on its boundaries, and let z_0 lie in D. Note that $f(z)$ is not necessarily analytic

Contour C (solid line)

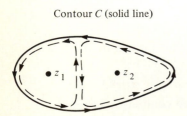

Figure 4.4–7

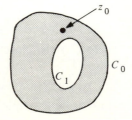

Figure 4.4–8

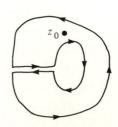

Figure 4.4–9

inside C_1. Show that

$$\frac{1}{2\pi i}\oint_{C_0}\frac{f(z)}{z-z_0}\,dz = f(z_0) + \frac{1}{2\pi i}\oint_{C_1}\frac{f(z)}{(z-z_0)}\,dz.$$

This is the Cauchy integral formula for doubly connected domains. *Hint:* Do the integration $(1/2\pi i)\oint f(z)/(z-z_0)\,dz$ around the simple closed contour shown in Fig. 4.4–9. What portions of the integral cancel?

b) Use the preceding result to show that

$$\frac{1}{2\pi i}\oint_{|z|=2}\frac{dz}{(z-1)\sin z} = \frac{1}{\sin 1} + \frac{1}{2\pi i}\oint_{|z|=1/2}\frac{dz}{(z-1)\sin z}.$$

c) Let D be an n-tuply connected domain bounded by the closed contours $C_0, C_1, \ldots, C_{n-1}$ as shown in Fig. 4.4–10. Let $f(z)$ be analytic in D and on its boundaries, and let z_0 lie in D. Show that

$$\frac{1}{2\pi i}\oint_{C_0}\frac{f(z)}{(z-z_0)}\,dz = f(z_0) + \frac{1}{2\pi i}\oint_{C_1}\frac{f(z)}{(z-z_0)}\,dz$$

$$+ \cdots + \frac{1}{2\pi i}\oint_{C_{n-1}}\frac{f(z)}{(z-z_0)}\,dz.$$

This is the Cauchy integral formula for n-tuply connected domains.

10. Evaluate the following integrals. *Hint:* See Eqs. (4.4–8) and (4.4–10).

   a) $(1/2\pi)\int_0^{2\pi}e^{(e^{i\theta})}\,d\theta$ d) $\int_0^{2\pi}\mathrm{Log}\,(\cos\theta + 5/4)\,d\theta$

   b) $\int_0^{2\pi}\cos(\cos\theta + i\sin\theta)\,d\theta$ *Hint:* $\cos\theta + 5/4 = |1 + 1/2e^{i\theta}|^2$.

   c) $\int_{-\pi}^{\pi}\cos(\cos\theta)\cosh(\sin\theta)\,d\theta$

11. a) Let $u(x, y)$ be a harmonic function. Let u_0 be the value of u at the center of the circle, of radius r, shown in Fig. 4.4–11. The values of u at four equally spaced points on the circumference are u_1, u_2, u_3, u_4. Refer to Eq. (4.4–10a) and use an approximation to the integral to show that

$$u_0 \approx \frac{u_1 + u_2 + u_3 + u_4}{4}.$$

b) Use a pocket calculator to evaluate the harmonic function $e^x\cos y$ at the four points $(1.1, 1)$, $(0.9, 1)$, $(1, 1.1)$, and $(1, 0.9)$. Compare the average of these results with $e^x\cos y$ evaluated at $(1, 1)$.

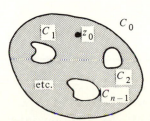

Figure 4.4–10

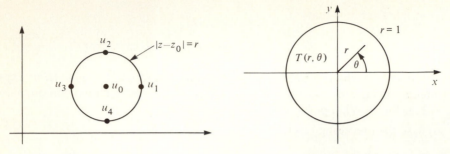

Figure 4.4–11   **Figure 4.4–12**

12. a) Consider the region $|z - (1 + i)| \leq 1$ and the function $f(z) = z^2$. Where in the region do the maximum and minimum values of $|f(z)|$ occur? Do these points lie on the boundary of the region?

    b) Repeat part (a) but use $|z - i| \leq 1/2$ as the region and $1/z$ as the function.

13. Let $f(z)$ be a nonconstant function that is continuous and nonzero throughout a closed bounded region R. Let $f(z)$ be analytic at every interior point of R. Show that the minimum value of $|f(z)|$ in R must occur on the boundary of R. *Hint:* Consider $g(z) = 1/f(z)$ and recall the maximum modulus theorem.

14. Consider the region $|z - 1| \leq 2$. Let $f(z) = z - i$. Where do the maximum and minimum values of $|f(z)|$ occur in this region? Explain why one of your results does not lie on the boundary of the region. Give the numerical values of the maximum and minimum values of $|f(z)|$ in the region.

15. Let $u(x, y)$ be real, nonconstant, and continuous in a closed bounded region R. Let $u(x, y)$ be harmonic in the interior of R. Prove that the maximum value of $u(x, y)$ in this region occurs on the boundary. This is known as the *maximum principle*. *Hint:* Consider $F(z) = u(x, y) + iv(x, y)$, where v is the harmonic conjugate of u. Let $f(z) = e^{F(z)}$. Explain why $|f(z)|$ has its maximum value on the boundary. How does it follow that $u(x, y)$ has its maximum value on the boundary?

16. For $u(x, y)$ described in Exercise 15 show that the minimum value of this function occurs on the boundary. This is known as the *minimum principle*. *Hint:* Follow the suggestions given in Exercise 15 but show that $|f(z)|$ has its minimum value on the boundary.

17. A long cylinder of unit radius, shown in Fig. 4.4–12, is filled with a heat-conducting material. The temperature inside the cylinder is described by the harmonic function $T(r, \theta)$ (see Section 2.6). The temperature on the surface of the cylinder is known and is given by $\sin \theta \cos^2 \theta$. Since $T(r, \theta)$ is continuous for $0 \leq r \leq 1$, $0 \leq \theta \leq 2\pi$, we require that $T(1, \theta) = \sin \theta \cos^2 \theta$. Use the results derived in Exercises 15 and 16 to establish upper and lower bounds on the temperature inside the cylinder.

18. a) Use the extension of the Cauchy integral formula to show that $\oint e^{az}/(z^{n+1}) \, dz = a^n 2\pi i/n!$, where the integration is performed around $|z| = 1$.

b) Use the results of part (a) to show for a real $\int_0^{2\pi} e^{a\cos\theta} \cos(a\sin\theta - n\theta)\,d\theta$
$= 2\pi a^n/n!$, and $\int_0^{2\pi} e^{a\cos\theta} \sin(a\sin\theta - n\theta)\,d\theta = 0$.

4.5   INTRODUCTION TO DIRICHLET PROBLEMS—THE POISSON INTEGRAL FORMULA FOR THE CIRCLE AND HALF PLANE

In previous sections we have seen the close relationship that exists between harmonic functions and analytic functions. In this section we continue exploring this connection and, in so doing, will solve some physical problems whose solutions are harmonic functions.

An important type of mathematical problem, with physical application, is the Dirichlet problem. Here, an unknown function must be found that is harmonic within a domain and that also assumes preassigned values on the boundary of the domain.[†] For an example of such a problem refer to Exercise 17 of the previous section. In this exercise the temperature inside the cylinder $T(r, \theta)$ is a harmonic function. We know what the temperature is on the boundary. If we try to find $T(r, \theta)$ subject to the requirement that on the boundary it agrees with the given function $\sin\theta\cos^2\theta$, we are solving a Dirichlet problem.

Dirichlet problems, such as the one just discussed, in which the boundaries are of simple geometrical shape, are frequently solved by *separation of variables*, a technique discussed in most textbooks on partial differential equations. Another method, which can sometimes be used for such simple domains as well as for those of more complicated shapes, is *conformal mapping*. This subject is considered at some length in Chapter 8. Finally, an approach is discussed in this section that is applicable when the boundary of the domain is a circle or an infinite straight line.

The Dirichlet Problem for a Circle

The Cauchy integral formula is helpful in solving the Dirichlet problem when we have a circular boundary. Consider a circle of radius R whose center lies at the origin of the complex w-plane (see Fig. 4.5–1). Let $f(w)$ be a function that is analytic on and throughout the interior of this circle.

The variable z locates some arbitrary point inside the circle. Applying the Cauchy integral formula on this circular contour and using w as the variable of integration, we have

$$f(z) = \frac{1}{2\pi i} \oint \frac{f(w)}{w - z}\,dw. \tag{4.5–1}$$

Suppose we write $f(z) = U(x, y) + iV(x, y)$. We would like to use the preceding integral to obtain explicit expressions for U and V.

[†] The function being sought should be continuous in the region consisting of the domain and its boundary except that discontinuities are permitted at those boundary points where the given boundary condition is itself discontinuous.

We begin by considering the point z_1 defined by $z_1 = R^2/\bar{z}$. Note that

$$|z_1| = \frac{R^2}{|\bar{z}|} = \frac{R}{|z|}R.$$

Since $|z| < R$, the preceding shows that $|z_1| > R$, that is, the point z_1 lies outside the circle in Fig. 4.5–1. It is easy to show that $\arg z_1 = \arg z$. The function $f(w)/(w - z_1)$ is analytic in the w-plane on and inside the given circle. Hence, from the Cauchy integral theorem,

$$0 = \frac{1}{2\pi i} \oint \frac{f(w)}{w - z_1} dw = \frac{1}{2\pi i} \oint \frac{f(w)}{w - \dfrac{R^2}{\bar{z}}} dw. \tag{4.5–2}$$

Subtracting Eq. (4.5–2) from Eq. (4.5–1), we obtain

$$f(z) = \frac{1}{2\pi i} \oint f(w) \left[\frac{1}{w - z} - \frac{1}{w - \dfrac{R^2}{\bar{z}}} \right] dw$$

$$= \frac{1}{2\pi i} \oint f(w) \left[\frac{z - \dfrac{R^2}{\bar{z}}}{(w - z)\left(w - \dfrac{R^2}{\bar{z}}\right)} \right] dw. \tag{4.5–3}$$

Because we are integrating around a circular contour, we switch to polar coordinates. Let $w = Re^{i\phi}$ and $z = re^{i\theta}$. Thus $\bar{z} = re^{-i\theta}$. Along the path of integration $dw = Re^{i\phi}i\,d\phi$, and ϕ ranges from 0 to 2π. Rewriting the right side of Eq. (4.5–3), we have

$$f(r, \theta) = \frac{1}{2\pi i} \int_0^{2\pi} f(R, \phi) \left[\frac{re^{i\theta} - \dfrac{R^2}{re^{-i\theta}}}{(Re^{i\phi} - re^{i\theta})\left(Re^{i\phi} - \dfrac{R^2}{re^{-i\theta}}\right)} \right] Re^{i\phi}i\,d\phi$$

$$= \frac{1}{2\pi} \int_0^{2\pi} f(R, \phi) \left[\frac{\left(re^{i\theta} - \dfrac{R^2}{r}e^{i\theta}\right)Re^{i\phi}}{(Re^{i\phi} - re^{i\theta})\left(Re^{i\phi} - \dfrac{R^2}{r}e^{i\theta}\right)} \right] d\phi.$$

If we multiply the two terms in the denominator of the preceding integral together and then multiply numerator and denominator by $(-r/R)e^{-i(\theta + \phi)}$, we can show, with the aid of Euler's identity, that

$$f(r, \theta) = \frac{1}{2\pi} \int_0^{2\pi} \frac{f(R, \phi)(R^2 - r^2)\,d\phi}{R^2 + r^2 - 2Rr\cos(\phi - \theta)}. \tag{4.5–4}$$

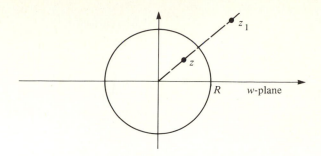

Figure 4.5–1

The analytic function $f(z)$ will now be represented in terms of its real and imaginary parts U and V. Thus $f(R, \phi) = U(R, \phi) + iV(R, \phi)$, $f(r, \theta) = U(r, \theta) + iV(r, \theta)$ and Eq. (4.5–4) becomes

$$U(r, \theta) + iV(r, \theta) = \frac{1}{2\pi} \int_0^{2\pi} \frac{[U(R, \phi) + iV(R, \phi)][R^2 - r^2]\, d\phi}{R^2 + r^2 - 2Rr\cos(\phi - \theta)}. \quad (4.5–5)$$

By equating the real parts on either side of this equation, we obtain the following formula.

POISSON INTEGRAL
FORMULA (FOR INTERIOR
OF A CIRCLE)
$$U(r, \theta) = \frac{1}{2\pi} \int_0^{2\pi} \frac{U(R, \phi)(R^2 - r^2)\, d\phi}{R^2 + r^2 - 2Rr\cos(\phi - \theta)} \quad (4.5–6)$$

A corresponding expression relates $V(r, \theta)$ and $V(R, \phi)$ and is obtained by our equating imaginary parts in Eq. (4.5–5).

Equation (4.5–6), the Poisson integral formula, is important. The formula yields the value of the harmonic function $U(r, \theta)$ everywhere inside a circle of radius R, provided we know the values $U(R, \phi)$ assumed by U on the circumference of the circle.

Since we required that $f(z)$ be analytic inside, and on, the circle of radius R, the reader must assume that the function $U(R, \phi)$ in Eq. (4.5–6) is continuous. In fact, this condition can be relaxed[†] to allow $U(R, \phi)$ to have a finite number of finite "jump" discontinuities. The Poisson integral formula will remain valid.

Example 1

An electrically conducting tube of unit radius is separated into two halves by means of infinitesimal slits. The top half of the tube ($R = 1, 0 < \phi < \pi$) is maintained at an electrical potential of 1 volt while the bottom half ($R = 1, \pi < \phi < 2\pi$) is at -1 volts. Find the potential at an arbitrary point r, θ inside the tube (see Fig. 4.5–2).

[†]See R. V. Churchill, J. W. Brown, R. F. Verhey, *Complex Variables and Applications* (New York: McGraw-Hill, 1974), sec. 100.

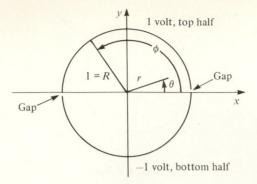

Figure 4.5–2

Solution

Since the electrostatic potential is a harmonic function (see Section 2.6), the Poisson integral formula is applicable. From Eq. (4.5–6), with $R = 1$, we have

$$U(r, \theta) = \frac{1}{2\pi} \int_0^{\pi} \frac{1 - r^2 \, d\phi}{1 + r^2 - 2r \cos(\phi - \theta)} - \frac{1}{2\pi} \int_{\pi}^{2\pi} \frac{1 - r^2 \, d\phi}{1 + r^2 - 2r \cos(\phi - \theta)}.$$

$$(4.5\text{–}7)$$

In each integral we make the change of variables $x = \phi - \theta$; from a standard table of integrals we find the formula

$$\int \frac{dx}{a + b \cos x} = \frac{2}{\sqrt{a^2 - b^2}} \tan^{-1}\left[\frac{\sqrt{a^2 - b^2} \tan(x/2)}{a + b} \right]. \qquad (4.5\text{–}8)$$

Using this formula in Eq. (4.5–7) with $a = 1 + r^2$, $b = -2r$, we obtain

$$U(r, \theta) = \frac{1}{\pi}\left[2 \tan^{-1}\left(\frac{1 + r}{1 - r} \tan\left(\frac{\pi}{2} - \frac{\theta}{2} \right) \right) - \tan^{-1}\left(\frac{1 + r}{1 - r} \tan\left(\pi - \frac{\theta}{2} \right) \right) \right.$$

$$\left. - \tan^{-1}\left(\frac{1 + r}{1 - r} \tan\left(-\frac{\theta}{2} \right) \right) \right]. \qquad (4.5\text{–}9)$$

Since the arctangent is a multivalued function, some care must be taken in applying this formula. Recalling that the values assumed by U on the boundaries are ± 1, we can use physical reasoning[†] to conclude that $-1 \leq U(r, \theta) \leq 1$ when $r \leq 1$. Moreover, the values of the arctangents must be chosen so that $U(r, \theta)$ is continuous for all $r < 1$, and $U(1, \theta)$ is discontinuous only at the slits $\theta = 0$ and $\theta = \pi$. ◀

[†] This same conclusion could also be reached through the maximum and minimum principles (Exercises 15 and 16, Section 4.4) if the given potential along the boundary $R = 1$ had been continuous.

The Dirichlet Problem for a Half Plane (the Infinite Line Boundary)

As in the case of the circle, we will state our Dirichlet problem in the w-plane. Our problem is to find a function $\phi(u, v)$ that is harmonic in the upper half space (the domain $v > 0$). In addition, $\phi(u, v)$ must satisfy a prescribed boundary condition $\phi(u, 0)$ on the line $v = 0$.

Let $f(w) = \phi(u, v) + i\psi(u, v)$ be a function that is analytic for $v \geq 0$. Consider the closed semicircle C (see Fig. 4.5–3) whose base extends along the u-axis from $-R$ to $+R$. Let z be a point inside this semicircle. Then, from the Cauchy integral formula,

$$f(z) = \frac{1}{2\pi i} \oint_C \frac{f(w)}{w - z} dw, \tag{4.5-10}$$

where the integral is taken along the base and arc of the given contour. Now, since z lies inside the semicircle, observe that $\bar{z}$ must lie in the space $v < 0$ and is therefore outside the semicircle. Hence, the function $f(w)/(w - \bar{z})$ is analytic on and interior to the contour C. Thus, from the Cauchy–Goursat theorem,

$$0 = \frac{1}{2\pi i} \oint_C \frac{f(w)}{(w - \bar{z})} dw. \tag{4.5-11}$$

Let us subtract Eq. (4.5–11) from Eq. (4.5–10):

$$f(z) = \frac{1}{2\pi i} \oint_C f(w)\left(\frac{1}{w - z} - \frac{1}{w - \bar{z}}\right) dw = \frac{1}{2\pi i} \oint_C \frac{(z - \bar{z})f(w)}{(w - z)(w - \bar{z})} dw. \tag{4.5-12}$$

We break the integral along C into two parts: along the base ($v = 0$, $-R \leq u \leq R$), which we symbolize with $\longrightarrow$; and along the arc of radius R, which we symbolize with $\frown$. Thus

$$f(z) = \frac{1}{2\pi i} \int_{\longrightarrow} \frac{(z - \bar{z})f(w)}{(w - z)(w - \bar{z})} dw + \frac{1}{2\pi i} \int_{\frown} \frac{(z - \bar{z})f(w)}{(w - z)(w - \bar{z})} dw. \tag{4.5-13}$$

Consider the Cartesian representations $z = x + iy$, $\bar{z} = x - iy$, and $w = u + iv$. We see that $z - \bar{z} = 2iy$ and that $w = u$ along the base. Hence, the first integrand on the

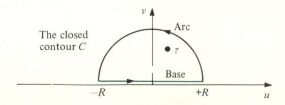

Figure 4.5–3

right in the preceding equation can be rewritten as

$$\frac{(z - \bar{z})f(w)}{(w - z)(w - \bar{z})} = \frac{2iyf(u)}{[u - (x + iy)][u - (x - iy)]} = \frac{2iyf(u)}{(u - x)^2 + y^2},$$

so that Eq. (4.5–13) becomes

$$f(z) = \frac{y}{\pi} \int_{-R}^{+R} \frac{f(u)\,du}{(u - x)^2 + y^2} + \frac{y}{\pi} \int_{\curvearrowright} \frac{f(w)\,dw}{(w - z)(w - \bar{z})}. \qquad (4.5–14)$$

In Exercise 8 of this section we discover that, as the radius R of the arc tends to infinity, the value of the integral along the arc in Eq. (4.5–14) goes to zero. The proof requires our assuming the existence of a constant m such that $|f(w)| \le m$ for all $\operatorname{Im} w \ge 0$, that is, $|f(w)|$ is bounded in the upper half plane. Passing to the limit $R \to \infty$, we find that Eq. (4.5–14) simplifies to

$$f(z) = \frac{y}{\pi} \int_{-\infty}^{+\infty} \frac{f(u)\,du}{(u - x)^2 + y^2}. \qquad (4.5–15)$$

If $f(w)$ is known on the whole real axis of the w-plane, (that is, along $w = u$), this formula will yield the value of the function at any arbitrary point $w = z$, provided $\operatorname{Im} z > 0$.

Let us now rewrite $f(z)$ and $f(w)$ explicitly in terms of real and imaginary parts. With $f(z) = \phi(x, y) + i\psi(x, y)$ and $f(w) = \phi(u, v) + i\psi(u, v)$ we obtain from Eq. (4.5–15)

$$\phi(x, y) + i\psi(x, y) = \frac{y}{\pi} \int_{-\infty}^{+\infty} \frac{\phi(u, 0) + i\psi(u, 0)}{(u - x)^2 + y^2}\,du.$$

Equating the real parts in this equation, we arrive at the following formula.

POISSON INTEGRAL FORMULA
(FOR THE UPPER HALF PLANE) $\qquad \phi(x, y) = \dfrac{y}{\pi} \displaystyle\int_{-\infty}^{+\infty} \dfrac{\phi(u, 0)\,du}{(u - x)^2 + y^2} \qquad (4.5–16)$

A corresponding equation, relating $\psi(x, y)$ and $\psi(u, 0)$ is obtained if we equate imaginary parts.

Equation (4.5–16), called the Poisson Integral Formula for the Upper Half Plane, will yield the value of a harmonic function $\phi(x, y)$ anywhere in the upper half plane provided ϕ is already completely known over the entire real axis. It can be shown that this is the only solution to the Dirichlet problem that is *bounded* in the upper half plane. Without this restriction other solutions can be found.

In our derivation we assumed that $\phi(u, v)$ is the real part of a function $f(u, v)$, which is analytic for $\operatorname{Im} v \ge 0$. This would require that $\phi(u, 0)$ in Eq. (4.5–16) be continuous for $-\infty < u < \infty$. Actually, this requirement can be relaxed to permit $\phi(u, 0)$ to have a finite number of finite jumps. We then can still use Eq. (4.5–16).

Example 2

As indicated in Fig. 4.5–4, the upper half space $\operatorname{Im} w > 0$ is filled with a heat-conducting material. The boundary $v = 0$, $u > 0$ is maintained at a temperature of 0

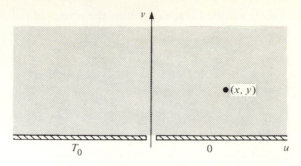

Figure 4.5–4

while the boundary $v = 0$, $u < 0$ is kept at temperature T_0. Find the steady state distribution of temperature $\phi(x, y)$ throughout the conducting material.

Solution

Since, as shown in Section 2.6, the temperature is a harmonic function, the Poisson integral formula is directly applicable. We have $\phi(u, 0) = T_0$, $u < 0$; and $\phi(u, 0) = 0$, $u > 0$. Thus

$$\phi(x, y) = \frac{y}{\pi} \int_{-\infty}^{0} \frac{T_0 \, du}{(u - x)^2 + y^2} + \frac{y}{\pi} \int_{0}^{\infty} \frac{0 \, du}{(u - x)^2 + y^2}.$$

The second integral is zero. In the first we make the change of variables $p = x - u$. Thus

$$\phi(x, y) = \frac{T_0 y}{\pi} \int_{x}^{\infty} \frac{dp}{p^2 + y^2} = \frac{T_0}{\pi} \tan^{-1} \frac{p}{y} \Big|_{x}^{\infty} = \frac{T_0}{\pi} \left[\frac{\pi}{2} - \tan^{-1} \frac{x}{y} \right].$$

$$(4.5–17)$$

From the trigonometric identity $\tan^{-1} s = \pi/2 - \tan^{-1}(1/s)$ we see that the expression in the brackets on the right side of Eq. (4.5–17) is $\tan^{-1}(y/x) = \theta$, where θ is the polar angle associated with the point (x, y). From physical considerations we require that $0 \le \phi(x, y) \le T_0$, that is, the maximum and minimum temperatures are

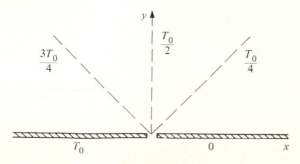

Figure 4.5–5

on the boundary. This is satisfied if we take θ as the principal polar angle. Thus

$$\phi(x, y) = \frac{T_0}{\pi}\theta, \qquad 0 \le \theta \le \pi.$$

Some surfaces on which the temperature displays constant values are exhibited as rays in Fig. 4.5–5. ◀

EXERCISES

1. a) In Example 1 of this section use physical intuition to guess the (constant) value of the potential U along the ray $0 \le r < 1$, $\theta = 0$. Verify that Eq. (4.5–9) yields this numerical result.

   b) Verify that Eq. (4.5–9) does satisfy the following boundary conditions.

$$\lim_{r \to 1} U(r, \theta) = \begin{cases} 1, & 0 < \theta < \pi \\ -1, & \pi < \theta < 2\pi \end{cases}$$

2. a) The temperature of the surface of a cylinder of radius 5 is maintained as shown in Fig. 4.5–6. Show that the steady state temperature inside the cylinder $U(r, \theta)$ is given by

$$U(r, \theta) = \frac{100}{\pi}\left[\tan^{-1}\left(\frac{5 + r}{5 - r} \tan\left(\frac{\pi}{2} - \frac{\theta}{2} \right) \right) + \tan^{-1}\left(\frac{5 + r}{5 - r} \tan\frac{\theta}{2} \right) \right].$$

   b) Verify that the following boundary conditions are satisfied by the preceding formula.

$$\lim_{r \to 5} U(r, \theta) = \begin{cases} 100, & 0 < \theta < \pi \\ 0, & \pi < \theta < 2\pi \end{cases}$$

   c) Plot $U(r, \pi/2)$ for $0 < r \le 5$. Use a pocket calculator for the numerical data.

3. Use the Poisson integral formula for the circle to show that if the electrostatic potential on the surface of any cylinder is constant and equal to V_0, then the potential everywhere inside is equal to V_0.

4. The purpose of this exercise is to obtain a formula for a function that is harmonic in the unbounded domain external to a circle. The function is required

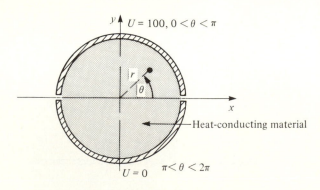

Figure 4.5–6

to achieve prescribed values on the circumference of the circle. This is known as the external Dirichlet problem for a circle.

a) Consider a function $f(w)$ that is analytic at all points in the w-plane that satisfy $|w| \geq R$. Place two circles, as shown in Fig. 4.5–7, in the w-plane centered at $w = 0$. Their radii are R and R', with $R < R'$. Let $z = re^{i\theta}$ be a point lying within the annular domain formed by the circles. Thus $R < r < R'$. Show that

$$f(z) = \frac{1}{2\pi i} \oint_{|w| = R'} \frac{f(w)}{w - z}\,dw + \frac{1}{2\pi i} \oint_{|w| = R} \frac{f(w)}{w - z}\,dw.$$

Note the direction of integration around the two circles. *Hint:* See Exercise 9a in Section 4.4.

b) Let $z_1 = R^2/\bar{z}$. Note that this point lies inside the inner circle. Show that

$$0 = \frac{1}{2\pi i} \oint_{|w| = R'} \frac{f(w)}{w - z_1}\,dw + \frac{1}{2\pi i} \oint_{|w| = R} \frac{f(w)}{w - z_1}\,dw.$$

c) Subtract the formula of part (b) from that derived in part (a) and show that

$$f(z) = \frac{1}{2\pi i} \oint_{|w| = R'} \frac{f(w)(z - R^2/\bar{z})}{(w - z)(w - R^2/\bar{z})}\,dw$$

$$+ \frac{1}{2\pi i} \oint_{|w| = R} \frac{f(w)(z - R^2/\bar{z})}{(w - z)(w - R^2/\bar{z})}\,dw.$$

d) Assume that $|f(w)| \leq m$ (a constant) when $|w| > R$. Let $R' \to \infty$. Show that in the limit the integral around $|w| = R'$ goes to zero. *Hint:* Use the ML inequality.

e) Rewrite the remaining integral in part (c) by using polar coordinates $z = re^{i\theta}$, $w = Re^{i\phi}$. Put $f(z) = U(r, \theta) + iV(r, \theta)$. Show that

$$U(r, \theta) = \frac{1}{2\pi} \int_0^{2\pi} \frac{U(r, \phi)(r^2 - R^2)\,d\phi}{R^2 + r^2 - 2Rr\cos(\phi - \theta)}, \qquad r > R. \quad (4.5\text{–}18)$$

Hint: Study the derivation of the Poisson integral formula for the interior of a circle.

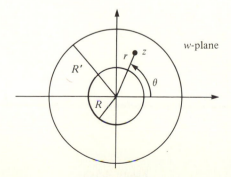

Figure 4.5–7

5. In this exercise we have an external Dirichlet problem with a circular boundary.

   a) Consider the configuration shown in Exercise 2. Assume that the temperature distribution along the cylindrical surface is the same as was given in that exercise but that now the region external to the cylinder is filled with a heat-conducting material. Use Eq. (4.5–18), derived in Exercise 4 to show that $U(r, \theta)$, the temperature distribution, is given for $r > 5$ by

$$U(r, \theta) = \frac{100}{\pi}\left[\tan^{-1}\left(\frac{5 + r}{r - 5}\tan\left(\frac{\pi}{2} - \frac{\theta}{2}\right)\right) + \tan^{-1}\left(\frac{5 + r}{r - 5}\tan\frac{\theta}{2}\right)\right].$$

   b) Verify that, for suitable choices of the branches of the arctangents, the expression derived in part (a) fulfills the boundary conditions

$$\lim_{r \to 5} U(r, \theta) = 100, \quad 0 < \theta < \pi, \quad \text{and} \quad \lim_{r \to 5} U(r, \theta) = 0, \quad \pi < \theta < 2\pi$$

   c) Plot $U(r, \pi/2)$ for $5 \le r \le 50$. What temperature is created at $r = \infty$ by the configuration?

6. a) An electrically conducting metal sheet is perpendicular to the y-axis and passes through $y = 0$, as shown in Fig. 4.5–8. The right half of the sheet, $x > 0$, is maintained at an electrical potential of V_0 volts while the left half, $x < 0$, is maintained at a voltage $-V_0$. Show that in the half space, $y \ge 0$, the electrostatic potential is given by

$$\phi(x, y) = V_0 - \frac{2V_0}{\pi}\tan^{-1}\frac{y}{x} = V_0 - \frac{2V_0}{\pi}\text{Im}(\text{Log } z),$$

   where $0 \le \tan^{-1} \le \pi$.

   b) Sketch the equipotential lines (or surfaces) on which $\phi(x, y) = V_0/2$, $\phi(x, y) = 0$, $\phi(x, y) = -V_0/2$.

   c) Find the components of the electric field E_x and E_y at $x = 1$, $y = 1$, and draw a vector representing the field at this point (see Section 2.6).

7. a) The surface $y = 0$ is maintained at an electrostatic potential $V(x)$ described by

$$-\infty < x < -h, \quad V(x) = 0; \quad -h < x < h, \quad V(x) = V_0;$$
$$h < x < \infty, \quad V(x) = 0.$$

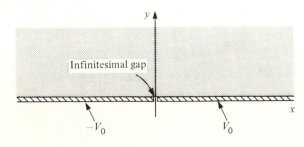

Figure 4.5–8

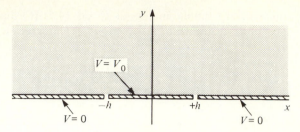

Figure 4.5–9

This potential distribution is shown in Fig. 4.5–9. Show that the electrostatic potential in the space $y > 0$ is given by

$$\phi(x, y) = \frac{V_0}{\pi} \left[-\tan^{-1} \frac{x - h}{y} + \tan^{-1} \frac{x + h}{y} \right].$$

b) Show that the preceding expression can be rewritten as

$$\phi(x, y) = \frac{V_0}{\pi} \left[\tan^{-1} \frac{y}{x - h} - \tan^{-1} \frac{y}{x + h} \right]$$

$$= \frac{V_0}{\pi} \operatorname{Im} \left[\operatorname{Log}(z - z_0) - \operatorname{Log}(z + z_0) \right],$$

where $z_0 = h$, $z = x + iy$.

c) Show that along the line $x = 0$, we have, when $y \gg h$,

$$\phi(0, y) \approx \frac{V_0}{\pi} \frac{2h}{y}.$$

Hint: For small arguments $\tan^{-1} w \approx w$.

8. Complete the proof of the Poisson integral formula for the upper half plane by showing that the integral over the arc (of radius R) in Eq. (4.5–14) goes to zero as $R \to \infty$. *Hint:* Explain why $|w - z| \geq |w| - |z|$ and $|w - \bar{z}| \geq |w| - |\bar{z}|$ in the integrand. We must assume that $|f(w)| \leq m$ for $\operatorname{Im} w \geq 0$. Now explain why $|f(w)/[(w - z)(w - \bar{z})]| \leq m/(R - |z|)^2$ on the path of integration. Calling the right side of this inequality M, show that the magnitude of the integral on the arc is $\leq M\pi R$ if we ignore the factor y/π in Eq. (4.5–14). Allow $R \to \infty$.

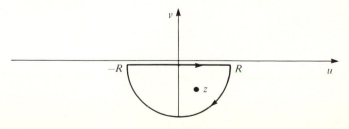

Figure 4.5–10

9. Derive a Poisson integral formula analogous to Eq. (4.5–16) that applies in the *lower* half plane. *Hint:* Begin with the contour of integration shown in Fig. 4.5–10. Answer:

$$\phi(x, y) = -\frac{y}{\pi} \int_{-\infty}^{+\infty} \frac{\phi(u,0)\, du}{(u - x)^2 + y^2}, \qquad y < 0.$$

APPENDIX TO CHAPTER 4
GREEN'S THEOREM IN THE PLANE[†]

Let us prove our theorem for a simple closed contour C that has this property: If a straight line is drawn parallel to either the x- or y-axis, it will intersect C at two points at most. Such a curve is shown in Fig. A4–1. The points A and B are the pair of points on C having the smallest and largest x-coordinates. These coordinates are a and b, respectively. Now consider

$$\iint_R \frac{-\partial P}{\partial y}\, dx\, dy,$$

where the integral is taken over the region R consisting of the contour C and its interior. The function $P(x, y)$ is assumed to be continuous and to have continuous first partial derivatives in R.

The contour C creates two distinct paths connecting A and B. They are given by the equations $y = g(x)$ and $y = f(x)$ (see Fig. A4–1). Thus

$$\iint_R \frac{-\partial P}{\partial y}\, dx\, dy = -\int_{x=a}^{x=b}\left[\int_{y=f(x)}^{y=g(x)} \frac{\partial P}{\partial y}\, dy\right] dx = -\int_a^b P(x, y)\Big|_{y=f(x)}^{y=g(x)} dx$$

$$= \int_a^b [P(x, f(x)) - P(x, g(x))]\, dx$$

$$= \int_a^b P(x, f(x))\, dx + \int_b^a P(x, g(x))\, dx.$$

This final pair of integrals, one from a to b, the other from b to a, together form the line integral $\int P(x, y)\, dx$ taken around the contour C in the positive (counterclock-

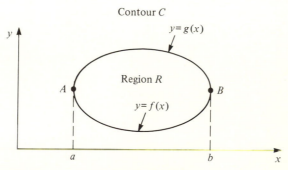

Contour C

$y = g(x)$

Region R

$y = f(x)$

A B

a b

Figure A4–1

[†]See Section 4.2.

wise) direction. Thus

$$\iint_R -\frac{\partial P}{\partial y}\, dx\, dy = \oint_C P(x,\, y)\, dx. \tag{A4-1}$$

In a similar way (refer to Fig. A4–2) we have for a function $Q(x,\, y)$, which has the same properties of continuity as $P(x,\, y)$:

$$\iint_R \frac{\partial Q}{\partial x}\, dx\, dy = \int_e^d [Q(n(y),\, y) - Q(m(y),\, y)]\, dy$$

$$= \int_e^d Q(n(y),\, y)\, dy + \int_d^e Q(m(y),\, y)\, dy = \oint_C Q(x,\, y)\, dy. \tag{A4-2}$$

Adding Eqs. (A4–1) and (A4–2) we obtain our desired result:

$$\oint_C P(x,\, y)\, dx + Q(x,\, y)\, dy = \iint_R \left(\frac{\partial Q}{\partial x} - \frac{\partial P}{\partial y} \right) dx\, dy.$$

The condition that straight lines drawn parallel to the x- or y-axes intersect C at two points, at most, is easily relaxed. A slightly more complicated proof is required.

The following theorem, related to Green's theorem, is of use in complex variable theory. It enables one to prove a converse of the Cauchy–Goursat theorem (see Exercise 3, Section 4.2).

Theorem 13

Let $P(x,\, y)$, $Q(x,\, y)$, $\partial P/\partial y$ and $\partial Q/\partial x$ be continuous in a simply connected domain D. Suppose $\oint P\, dx + Q\, dy = 0$ around every simple closed contour in D. Then, $\partial Q/\partial x = \partial P/\partial y$ in D. ∎

To prove this theorem suppose that $\partial Q/\partial x - \partial P/\partial y > 0$ at the point $x_0,\, y_0$ in D. Then, since both these derivatives are continuous, we can find in D a circle C centered at $x_0,\, y_0$ such that $\partial Q/\partial x - \partial P/\partial y > 0$ inside and on C. Applying Green's theorem to this circle, we have

$$\oint_C P\, dx + Q\, dy = \iint_{\text{interior of } C} \left(\frac{\partial Q}{\partial x} - \frac{\partial P}{\partial y} \right) dx\, dy. \tag{A4-3}$$

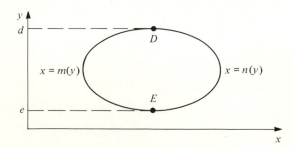

Figure A4–2

Since the integrand on the right is positive, the integral on the right must result in a positive number. The integral on the left is, by hypothesis, zero. Hence, we have obtained a contradiction.

Assuming $\partial Q / \partial x - \partial P / \partial y < 0$, we can go through a similar argument and also obtain a contradiction. Thus since $\partial Q / \partial x$ is neither less than nor greater than $\partial P / \partial y$, we have, at (x_0, y_0), $\partial Q / \partial x = \partial P / \partial y$.

CHAPTER 5 INFINITE SERIES INVOLVING A COMPLEX VARIABLE

5.1 INTRODUCTION AND REVIEW OF REAL SERIES

The reader doubtless already knows something about infinite series involving functions of a real variable. For example, the equations

$$e^x = \sum_{n=0}^{\infty} \frac{x^n}{n!} = 1 + x + \frac{x^2}{2!} + \frac{x^3}{3!} + \cdots \tag{5.1–1}$$

and

$$\frac{1}{1-x} = \sum_{n=0}^{\infty} \frac{(x+1)^n}{2^{n+1}} = \frac{1}{2} + \frac{(x+1)}{4} + \frac{(x+1)^2}{8} + \cdots \tag{5.1–2}$$

should look at least slightly familiar. The infinite sums appearing here are called power series because each term is of the form $(x - x_0)^n$, where x_0 is a real constant and $n \geq 0$ is an integer. When the real variable x assumes certain allowable values, the infinite sums correctly represent the functions on the left.

In this chapter we will learn to obtain power series representations of functions of a complex variable z. These series will contain terms of the form $(z - z_0)^n$ instead of $(x - x_0)^n$. Here z_0 is a complex constant. We are interested in such series for several reasons. First, there is a close connection between the analyticity of a function and its ability to be represented by a power series. Secondly, just as for real series, complex series are useful for numerical approximation. Without the benefit of either tables or a pocket calculator we could use the first three terms in the sum in Eq. (5.1–1) to establish that $e^{0.2} \doteq 1 + 0.2 + 0.02 = 1.22$, a result accurate to better than 0.15%. Even better accuracy is obtained with more terms. Similarly, without resorting to Eq. (3.1–1) we can use a power series in the variable z to obtain a good approximation to $e^{0.2 + 0.1i}$. Power series are used to obtain numerical approximations for the value of an integral that cannot be found in closed form. For example, to evaluate $\int_0^{0.2} (e^x - 1)/x \, dx$ we replace e^x by perhaps the first three terms in Eq. (5.1–1). Having done so, we find that we must instead determine $\int_0^{0.2} (1 + x/2) \, dx$, which is an easy matter. Given a series expansion for e^z, we could proceed in a similar manner to find $\int_0^{0.2+0.1i}(e^z - 1)/z\, dz$ taken along some path. Power series are also used extensively in the solution of differential equations.

Toward the end of this chapter we will consider Laurent series, which contain $(z - z_0)$ raised to positive and negative powers. These series lead directly to the subject of residues (see Chapter 6). Residues are enormously useful in the rapid evaluation of contour integrals taken along closed contours.

A power series expansion of the real function $f(x)$ takes the form

$$f(x) = \sum_{n=0}^{\infty} c_n(x - x_0)^n = c_0 + c_1(x - x_0) + c_2(x - x_0)^2 + \cdots . \quad (5.1\text{--}3)$$

Here, c_0, c_1, etc. are called the coefficients of the expansion. For finding the coefficients the reader probably recalls the straightforward procedure

$$c_n = \frac{f^{(n)}(x_0)}{n!}. \quad (5.1\text{--}4)$$

When coefficients obtained in this way are used in Eq. (5.1–3), we say that Eq. (5.1–3) is the Taylor series expansion of $f(x)$ about x_0. The coefficients in Eqs. (5.1–1) and (5.1–2) can be derived through the use of Eq. (5.1–4). However, attempting the expansion

$$x^{1/2} = \sum_{n=0}^{\infty} c_n x^n, \quad (5.1\text{--}5)$$

we land in trouble since $x^{1/2}$ does not possess a first- or higher-order derivative at $x = 0$. Thus not all functions of x have a Taylor series expansion.

Let us consider another difficulty posed by Taylor series. If in Eq. (5.1–2) we substitute $x = -1/2$ in the series as well as in the function on the left, we have

$$\frac{2}{3} = \frac{1}{2} + \frac{1}{8} + \frac{1}{32} + \frac{1}{128} + \cdots . \quad (5.1\text{--}6)$$

Adding the first four terms on the right and getting 0.664, we become convinced that the infinite sum approaches $2/3 = 0.6666\ldots$. However, with $x = 2$ on both sides in Eq. (5.1–2) we obtain

$$-1 = \frac{1}{2} + \frac{3}{4} + \frac{9}{8} + \cdots . \quad (5.1\text{--}7)$$

Clearly, the infinite sum will not yield the numerical value -1. What we have seen—that there are values of x for which the series is not valid—is typical of Taylor series for functions of real variables.

The problems just discussed also occur when we study power series expansions of functions of a complex variable. We will thus be concerned with the question of which functions $f(z)$ can be represented by a power series, how the series is obtained, and for what values of z the series actually does represent or "converge to" $f(z)$. We must also carefully consider the meaning of the term "converge" when applied to series generally. In studying these matters we will also achieve a better understanding of the behavior of real power series.

EXERCISES[†]

1. Use Eq. (5.1–4) to obtain the indicated Taylor series. Give the general coefficient c_n and write out the first few terms explicitly.

   a) $\sin x$, about $x = 0$

   b) $\cos x$, about $x = 0$

   c) e^x, about $x = 1$

   d) $\dfrac{1}{1-x}$, about $x = 0$

   e) $\dfrac{1}{1-2x}$, about $x = 1$

2. Review the "ratio test" in an elementary calculus book and use this test to establish those values of x for which the series in Exercise 1 are convergent. For parts (d) and (e) above it will also be helpful to review the "nth term test" to establish behavior at the endpoints of the interval of convergence.[‡]

5.2 CONVERGENCE AND UNIFORM CONVERGENCE OF COMPLEX SERIES

Before we enter a discussion of power series involving a complex variable we must try to define precisely the term "convergent" as applied to series in general. The definition is also applicable to real series. In addition, we should know something about the properties of convergent series and how to test for convergence.

Let

$$u_1(z) + u_2(z) + \cdots = \sum_{j=1}^{\infty} u_j(z) \tag{5.2–1}$$

be a series with an infinite number of terms in which the members $u_1(z), u_2(z), \ldots$ are functions of the complex variable z. A function $u_j(z)$ is assumed to exist for each positive[§] value of the integer index j. Examples of such series are $e^z + e^{2z} + e^{3z} + \cdots$ and $(z-1) + (z-1)^2 + (z-1)^3 + \cdots$. We define

$$S_1(z) = u_1(z),$$
$$S_2(z) = u_1(z) + u_2(z), \tag{5.2–2}$$
$$S_3(z) = u_1(z) + u_2(z) + u_3(z),$$

and so forth.

A sum

$$S_n(z) = \sum_{j=1}^{n} u_j(z) \tag{5.2–3}$$

involving the first n terms in Eq. (5.2–1) is called the *nth partial sum* of the infinite

---

[†] These exercises are primarily for review.

[‡] The general (nth term) of a series must go to zero as $n \to \infty$ if the series converges.

[§] Sometimes we will use series that begin with $u_0(z)$ or $u_N(z)$, where $N \neq 1$. The discussion presented here applies substantially unchanged to series of this sort. If necessary, such series can be reindexed to begin with $j = 1$.

series. Partial sums can be arranged to create a sequence of functions of the form

$$S_1(z), S_2(z), S_3(z), \ldots, S_n(z), \ldots . \tag{5.2-4}$$

For example, if the series shown in Eq. (5.2–1) were $\sum_{j=1}^{\infty} e^{jz}$, then the sequence of partial sums would be $e^z, e^z + e^{2z}, e^z + e^{2z} + e^{3z}, \ldots .$

Let us assume that the sequence of partial sums in Eq. (5.2–4) has a limit $S(z)$ as $n \to \infty$. Mathematically we can define this as stated below.

Definition   Limit of a sequence

For the sequence of functions $S_1(z), S_2(z), \ldots, S_n(z)$ we say that there is a limit $S(z)$ as $n \to \infty$ if, given a number $\varepsilon > 0$, we determine that there exists an integer N such that

$$|S(z) - S_n(z)| < \varepsilon, \quad \text{for all } n > N. \tag{5.2-5}$$

We then write $\lim_{n \to \infty} S_n(z) = S(z)$. ∎

A sequence may or may not have a limit. Typically, a limit will exist for some values of z but not for others. The concept of a limit of a sequence is not confined to sequences formed from the partial sums of a series. However, when $S_1(z), S_2(z), \ldots$ are partial sums, we can make the following definition.

Definition   Ordinary convergence

If the sequence of partial sums in Eq. (5.2–2) formed from the infinite series in Eq. (5.2–1) has limit $S(z)$ as $n \to \infty$, we say that the infinite series in Eq. (5.2–1) converges and has sum $S(z)$. The set of all values of z for which the series converges is called its *region of convergence*. For z lying in this region we write

$$S(z) = \sum_{j=1}^{\infty} u_j(z). \tag{5.2-6}$$

If the sequence of partial sums does not have a limit, we say that the series in Eq. (5.2–1) diverges. ∎

In informal language the series in Eq. (5.2–1) "converges to $S(z)$" if in Eq. (5.2–2) the succession of approximations to the whole series shown in Eq. (5.2–1) yields $S(z)$. Equation (5.2–5) says that we can make the difference between the sum of the series and the partial sum of the series, $S_n(z)$, be smaller than any preassigned number ε by taking enough terms in the partial sum. We must make n, the number of terms used to form the partial sum, exceed a number N. Typically, N depends on ε and z.

Occasionally we will deal with series of the form $\sum_{j=1}^{\infty} c_j$, where $c_1, c_2, \ldots$ are complex constants. The definition of convergence of such series is identical to that just given for a series of functions, except that the sum and partial sums are independent of z. The convergence or divergence of such series likewise is independent of z.

Example 1

Show that

$$\sum_{j=1}^{\infty} z^{j-1} = 1 + z + z^2 + \cdots = S(z) = \frac{1}{1-z}, \qquad |z| < 1. \qquad (5.2-7)$$

Solution

This result should look plausible since, if we replace z by x in Eq. (5.2–7), we obtain a familiar real geometric series and its sum.

The nth partial sum of our series is $S_n(z) = 1 + z + z^2 + \cdots + z^{n-1}$. Notice that $S_n(z) - zS_n(z) = (1 + z + \cdots + z^{n-1}) - (z + z^2 + \cdots + z^n) = 1 - z^n$, so that $S_n(z)(1 - z) = 1 - z^n$ or, for $z \neq 1$,

$$S_n(z) = \frac{1 - z^n}{1 - z} = 1 + z + z^2 + \cdots + z^{n-1}. \qquad (5.2-8)$$

Since the sum in Eq. (5.2–7) is $S(z) = 1/(1 - z)$, we have

$$|S(z) - S_n(z)| = \left| \frac{1 - (1 - z^n)}{1 - z} \right| = \frac{|z|^n}{|1 - z|}. \qquad (5.2-9)$$

Referring to Eq. (5.2–5), we require for convergence that

$$\frac{|z|^n}{|1 - z|} < \varepsilon, \quad \text{for } n > N, \qquad (5.2-10)$$

or that

$$\left| \frac{1}{z} \right|^n > \frac{1}{\varepsilon|1 - z|}.$$

Taking logarithms of the preceding, we obtain

$$n \operatorname{Log} \left| \frac{1}{z} \right| > \operatorname{Log} \frac{1}{\varepsilon|1 - z|}. \qquad (5.2-11)$$

Inside the disc $|z| < 1$ we have $|1/z| > 1$ and $\operatorname{Log}|1/z| > 0$. Inside this disc $|z - 1|$ is less than 2. Assuming $\varepsilon < 1/2$, we then have

$$\frac{1}{\varepsilon|z - 1|} > 1 \quad \text{and} \quad \operatorname{Log} \frac{1}{\varepsilon|z - 1|} > 0.$$

Both sides of Eq. (5.2–11) are thus positive and the inequality can be rearranged as

$$n > \frac{\operatorname{Log} \varepsilon|1 - z|}{\operatorname{Log}|z|}. \qquad (5.2-12)$$

If we choose N as an integer that equals or exceeds the right side of Eq. (5.2–12) and take $n > N$, then Eq. (5.2–10) is satisfied. Hence, $|S(z) - S_n(z)|$ in Eq. (5.2–9) will be $< \varepsilon$. Thus Eq. (5.2–5) is satisfied and, according to our definition of convergence, Eq. (5.2–7) has been proved. ◀

Example 2

Use the known sum of a geometric series, shown in Eq. (5.2–7), to sum $\sum_{n=0}^{\infty} e^{inz} = 1 + e^{iz} + e^{i2z} + \cdots$. State the region of convergence.

Solution

We replace z by e^{iz} in Eq. (5.2–7) and obtain

$$1 + e^{iz} + e^{i2z} + \cdots = \frac{1}{1 - e^{iz}}, \qquad |e^{iz}| < 1.$$

Now

$$|e^{iz}| = |e^{i(x+iy)}| = |e^{ix}e^{-y}| = |e^{ix}||e^{-y}| = e^{-y}.$$

To justify the final step on the right, recall that $|e^{ix}| = 1$ and that e^{-y} is not negative. The requirement for convergence of our given series $|e^{iz}| < 1$ now becomes $e^{-y} < 1$. A sketch of e^{-y} against y reveals that the inequality is satisfied only if $y > 0$, that is, Im $z > 0$.   ◀

Often the functions $u_j(z)$ in an infinite series appear in the form

$$u_j(z) = R_j(x, y) + iI_j(x, y), \tag{5.2–13}$$

where R_j and I_j are the real and imaginary parts of u_j. Thus

$$\sum_{j=1}^{\infty} u_j(z) = \sum_{j=1}^{\infty} R_j(x, y) + iI_j(x, y). \tag{5.2–14}$$

Theorem 1

The convergence of both the real series $\sum_{j=1}^{\infty} R_j(x, y)$ and $\sum_{j=1}^{\infty} I_j(x, y)$ is a necessary and sufficient condition for the convergence of $\sum_{j=1}^{\infty} u_j(z)$, where $u_j(z) = R_j(x, y) + iI_j(x, y)$. If $\sum_{j=1}^{\infty} R_j(x, y)$ and $\sum_{j=1}^{\infty} I_j(x, y)$ converge to the functions $R(x, y)$ and $I(x, y)$, respectively, then $\sum_{j=1}^{\infty} u_j(z)$ converges to $S(z) = R(x, y) + iI(x, y)$.   ∎

The rather simple proofs will not be presented here.

We should recall that two convergent real series can be added term by term. The resulting series converges to a function obtained by adding the sums of the two original series. Convergent complex series can also be added in this way. Subtraction of series is also performed in an analogous manner.

The nth term test, derived for real series in elementary calculus, also applies to complex series, and is described by the following theorem.

Theorem 2 nth term test

The series $\sum_{n=1}^{\infty} u_n(z)$ diverges if

$$\lim_{n \to \infty} u_n(z) \neq 0 \tag{5.2–15a}$$

or, equivalently, if

$$\lim_{n \to \infty} |u_n(z)| \neq 0. \tag{5.2–15b}$$ ∎

Loosely speaking, if the terms of a series do not ultimately start shrinking to zero, then the series cannot converge. Notice that the phrase "only if" does not appear in this theorem; there are divergent series whose nth term goes to zero as $n \to \infty$. The test can, however, be used to establish the divergence of some series, as Exercise 1 in this section demonstrates.

The notions of absolute and conditional convergence that apply to real series also apply to complex series, as we see from the following definition.

Definition   Absolute and conditional convergence

The series $\sum_{j=1}^{\infty} u_j(z)$ is called *absolutely convergent* if the series $\sum_{j=1}^{\infty} |u_j(z)|$ is convergent. If the former series converges but the latter series diverges, we say that $\sum_{j=1}^{\infty} u_j(z)$ is *conditionally convergent.*   ∎

In words, a series is absolutely convergent if the series formed by taking the magnitude of each of its terms is convergent. An absolutely convergent complex series has some useful properties that we will now list without proof. In each case the proof is similar to that for a corresponding property of absolutely convergent real series. The reader is referred to standard texts on the calculus of real variables.

Theorem 3

An absolutely convergent series converges in the ordinary sense.   ∎

Theorem 4

The sum of an absolutely convergent series is independent of the order in which the terms are added.   ∎

Theorem 5

Two absolutely convergent series can be multiplied together as if they were polynomials. The sum of the resulting series is the product of the sums of the two original series.   ∎

The ratio test, which is used to establish the absolute convergence of a real infinite series, also applies to complex series.

Theorem 6   Ratio test

For the series $\sum_{j=1}^{\infty} u_j(z)$ consider

$$\Gamma = \lim_{j \to \infty} \left| \frac{u_{j+1}(z)}{u_j(z)} \right|; \qquad (5.2-16)$$

then,

a) the series converges absolutely if $\Gamma < 1$,

b) the series diverges if $\Gamma > 1$,

c) Eq. (5.2–16) provides no information about the series if the indicated limit fails to exist or if $\Gamma = 1$.   ∎

Example 3

Use the ratio test to investigate the convergence of

$$\sum_{j=1}^{\infty} (-1)^j j 2^{j+1} z^{2j} = -4z^2 + 16z^4 - 48z^6 + \cdots.$$

Solution

$$u_j = (-1)^j j 2^{j+1} z^{2j} \quad \text{and} \quad u_{j+1} = (-1)^{j+1}(j+1)2^{j+2} z^{2(j+1)}.$$

Thus

$$\left| \frac{u_{j+1}}{u_j} \right| = \left| \frac{(-1)^{j+1}(j+1)2^{j+2} z^{2j+2}}{(-1)^j j 2^{j+1} z^{2j}} \right| = \left| \frac{j+1}{j} 2z^2 \right|.$$

Following Eq. (5.2–16), we, in the preceding equation, allow $j \to \infty$, notice that $(j+1)/j = 1 + 1/j$ becomes 1 in the limit, and find that

$$\Gamma = 2|z^2|.$$

Now, we use part (a) of Theorem 6 and set $\Gamma < 1$. This requires that

$$2|z^2| < 1 \quad \text{or} \quad |z| < \frac{1}{\sqrt{2}}.$$

Thus our series converges absolutely if z lies inside a circle of radius $1/\sqrt{2}$ centered at the origin. Using part (b) of the same theorem we readily show that the series diverges for $|z| > 1/\sqrt{2}$, that is, when z lies outside the circle just mentioned.

On $|z| = 1/\sqrt{2}$ we have $\Gamma = 1$, which provides no information about convergence. However, observe that on $|z| = 1/\sqrt{2}$ we have

$$|u_j(z)| = j 2^{j+1} \left(\frac{1}{\sqrt{2}} \right)^{2j} = j \frac{2^{j+1}}{2^j} = j2.$$

Clearly, as $j \to \infty$, we do not have $|u_j| \to 0$. Thus according to Theorem 2, the series diverges on $|z| = 1/\sqrt{2}$. ◀

In Example 1 we showed that $\sum_{j=1}^{\infty} z^{j-1} = 1/(1-z)$ for $|z| < 1$. We found the number N that would render the magnitude of the difference between the nth partial sum $S_n(z)$ and the sum $S(z) = 1/(1-z)$ less than ε if $n > N$. N (see Eq. 5.2–12) was found to depend on both ε and z.

Suppose, however, that in the course of establishing the convergence of a series we find, when z lies in some region R, an expression for ε that is independent of z. Such a series has special properties and is called *uniformly convergent* in R. More precisely:

Definition   Uniform convergence

The series $\sum_{j=1}^{\infty} u_j(z)$ whose nth partial sum is $S_n(z)$ is said to converge uniformly to $S(z)$ in a region R if, for any $\varepsilon > 0$, there exists a number N independent of z so that for all z in R

$$|S(z) - S_n(z)| < \varepsilon, \quad \text{for all } n > N. \quad \blacksquare \qquad (5.2\text{–}17)$$

There are various ways to show that a series is uniformly convergent in a region. In Exercise 5 of this section, for example, the reader will encounter one method. The series $\sum_{j=1}^{\infty} z^{j-1}$ is shown to be uniformly convergent in the disc $|z| \leq r$ (where $r < 1$) since we are able to find the required value of N in Eq. (5.2–17) that depends on only r and ε. This approach is time consuming. It is often easier to establish uniform convergence with the Weierstrass M test, which is described as follows:

Theorem 7 Weierstrass M test

Let $\sum_{j=1}^{\infty} M_j$ be a convergent series whose terms $M_1, M_2, \ldots$ are all positive constants. The series $\sum_{j=1}^{\infty} u_j(z)$ converges uniformly in a region R if

$$|u_j(z)| \leq M_j, \quad \text{for all } z \text{ in } R. \quad \blacksquare \qquad (5.2\text{–}18)$$

The test asserts that a series of complex functions $u_1(z) + u_2(z) + \cdots$ is uniformly convergent in a region R if there exists a convergent series of positive constants $M_1 + M_2 + \cdots$ each of whose terms $M_1, M_2, \ldots$ equals or exceeds the magnitude of the corresponding term $|u_1|, |u_2|, \ldots$ throughout R.

If Eq. (5.2–18) is satisfied, then $\sum_{j=1}^{\infty} u_j(z)$ is also absolutely convergent in R. This follows from the "comparison test" that the reader encountered for real series; it also applies to complex series.

The proof of the M test, Theorem 7, is not difficult provided one has first made a small study of what is called the Cauchy convergence criterion. We will not take this detour. The interested reader is referred to more advanced texts on complex variables for a proof.

Example 4

Use the M test to show that $\sum_{j=1}^{\infty} z^{j-1}$ is uniformly convergent in the disc $|z| \leq 3/4$.

Solution

From Eq. (5.2–7) with $z = 3/4$ or from a previous knowledge of real geometric series, we know that if $M_j = (3/4)^{j-1}$, then

$$\sum_{j=1}^{\infty} M_j = 1 + \frac{3}{4} + \left(\frac{3}{4}\right)^2 + \cdots = \frac{1}{1 - \dfrac{3}{4}}. \qquad (5.2\text{–}19)$$

Now, with $u_j = z^{j-1}$ we have the given series:

$$\sum_{j=1}^{\infty} u_j = \sum_{j=1}^{\infty} z^{j-1} = 1 + z + z^2 + \cdots. \qquad (5.2\text{–}20)$$

If $|z| \leq 3/4$, then the magnitude of each term of the series in Eq. (5.2–20) is less than or equal to the corresponding term in Eq. (5.2–19), for example, $|z^2| \leq (3/4)^2$, $|z^3| \leq (3/4)^3$, etc. so that $|u_j| \leq M_j$ and the M test is satisfied in the given region.

Comment: We can, by an identical argument, show that $\sum_{j=1}^{\infty} z^{j-1}$ is uniformly convergent in any circular region $|z| \leq r$ provided $r < 1$. The proof involves replacing 3/4 by r in the preceding example. ◄

We have dwelt on uniform convergence because series with this feature have some useful properties, which we will now list. The scope of this text does not allow for a derivation of all these properties. Most are derived as part of the exercises at the end of this section, and the reader is referred to more advanced texts for justification of the others.

Theorem 8

Let $\sum_{j=1}^{\infty} u_j(z)$ converge uniformly in a region R to $S(z)$. Let $f(z)$ be bounded in R, that is, $|f(z)| \leq k$ (k is constant) throughout R. Then, in R,

$$\sum_{j=1}^{\infty} f(z)u_j(z) = f(z)u_1(z) + f(z)u_2(z) + \cdots = f(z)S(z).$$

The series converges uniformly to $f(z)S(z)$. ∎

Thus since $\sum_{j=1}^{\infty} z^{j-1}$ converges uniformly for $|z| \leq r$, where $r < 1$, $\sum_{j=1}^{\infty} e^z z^{j-1}$ converges uniformly in the same region. (Recall that since e^z is continuous in the disc $|z| \leq r$ it must be bounded in this region.)

Theorem 9

Let $\sum_{j=1}^{\infty} u_j(z)$ be a series converging uniformly to $S(z)$ in R. If all the functions $u_1(z), u_2(z),\ldots$ are continuous in R, then so is the sum $S(z)$. ∎

For example, all the terms in the series $1 + z + z^2 + \cdots$ are continuous in the z-plane. This series is known to be uniformly convergent if $|z| \leq r$, ($r < 1$). Thus the sum must be continuous in $|z| \leq r$. A glance at the sum $1/(1 - z)$ (see Example 1 in this section) reveals this to be true.

Theorem 10   Term-by-term integration

Let $\sum_{j=1}^{\infty} u_j(z)$ be a series that is uniformly convergent to $S(z)$ in R and let all the terms $u_1(z), u_2(z),\ldots$ be continuous in R. If C is a contour in R, then

$$\int_C S(z)\, dz = \sum_{j=1}^{\infty} \int_C u_j(z)\, dz = \int_C u_1(z)\, dz + \int_C u_2(z)\, dz + \cdots, \quad (5.2\text{--}21)$$

that is, when a uniformly convergent series of continuous functions is integrated term by term the resulting series has a sum that is the integral of the sum of the original series. ∎

To illustrate this theorem we again consider

$$\frac{1}{1 - z} = 1 + z + z^2 + \cdots, \qquad |z| \leq r, \quad \text{and} \quad r < 1.$$

We integrate this series term by term along a contour C that lies entirely inside the disc $|z| \leq r$. The contour is assumed to connect the points $z = 0$ and $z = z'$. Thus from Eq. (5.2–21)

$$\int_0^{z'} \frac{1}{1-z}\, dz = \int_0^{z'} dz + \int_0^{z'} z\, dz + \int_0^{z'} z^2\, dz + \cdots . \tag{5.2–22}$$

The left side involves an integrand $1/(1-z)$, which is the derivative of an analytic branch of a multivalued function. Such integrations were considered in Section 4.3. We have

$$\int_0^{z'} \frac{1}{1-z}\, dz = -\text{Log}\,(1-z)\Big|_0^{z'} = \text{Log}\left(\frac{1}{1-z'}\right), \tag{5.2–23}$$

where we have elected to use the principal branch of $\log(1-z)$ since it is analytic in the disc under consideration. The preceding result can be used on the left in Eq. (5.2–22); the integrations on the right in Eq. (5.2–22) are readily performed. We have, finally,

$$\text{Log}\,\frac{1}{(1-z')} = z' + \frac{(z')^2}{2} + \frac{(z')^3}{3} + \cdots = \sum_{j=1}^{\infty} \frac{(z')^j}{j}, \qquad |z'| \le r, \qquad r < 1. \tag{5.2–24}$$

As a practical matter, the restriction on z' in this equation can be written simply $|z'| < 1$.

As shown in the exercises, Theorem 10 can be used to establish the following theorem.

Theorem 11   Analyticity of the sum of a series

If $\sum_{j=1}^{\infty} u_j(z)$ converges uniformly to $S(z)$ for all z in R and if $u_1(z), u_2(z), \ldots$ are all analytic in R, then $S(z)$ is analytic in R.   ∎

The preceding theorem guarantees the existence of the derivative of the sum of a uniformly convergent series of analytic functions. We have a way to arrive at this derivative:

Theorem 12   Term-by-term differentiation

Let $\sum_{j=1}^{\infty} u_j(z)$ converge uniformly to $S(z)$ in a region R. If $u_1(z), u_2(z), \ldots$ are all analytic in R, then at any interior point of this region

$$\frac{dS}{dz} = \sum_{j=1}^{\infty} \frac{du_j(z)}{dz} . \quad ∎ \tag{5.2–25}$$

The theorem states that when a uniformly convergent series of analytic functions is differentiated term by term, we obtain the derivative of the sum of the original series.

We illustrate the preceding with our geometric series. Since $1/(1-z) = \sum_{j=1}^{\infty} z^{j-1} = 1 + z + z^2 + \cdots$, where convergence is uniform for $|z| \le r$ (with $r < 1$), we have upon differentiation

$$\frac{d}{dz}\frac{1}{1-z} = \frac{1}{(1-z)^2} = \frac{d}{dz}(1 + z + z^2 + \cdots) = 1 + 2z + 3z^2 + \cdots ,$$

or

$$\frac{1}{(1 - z)^2} = \sum_{j=1}^{\infty} jz^{j-1}, \qquad |z| < r, \qquad r < 1. \qquad (5.2\text{--}26)$$

EXERCISES

1. Use the nth term test to show that

   a) the series of Example 1, $\sum_{n=1}^{\infty} z^{n-1}$, diverges for $|z| \geq 1$;

   b) $\displaystyle\sum_{n=1}^{\infty} e^{nz}$, diverges for Re $z \geq 0$;

   c) $\displaystyle\sum_{j=1}^{\infty} \frac{1}{(z-1)^j}$, diverges for $|z - 1| \leq 1$;

   d) $\displaystyle\sum_{j=1}^{\infty} \text{Log}\, \frac{z}{j}$, diverges for all z;

   e) $\displaystyle\sum_{n=1}^{\infty} \left(1 + \frac{1}{n}\right)^{-n}$, diverges. *Hint:* Look up the definition of e.

2. Use the ratio test to establish the absolute convergence of each of the following series in the indicated domain.

   a) $\displaystyle\sum_{n=1}^{\infty} e^{nz}$,   Re $z < 0$ (See also Exercise 1b.)

   b) $\displaystyle\sum_{j=1}^{\infty} \frac{1}{(z-1)^j}$, $|z - 1| > 1$ (See also Exercise 1c.)

   c) $\displaystyle\sum_{n=1}^{\infty} \frac{z^n}{n!}$, $|z| < \infty$

3. Use the fifth partial sum (that is, the first five terms) in Eq. (5.2–24) to find an approximate value of Log 2. Sum the terms with a pocket calculator. Compare your result, to three significant digits, with the exact value.

4. Use the Weierstrass M test to establish the uniform convergence of the following series in the indicated regions.

   a) $\displaystyle\sum_{n=0}^{\infty} \frac{z^n e^{in^2 x}}{n!}$, for $|z| \leq 1$

   Hint: Look up the infinite series for the number e.

   b) $\displaystyle\sum_{n=1}^{\infty} \frac{(-z)^n \left(1 + \dfrac{1}{n}\right)}{n!}$, for $|z| \leq 1$

   See the hint in part (a).

   c) $\displaystyle\sum_{j=1}^{\infty} \frac{z^j}{\left(j + \dfrac{1}{j}\right)}$, $|z| \leq 0.99$

Hint: Use Eq. (5.2–24) with an appropriate value of z.

d) $\sum_{j=1}^{\infty} \left(j - \frac{1}{j} \right) z^{j-1}$, $|z| \leq 0.99$.

Hint: See Eq. (5.2–26).

5. In this exercise we show that the series $\sum_{j=1}^{\infty} z^{j-1}$ converges uniformly to $1/(1-z)$ in the disc $|z| \leq r$, where $r < 1$. The proof requires that we obtain a value for N satisfying Eq. (5.2–17).

a) Explain why $\text{Log}(1/|z|) \geq \text{Log}(1/r)$ in the disc.

b) Prove $1/|1-z| \leq 1/(1-r)$ and $\text{Log}[1/(\varepsilon|1-z|)] \leq \text{Log}[1/(\varepsilon(1-r))]$ for $|z| \leq r$. Take $\varepsilon > 0$.

c) Assume that $0 < \varepsilon < 1/2$. Show that in the disc $|z| \leq r$ we have

$$\frac{\text{Log}\,\dfrac{1}{\varepsilon|1-z|}}{\text{Log}\,\dfrac{1}{|z|}} \leq \frac{\text{Log}\,\dfrac{1}{\varepsilon(1-r)}}{\text{Log}\,\dfrac{1}{r}}.$$

d) Observe that the left side of the preceding inequality is equal to the right side of Eq. (5.2–12). Explain why we can take N as any integer greater than or equal to

$$\frac{\text{Log}\,\dfrac{1}{\varepsilon(1-r)}}{\text{Log}\,\dfrac{1}{r}},$$

and Eq. (5.2–17) will be satisfied. Observe that N is independent of z.

6. We will prove Theorem 10, that is, establish that in a region R

$$\sum_{j=1}^{\infty} \int_C u_j(z)\, dz = \int_C S(z)\, dz,$$

where $\sum_{j=1}^{\infty} u_j(z)$ is assumed to converge uniformly to $S(z)$. From the definition of convergence we must prove that, given $\varepsilon_1 > 0$, there exists a number N such that

$$\left| \int_C S(z)\, dz - \sum_{j=1}^{n} \int_C u_j(z)\, dz \right| < \varepsilon_1 \quad \text{for } n > N. \tag{5.2–27}$$

a) Notice that for a finite sum

$$\sum_{j=1}^{n} \int_C u_j(z)\, dz = \int_C \sum_{j=1}^{n} u_j(z)\, dz$$

since u_1, u_2, etc. are assumed to be continuous (see Sec. 4.1). Explain why the following is true.

$$\left| \int_C S(z)\, dz - \sum_{j=1}^{n} \int_C u_j(z)\, dz \right| = \left| \int_C S(z)\, dz - \int_C \sum_{j=1}^{n} u_j(z)\, dz \right|$$

$$= \left| \int_C \left[S(z) - \sum_{j=1}^{n} u_j(z) \right] dz \right|$$

b) Given $\varepsilon > 0$, determine that there exists an N such that

$$\left| \int_C \left[S(z) - \sum_{j=1}^{n} u_j(z) \right] dz \right| < \varepsilon L, \quad \text{for } n > N, \qquad (5.2\text{–}28)$$

where L is the length of C. (Recall the definition of a uniformly convergent series and use the ML inequality.) Now observe that if we take $\varepsilon = \varepsilon_1/L$ in Eq. (5.2–28) we have proved Eq. (5.2–27).

7. Use Theorem 10 together with Morera's theorem (see Exercise 3, Section 4.2) to prove Theorem 11.

8. Prove Theorem 12. *Hint:* Consider the series $\sum_{j=1}^{\infty} u_j(z') = S(z')$, where convergence is uniform in a region R of the z'-plane. Let z be any point in R except a boundary point. Consider a simple closed contour C lying in R and enclosing z. Then, dividing the preceding series by $(z' - z)^2$ and invoking Theorem 8, we have

$$\frac{S(z')}{2\pi i (z' - z)^2} = \frac{u_1(z')}{2\pi i (z' - z)^2} + \frac{u_2(z')}{2\pi i (z' - z)^2} + \frac{u_3(z')}{2\pi i (z' - z)^2} + \cdots,$$

where z' is now assumed to lie on C. Note that $1/(z' - z)$ is bounded on C. Integrate the left side of the preceding equation around C, and make a term-by-term integration of the right side around the same contour. Use Theorems 8 and 10 for justification. Evaluate each integral by using an extension of the Cauchy integral formula and thus obtain Eq. (5.2–25).

5.3   POWER SERIES AND TAYLOR SERIES

As we noted earlier, a power series is a sum of the form $\sum_{n=0}^{\infty} c_n(z - z_0)^n$. Part of our task in this section is to see when the theorems of the previous section, especially those on uniform convergence, are applicable to power series. The series notation $\sum_{j=1}^{\infty} u_j(z)$ used in the previous section can be used to generate a power series with the substitution

$$u_j(z) = c_{j-1}(z - z_0)^{j-1}. \qquad (5.3\text{–}1)$$

We begin by discussing two theorems that apply specifically to power series.

Theorem 13

If $\sum_{n=0}^{\infty} c_n(z - z_0)^n$ converges when $z = z_1$, then this series converges for all z satisfying $|z - z_0| < |z_1 - z_0|$. The convergence is absolute for these values of z. ∎

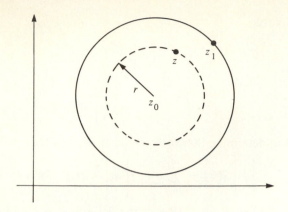

Figure 5.3–1

To understand this theorem, imagine a circle in the z-plane centered at z_0, as shown in Fig. 5.3–1. Suppose the given series is known to converge for $z = z_1$. Then the series will converge for any z lying within the solid circle in the figure.

The proof of Theorem 13, which involves a comparison test, is not difficult; in fact, it will not be presented here because it is sufficiently similar to the proof to be presented for Theorem 14.

Theorem 14   Uniform convergence and analyticity of power series

If $\sum_{n=0}^{\infty} c_n(z - z_0)^n$ converges when $z = z_1$, then the series converges uniformly for all z in the disc $|z - z_0| \leq r$, where $r < |z_1 - z_0|$. The sum of the series is an analytic function for $|z - z_0| \leq r$. ∎

We assume in this theorem, as we do in the following proof, that $z_1 \neq z_0$; therefore, the distance $|z_1 - z_0|$ is nonzero. The theorem asserts that if the power series converges at z_1 in Fig. 5.3–1, then the series converges to an analytic function on and inside the broken circle shown in this figure.

The proof of Theorem 14 involves the Weierstrass M test. To begin, we consider the convergent series

$$\sum_{n=0}^{\infty} c_n(z_1 - z_0)^n = c_0 + c_1(z_1 - z_0) + c_2(z_1 - z_0)^2 + \cdots. \qquad (5.3\text{–}2)$$

For a convergent series of constants, such as the preceding one, we can find a number m that equals or exceeds the magnitude of any of the terms. Thus

$$|c_n(z_1 - z_0)^n| \leq m, \qquad n = 0, 1, 2, \ldots. \qquad (5.3\text{–}3)$$

Now consider the original series

$$\sum_{n=0}^{\infty} c_n(z - z_0)^n = c_0 + c_1(z - z_0) + c_2(z - z_0)^2 + \cdots, \qquad (5.3\text{–}4)$$

where we take $|z - z_0| \leq r$ and $r < |z_1 - z_0|$. Notice that the terms in Eq. (5.3–4) can be written

$$c_n(z - z_0)^n = c_n(z_1 - z_0)^n \left(\frac{z - z_0}{z_1 - z_0} \right)^n.$$

Taking magnitudes yields

$$|c_n(z - z_0)^n| = |c_n(z_1 - z_0)^n| \left| \frac{z - z_0}{z_1 - z_0} \right|^n. \tag{5.3-5}$$

Let $p = r/|z_1 - z_0|$, where, by hypothesis, $p < 1$. Since $|z - z_0| < r$, we have

$$\left| \frac{z - z_0}{z_1 - z_0} \right| < p. \tag{5.3-6}$$

Simultaneously applying this inequality, as well as Eq. (5.3–3), to the right side of Eq. (5.3–5), we obtain

$$|c_n(z - z_0)^n| < mp^n. \tag{5.3-7}$$

Let $M_n = mp^n$. From Eq. (5.3–7) we have

$$|c_n(z - z_0)^n| < M_n. \tag{5.3-8}$$

The summation

$$\sum_{n=0}^{\infty} M_n = \sum_{n=0}^{\infty} mp^n = m \sum_{n=0}^{\infty} p^n, \qquad p < 1 \tag{5.3-9}$$

involves a convergent geometric series of real constants (see Eq. 5.2–7).

The inequality shown in Eq. (5.3–8), the convergence of Eq. (5.3–9), and Theorem 7 together guarantee the uniform convergence $\sum_{n=0}^{\infty} c_n(z - z_0)^n$ for $|z - z_0| \leq r$. Because the individual terms $c_n(z - z_0)^n$ in this series are each analytic functions, it follows (see Theorem 11) that the sum of this series is an analytic function in $|z - z_0| \leq r$. The proof of Theorem 14 is complete.

Now consider all the possible values of z for which $\sum_{n=0}^{\infty} c_n(z - z_0)^n$ is convergent. Suppose we find that value of z lying farthest from z_0 for which this series converges. Calling this value z_2 and taking $|z_2 - z_0| = \rho$, we see from Theorem 13 that $|z - z_0| < \rho$ describes the largest disc centered at z_0 within which our power series is convergent. By Theorem 14 this series converges uniformly to an analytic function on and inside any circle centered at z_0 whose radius is less than ρ. A circle such as the one just described is known as the *circle of convergence* of a power series.

Definition Circle of convergence

The largest circle centered at z_0 inside which the series $\sum_{n=0}^{\infty} c_n(z - z_0)^n$ converges everywhere is called the *circle of convergence* of this series. The radius ρ of the circle is called the *radius of convergence* of the series. The center of the circle z_0 is called the *center of expansion* of the series. ∎

It is possible for the radius ρ to be as large as infinity in certain cases.

We know from Theorem 14 that the sum of a uniformly convergent power series is an analytic function. Is an analytic function always representable by a power series? The answer is yes, as we will see by proving the following theorem.

Theorem 15 Taylor series

Let $f(z)$ be analytic at z_0. Let C be the largest circle centered at z_0, inside which $f(z)$ is everywhere analytic, and let a be the radius of C. Then there exists a power series $\sum_{n=0}^{\infty} c_n(z - z_0)^n$, which converges to $f(z)$ in C; that is,

$$f(z) = \sum_{n=0}^{\infty} c_n(z - z_0)^n, \qquad |z - z_0| < a, \qquad (5.3\text{--}10)$$

where

$$c_n = \frac{f^{(n)}(z_0)}{n!}. \qquad (5.3\text{--}11)$$

This power series is called the Taylor series expansion of $f(z)$ about z_0. In the special case $z_0 = 0$ we call the Taylor series a Maclaurin series. ■

Notice that the preceding theorem makes no guarantees concerning the convergence of the series to $f(z)$ when z lies on C. Here, each series, and each value of z on C, must be examined on an individual basis.

For simplicity we will prove the theorem by making an expansion about $z_0 = 0$, and we will then indicate how to extend our work to the case $z_0 \neq 0$.

We construct a circle C in Fig. 5.3–2. Its center is at the origin and its radius, which we call a, is the distance from the origin to the closest singularity of $f(z)$. We assume that the origin is not a singular point. Let z_1 lie within this contour. We enclose z_1 by a second circle C' centered at the origin but having a radius less than that of C. Since the radius of C' is b, we have $|z_1| < b < a$. By the Cauchy integral formula

$$f(z_1) = \frac{1}{2\pi i} \oint_{C'} \frac{f(z)}{(z - z_1)} dz = \frac{1}{2\pi i} \oint_{C'} \frac{f(z)\, dz}{z\left(1 - \dfrac{z_1}{z}\right)}. \qquad (5.3\text{--}12)$$

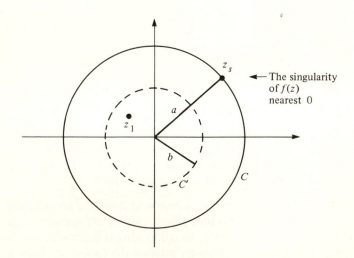

The singularity of $f(z)$ nearest 0

Figure 5.3–2

Now consider

$$\frac{1}{1 - \frac{z_1}{z}} = 1 + \frac{z_1}{z} + \left(\frac{z_1}{z}\right)^2 + \left(\frac{z_1}{z}\right)^3 + \cdots. \qquad (5.3\text{–}13)$$

This is just the series of Example 4 of Section 5.2 with z replaced by z_1/z. According to that example, the series of Eq. (5.3–13) is uniformly convergent when

$$\left|\frac{z_1}{z}\right| \le r, \quad \text{where } r < 1. \qquad (5.3\text{–}14)$$

If z is confined to the contour C' (see Fig. 5.3–2), we observe that $|z_1/z| < 1$, and we can readily find a value of r satisfying Eq. (5.3–14).

The function $f(z)/z$ is bounded on C'. By invoking Theorem 8 we can formally multiply both sides of Eq. (5.3–13) by $f(z)/z$ and obtain

$$\frac{f(z)}{z\left(1 - \frac{z_1}{z}\right)} = f(z)\frac{1}{z} + f(z)\frac{z_1}{z^2} + f(z)\frac{z_1^2}{z^3} + \cdots, \qquad (5.3\text{–}15)$$

which is uniformly convergent in some region containing C'.

Notice that the right side of Eq. (5.3–15) is a series expansion of the integrand in Eq. (5.3–12). Each term of this series is continuous on C'. Thus from Theorem 10 a term-by-term integration of the series in Eq. (5.3–15) is possible around C'. Therefore from Eqs. (5.3–12) and (5.3–15)

$$f(z_1) = \frac{1}{2\pi i}\oint_{C'} \frac{f(z)}{z}\, dz + \frac{z_1}{2\pi i}\oint_{C'} \frac{f(z)}{z^2}\, dz + \frac{z_1^2}{2\pi i}\oint_{C'} \frac{f(z)}{z^3}\, dz + \cdots. \qquad (5.3\text{–}16)$$

Since z_1 is a fixed point, it was brought out from under each integral sign.

There are an infinite number of integrals on the right in Eq. (5.3–16), each of which can be evaluated with the extended Cauchy integral formula. With $z_0 = 0$ in Eq. (4.4–16) we have

$$\frac{1}{2\pi i}\oint_{C'} \frac{f(z)}{z^{n+1}}\, dz = \frac{f^{(n)}(0)}{n!}, \quad n = 0, 1, 2, \ldots. \qquad (5.3\text{–}17)$$

Rewriting Eq. (5.3–16) with this formula we obtain

$$f(z_1) = \sum_{n=0}^{\infty} c_n z_1^n = c_0 + c_1 z_1 + c_2 z_1^2 + \cdots, \qquad (5.3\text{–}18)$$

for $|z_1| < b < a$, where

$$c_n = \frac{f^{(n)}(0)}{n!}. \qquad (5.3\text{–}19)$$

Replacing what is now the dummy variable z_1 in Eq. (5.3–18) by z, we find that we have derived Eqs. (5.3–10) and (5.3–11) for the special case $z_0 = 0$. The constraint on $|z_1|$ now becomes $|z| < b < a$, and because b can be made arbitrarily close to a, we will write this $|z| < a$. Thus with $z_0 = 0$, z is constrained to lie inside the largest circle, centered at the origin within which $f(z)$ is analytic.

The more general result $z_0 \neq 0$ described in Theorem 15, is obtained by a derivation much like the one just given. The contours C and C' in Fig. 5.3-2 become circles centered at z_0. Again, z_1 lies inside C'. Equation (5.3–12) still holds, but the integrand is now written

$$\frac{f(z)}{z - z_1} = \frac{f(z)}{(z - z_0)\left[1 - \frac{(z_1 - z_0)}{(z - z_0)}\right]},$$

and a series expansion is made in powers of $(z_1 - z_0)/(z - z_0)$. The reader can supply the additional details in Exercise 2 of this section.

Theorem 15 is enormously useful. It tells us that any function $f(z)$, analytic at z_0, is the sum of a power series, called a Taylor series, containing powers of $(z - z_0)$. In Eq. (5.3–11) the theorem tells us how to obtain the coefficients for the Taylor series. Since all the derivatives of an analytic function exist (see Section 4.4), all the coefficients are defined. The procedure for getting the coefficients is identical to that used in Taylor expansions of functions of a real variable (see Eq. 5.1–4). Finally, the theorem guarantees that as long as z lies within a certain circle, centered at z_0, the Taylor series converges to $f(z)$. The radius of this circle is precisely the distance from z_0 to the nearest singularity of $f(z)$ in the complex plane. Some examples of Taylor series follow.

Example 1

Expand e^z in (a) a Maclaurin series and (b) a Taylor series about $z = i$.

Solution

Part (a):

$$e^z = c_0 + c_1 z + c_2 z^2 + \cdots.$$

From Eq. (5.3–11), with $z_0 = 0$,

$$c_n = \frac{\dfrac{d^n}{dz^n} e^z \Big|_{z=0}}{n!} = \frac{e^z}{n!}\Big|_{z=0} = \frac{1}{n!}.$$

Thus

$$e^z = \sum_{n=0}^{\infty} \frac{1}{n!} z^n. \tag{5.3–20}$$

Because e^z is analytic for all finite z, Theorem 15 guarantees that the series in Eq. (5.3–20) converges to e^z everywhere in the complex plane. Putting $z = x$ in Eq. (5.3–20) yields a familiar expansion for the real function e^x.

Part (b):

$$e^z = c_0 + c_1(z - i) + c_2(z - i)^2 + \cdots.$$

From Eq. (5.3–11), with $z_0 = i$,

$$c_n = \frac{\dfrac{d^n}{dz^n} e^z \Big|_{z=i}}{n!} = \frac{e^z}{n!}\Big|_{z=i} = \frac{e^i}{n!};$$

thus

$$e^z = \sum_{n=0}^{\infty} \frac{e^i}{n!}(z - i)^n.$$

The series representation is again valid throughout the complex plane. ◄

Other useful Maclaurin series expansions besides Eq. (5.3–20) are

$$\sin z = z - \frac{z^3}{3!} + \frac{z^5}{5!} + \cdots,$$

$$\cos z = 1 - \frac{z^2}{2!} + \frac{z^4}{4!} + \cdots,$$

$$\sinh z = z + \frac{z^3}{3!} + \frac{z^5}{5!} + \cdots,$$ (5.3–21)

$$\cosh z = 1 + \frac{z^2}{2!} + \frac{z^4}{4!} + \cdots.$$

Because the four preceding functions on the left are analytic for $|z| < \infty$, the series representations are valid throughout the z-plane.

The question of convergence is of consequence in the following example.

Example 2

Expand

$$f(z) = \frac{1}{(1 - z)}$$

in the Taylor series $\sum_{n=0}^{\infty} c_n(z + 1)^n$. For what values of z must the series converge to $f(z)$?

Solution

The expansion is about $z_0 = -1$. From Eq. (5.3–11), with $f(z) = 1/(1 - z)$, we find $c_0 = 1/2$, $c_1 = 1/4$, and, in general, $c_n = 1/2^{n+1}$. Thus

$$f(z) = \frac{1}{1 - z} = \sum_{n=0}^{\infty} \frac{1}{2^{n+1}}(z + 1)^n.$$ (5.3–22)

To study the validity of this series representation we must see where the singularities of $1/(1 - z)$ lie in the complex plane and determine which one lies closest to $z_0 = -1$.

Since $f(z)$ is analytic except at $z = 1$, Theorem 15 *guarantees* that the series in Eq. (5.3–22) will converge to $f(z)$ for all z lying inside a circle centered at -1 having radius 2. We will soon see that it is impossible for the series to converge to $f(z)$ outside this circle. ◄

Given any analytic function $f(z)$ we know that it can be represented in a Taylor series about the point z_0. Might there be some other power series using powers of $(z - z_0)$ that converges to $f(z)$ in a neighborhood of z_0? The answer is no. Let

$$f(z) = b_0 + b_1(z - z_0) + b_2(z - z_0)^2 + \cdots, \qquad |z - z_0| \le r. \quad (5.3\text{--}23)$$

We can show that this must be the Taylor expansion of $f(z)$ about z_0. Invoking Theorem 14 and Theorem 12, we differentiate Eq. (5.3–23) and find that

$$f'(z) = b_1 + 2b_2(z - z_0) + \cdots, \qquad |z - z_0| < r. \qquad (5.3\text{--}24)$$

This series can be differentiated again:

$$f''(z) = 2b_2 + 3 \cdot 2b_3(z - z_0) + \cdots. \qquad (5.3\text{--}25)$$

and again.

Setting $z = z_0$ in Eqs. (5.3–23), (5.3–24), and (5.3–25) we get

$$b_0 = f(z_0), \qquad b_1 = f'(z_0), \qquad b_2 = \frac{f''(z_0)}{2},$$

and, in general, one readily shows that

$$b_n = \frac{f^{(n)}(z_0)}{n!}, \qquad n = 0, 1, 2, \ldots.$$

But these coefficients are precisely those used in the Taylor expansion (see Eq. 5.3–11). Thus we conclude:

Theorem 16

The Taylor series expansion about z_0 of a function $f(z)$ is the *only* power series that will converge to $f(z)$ everywhere in a circle centered at z_0. ∎

According to Theorem 15, when a function $f(z)$ is expanded in a Taylor series about z_0, the resulting series *must* converge to $f(z)$ whenever z resides within a certain circle centered at z_0.

Theorem 15 explains how to find the radius of the circle. We can prove that this is the largest circle, centered at z_0, in which the Taylor series converges to $f(z)$.

Let z_s be that singularity of $f(z)$ lying closest to z_0. The circle shown in solid line in Fig. 5.3–3 is the one described in Theorem 15. Assume that the Taylor expansion of $f(z)$ converges to $f(z)$ in the disc $|z - z_0| < \alpha$, where $\alpha > a = |z_s - z_0|$. We thus have a power series that converges in the disc $|z - z_0| < \alpha$. Now, according to Theorem 14, such a power series converges to a function that is analytic throughout a disc that is larger than, and contains, the disc of radius a shown in Fig. 5.3–3. This larger disc contains the point z_s, where $f(z)$ is known to be nonanalytic.

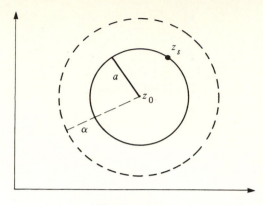

Figure 5.3–3

Thus our assumption that the Taylor expansion converges to $f(z)$ in a circle larger than $|z - z_s| = a$ must be false. To summarize:

Theorem 17

Let $f(z)$ be expanded in a Taylor series about z_0. The largest circle within which this series converges to $f(z)$ at each point is $|z - z_0| = a$, where a is the distance from z_0 to the nearest singular point of $f(z)$. ■

Notice that this theorem *does not* assert that the Taylor series fails to converge outside $|z - z_0| = a$. It asserts that this is the largest circle throughout which the series converges to $f(z)$.

The circle in which the Taylor series $\sum_{n=0}^{\infty}[f^{(n)}(z_0)/n!](z - z_0)^n$ is everywhere convergent to $f(z)$ and the circle throughout which this series converges are not necessarily the same. The second circle could be larger. This fact is considered in Exercise 8 of this section. It can be shown, however, that when the singularity z_s lying nearest z_0 is one where $|f(z)|$ becomes infinite, then the two circles are identical. This is the case in most of the examples that we will consider.

Example 3

Without actually obtaining the Taylor series give the largest circle throughout which the indicated expansion is valid.

$$f(z) = \frac{1}{z^2 + 1} = \sum_{n=0}^{\infty} c_n(z - 2)^n \qquad (5.3\text{--}26)$$

Solution

Convergence takes place within a circle centered at $z = 2$, as shown in Fig. 5.3–4. The singularities of $f(z)$ lie at $\pm i$. The nearest singularity to $z = 2$ is, in this case, either $+i$ or $-i$. They are equally close. The distance from $z = 2$ to these points is $\sqrt{5}$. Thus the Taylor series converges to $f(z)$ throughout the circular domain $|z - 2| < \sqrt{5}$. ◄

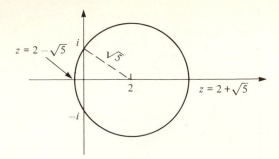

Figure 5.3–4

Example 4

Consider the real Taylor series expansion

$$\frac{1}{x^2 + 1} = \sum_{n=0}^{\infty} c_n(x - 2)^n. \qquad (5.3\text{-}27)$$

Determine the largest interval along the x-axis inside which the series converges to $1/(x^2 + 1)$.

Solution

By requiring z to be a real variable ($z = x$) in Eq. (5.3–26), we will obtain the series of the present problem. If z is a real variable and if, in addition, the series on the right in Eq. (5.3–26) is to converge to the function on the left, not only must z lie on the real axis in Fig. 5.3–4, it must be inside the indicated circle as well. Thus we require $2 - \sqrt{5} < x < 2 + \sqrt{5}$ for convergence. Whether the series in Eq. (5.3–27) will converge to $1/(x^2 + 1)$ at either of the endpoints of this interval, that is, $x = 2 \pm \sqrt{5}$, cannot be determined from Theorem 17. ◀

EXERCISES

1. Derive the Maclaurin series expansions in Eq. (5.3–21).

2. Theorem 15 on Taylor series was derived for expansions about the origin $z_0 = 0$. Follow the suggestions given in that derivation and give a proof valid for any z_0.

3. Use Eq. (5.3–11) to obtain the first few terms of the indicated Taylor series expansions. Give an explicit formula for the nth coefficient c_n. State the largest circle within which your expansion is valid.

a) Log z,   about 1,

b) $\dfrac{1}{z^2}$,   about 1

c) $\dfrac{1}{z}$,   about i

d) $\dfrac{1}{z + 1}$,   about 2

e) sin z,   about $-i$

f) cosh z,   about $1 + i$

4. a) Explain why $z^{1/2}$ (principal branch) cannot be expanded in a Maclaurin series. Also, explain why the series expansion sought in Eq. (5.1–5) does not exist.

   b) Explain whether this same branch of $z^{1/2}$ can be expanded in a Taylor series about 1. If so, find the first three terms and state the circle within which the expansion is valid.

5. Without actually obtaining the series determine the center and radius of the circle within which the indicated Taylor series converges to the given function.

   a) $\tan z = \sum\limits_{n=0}^{\infty} c_n z^n$

   b) $\tan z = \sum\limits_{n=0}^{\infty} c_n \left(z - \dfrac{\pi}{8} \right)^n$

   c) $\dfrac{1}{e^z - e}$,   expanded about $z = 0$

   d) $\dfrac{1}{\cos \dfrac{z}{2} \cosh z}$,   expanded about $z = 0$

   e) $\dfrac{1}{(z - 2) \cos z}$,   expanded about $z = -i$

   f) $\dfrac{1}{(z + 4)(z^2 + z + 1)}$,   expanded about $z = 1$

6. Without obtaining the series determine the interval along the x-axis for which the indicated real Taylor series converges to the given real function. Convergence at the endpoints of the interval need not be considered.

   a) $\dfrac{1}{1 - x}$,   expanded about $x = -1$

   How do your findings confirm Eqs. (5.1–2), (5.1–6), and (5.1–7)?

   b) $\dfrac{1}{x^2 - x + 1}$,   expanded about $x = 1$

   c) $\dfrac{1}{\sin x}$,   expanded about $x = \dfrac{3}{4}$

7. a) Use a Maclaurin series expansion to establish the binomial expansion for an integer $N \geq 0$.

$$(1 + z)^N = \sum_{n=0}^{N} \frac{N! z^n}{n!(N - n)!}, \quad \text{all } z$$

   b) By substituting z/w for z in the preceding equation and then multiplying both sides by w^N, show that

$$(w + z)^N = \sum_{n=0}^{N} \frac{N! w^{N-n} z^n}{n!(N - n)!}.$$

   This is a more general form of the binomial expansion.

8. Let a function $f(z)$ be expanded in a Taylor series about z_0. The circle, centered at z_0, within which this series converges is in certain cases larger than, but concentric with, the circle in which the series converges to $f(z)$. We will investigate this possibility in one particular case.

a) Let $f(z) = \mathrm{Log}\,(z - 1/\sqrt{2}\,)$, where the principal branch of the logarithm is used. Thus $f(z)$ is defined by means of the branch cut $y = 0$, $x \leq 1/\sqrt{2}$.
Show that

$$f(z) = \sum_{n=0}^{\infty} c_n \left(z - \frac{i}{\sqrt{2}} \right)^n,$$

where

$$c_0 = \frac{i3\pi}{4} \quad \text{and} \quad c_n = \frac{(-1)^{n+1} e^{-i(3\pi/4)n}}{n}, \qquad n \neq 0.$$

b) What is the radius of the largest circle centered at $i/\sqrt{2}$ within which the series of part (a) converges to $f(z)$? *Hint:* Draw the branch cut mentioned in part (a).

c) Use the ratio test to establish that the series of part (a) converges inside $|z - i/\sqrt{2}\,| = 1$ and diverges outside this circle. Compare this circle with the one in part (b).

9. This problem investigates Taylor series with a *remainder*. In numerical calculations we often use the first n terms of a Taylor series instead of the entire (infinite) expansion. The difference between the sum $f(z)$ of the infinite series and the sum of the finite series actually used constitutes an error and is called the *remainder*. The size of the remainder is of interest. An upper bound for its magnitude can be determined with the help of Taylor's theorem. We will derive this theorem for the special case of an expansion about the origin.

a) Refer to the proof of Theorem 15 and to Fig. 5.3–2. Use Eq. (5.3–12) to show that

$$f(z_1) = \frac{1}{2\pi i} \oint_{C'} f(z) \left[\frac{1}{z} + \frac{z_1}{z^2} + \cdots + \frac{z_1^{n-1}}{z^n} + \left(\frac{z_1}{z} \right)^n \frac{1}{(z - z_1)} \right] dz.$$

Hint: Refer to Eq. (5.2–8), which implies that

$$\frac{1}{1 - z} = 1 + z + z^2 + \cdots + z^{n-1} + \frac{z^n}{1 - z},$$

and replace z by z_1/z.

b) Use the expression for $f(z_1)$ given in part (a) to show, after integration, that

$$f(z_1) = f(0) + f'(0)z_1 + \frac{f''(0)}{2} z_1^2 + \cdots + \frac{f^{(n-1)}(0)}{(n-1)!} z_1^{n-1} + R_n, \quad (5.3\text{–}28)$$

where

$$R_n = \frac{z_1^n}{2\pi i} \oint_{C'} \frac{f(z)}{z^n(z - z_1)} \, dz. \qquad (5.3\text{–}29)$$

Comparing Eqs. (5.3–28) and (5.3–18) we see that R_n is the remainder representing the difference between $f(z_1)$ and the first n terms of its Maclaurin expansion.

c) We can place an upper bound on the remainder in Eq. (5.3–29). Assume $|f(z)| \leq m$ everywhere on $|z| = b$ (the contour C'). Use the ML inequality to show that

$$|R_n| \leq \left| \frac{z_1}{b} \right|^n \frac{mb}{b - |z_1|}. \tag{5.3–30}$$

Hint: Note that for z lying on contour C'

$$\frac{1}{|z - z_1|} \leq \frac{1}{b - |z_1|}.$$

Why?

In passing, we notice that since $|z_1/b| < 1$, the remainder R_n in Eq. (5.3–30) tends to zero as $n \to \infty$. Using this limit we find that the right side of Eq. (5.3–28) becomes the Maclaurin series of $f(z_1)$. This constitutes a derivation of the Maclaurin expansion shown in Eq. (5.3–18) not requiring the use of uniform convergence. A similar derivation applies for the Taylor expansion.

d) Suppose we wish to determine the approximate value of $\cosh i$ by the finite series $i^0 + i^2/2! + \cdots + i^{10}/10!$ (see Eq. 5.3–21). Taking the contour C' in Eq. (5.3–29) as $|z| = 2$, show by using Eq. (5.3–30) that the error made cannot exceed $(\cosh 2)/2^{10} \doteq 3.67 \times 10^{-3}$. *Hint:* Observe that $|\cosh z| \leq \cosh x$ and use this fact to find m.

5.4   TECHNIQUES FOR OBTAINING TAYLOR SERIES EXPANSIONS

Equation (5.3–11) allows us, in principle, to obtain the coefficients for the Taylor series expansion of any analytic function. Alternatively, there are sometimes short-cuts that can be applied that will relieve us of the tedium of taking high-order derivatives to obtain the coefficients. We will explore some of these techniques in this section.

Term-by-term Differentiation and Integration

In our derivation of Theorem 16 we observed that the Taylor series of $f(z)$ can be differentiated term by term. If the original series converged to $f(z)$ inside a circle having center z_0 and radius r, the series obtained through differentiation converges to $f'(z)$ inside this circle. The procedure can be repeated indefinitely to yield a series for any derivative of $f(z)$.

Finally, Theorem 10 permits us to integrate the Taylor series for $f(z)$ term by term along a path lying inside the circle of radius r. The resulting series converges to the integral of $f(z)$ taken along this path.

Example 1

Use term-by-term differentiation and the expansion of $1/z$ about $z = 1$ to obtain the expansion of $1/z^2$ about the same point.

Solution

From Theorem 15, with $f(z) = 1/z$ and the use of Eqs. (5.3–10) and (5.3–11), we find that

$$\frac{1}{z} = 1 - (z - 1) + (z - 1)^2 + \cdots, \qquad |z - 1| < 1. \qquad (5.4–1)$$

Differentiating both sides with respect to z and multiplying by (-1), we obtain

$$\frac{1}{z^2} = 1 - 2(z - 1) + 3(z - 1)^2 + \cdots = \sum_{n=0}^{\infty} (-1)^n (n + 1)(z - 1)^n, \quad (5.4–2)$$

valid for $|z - 1| < 1$. ◄

Term-by-term integration is illustrated in the following example.

Example 2

Obtain the Maclaurin expansion of

$$S_i(z) = \int_0^z f(z') \, dz', \qquad (5.4–3)$$

where

$$f(z') = \frac{\sin z'}{z'}, \qquad z' \neq 0, \qquad (5.4–4a)$$

$$f(0) = 1, \qquad z' = 0. \qquad (5.4–4b)$$

The function $S_i(z)$ is called the *sine integral* and cannot be evaluated in terms of elementary functions.

Solution

From Eq. (5.3–21) we have

$$\sin z' = z' - \frac{(z')^3}{3!} + \frac{(z')^5}{5!} + \cdots,$$

so that

$$\frac{\sin z'}{z'} = 1 - \frac{(z')^2}{3!} + \frac{(z')^4}{5!} + \cdots.$$

Notice that this series converges to $f(z')$ in Eq. (5.4–4) for $z' \neq 0$ and $z' = 0$. We now integrate as follows:

$$\int_0^z \frac{\sin z'}{z'} \, dz' = \int_0^z dz' + \int_0^z \frac{-(z')^2}{3!} \, dz' + \int_0^z \frac{(z')^4}{5!} \, dz' + \cdots$$

$$= z - \frac{z^3}{3 \cdot 3!} + \frac{z^5}{5 \cdot 5!} + \cdots.$$

Thus

$$S_i(z) = \sum_{n=0}^{\infty} c_n z^{(2n+1)}, \qquad (5.4\text{--}5)$$

where

$$c_n = \frac{(-1)^n}{(2n+1)!(2n+1)}.$$

The expansion is valid throughout the z-plane. ◀

Example 3

Use the Maclaurin series for e^z to obtain the Maclaurin series for e^{z^2}.

Solution

From Eq. (5.3–20)

$$e^z = 1 + z + \frac{z^2}{2!} + \frac{z^3}{3!} + \cdots, \qquad |z| < \infty.$$

Now, substituting z^2 for z we have

$$e^{z^2} = 1 + z^2 + \frac{z^4}{2!} + \frac{z^6}{3!} + \cdots, \qquad |z|^2 < \infty \quad \text{or} \quad \text{all finite } z. \quad ◀$$

Multiplication and Division of Series

Because of the absolute convergence of Taylor series (see Theorem 13) Taylor series expansions about z_0 of two functions can be multiplied[†] as if they were polynomials. The series are multiplied term by term; then, like powers of $(z - z_0)$ are combined. This is known as the Cauchy product of the two series. The result is a Taylor expansion of the product of the sums of the two original series. The procedure is readily extended to finding the Taylor series expansion of the product of more than two functions.

Multiplication of series is often a tedious procedure, especially if we want a general formula for the nth coefficient in the resulting series. However, if we need only the first few terms in the result, it is easy to use.

Example 4

Obtain the first three nonzero terms in the Maclaurin expansion of

$$f(z) = \frac{e^z}{1 - z} \sin z.$$

Solution

From Eqs. (5.3–20), (5.3–21), and (5.2–7) we obtain the Maclaurin series for e^z, $\sin z$, and $1/(1 - z)$, respectively. The first two expansions are valid for all finite z while the third is restricted to $|z| < 1$. Thus

$$f(z) = \left(1 + z + \frac{z^2}{2!} + \cdots\right)\left(z - \frac{z^3}{3!} + \frac{z^5}{5!} + \cdots\right)(1 + z + z^2 + \cdots), \qquad |z| < 1.$$

[†]See K. Knopp, *Infinite Sequences and Series* (New York: Dover, 1956), sec. 3.6.

We now multiply the first two series and obtain

$$f(z) = \left(z + z^2 + \left(\frac{1}{2!} - \frac{1}{3!} \right) z^3 + \cdots \right) \left(1 + z + z^2 + \cdots \right).$$

Note that $1/2! - 1/3! = 1/3$.

The two preceding series are now multiplied with the result that

$$f(z) = \left(z + (1 + 1)z^2 + (1 + 1 + 1/3)z^3 + \cdots \right).$$

Hence

$$\frac{e^z \sin z}{1 - z} = z + 2z^2 + \frac{7}{3}z^3 + \cdots, \qquad |z| < 1. \blacktriangleleft \tag{5.4-6}$$

Suppose $f(z)$ and $g(z)$ are both analytic at z_0. Then, if $g(z_0) \neq 0$, the quotient

$$h(z) = \frac{f(z)}{g(z)}$$

is analytic at z_0 and can be expanded in a Taylor series about this point. In Exercise 11 of this section we justify (to some extent) obtaining this expansion by a long division of the Taylor series for $f(z)$ and $g(z)$. An example of this method follows.

Example 5

Obtain the Maclaurin expansion of $\tan z = \sin z / \cos z$ from the Maclaurin series of $\sin z$ and $\cos z$.

Solution

From Eq. (5.3–21) we have

$$\sin z = z - \frac{z^3}{3!} + \frac{z^5}{5!} + \cdots,$$

$$\cos z = 1 - \frac{z^2}{2!} + \frac{z^4}{4!} + \cdots.$$

We divide these series as follows:

$$
\begin{array}{r}
z + \left(\dfrac{1}{2!} - \dfrac{1}{3!} \right) z^3 + \cdots = \tan z \\[2mm]
\hline
\end{array}
$$

$$1 - \frac{z^2}{2!} + \frac{z^4}{4!} - \cdots \;\bigg)\; z - \frac{z^3}{3!} + \frac{z^5}{5!} - \cdots$$

$$z - \frac{z^3}{2!} + \frac{z^5}{4!}$$

$$\overline{\left(\frac{1}{2!} - \frac{1}{3!} \right) z^3 + \left(\frac{1}{5!} - \frac{1}{4!} \right) z^5 + \cdots}$$

$$\left(\frac{1}{2!} - \frac{1}{3!} \right) z^3 - \left(\frac{1}{2!} - \frac{1}{3!} \right) \frac{z^5}{2!}.$$

We have obtained the first two terms of the desired Maclaurin series. Because cos z becomes zero at $\pm\pi/2$, the infinite expansion is valid for $|z| < \pi/2$. ◀

The Method of Partial Fractions

Consider a rational algebraic function

$$f(z) = \frac{P(z)}{Q(z)},$$

where P and Q are polynomials in z. If $Q(z_0) \neq 0$, then $f(z)$ has a Taylor expansion about z_0. The coefficients in this series can, in principle, be obtained through differentiation of $f(z)$, but the process is often tedious.

When the degree of Q (its highest power of z) exceeds the degree of P, the use of partial fractions provides a systematic procedure for obtaining the coefficients in the Taylor series. When the degree of P equals or exceeds that of Q, the method to be presented also helps, provided we first perform a simple division (see Exercise 13 in this section).

The reader should review the techniques learned in elementary calculus for decomposing real rational functions into partial fractions. The method works equally well for complex functions—the algebraic manipulations are the same. The form of partial fraction expansions is governed by the following rules.

Rule I   Nonrepeated factors

Let $P(z)/Q(z)$ be a rational function where the polynomial $P(z)$ is of lower degree than the polynomial $Q(z)$. If $Q(z)$ can be factored into the form

$$Q(z) = C(z - a_1)(z - a_2)\ldots(z - a_n), \tag{5.4-7}$$

where $a_1, a_2, \ldots$ are all different constants and C is a constant, then

$$\frac{P(z)}{Q(z)} = \frac{A_1}{(z - a_1)} + \frac{A_2}{(z - a_2)} + \cdots + \frac{A_n}{(z - a_n)}, \tag{5.4-8}$$

where $A_1, A_2, \ldots$ are constants. Equation (5.4-8), called the partial fraction expansion of $P(z)/Q(z)$, is valid for all $z \neq a_j$ ($j = 1, 2, \ldots, n$). ■

This rule does not apply to the function $z/[(z + 1)(z^2 - 1)]$ since $Q(z) = (z + 1)^2(z - 1)$. The first factor here appears raised to the second power and not to the first as required by Eq. (5.4-7). Instead, we use the following rule.

Rule II   Repeated factors

Let $Q(z)$ be factored as in Eq. (5.4-7), except that $(z - a_1)$ appears raised to the m_1 power, $(z - a_2)$ to the m_2 power, etc. Then, $P(z)/Q(z)$ can be decomposed as in Eq. (5.4-8), except that for each factor of $Q(z)$ of the form

$(z - a_j)^{m_j}$, where $m_j \geq 2$, we replace $A_j/(z - a_j)$ in Eq. (5.4–8) by

$$\frac{A_{j1}}{(z - a_j)} + \frac{A_{j2}}{(z - a_j)^2} + \cdots + \frac{A_{jm_j}}{(z - a_j)^{m_j}}. \qquad \blacksquare$$

Thus Rule I tells us that

$$\frac{z}{(z - 1)(z + 1)} = \frac{A_1}{(z - 1)} + \frac{A_2}{(z + 1)},$$

whereas Rule II establishes

$$\frac{z}{(z - 1)^2(z + 1)} = \frac{A_{11}}{(z - 1)} + \frac{A_{12}}{(z - 1)^2} + \frac{A_2}{(z + 1)}.$$

The utility of partial fractions in the generation of series and various procedures for obtaining the coefficients are illustrated in Examples 6 and 7.

First, however, for future reference we state four Maclaurin expansions:

$$\frac{1}{1 - w} = 1 + w + w^2 + \cdots, \qquad |w| < 1; \qquad (5.4\text{–}9a)$$

$$\frac{1}{1 + w} = 1 - w + w^2 - w^3 + \cdots, \qquad |w| < 1; \qquad (5.4\text{–}9b)$$

$$\frac{1}{(1 - w)^2} = 1 + 2w + 3w^2 + \cdots, \qquad |w| < 1; \qquad (5.4\text{–}9c)$$

$$\frac{1}{(1 + w)^2} = 1 - 2w + 3w^2 - \cdots, \qquad |w| < 1. \qquad (5.4\text{–}9d)$$

Equation (5.4–9a) is actually Eq. (5.2–7) with z replaced by w. Equation (5.4–9c) is similarly derived from Eq. (5.2–26). Equations (5.4–9b) and (5.4–9d) are obtained if we substitute $-w$ for w in Eqs. (5.4–9a) and (5.4–9c), respectively. A general expansion for $1/(1 \pm w)^N$, which one sometimes needs, can be obtained from Exercise 1(b) in this section.

Example 6

Expand

$$f(z) = \frac{z}{z^2 - z - 2} = \frac{z}{(z + 1)(z - 2)}$$

in a Taylor series about the point $z = 1$.

Solution

We have, from Rule I,

$$\frac{z}{(z + 1)(z - 2)} = \frac{a}{(z + 1)} + \frac{b}{(z - 2)}. \qquad (5.4\text{–}10)$$

(It is simpler to use a and b here instead of the subscript notation A_1 and A_2.)

Clearing the fractions in Eq. (5.4–10) yields

$$z = a(z - 2) + b(z + 1).$$

We can find a and b by letting z in the above equation equal -1 and then 2. This type of procedure is useful and can be generalized to yield the required partial fractions whenever $Q(z)$ has any number of nonrepeated factors (see Exercise 10 in this section). For another approach we rearrange the previous equation as

$$z = (a + b)z + (-2a + b).$$

The coefficients of like powers of z on each side of the equation must be in agreement. We then have

$$1 = a + b, \quad (z^1 \text{ coefficients});$$

$$0 = (-2a + b), \quad (z^0 \text{ coefficients});$$

whose solution is $a = 1/3$, $b = 2/3$.

Thus from Eq. (5.4–10)

$$\frac{z}{(z + 1)(z - 2)} = \frac{1/3}{(z + 1)} + \frac{2/3}{(z - 2)}. \tag{5.4–11}$$

To expand $z/[(z + 1)(z - 2)]$ in powers of $(z - 1)$ we expand each fraction on the right in Eq. (5.4–11) in these powers. Thus

$$\frac{1/3}{(z + 1)} = \frac{1/3}{(z - 1) + 2} = \frac{1/6}{1 + \dfrac{(z - 1)}{2}} = 1/6\left[1 - \frac{(z - 1)}{2} + \frac{(z - 1)^2}{4} - \cdots\right],$$

$$\text{for } |z - 1| < 2. \tag{5.4–12}$$

The preceding series is obtained with the substitution $w = (z - 1)/2$ in Eq. (5.4–9b). The requirement $|z - 1| < 2$ is identical to the constraint $|w| < 1$.

Similarly,

$$\frac{2/3}{(z - 2)} = \frac{2/3}{(z - 1) - 1} = \frac{-2/3}{1 - (z - 1)} = -\frac{2}{3}\left[1 + (z - 1) + (z - 1)^2 + \cdots\right],$$

$$\text{for } |z - 1| < 1, \tag{5.4–13}$$

where we have used Eq. (5.4–9a) and taken $w = z - 1$. The series in Eqs. (5.4–12) and (5.4–13) are now substituted in the right side of Eq. (5.4–11). Thus

$$\frac{z}{(z + 1)(z - 2)} = \frac{1}{6}\underbrace{\left[1 - \frac{(z - 1)}{2} + \frac{(z - 1)^2}{4} - \cdots\right]}_{|z-1|<2}$$

$$-\frac{2}{3}\underbrace{\left[1 + (z - 1) + (z - 1)^2 + \cdots\right]}_{|z-1|<1}.$$

In the domain $|z - 1| < 1$ *both* series converge and their terms can be combined.

Thus

$$\frac{z}{(z+1)(z-2)} = \left(\frac{1}{6} - \frac{2}{3}\right) + \left(-\frac{1}{12} - \frac{2}{3}\right)(z-1) + \left(\frac{1}{24} - \frac{2}{3}\right)(z-1)^2 + \cdots,$$

or

$$\frac{z}{(z+1)(z-2)} = \sum_{n=0}^{\infty} c_n(z-1)^n, \qquad |z-1| < 1, \qquad (5.4\text{--}14)$$

where

$$c_n = \frac{1}{6}\left(-\frac{1}{2}\right)^n - \frac{2}{3}.$$

We could have obtained the constraint $|z - 1| < 1$ by studying the location of the singularities of $z/[(z+1)(z-2)]$ to see which one ($z = 2$) lies closest to $(1,0)$. ◀

Example 7

Expand

$$f(z) = \frac{z}{(z+1)^2(z-2)}$$

in a Maclaurin series.

Solution

Since $(z + 1)$ is raised to the second power, we must follow Rule II and seek a partial fraction expansion of the form

$$\frac{z}{(z+1)^2(z-2)} = \frac{A}{(z+1)} + \frac{B}{(z+1)^2} + \frac{C}{(z-2)}. \qquad (5.4\text{--}15)$$

Clearing fractions we obtain

$$z = A(z+1)(z-2) + B(z-2) + C(z+1)^2, \qquad (5.4\text{--}16)$$

or

$$z = (A+C)z^2 + (-A+B+2C)z + (-2A-2B+C). \qquad (5.4\text{--}17)$$

By putting $z = 2$ and then $z = -1$ in Eq. (5.4–16), we discover that $C = 2/9$ and $B = 1/3$. Note that z^2 does not appear on the left in Eq. (5.4–17), which means z^2 must not appear on the right; hence, $A = -C = -2/9$. Thus from Eq. (5.4–15)

$$\frac{z}{(z+1)^2(z-2)} = \frac{-2/9}{z+1} + \frac{1/3}{(z+1)^2} + \frac{2/9}{z-2}. \qquad (5.4\text{--}18)$$

We now expand each fraction in powers of z. Taking $w = z$, we have from Eq. (5.4–9b)

$$\frac{-2/9}{1+z} = -\frac{2}{9}\left[1 - z + z^2 - \cdots\right], \qquad |z| < 1,$$

and from Eq. (5.4–9d)

$$\frac{1/3}{(1+z)^2} = \frac{1}{3}\left[1 - 2z + 3z^2 - 4z^3 + \cdots\right], \qquad |z| < 1.$$

With $w = z/2$ in Eq. (5.4–9a) we obtain

$$\frac{2/9}{z-2} = \frac{-1/9}{1-z/2} = -\frac{1}{9}\left[1 + \frac{z}{2} + \frac{z^2}{4} + \cdots\right], \qquad |z| < 2.$$

The substitution of the three preceding series on the right in Eq. (5.4–18) yields

$$\frac{z}{(z-1)^2(z-2)} = -\frac{2}{9}\underbrace{\left[1 - z + z^2 - \cdots\right]}_{|z|<1} + \frac{1}{3}\underbrace{\left[1 - 2z + 3z^2 - \cdots\right]}_{|z|<1}$$

$$-\frac{1}{9}\underbrace{\left[1 + \frac{z}{2} + \frac{z^2}{4} + \cdots\right]}_{|z|<2}.$$

Inside $|z| = 1$ we can add the three series together and obtain

$$\frac{z}{(z-1)^2(z-2)} = \sum_{n=0}^{\infty} c_n z^n, \qquad |z| < 1, \tag{5.4–19}$$

where

$$c_n = (-1)^{n+1}\frac{2}{9} + \frac{(-1)^n}{3}(n+1) - \frac{1}{9}\left(\frac{1}{2}\right)^n. \qquad \blacktriangleleft$$

EXERCISES

1. a) Differentiate the series of Eq. (5.4–2) to show that

$$\frac{1}{z^3} = 1 - \frac{3 \cdot 2}{2}(z-1) + \frac{4 \cdot 3}{2}(z-1)^2 - \frac{5 \cdot 4}{2}(z-1)^3 + \cdots,$$

$$|z - 1| < 1.$$

   b) Use the series in Eq. (5.2–7), and successive differentiation, to show that for $N \geq 0$

$$\frac{1}{(1-z)^N} = \sum_{n=0}^{\infty} c_n z^n, \qquad c_n = \frac{(N-1+n)!}{n!(N-1)!}, \qquad |z| < 1.$$

2. a) Show that

$$\frac{1}{1+z^2} = 1 - z^2 + z^4 - z^6 + \cdots, \qquad |z| < 1.$$

   b) Integrate the above series, and its sum, along a contour from the origin to an arbitrary point z to show that

$$\tan^{-1}z = \sum_{n=0}^{\infty}(-1)^n\frac{z^{2n+1}}{(2n+1)}, \qquad |z| < 1.$$

3. a) By considering the first and second derivatives of the geometric series in Eq. (5.2–7) find a formula for the sum $\sum_{n=1}^{\infty} n^2 z^n$ for $|z| < 1$.

   b) Use your result to evaluate $\sum_{n=1}^{\infty} n^2/2^n$.

4. The Fresnel integrals $C(P)$ and $S(P)$ are used in optics and in the design of microwave antennas. They are defined by

$$C(P) = \int_0^P \cos\left(\frac{\pi t^2}{2}\right) dt$$

and

$$S(P) = \int_0^P \sin\left(\frac{\pi t^2}{2}\right) dt,$$

where P is a real number. Notice that

$$F(P) = C(P) + iS(P) = \int_0^P e^{i\pi/2 t^2} dt.$$

Why is this so?

a) Expand the preceding integrand in a Maclaurin series and integrate to show that

$$C(P) + iS(P) = \sum_{n=0}^{\infty} \frac{\left(\frac{i\pi}{2}\right)^n P^{2n+1}}{n!(2n+1)}.$$

b) Allow P to assume the values 0, 0.2, 0.4, 0.6, and 0.8. Use the first four terms of your series and a pocket calculator to compute the corresponding values of $F(P)$. Plot these values of $F(P)$ as points in a complex plane whose real axis is $C(P)$ and whose imaginary axis is $S(P)$. Connect the points by a smooth curve. This locus is, approximately, a portion of the "Cornu spiral."[†] The entire spiral can be used to determine $F(P) = C(P) + iS(P)$ for any real number P.

5. Use the first three terms in Eq. (5.4–6) to obtain an approximate value of $\int_0^{0.2i} (e^z \sin z)/(1 - z)\, dz$.

6. Find an approximate value of $\int_0^{0.2+0.1i} (e^z - 1)/z\, dz$ by replacing the numerator of the integrand by the first four terms of its Maclaurin series.

7. Use the method of partial fractions to obtain the Taylor series expansion of the following functions about the indicated point. State where each expansion is valid.

   a) $\dfrac{z}{(z+1)(z-1)}$, about $z = 0$ c) $\dfrac{1}{(z-1)^2}$, about $z = -1$

   b) $\dfrac{z}{(z+1)(z-1)}$, about $z = 3$ d) $\dfrac{1}{(z-1)^3}$, about $z = -1$

[†]See B. Rossi, *Optics* (Reading, Mass.: Addison-Wesley, 1959), p.187.

e) $\dfrac{1}{(z-3)^2 z}$, about $z = 1$ g) $\dfrac{1}{(z)(z-2)(z-3)}$, about $z = -1$

f) $\dfrac{1}{(z)(z-2)}$, about $z = i$

8. Use the result of Exercise 7(g) to obtain the numerical value of the tenth derivative of $1/[(z)(z-2)(z-3)]$ at $z = -1$.

9. Expand $1/[(1-w)^2(1+w)^2]$ in a Maclaurin series in powers of w by

  a) multiplication of the series in Eqs. (5.4–9c) and (5.4–9d).
  b) a partial fraction decomposition and use of all the equations in (5.4–9).

10. Assume $P(z)/Q(z)$ satisfies the requirements of Rule I for partial fractions. Thus $Q(z)$ has no repeated factors and

$$\frac{P(z)}{Q(z)} = \frac{P}{C(z-a_1)(z-a_2)\ldots(z-a_n)}$$

$$= \frac{A_1}{z-a_1} + \frac{A_2}{z-a_2} + \cdots + \frac{A_n}{z-a_n}.$$

  a) Multiply both sides of the preceding equation by $Q(z)$ and cancel common factors to show that

$$P(z) = A_1(z-a_2)(z-a_3)\cdots(z-a_n) + A_2(z-a_1)(z-a_3)\cdots(z-a_n)$$
$$+ \cdots + A_n(z-a_1)(z-a_2)\cdots(z-a_{n-1}).$$

  b) Show how to obtain any coefficient A_j ($j = 1, 2, \ldots, n$) by setting $z = a_j$ in the previous equation.

  c) Show that the result obtained in part (b) is identical to

$$A_j = \lim_{z \to a_j}\left[(z-a_j)\frac{P(z)}{Q(z)}\right] = \frac{P(a_j)}{Q'(z_j)}.$$

  Hint: Consider L'Hôpital's rule.

  d) Expand $z/[(z^2+1)(z-2)]$ in partial fractions by using the results of part (b) or (c).

11. Let $h(z) = f(z)/g(z)$ where

$$f(z) = a_0 + a_1 z + a_2 z^2 + \cdots,$$
$$g(z) = b_0 + b_1 z + b_2 z^2 + \cdots, \qquad b_0 \neq 0.$$

We want the Maclaurin expansion $h(z) = c_0 + c_1 z + c_2 z^2 + \cdots$.

  a) In the given quotient replace f, g, and h by their Maclaurin series. Now clear of fractions by cross multiplying, and show by equating coefficients of corresponding powers of z that

$$a_0 = c_0 b_0, \qquad a_1 = c_1 b_0 + c_0 b_1, \qquad a_2 = c_2 b_0 + c_1 b_1 + c_0 b_2.$$

  b) Solve these three equations for the three unknowns c_0, c_1, and c_2 in terms of the a_j and b_j coefficients.

c) Show that the same values of c_0, c_1, and c_2 are obtained through the formal long division

$$b_0 + b_1 z + b_2 z^2 + \cdots \overline{\smash{)}\, a_0 + a_1 z + a_2 z^2 + \cdots} \; .$$

The preceding procedure will, of course, work if we have expansions in powers of $(z - z_0)$ rather than z.[†]

12. Show that

$$\sec z = \frac{1}{\cos z} = 1 + \frac{z^2}{2} + \frac{5z^4}{24} + \cdots$$

by either long division of a Maclaurin series or by the method shown in Exercise 11(a). Where is the Maclaurin series expansion of $\sec z$ valid?

13. Expand $(z^3 + 4z^2 + 2z - 6)/(z^2 + 5z + 6)$ in a Taylor series about $z = -1$. *Hint:* This is an "improper" fraction. Show by long division that it can be rewritten

$$(z - 1) + \frac{z}{(z + 3)(z + 2)} = (z + 1) - 2 + \frac{z}{(z + 3)(z + 2)} \, .$$

Now apply the method of partial fractions to the remaining fraction.

14. The Bernoulli numbers B_0, B_1, $B_2, \ldots$ are defined by

$$B_n = n! c_n,$$

where

$$f(z) = \left\{ \begin{matrix} \dfrac{z}{e^z - 1}, & z \neq 0 \\ 1, & z = 0 \end{matrix} \right\} = \sum_{n=0}^{\infty} c_n z^n.$$

Note that $f(z)$ is analytic at $z = 0$ since, for all z,

$$\frac{z}{e^z - 1} = \frac{z}{z + \dfrac{z^2}{2!} + \dfrac{z^3}{3!} + \cdots} = \frac{1}{1 + \dfrac{z}{2!} + \dfrac{z^2}{3!} + \cdots} \, .$$

Perform long division on the right-hand quotient to show that $B_0 = 1$, $B_1 = -1/2$, $B_2 = 1/6$.

15. Consider a power series whose circle of convergence C (see Fig. 5.4–1) has center z_0 and radius r_0. The sum of this power series is a function $f(z)$ defined in C. Let us expand $f(z)$ in a Taylor series about the point z_1 lying inside C. A new power series that will converge within a circle of radius r_1 and center z_1 is obtained. This new circle may, in some cases, extend outside the circle C in which $f(z)$ is defined. If this is the case, the analytic function, which is the sum of the new power series in the domain $|z - z_0| < r_1$, is called the *analytic continuation* of $f(z)$. A new function has been obtained that is analytic in a domain extending beyond the original domain in which $f(z)$ is defined.

[†]For additional information on division of series see K. Knopp, *Infinite Sequences and Series* (New York: Dover, 1956), sec. 4.3.

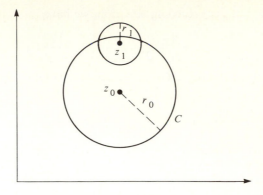

Figure 5.4–1

a) Consider the power series $1 - z + z^2 - z^3 + \cdots$. To what sum $f(z)$ does this series converge for $|z| < 1$?

b) Expand the function $f(z)$ found in part (a) in a power series about $z = 1/2$.

c) Make a sketch or describe the circle in which the series of part (b) converges. Does this circle extend outside the circle in which the series of part (a) converges?

d) What is the sum of the series obtained in part (b)? This sum is an analytic continuation of $f(z)$ in part (a) outside the circle $|z| = 1$.

5.5 LAURENT SERIES

A Taylor series expansion of a function $f(z)$ contains only terms of the form $(z - z_0)$ raised to positive integer powers. The Laurent series, which is related to the Taylor series, is defined as follows:

Definition Laurent series

The Laurent series expansion of a function $f(z)$ is an expansion of the form

$$f(z) = \sum_{n=-\infty}^{\infty} c_n(z - z_0)^n = \cdots + c_{-2}(z - z_0)^{-2} + c_{-1}(z - z_0)^{-1}$$

$$+ c_0 + c_1(z - z_0) + \cdots. \quad \blacksquare \quad (5.5–1)$$

Thus a Laurent expansion, unlike a Taylor expansion, can contain one or more terms with $(z - z_0)$ raised to a negative power. It can also contain positive powers of $(z - z_0)$.

Examples of Laurent series are often obtained from some simple manipulations on Taylor series. Thus

$$e^u = 1 + u + \frac{u^2}{2!} + \cdots, \quad \text{all finite } u.$$

Putting $u = (z - 1)^{-1}$ in the preceding equation, we have

$$e^{1/(z-1)} = 1 + (z - 1)^{-1} + \frac{(z - 1)^{-2}}{2!} + \frac{(z - 1)^{-3}}{3!} + \cdots, \qquad z \neq 1.$$

We can reverse the order of the terms on the right to comply with the form of Eq. (5.5–1) and obtain

$$e^{1/(z-1)} = \cdots + \frac{(z - 1)^{-3}}{3!} + \frac{(z - 1)^{-2}}{2!} + \frac{(z - 1)^{-1}}{1!} + 1, \qquad z \neq 1. \quad (5.5-2)$$

This is a Laurent series with no positive powers of $(z - 1)$. Multiplying both sides of Eq. (5.5–2) by $(z - 1)^2$, we have

$$(z - 1)^2 e^{1/(z-1)} = \cdots + \frac{(z - 1)^{-1}}{3!} + \frac{1}{2!} + (z - 1) + (z - 1)^2, \qquad z \neq 1.$$

$$(5.5-3)$$

This is a Laurent series with both negative and positive powers of $(z - 1)$.

A knowledge of Laurent series is necessary for an understanding of the calculus of residues. Residues, treated in Chapter 6, are an invaluable tool in the evaluation of many types of integrals.

What kinds of functions can be represented by Laurent series, and in what region of the complex plane will the representation be valid? The answer is contained in the following theorem, which we will soon prove.

Theorem 18   Laurent's theorem

Let $f(z)$ be analytic in D, an annular domain $r_1 < |z - z_0| < r_2$. If z lies in D, $f(z)$ can be represented by the Laurent expansion

$$f(z) = \sum_{n=-\infty}^{\infty} c_n(z - z_0)^n = \cdots + c_{-2}(z - z_0)^{-2} + c_{-1}(z - z_0)^{-1}$$

$$+ c_0 + c_1(z - z_0) + c_2(z - z_0)^2 + \cdots. \quad (5.5-4)$$

The coefficients are given by

$$c_n = \frac{1}{2\pi i} \oint_C \frac{f(z)}{(z - z_0)^{n+1}} \, dz, \quad (5.5-5)$$

where C is any simple closed contour lying in D and enclosing the inner boundary $|z - z_0| = r_1$. The series is uniformly convergent in any annular region centered at z_0 and lying in D. ∎

The theorem asserts that if $f(z)$ is analytic in a "washer-shaped" domain, like the one shown in Fig. 5.5–1, then it can be represented by a Laurent series throughout this domain. The coefficients can be found by a line integral (see Eq. 5.5–5) taken around a loop C, such as the one shown in the figure.

For simplicity we consider a proof in which z_0 is zero; that is, we seek an expansion in an annulus centered at the origin. The annulus, having inner and outer radii r_1 and r_2, is shown in Fig. 5.5–2(a).

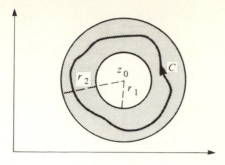

Figure 5.5-1

Now, using the contour C', which lies in the annulus, we apply the Cauchy integral formula. Observe that C' encloses the point z_1 and that $f(z)$ is analytic on and inside this contour. Thus

$$f(z_1) = \frac{1}{2\pi i} \oint_{C'} \frac{f(z)}{(z - z_1)} \, dz. \tag{5.5-6}$$

The portions of the preceding integral taken along the contiguous lines l_1 and l_2 (see Fig. 5.5-2b) cancel because of the opposite directions of integration. Thus Eq. (5.5-6) becomes

$$f(z_1) = I_A + I_B, \tag{5.5-7}$$

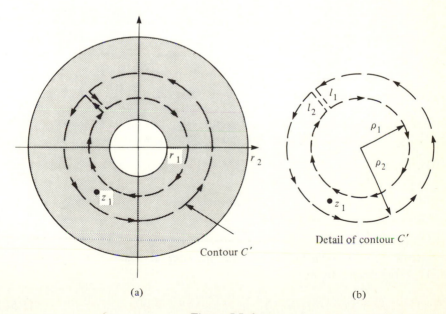

Contour C'

Detail of contour C'

(a)

(b)

Figure 5.5-2

where

$$I_A = \frac{1}{2\pi i} \oint_{|z|=\rho_2} \frac{f(z)}{(z-z_1)} \, dz \tag{5.5-8}$$

and

$$I_B = \frac{1}{2\pi i} \oint_{|z|=\rho_1} \frac{f(z)}{(z-z_1)} \, dz. \tag{5.5-9}$$

The integral I_A is dealt with in the same manner as the integral in Eq. (5.3–12) in our derivation of the Taylor series. We have

$$I_A = \frac{1}{2\pi i} \oint_{|z|=\rho_2} \frac{f(z)}{z\left(1 - \frac{z_1}{z}\right)} \, dz = \frac{1}{2\pi i} \oint_{|z|=\rho_2} \frac{f(z)}{z}\left(1 + \frac{z_1}{z} + \left(\frac{z_1}{z}\right)^2 + \cdots\right) dz$$

$$= \sum_{n=0}^{\infty} c_n z_1^n, \tag{5.5-10}$$

where

$$c_n = \frac{1}{2\pi i} \oint_{|z|=\rho_2} \frac{f(z)}{z^{n+1}} \, dz, \qquad n = 0, 1, 2, \ldots. \tag{5.5-11}$$

In Eq. (5.5–10) we require that $|z_1/z| < 1$ or $|z_1| < \rho_2$ (since $|z| = \rho_2$).

In the integral I_B (see Eq. 5.5–9) we reverse the direction of integration and compensate with a minus sign in the integrand. Thus

$$I_B = \frac{1}{2\pi i} \oint_{|z|=\rho_1} \frac{f(z)}{(z_1-z)} \, dz = \frac{1}{2\pi i} \oint_{|z|=\rho_1} \frac{f(z)}{z_1\left(1 - \frac{z}{z_1}\right)} \, dz. \tag{5.5-12}$$

Now

$$\frac{1}{1 - \frac{z}{z_1}} = 1 + \frac{z}{z_1} + \left(\frac{z}{z_1}\right)^2 + \cdots, \quad \text{if } \left|\frac{z}{z_1}\right| < 1 \quad \text{or} \quad |z| < |z_1|.$$

This series converges uniformly in a region containing $|z| = \rho_1$ (since $|z| = \rho_1 < |z_1|$). Using this series in Eq. (5.5–12) and integrating we obtain

$$I_B = \frac{1}{2\pi i} \oint_{|z|=\rho_1} \frac{f(z)}{z_1}\left(1 + \frac{z}{z_1} + \left(\frac{z}{z_1}\right)^2 + \cdots\right) dz = \frac{z_1^{-1}}{2\pi i} \oint_{|z|=\rho_1} f(z) \, dz$$

$$+ \frac{z_1^{-2}}{2\pi i} \oint_{|z|=\rho_1} z f(z) \, dz + \frac{z_1^{-3}}{2\pi i} \oint_{|z|=\rho_1} z^2 f(z) \, dz + \cdots. \tag{5.5-13}$$

We have moved the constant z_1 outside the integral signs. We may rewrite Eq. (5.5–13) more succinctly as

$$I_B = \sum_{n=-\infty}^{-1} c_n z_1^n, \qquad |z_1| > \rho_1, \tag{5.5-14}$$

where

$$c_n = \frac{1}{2\pi i} \oint_{|z|=\rho_1} z^{-n-1} f(z) \, dz = \frac{1}{2\pi i} \oint_{|z|=\rho_1} \frac{f(z)}{z^{n+1}} \, dz, \qquad n = -1, -2, \ldots .$$

(5.5-15)

Let us compare Eqs. (5.5-11) and (5.5-15). The former equation gives the coefficients for a series representation of the integral I_A (see Eq. 5.5-10). The index n in Eq. (5.5-11) is zero or positive. Equation (5.5-15) gives the coefficients for a series expansion of the integral I_B (see Eq. 5.5-14). The index n in Eq. (5.5-15) is negative. Both Eq. (5.5-11) and Eq. (5.5-15) are identical in form except that the paths of integration are circles of different radii. Observe, however, that $f(z)/z^{n+1}$ is analytic throughout the annular domain $r_1 < |z| < r_2$. Thus, by the principle of deformation of contours (see Section 4.2), we can perform the integrations in both Eq. (5.5-11) and Eq. (5.5-15) around any simple closed contour C lying in this domain and encircling the inner boundary $|z| = r_1$. The same contour can be used in both formulas.

Substituting series expansions, as shown in Eqs. (5.5-14) and (5.5-10), for the two integrals on the right in Eq. (5.5-7), we have

$$f(z_1) = \underbrace{\sum_{n=0}^{\infty} c_n z_1^n}_{|z_1| < \rho_2} + \underbrace{\sum_{n=-\infty}^{-1} c_n z_1^n}_{|z_1| > \rho_1},$$

(5.5-16)

where

$$c_n = \frac{1}{2\pi i} \oint_C \frac{f(z)}{z^{n+1}} \, dz, \qquad n = 0, \pm 1, \pm 2, \ldots .$$

(5.5-17)

We can rewrite Eq. (5.5-16) as a single summation,

$$f(z_1) = \sum_{n=-\infty}^{+\infty} c_n z_1^n,$$

(5.5-18)

that is valid when z_1 satisfies $\rho_1 < |z_1| < \rho_2$. Since ρ_1 can be brought arbitrarily close to r_1 and ρ_2 arbitrarily close to r_2 (see Fig. 5.5-2) this restriction can be relaxed to $r_1 < |z_1| < r_2$. Replacing z_1 by z in Eq. (5.5-18), we find that we have derived Eq. (5.5-4) for the special case $z_0 = 0$. A derivation valid for an arbitrary z_0 is developed in Exercise 6 of this section.

The M test (see Theorem 7) can be used to study the uniform convergence of each series on the right in Eq. (5.5-16). It is easily established that the overall series shown in Eq. (5.5-18) is uniformly convergent in any closed annular region contained in, and concentric with, the domain $r_1 < |z| < r_2$ (see Fig. 5.5-2). Just as is the case for a Taylor series, the uniform convergence of a Laurent series permits term-by-term integration and differentiation. New series are obtained that converge to the integral and derivative, respectively, of the sum of the original series.

The imaginative reader may compare Eq. (5.5-15) with the extended Cauchy integral formula (see Eq. 4.4-16) and conclude that the coefficients for our Laurent series with $z_0 = 0$ are given by

$$c_n = \frac{f^{(n)}(0)}{n!}, \quad \text{for } n \geq 0. \tag{5.5-19}$$

In fact, this very step was taken in the derivation of the Taylor series (see Eq. 5.3–19). This manuever is not permitted here. The Cauchy integral formula and its extension apply only when $f(z)$ in Eq. (5.5–15) is analytic not only on C but throughout its interior. We have made no such assumption concerning $f(z)$. Our derivation admits the possibility of $f(z)$ having singularities inside the circle $|z| = r_1$ in Fig. 5.5–2. If $f(z)$ were analytic throughout the disc $|z| \leq r_1$, as well as in the annulus $r_1 < |z| < r_2$ shown in Fig. 5.5–2, then $c_0, c_1, c_2, \ldots$ would indeed be given by Eq. (5.3–19). Moreover, according to Eq. (5.5–15) and the Cauchy integral theorem, the coefficients $c_{-1}, c_{-2}, \ldots$ would be zero. A special kind of Laurent series, namely, a Taylor expansion of $f(z)$, is obtained. The preceding discussion is easily altered to deal with Eq. (5.5–5); in other words, that integral, in general, is *not* $f^n(z_0)/n!$.

Although the coefficients in a Laurent expansion can, in principle, be derived from Eq. (5.5–5), this formula is rarely used. In practice, the coefficients are obtained from manipulations involving known series, as in the derivation of Eq. (5.5–2), and with partial fraction decompositions. The techniques are illustrated in the following examples. A useful corollary to Theorem 18, which we will not prove, is that in a given annulus the Laurent series for a function is unique. Hence, if we find a Laurent expansion of $f(z)$ valid in an annular domain, we have found the only Laurent expansion of $f(z)$ in this domain. We will learn from Examples 2 and 3 in this section, that a given $f(z)$ can have different Laurent expansions valid in different annuli sharing the same center z_0.

A discussion of Laurent series and of analytic functions sometimes involves the notion of an isolated singular point, which is defined as follows:

Definition   Isolated singular point

The point z_p is an *isolated singular point* of $f(z)$ if $f(z)$ is not analytic at z_p but is analytic in a deleted neighborhood of z_p. ∎

For example, $1/[(z - 1)(z - 2)^3]$ has isolated singular points at $z = 1$ and $z = 2$.

Let a function $f(z)$ be expanded in a Laurent series involving powers of $(z - z_0)$, where z_0 happens to be an isolated singular point of $f(z)$. We can find a series that converges to $f(z)$ in an annular domain centered at z_0. The inner radius r_1 of the domain can be made arbitrarily small but cannot be made zero since the point z_0 must, according to Laurent's theorem, be excluded from the domain. A series representation in such a domain (a disc with the center removed) is valid in a deleted neighborhood of z_0. An instance of this occurs in Example 3 below. In our work on residues in the next chapter we will be especially concerned with such series.

Finally, if $f(z)$ is analytic at all points in the z-plane lying outside some circle, then it is possible to find a Laurent expansion for $f(z)$ valid in an annulus whose outer radius r_2 is infinitely large. Equation (5.5–2) shows this possibility. Here we have a Laurent series expansion that is valid in a deleted neighborhood of $z = 1$. The outer radius of the domain in which the expansion holds is infinite.

Example 1

Expand

$$f(z) = \frac{1}{(z-3)}$$

in a Laurent series in powers of $(z-1)$. State the domain in which the series converges to $f(z)$.

Solution

Noting that $f(z)$ has its only singularity at $z = 3$, we see that a Taylor series representation of $f(z)$ is valid in the domain $|z - 1| < 2$ (see Fig. 5.5–3). According to Theorem 18, with $z_0 = 1$, we can represent $f(z)$ in a Laurent series in the domain $|z - 1| > 2$. We proceed by recalling Eq. (5.4–9a).

$$\frac{1}{1-w} = 1 + w + w^2 + \cdots, \qquad |w| < 1. \tag{5.5-20}$$

Now

$$\frac{1}{z-3} = \frac{1}{(z-1)-2} = \frac{1/(z-1)}{1-2/(z-1)}. \tag{5.5-21}$$

Comparing Eqs. (5.5–20) and (5.5–21) and taking $w = 2/(z-1)$, we obtain our Laurent series. Thus

$$\frac{1}{z-3} = \frac{1}{(z-1)}\left[1 + \frac{2}{(z-1)} + \frac{4}{(z-1)^2} + \cdots\right]$$

$$= (z-1)^{-1} + 2(z-1)^{-2} + 4(z-1)^{-3} + \cdots.$$

The condition $|w| < 1$ in Eq. (5.5–20) becomes $|2/(z-1)| < 1$ or $|z - 1| > 2$. We anticipated that our Laurent series would be valid in this domain. ◀

Example 2

Expand

$$f(z) = \frac{1}{(z+1)(z+2)}$$

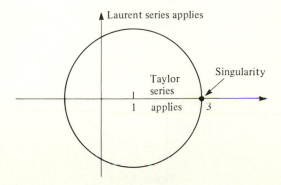

Figure 5.5–3

in a Laurent series in powers of $(z - 1)$ valid in an annular domain containing the point $z = 7/2$. State the domain in which the series converges to $f(z)$.

Solution

From Theorem 18 we know that a Laurent series in powers of $(z - 1)$ is capable of representing $f(z)$ in annular domains centered at $z_0 = 1$. Refer to Fig. 5.5–4. Since $f(z)$ has singularities at -2 and -1, we see that one such domain is D_1 defined by $2 < |z - 1| < 3$, while another is D_2 given by $|z - 1| > 3$. A Taylor series representation is also available in the domain D_3 described by $|z - 1| < 2$. Since $z = 7/2$ lies in D_1, it is the Laurent expansion valid in this domain that we seek.

We break $f(z)$ into partial fractions. Thus

$$\frac{1}{(z + 1)(z + 2)} = \frac{1}{(z + 1)} - \frac{1}{(z + 2)}. \tag{5.5–22}$$

Because we wish to generate powers of $(z - 1)$, we rewrite the first fraction as

$$\frac{1}{(z + 1)} = \frac{1}{(z - 1) + 2} = \frac{1/2}{1 + (z - 1)/2} \tag{5.5–23}$$

or, alternatively, as

$$\frac{1}{z + 1} = \frac{1}{(z - 1) + 2} = \frac{1/(z - 1)}{1 + 2/(z - 1)}. \tag{5.5–24}$$

Recall now Eq. (5.4–9b):

$$\frac{1}{1 + w} = 1 - w + w^2 - w^3 + \cdots, \qquad |w| < 1.$$

With $w = (z - 1)/2$ we expand Eq. (5.5–23) to obtain

$$\frac{1}{z + 1} = \frac{1}{2}\left[1 - \frac{(z - 1)}{2} + \frac{(z - 1)^2}{4} + \cdots \right], \quad \text{if } \left| \frac{z - 1}{2} \right| < 1 \quad \text{or} \quad |z - 1| < 2.$$

$$\tag{5.5–25}$$

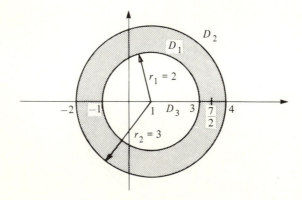

Figure 5.5–4

Taking $w = 2/(z - 1)$, we can expand Eq. (5.5–24) as follows:

$$\frac{1}{z + 1} = \frac{1}{(z - 1)}\left[1 - \frac{2}{(z - 1)} + \frac{4}{(z - 1)^2} - \cdots\right] = (z - 1)^{-1} - 2(z - 1)^{-2}$$

$$+ 4(z - 1)^{-3} - \cdots, \quad \text{if } \left|\frac{2}{z - 1}\right| < 1 \quad \text{or} \quad |z - 1| > 2. \tag{5.5–26}$$

We have expressed $1/(z + 1)$ as a Taylor series and a Laurent series both in powers of $(z - 1)$. A similar procedure can be applied to the remaining partial fraction in Eq. (5.5–22). Thus, with $w = (z - 1)/3$,

$$-\frac{1}{(z + 2)} = \frac{-1}{(z - 1) + 3} = \frac{-1/3}{1 + \left(\dfrac{z - 1}{3}\right)}$$

$$= -\frac{1}{3}\left[1 - \frac{(z - 1)}{3} + \frac{(z - 1)^2}{9} - \cdots\right], \quad |z - 1| < 3, \tag{5.5–27}$$

and, with $w = 3/(z - 1)$,

$$-\frac{1}{z + 2} = \frac{-1}{(z - 1) + 3} = \frac{-1/(z - 1)}{1 + 3/(z - 1)}$$

$$= -\frac{1}{(z - 1)}\left[1 - \frac{3}{(z - 1)} + \frac{9}{(z - 1)^2} - \cdots\right]$$

$$= -(z - 1)^{-1} + 3(z - 1)^{-2} - 9(z - 1)^{-3} + \cdots, \quad |z - 1| > 3. \tag{5.5–28}$$

In the domain D_1 the series in Eqs. (5.5–25) and (5.5–28) are of no use. However, the series in Eqs. (5.5–26) and (5.5–27) converge to their respective functions in this domain. Using these equations, we replace each fraction on the right in Eq. (5.5–22) by a series and obtain

$$\frac{1}{(z + 1)(z + 2)} = \underbrace{(z - 1)^{-1} - 2(z - 1)^{-2} + 4(z - 1)^{-3} - \cdots}_{|z-1|>2}$$

$$\underbrace{-\frac{1}{3} + \frac{1}{9}(z - 1) - \frac{1}{27}(z - 1)^2 + \cdots}_{|z-1|<3}, \tag{5.5–29}$$

which, when written more succinctly, reads

$$\frac{1}{(z + 1)(z + 2)} = \sum_{n=-\infty}^{+\infty} c_n(z - 1)^n, \tag{5.5–30}$$

where

$$c_n = \left(-\frac{1}{3}\right)^{n+1}, \quad n \geq 0, \tag{5.5–31a}$$

and

$$c_n = (-1)^{n+1} 2^{-n-1}, \qquad n \le -1. \tag{5.5-31b}$$

We see from Eq. (5.5–29) that the Laurent series in Eq. (5.5–30) is composed of two series that simultaneously converge to their respective partial fractions only in the annulus $2 < |z - 1| < 3$.

Comment: A Laurent series expansion of $f(z)$ in the domain $|z - 1| > 3$, that is, D_2, is possible. We represent the partial fractions in Eq. (5.2–22) by the series shown in Eqs. (5.5–26) and (5.5–28). Both are valid in D_2. Adding these series we have

$$\frac{1}{(z+1)(z+2)} = \sum_{n=-\infty}^{-2} c_n (z-1)^n, \qquad |z-1| > 3,$$

where

$$c_n = (-1)^n [3^{-n-1} - 2^{-n-1}], \qquad n = -2, -3, \ldots . \qquad \blacktriangleleft$$

Example 3

Expand

$$f(z) = \frac{1}{(z)(z-1)}$$

in a Laurent series that is valid in a deleted neighborhood of $z = 1$. State the domain throughout which the series is valid.

Solution

Observe that $f(z)$ has singularities at $z = 1$ and $z = 0$. The annulus $0 < |z - 1| < 1$ is the largest deleted neighborhood of $z = 1$ that excludes both singularities of $f(z)$. Hence, we take $z_0 = 1$ in Theorem 18 and seek a Laurent expansion in powers of $(z - 1)$. Decomposing $f(z)$ into partial fractions, we obtain

$$\frac{1}{(z)(z-1)} = -\frac{1}{z} + \frac{1}{z-1}. \tag{5.5-32}$$

This equality breaks down at $z = 0$ and $z = 1$. The second fraction, $(z - 1)^{-1}$, is already expanded in powers of $(z - 1)$. It is a one-term Laurent series. No other expansion of this fraction in powers of $(z - 1)$ is possible. For the fraction $-1/z$ we have the choice of two series containing powers of $(z - 1)$. Thus

$$-\frac{1}{z} = \frac{-1}{1 + (z-1)} = -\left(1 - (z-1) + (z-1)^2 - \cdots\right), \qquad |z-1| < 1, \tag{5.5-33}$$

and

$$-\frac{1}{z} = \frac{-1/(z-1)}{1 + 1/(z-1)} = -(z-1)^{-1}\left(1 - \frac{1}{(z-1)} + \frac{1}{(z-1)^2} - \cdots\right)$$

$$= -(z-1)^{-1} + (z-1)^{-2} - (z-1)^{-3} + \cdots, \qquad |z-1| > 1. \tag{5.5-34}$$

Using Eq. (5.5–33) on the right in Eq. (5.5–32) to represent $-1/z$, we get

$$\frac{1}{(z)(z-1)} = \underbrace{-1 + (z-1) - (z-1)^2 + \cdots}_{|z-1|<1} + \underbrace{(z-1)^{-1}}_{z\neq1},$$

or, more neatly,

$$\frac{1}{(z)(z-1)} = \sum_{n=-1}^{\infty} (-1)^{n+1}(z-1)^n, \qquad 0 < |z-1| < 1.$$

Comment: Had we used Eq. (5.5–34) instead of Eq. (5.5–33) to represent $-1/z$ on the right in Eq. (5.5–32), we would have obtained the Laurent expansion

$$\frac{1}{z(z-1)} = (z-1)^{-2} - (z-1)^{-3} + (z-1)^{-4} - \cdots.$$

This expansion is valid in the same annulus as the series in Eq. (5.5–34), that is, $|z-1| > 1$, which is not the required deleted neighborhood of $z = 1$. ◀

EXERCISES

1. Obtain the indicated Laurent expansions of the following functions in an annular domain whose outer radius is infinite. State the domain in which the series is valid.

   a) $\dfrac{1}{z+3}$, in powers of $(z-1)$ b) $\dfrac{1}{z+i}$, in powers of $(z-2)$

2. a) For $f(z) = 1/[(z-1)(z+1)]$ how many distinct series expansions are possible that employ integer powers of $(z-2)$? Find each series and state the circle or annulus within which each is valid. Which one of these series is a Taylor series?

   b) Expand $1/[(z)(z+1)(z-2)]$ in three different Laurent series involving powers of z. State the domain in which each series applies.

3. For the following functions find the Laurent expansion valid in an annular domain that contains the point $(6, 0)$. The center of the annulus is $(3, 0)$. In what domain is your series valid?

   a) $\dfrac{z}{(z-2)(z+1)}$ b) $\dfrac{1}{(z-1)(z+2)}$

4. For the following functions find the Laurent series that is valid in a deleted neighborhood of the specified singular point. In what domain is the series valid?

   a) $\dfrac{1}{(z)(z+3)}$, $z = 0$ d) $z^2 e^{1/z^2}$, $z = 0$

   b) $\dfrac{1}{(z)(z+3)}$, $z = -3$ e) $\dfrac{e^z}{z-2}$, $z = 2$

   c) $\dfrac{\sin z}{z^2}$, $z = 0$ f) $\dfrac{\cos z}{(z-2)^2}$, $z = 2$

5. The exponential integral $E_1(a)$ is defined by the improper integral

$$E_1(a) = \int_a^\infty \frac{e^{-x}}{x}\, dx, \qquad a > 0.$$

Thus

$$E_1(a) - E_1(b) = \int_a^b \frac{e^{-x}}{x}\, dx.$$

Use a Laurent expansion for e^{-z}/z and a term-by-term integration to show that

$$E_1(a) - E_1(b) = \log\frac{b}{a} - (b - a) + \frac{b^2 - a^2}{(2!)(2)} - \frac{(b^3 - a^3)}{(3!)(3)} + \cdots .$$

6. A derivation of Laurent's theorem was given for the case $z_0 = 0$. Derive this theorem for the more general case where z_0 is not necessarily 0. *Hint:* Redraw Fig. 5.5–2 with all circles centered at $z_0 \neq 0$. Notice that

$$f(z_1) = \frac{1}{2\pi i} \oint_{|z-z_0|=\rho_2} \frac{f(z)\, dz}{(z - z_0) - (z_1 - z_0)}$$

$$+ \frac{1}{2\pi i} \oint_{|z-z_0|=\rho_1} \frac{f(z)\, dz}{(z_1 - z_0) - (z - z_0)} .$$

Now expand each integral in either positive or negative powers of $(z_1 - z_0)$.

7. A Laurent series expansion is sometimes obtained by long division of known series. This technique was applied in Section 5.4 to obtain Taylor expansions (see Exercises 11 and 12, Section 5.4).

a) Show that

$$\operatorname{cosec} z = \frac{1}{\sin z} = \frac{1}{z} + \frac{z}{6} + \frac{7}{360}z^3 + \cdots , \qquad 0 < |z| < \pi$$

by performing a long division of the Maclaurin series for $\sin z$.

b) Show that

$$\cot z = \frac{1}{z} - \frac{z}{3} - \frac{z^3}{45} - \cdots , \qquad 0 < |z| < \pi$$

by dividing series for $\cos z$ and $\sin z$.

8. a) Explain why Log z (principal branch) cannot be expanded in a Laurent series involving powers of z.

b) Explain why the principal branch of $1/(z)^{1/2}$ cannot be expanded in a Laurent series valid in a deleted neighborhood of $z = 0$.

9. One way of defining the Bessel functions of the first kind is by means of an integral:

$$J_n(w) = \frac{1}{2\pi} \int_{-\pi}^{+\pi} \cos(n\theta - w \sin\theta)\, d\theta,$$

where n is an integer. The number n is called the order of the Bessel function. There is a connection between this integral and the coefficients of z in a Laurent expansion of $e^{w(z - 1/z)/2}$.

Let

$$e^{w(z-1/z)/2} = \sum_{n=-\infty}^{+\infty} c_n z^n, \qquad |z| > 0. \tag{5.5-35}$$

Show using Laurent's theorem that

$$c_n = J_n(w). \tag{5.5-36}$$

Hint: Refer to Eq. (5.5–5). Take as a contour $|z| = 1$. Make a change of variables to polar coordinates ($z = e^{i\theta}$). Then use Euler's identity and symmetry to argue that a portion of your result is zero.

The expression $e^{w(z-1/z)/2}$ is called a *generating function* for these Bessel functions.

10. Refer to Eqs. (5.5–35) and (5.5–36). Show that

$$J_n(w) = \sum_{k=0}^{\infty} \frac{(-1)^k (w/2)^{n+2k}}{k!(n+k)!}.$$

Hint: The left side of Eq. (5.5–35) is $e^{wz/2}e^{-w/2z}$. Multiply the Maclaurin series for the first term by a Laurent series for the second term.

6 RESIDUES AND THEIR USE IN INTEGRATION

6.1 DEFINITION OF THE RESIDUE

In the previous chapter we devoted nearly equal attention to Laurent series and Taylor series. Although some uses were demonstrated for Taylor series (for example, in the numerical evaluation of integrals and transcendental functions), almost none were demonstrated for Laurent series. In this chapter we will remedy this imbalance and show that a Laurent series (in particular, one of its terms) can be used in the evaluation of certain kinds of integrals.

Let z_0 be an isolated singular point of the analytic function $f(z)$. Consider any simple closed contour enclosing z_0 and no other singularity of $f(z)$. The integral $(1/2\pi i) \oint f(z)\, dz$ taken around C is typically nonzero. However, by the principle of deformation of contours (see Section 4.2) its value is independent of the precise shape of C, that is, all closed curves that contain z_0 and no other singular point of $f(z)$ will lead to the same value for the integral. This leads us to create the following definition.

Definition Residue

Let $f(z)$ be analytic on a simple closed contour C and at all points interior to C except for the point z_0. Then the residue of $f(z)$ at z_0, written $\text{Res}[f(z), z_0]$, is defined by

$$\text{Res}[f(z), z_0] = \frac{1}{2\pi i} \oint_C f(z)\, dz. \quad \blacksquare \qquad (6.1\text{–}1)$$

The connection between $\text{Res}[f(z), z_0]$ and a Laurent series for $f(z)$ will soon be apparent. Because z_0 is an isolated singular point of $f(z)$ a Laurent expansion

$$f(z) = \cdots + c_{-2}(z - z_0)^{-2} + c_{-1}(z - z_0)^{-1} + c_0 + c_1(z - z_0)$$

$$+ c_2(z - z_0)^2 + \cdots, \qquad 0 < |z - z_0| < R \qquad (6.1\text{–}2)$$

of $f(z)$ about z_0 is possible. This series converges to $f(z)$ at all points (except z_0) within a circle of radius R centered at z_0.

Now, to evaluate $\text{Res}[f(z), z_0]$ we take as C in Eq. (6.1–1) a circle of radius r centered at z_0. We choose $r < R$, which means that $f(z)$ in Eq. (6.1–1) can be represented by means of the Laurent series in Eq. (6.1–2), and a term-by-term integration is possible. Thus

$$\text{Res}[f(z), z_0] = \frac{1}{2\pi i} \oint_{|z-z_0|=r} \sum_{n=-\infty}^{\infty} c_n (z - z_0)^n \, dz$$

$$= \frac{1}{2\pi i} \sum_{n=-\infty}^{\infty} c_n \oint_{|z-z_0|=r} (z - z_0)^n \, dz. \tag{6.1–3}$$

It is now helpful to take note of Eq. (4.2–10).

$$\oint_{|z-z_0|=r} (z - z_0)^n \, dz = \begin{cases} 0, & n \neq -1 \\ 2\pi i, & n = -1 \end{cases} \tag{6.1–4}$$

Thus all the integrals on the right side of Eq. (6.1–3) have the value zero except the one for which $n = -1$. We then have

$$\text{Res}[f(z), z_0] = c_{-1}. \tag{6.1–5}$$

The result contained in this equation is extremely important and is summarized in the following theorem.

Theorem 1

The residue of the function $f(z)$ at the isolated singular point z_0 is the coefficient of $(z - z_0)^{-1}$ in the Laurent series representing $f(z)$ in the annulus $0 < |z - z_0| < R$. ∎

The term "residue," meaning "that which is left over," seems particularly appropriate when applied to Eqs. (6.1–1) and (6.1–5). When a valid Laurent series for $f(z)$ is used in Eq. (6.1–1) and the integration is performed term by term, all that remains is a particular coefficient in the series. An alternative derivation of Eq. (6.1–5) is given in Exercise 3 of this section.

Example 1

Let

$$f(z) = \frac{1}{(z)(z - 1)}.$$

Find, using the residue $(1/2\pi i) \oint_C f(z) \, dz$, where C is the contour shown in Fig. 6.1–1.

Solution

Note that $f(z)$ is analytic on C and at all points inside C except $z = 1$, which is an isolated singularity. Thus Theorem 1 is applicable. In Example 3 of Section 5.5 we

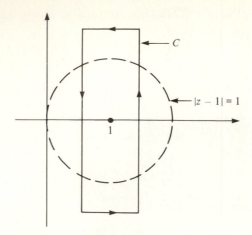

Figure 6.1–1

considered two Laurent series expansions of $f(z)$ that converge in annular regions centered at $z = 1$.

$$\frac{1}{(z)(z-1)} = (z-1)^{-1} - 1 + (z-1) - (z-1)^2 + \cdots, \qquad 0 < |z-1| < 1$$

$$(6.1\text{–}6)$$

$$\frac{1}{(z)(z-1)} = (z-1)^{-2} - (z-1)^{-3} + (z-1)^{-4} - \cdots, \qquad |z-1| > 1$$

$$(6.1\text{–}7)$$

The series of Eq. (6.1–6), which applies around the point $z = 1$, is of use to us here. The coefficient of $(z-1)^{-1}$ is 1, which means that

$$\text{Res}\left[\frac{1}{z(z-1)}, 1\right] = 1.$$

Thus

$$\frac{1}{2\pi i} \oint_C f(z)\, dz = 1.$$

Observe that the contour C lies in part outside the domain in which the Laurent series of Eq. (6.1–6) correctly represents $f(z)$ (see Fig. 6.1–1).

However, contour C can legitimately be deformed into a circle, centered at $z = 1$ and lying within the domain $0 < |z - 1| < 1$. It is around this circle that the term-by-term integration, leading to our final result, can be applied.

Comment: If we change the previous example so that the contour C now encloses only the singular point $z = 0$ (for example, C is $|z| = 1/2$), then our solution would require that we extract the residue at $z = 0$ from the expansion

$$\frac{1}{(z)(z-1)} = -z^{-1} - 1 - z - z^2 - \cdots, \qquad 0 < |z| < 1,$$

which the reader should confirm. The required residue, at $z = 0$, is -1. Thus for this new contour

$$\frac{1}{2\pi i} \oint_C \frac{dz}{z(z-1)} = -1. \qquad \blacktriangleleft$$

The preceding example could have been solved without recourse to residues. In fact, the Cauchy integral formula (see Section 4.4) could have yielded the answers more quickly. However, we will soon be dealing with more difficult integrations that can be performed only with residue calculus, as in the following example.

Example 2

Find $(1/2\pi i) \oint z \sin(1/z) \, dz$ integrated around $|z| = 2$.

Solution

The point $z = 0$ is an isolated singularity of $\sin(1/z)$ and lies inside the given contour of integration. We require a Laurent expansion of $z \sin(1/z)$ about this point. From Eq. (5.3–21a), with $1/z$ substituted for z, we can obtain

$$z \sin\left(\frac{1}{z}\right) = 1 - \frac{\left(\dfrac{1}{z}\right)^2}{3!} + \frac{\left(\dfrac{1}{z}\right)^4}{5!} - \cdots, \qquad |z| > 0.$$

Since the coefficient of z^{-1} in the preceding series is zero, we have

$$\text{Res}\left[\left(z \sin \frac{1}{z}\right), 0\right] = 0.$$

We see that a function can have a singularity at a point and possess a residue of zero there. The value of the given integral is thus zero. $\blacktriangleleft$

Thus far we have used residues to evaluate only those contour integrals whose path of integration encloses one isolated singularity of the integrand. Theorem 2 enables us to use residue calculus to evaluate integrals when more than one isolated singularity is enclosed.

Theorem 2   Residue theorem

Let C be a simple closed contour and let $f(z)$ be analytic on C and at all points inside C except for isolated singularities at $z_1, z_2, \ldots, z_n$. Then

$$\frac{1}{2\pi i} \oint_C f(z) \, dz = \text{Res}[f(z), z_1] + \text{Res}[f(z), z_2] + \cdots + \text{Res}[f(z), z_n],$$

which is more neatly written

$$\oint_C f(z) \, dz = 2\pi i \sum_{k=1}^{n} \text{Res}[f(z), z_k]. \quad \blacksquare \qquad (6.1\text{--}8)$$

Thus the integral of $f(z)$ around C is $2\pi i$ times the sum of the residues of $f(z)$ inside C.

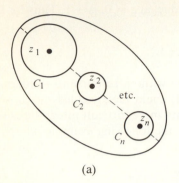
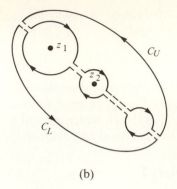

(a)   (b)

Figure 6.1–2

To prove the residue theorem, we first surround each of the singularities in C by circles $C_1, C_2, \ldots, C_n$ that intersect neither each other nor C (see Fig. 6.1–2a). A set of paths, illustrated with broken lines, is then drawn connecting $C, C_1, \ldots, C_n$ as shown. Two simple closed contours, C_U and C_L can then be formed as shown in Fig. 6.1–2(b). The function $f(z)$ is analytic on and inside C_U and C_L. Hence, from the Cauchy integral theorem,

$$\frac{1}{2\pi i} \oint_{C_U} f(z)\, dz = 0, \qquad \frac{1}{2\pi i} \oint_{C_L} f(z)\, dz = 0.$$

We now add these two expressions:

$$\frac{1}{2\pi i} \left[\oint_{C_U} f(z)\, dz + \oint_{C_L} f(z)\, dz \right] = 0. \qquad (6.1\text{–}9)$$

Note that those portions of the integral along C_U that take place along the paths illustrated with broken lines are exactly canceled by those portions of the integral along C_L that take place along the same path. Cancellation is due to the opposite directions of integration. What remains on the left side of Eq. (6.1–9) is the integral of $f(z)$ taken around C in the positive (counterclockwise) sense, plus the integrals around $C_1, C_2, \ldots, C_n$ in the negative direction. Hence

$$\frac{1}{2\pi i} \left[\oint_C f(z)\, dz + \oint_{C_1} f(z)\, dz + \oint_{C_2} f(z)\, dz + \cdots + \oint_{C_n} f(z)\, dz \right] = 0.$$

We can rearrange this as

$$\frac{1}{2\pi i} \oint_C f(z)\, dz = \frac{1}{2\pi i} \oint_{C_1} f(z)\, dz + \frac{1}{2\pi i} \oint_{C_2} f(z)\, dz + \cdots + \frac{1}{2\pi i} \oint_{C_n} f(z)\, dz,$$

$$(6.1\text{–}10)$$

where all the integrations are now performed in the positive sense. Each of the integrals on the right in Eq. (6.1–10) is taken around an isolated singularity and is numerically equal to the residue of $f(z)$ evaluated at that singularity. Hence

$$\frac{1}{2\pi i} \oint_C f(z)\, dz = \text{Res}[f(z), z_1] + \text{Res}[f(z), z_2] + \cdots + \text{Res}[f(z), z_n].$$

Multiplying both sides of the equation by $2\pi i$, we obtain Eq. (6.1–8). Note that in the summation of Eq. (6.1–8) we include residues at only those singularities inside C.

Example 3

Find $\oint_C 1/[(z)(z - 3)]\, dz$, where C is the circle $|z - 1| = 6$, by means of the residue theorem.

Solution

The contour C encloses the isolated singularities at $z = 1$ and $z = 0$. The residues of $1/[(z)(z - 1)]$ at 1 and 0 were derived in Example 1 and are 1 and -1, respectively. Applying Eq. (6.1–8), we see that since the sum of these residues is zero, the value of the given integral is zero. ◀

EXERCISES

1. Evaluate the following integrals by the method of residues.

   a) $\displaystyle\oint_{|z+1|=1} \frac{dz}{(z + 1)(z - 2)}$ b) $\displaystyle\oint_{|z|=3} \frac{dz}{(z + 1)(z - 2)}$

2. Evaluate the following integrals by the method of residues. Where necessary use Laurent expansions about singular points to obtain the residue.

   a) $\displaystyle\oint_{|z-2|=3} \sum_{n=0}^{\infty} \frac{1}{(z - 1)^{n-2}(n + 1)!}\, dz$

   b) $\displaystyle\oint_{|z-1|=2} ze^{1/z}\, dz$

   c) $\displaystyle\oint_{|z-2|=1} ze^{1/z}\, dz$

   d) $\displaystyle\frac{1}{2\pi i}\oint_{|z-i|=3} \frac{e^z - z}{z}\, dz$

   e) $\displaystyle\oint \frac{\sin z + z}{z^3}\, dz$

   around the square with corners at $x = \pm 1, y = \pm 1$

   f) $\displaystyle\oint_{|z|=2} \frac{z^{-4}}{\sin z}\, dz$

   Hint: See Exercise 7, Section 5.5.

   g) $\displaystyle\oint_{|z-1|=1/2} \frac{\text{Log } z}{(z - 1)^2}\, dz$

3. Show how the result in Eq. (6.1–5), which relates the residue to a particular coefficient in a Laurent series, can be derived through the use of Eq. (5.5–5) and the definition of the residue shown in Eq. (6.1–1).

4. Use residue calculus to show that if n is a nonnegative integer, then

$$\oint_C \left(z + \frac{1}{z}\right)^n dz = \begin{cases} \dfrac{2\pi i n!}{\left(\dfrac{n-1}{2}\right)!\left(\dfrac{n+1}{2}\right)!}, & (n \text{ odd}), \\ 0, & (n \text{ even}), \end{cases}$$

where C is any simple closed contour encircling the origin. *Hint:* Use the binomial theorem.

5. We wish to evaluate

$$\oint_{\substack{|z|=R \\ R>1}} (z^2 - 1)^{1/2}\, dz,$$

where we employ a branch of the integrand defined by a straight branch cut connecting $z = 1$ and $z = -1$, and $(z^2 - 1)^{1/2} > 0$ on the line $y = 0$, $x > 1$. Note that the singularities enclosed by the path of integration are not isolated.

a) Show that $(z^2 - 1)^{1/2} = z(1 - 1/z^2)^{1/2} = z - 1/(2z) + \cdots$, $|z| > 1$.
   Hint: Consider $(1 + w)^{1/2} = \sum_{n=0}^{\infty} c_n w^n$, $|w| < 1$, let $w = -1/z^2$.

b) Evaluate the given integral by a term-by-term integration of the Laurent series found in part (a).

6.2 ISOLATED SINGULARITIES

From the previous section it should be apparent that when $f(z)$ has an isolated singular point at z_0, it is useful to know the coefficient of $(z - z_0)^{-1}$ in the Laurent expansion about that point. In this section we do some preliminary work that will allow us, in Section 6.3, to often find this coefficient without obtaining the Laurent series.

Kinds of Isolated Singularities

Let $\sum_{n=-\infty}^{\infty} c_n(z - z_0)^n$ be the Laurent expansion of $f(z)$ about the isolated singular point z_0. We have

$$f(z) = \cdots + c_{-2}(z - z_0)^{-2} + c_{-1}(z - z_0)^{-1} + c_0 + c_1(z - z_0)^1 + \cdots .$$

Definition Principal part

That portion of the Laurent series containing only the negative powers of $(z - z_0)$ is known as the *principal part*. ∎

We now distinguish among three different kinds of principal parts.

a) A principal part with a *finite number* of nonzero terms. The principal part takes the form

$$c_{-N}(z - z_0)^{-N} + c_{-(N-1)}(z - z_0)^{-(N-1)} + \cdots + c_{-1}(z - z_0)^{-1},$$

where $c_{-N} \neq 0$. The most negative power of $(z - z_0)$ in the principal part is $-N$, where $N > 0$. Thus we arrive at the following definition.

Definition Pole of order N

A function whose Laurent expansion about a singular point z_0 has a principal part, in which the most negative power of $(z - z_0)$ is $-N$, is said to have a *pole of order N at z_0.* ■

The function $f(z) = 1/[(z)(z - 1)^2]$ has singularities at $z = 0$ and $z = 1$. The Laurent expansions about these points are

$$f(z) = z^{-1} + 2 + 3z + 4z^2 + \cdots, \qquad 0 < |z| < 1$$

and

$$f(z) = (z - 1)^{-2} - (z - 1)^{-1} + 1 - (z - 1) + (z - 1)^2 + \cdots,$$
$$0 < |z - 1| < 1.$$

The first series reveals that $f(z)$ has a pole of order 1 at $z = 0$, while the second shows a pole of order 2 at $z = 1$.

A function possessing a pole of order 1, at some point, is said to have a *simple pole* at that point.

  b) There are an *infinite number* of nonzero terms in the principal part. Unlike the case just described we are now unable to find in the principal part a nonzero term $C_{-N}(z - z_0)^{-N}$ containing the most negative power of $(z - z_0)$.

Definition Isolated essential singularity

A function, whose Laurent expansion about the isolated singular point z_0 contains an infinite number of nonzero terms in the principal part, is said to have an *isolated essential singularity* at z_0. ■

We will usually delete the word "isolated" and simply say essential singularity.

Transcendental functions defined in terms of exponentials (sine, cosh, etc.) exhibit this behavior when their arguments become infinite. For example, from Eq. (5.3–21), with z replaced by $1/z$, we have

$$\sin \frac{1}{z} = z^{-1} - \frac{z^{-3}}{3!} + \frac{z^{-5}}{5!} + \cdots, \qquad z \neq 0, \tag{6.2–1}$$

which shows that $\sin(1/z)$ has an essential singularity at $z = 0$.

Another example is $e^{1/(z-1)}/(z - 1)^2$. Using $e^u = 1 + u + u^2/2! + \cdots$, and putting $u = (z - 1)^{-1}$, we find that

$$\frac{e^{1/(z-1)}}{(z - 1)^2} = (z - 1)^{-2} + (z - 1)^{-3} + \frac{(z - 1)^{-4}}{2!} + \cdots, \qquad z \neq 1,$$

$$\tag{6.2–2}$$

which shows that the given function has an essential singularity at $z = 1$.

  c) There are functions possessing an isolated singular point z_0 such that a sought after Laurent expansion about z_0 will be found to have *no terms* in the principal part, that is, all the coefficients involving negative powers of

$(z - z_0)$ are zero. In fact, a Taylor series is obtained. In these cases it is found that the singularity exists because the function is undefined at z_0 or defined so as to create a discontinuity. By properly defining $f(z)$ at z_0 the singularity is removed.

Definition   Removable singular point

When a singularity of a function $f(z)$ at z_0 can be removed by suitably defining $f(z)$ at z_0, we say that $f(z)$ has a *removable singular point* at z_0. ■

One example of the preceding is $f(z) = \sin z/z$, which is undefined at $z = 0$. Since $\sin z = z - z^3/3! + z^5/5! - \cdots$, $|z| < \infty$, we have

$$\frac{\sin z}{z} = 1 - \frac{z^2}{3!} + \frac{z^4}{5!} - \cdots .$$

Because $\sin 0/0$ is undefined, the function on the left in this equation possesses a singular point at $z = 0$. The Taylor series on the right represents an analytic function everywhere inside its circle of convergence (the entire z-plane). The value of the series on the right, at $z = 0$, is 1. By defining $f(0) = 1$, we obtain a function

$$f(z) = \begin{cases} \dfrac{\sin z}{z}, & z \neq 0, \\ 1, & z = 0, \end{cases}$$

which is analytic for all z. The singularity of $f(z)$ at $z = 0$ has been removed by an appropriate definition of $f(0)$. This value could also have been obtained by evaluating $\lim_{z \to 0} \sin z/z$ with L'Hopital's rule.[†] Since we require continuity at $z = 0$, this limit must equal $f(0)$.

Establishing the Nature of the Singularity

When a function $f(z)$ possesses an essential singularity at z_0, the only means we have for obtaining its residue there is to use the Laurent expansion about this point and pick out the appropriate coefficient. Thus from Eq. (6.2–1) we see that $\text{Res}[\sin 1/z, 0] = 1$ (the coefficient of z^{-1}) while Eq. (6.2–2) shows that

$$\text{Res}\left[\frac{e^{1/(z-1)}}{(z-1)^2}, 1 \right] = 0,$$

which is the coefficient of $(z - 1)^{-1}$.

If however a function has a pole singularity at z_0, we need not obtain the entire Laurent expansion about z_0 in order to find the one coefficient in the series that we actually need. Provided we know that the singularity is a pole, there are a variety of techniques open to us. Furthermore, finding the residue is made easier by our first knowing the order of the pole. We will now find some rules for doing this.

[†]L'Hopital's rule is used for functions of a complex variable in the same way that it is used for real functions of a real variable (see Exercise 6, Section 2.3 for more information).

Let $f(z)$ have a pole of order N at $z = z_0$. Then

$$f(z) = c_{-N}(z - z_0)^{-N} + c_{-(N-1)}(z - z_0)^{-(N-1)} + \cdots + c_0 + c_1(z - z_0) + \cdots,$$
(6.2–3)

where $c_{-N} \neq 0$. Multiplying both sides by $(z - z_0)^N$, we have

$$(z - z_0)^N f(z) = c_{-N} + c_{-(N-1)}(z - z_0)$$
$$+ \cdots + c_0(z - z_0)^N + c_1(z - z_0)^{N+1} + \cdots. \quad (6.2–4)$$

From the preceding we have

$$\lim_{z \to z_0} \left[(z - z_0)^N f(z) \right] = c_{-N}. \quad (6.2–5)$$

Since $|(z - z_0)^N| \to 0$ as $z \to z_0$, Eq. (6.2–5) shows us that $\lim_{z \to z_0} |f(z)| \to \infty$, that is, if $f(z)$ has a pole at z_0 then $f(z)$ must have a limit of ∞ as $z \to z_0$.

The function $f(z) = \sinh z / z$ does not have a pole at $z = 0$. To see this we apply L'Hopital's rule to evaluate $\lim_{z \to 0} f(z)$. We obtain $\lim_{z \to 0} \cosh z / 1 = 1$. In fact, $f(z)$ has a removable singularity at $z = 0$.

A function $f(z)$ having an essential singularity at z_0 does not have a limit of ∞ as $z \to z_0$. For example, in the case of $e^{1/z}$, if we approach the origin along the line $y = 0$, $x > 0$, we find that $f(z) = e^{1/x}$ becomes unbounded as $x \to 0$. However, if we approach the origin along the line $x = 0$, we have $f(z) = e^{1/iy} = \cos(y^{-1}) - i \sin(y^{-1})$, which is a complex number of modulus 1 for all y.

Analytic branches of some multivalued functions, such as $\log z$ and $1/(z - 1)^{1/2}$ have moduli that become infinite at their singular points (in these examples at $z = 0$ and $z = 1$, respectively). However, these singular points are not poles but branch points. A function that "blows up" at a point does not necessarily have a pole there. However, a function that has a limit of ∞ at an *isolated* singular point does have a pole at that point. The following pair of rules, based on Eqs. (6.2–3) and (6.2–5), are useful in establishing the existence of a pole and its order. To establish the second rule we must multiply Eq. (6.2–3) by $(z - z_0)^n$.

Rule I

Let z_0 be an isolated singular point of $f(z)$. If $\lim_{z \to z_0} (z - z_0)^N f(z)$ exists and has a finite, nonzero value, then $f(z)$ has a pole of order N at z_0. ∎

Rule II

If N is the order of the pole of $f(z)$ at z_0, then

$$\lim_{z \to z_0} (z - z_0)^n f(z) = \begin{cases} 0, & n > N, \\ \infty, & n < N. \end{cases} \quad ∎ \quad (6.2–6)$$

Example 1

Discuss the singularities of

$$f(z) = \frac{z \cos z}{(z - 1)(z^2 + 1)^2(z^2 + 3z + 2)}.$$

Solution

This function possesses only isolated singularities, and they occur only where the denominator becomes zero. Factoring the denominator we have

$$f(z) = \frac{z \cos z}{(z - 1)(z + i)^2(z - i)^2(z + 2)(z + 1)}.$$

There is a pole of order 1 (simple pole) at $z = 1$ since

$$\lim_{z \to 1} [(z - 1)f(z)] = \lim_{z \to 1} \frac{(z - 1)z \cos z}{(z - 1)(z + i)^2(z - i)^2(z + 2)(z + 1)} = \frac{\cos 1}{24},$$

which is finite and nonzero.

There is a second-order pole at $z = -i$ since

$$\lim_{z \to -i} \left[(z + i)^2 f(z) \right] = \lim_{z \to -i} \frac{(z + i)^2 z \cos z}{(z - 1)(z + i)^2(z - i)^2(z + 2)(z + 1)}$$

$$= \frac{-i \cos(-i)}{8(2 - i)}.$$

Similarly, there is a pole of order 2 at $z = i$ and poles of order 1 at -2 and -1. ◀

Example 2

Discuss the singularities of

$$f(z) = \frac{e^z}{\sin z}.$$

Solution

Wherever $\sin z = 0$, that is, for $z = k\pi$, $k = 0, \pm 1, \pm 2, \ldots,$ $f(z)$ has isolated singularities.

Assuming these are simple poles, we evaluate $\lim_{z \to k\pi} [(z - k\pi)e^z / \sin z]$. This indeterminate form is evaluated from L'Hopital's rule and equals

$$\lim_{z \to k\pi} \frac{(z - k\pi)e^z + e^z}{\cos z} = \frac{e^{k\pi}}{\cos k\pi}.$$

Because this result is finite and nonzero, the pole at $z = k\pi$ is of first order.

Had this result been infinite, we would have recognized that the order of the pole exceeded 1, and we might have investigated $\lim_{z \to k\pi} (z - k\pi)^2 f(z)$, etc. On the other hand, had our result been zero, we would have concluded that $f(z)$ had a removable singularity at $z = k\pi$. ◀

Problems like the one discussed in Example 2, in which we must investigate the order of the pole for a function of the form $f(z) = g(z)/h(z)$, are so common that we will give them some special attention.

If $g(z)$ and $h(z)$ are analytic at z_0, with $g(z_0) \neq 0$ and $h(z_0) = 0$, then $f(z)$ will have an isolated singularity at z_0.[†] With these assumptions we expand $h(z)$ in a Taylor series about z_0.

[†] The zeros of an analytic function are isolated, that is, every zero has some neighborhood containing no other zero. Thus a zero appearing in a denominator creates an isolated singularity. See W. Kaplan, *Introduction to Analytic Functions* (Reading, Mass.: Addison-Wesley, 1966) pp. 93–94.

Definition Order of zero

Let

$$h(z) = a_N(z - z_0)^N + a_{N+1}(z - z_0)^{N+1} + \cdots.$$

The leading term, that is, the one containing the lowest power of $(z - z_0)$, is $a_N(z - z_0)^N$, where $a_N \neq 0$ and $N \geq 1$. We say that $h(z)$ has a *zero of order N* at z_0. ∎

Rewriting our expression for $f(z)$ by using the series for $h(z)$, we have

$$f(z) = \frac{g(z)}{a_N(z - z_0)^N + a_{N+1}(z - z_0)^{N+1} + \cdots}.$$

To show that this expression has a pole of order N at z_0, consider

$$\lim_{z \to z_0} \left[(z - z_0)^N f(z) \right] = \lim_{z \to z_0} \frac{(z - z_0)^N g(z)}{a_N(z - z_0)^N + a_{N+1}(z - z_0)^{N+1} + \cdots}$$

$$= \lim_{z \to z_0} \frac{g(z)}{a_N + a_{N+1}(z - z_0) + \cdots} = \frac{g(z_0)}{a_N}.$$

Since this limit is finite and nonzero, we may conclude the following rule.

Rule I Quotients

If $f(z) = g(z)/h(z)$, where $g(z)$ and $h(z)$ are analytic at z_0, and if $h(z_0) = 0$ and $g(z_0) \neq 0$, then the order of the pole of $f(z)$ at z_0 is identical to the order of the zero of $h(z)$ at this point. ∎

The preceding procedure can be modified to deal with the case $g(z_0) = 0$, $h(z_0) = 0$. Under these conditions if $\lim_{z \to z_0} f(z) = \lim_{z \to z_0} [g(z)/h(z)]$ is infinite, then $f(z)$ has a pole at z_0, whereas if the limit is finite, $f(z)$ has a removable singularity at z_0. L'Hopital's rule is often useful in finding the limit.

If there is a pole, and we want its order, we might expand both $g(z)$ and $h(z)$ in Taylor series about z_0. This would establish the order of the zeros of $g(z)$ and $h(z)$ at z_0. Then, as is shown in Exercise 3 of this section, the following rule applies.

Rule II Quotients

The order of the pole of $f(z) = g(z)/h(z)$ at z_0 is the order of the zero of $h(z)$ at this point less the order of the zero of $g(z)$ at the same point. ∎

The number found from this rule must be positive, otherwise there would be no pole.

Example 3

Find the order of the pole of $(z^2 + 1)/(e^z + 1)$ at $z = i\pi$.

Solution

With $g(z) = (z^2 + 1)$ and $h(z) = (e^z + 1)$ we verify that $g(i\pi) = -\pi^2 + 1 \neq 0$, and $h(i\pi) = e^{i\pi} + 1 = 0$.

To find the order of the zero of $(e^z + 1)$ at $z = i\pi$, we make the Taylor expansion

$$h(z) = e^z + 1 = c_0 + c_1(z - i\pi) + c_2(z - i\pi)^2 + \cdots.$$

Note that $c_0 = 0$ because $h(i\pi) = 0$. Since

$$c_1 = \frac{d}{dz}(e^z + 1)\bigg|_{z=i\pi} = -1,$$

which is nonzero, we see that $h(z)$ has a zero of order 1 at $z = i\pi$. Thus by our Rule I, $f(z)$ has a pole of order 1 at $z = i\pi$. ◀

Example 4

Find the order of the pole of

$$f(z) = \frac{\sinh z}{\sin^5 z}, \quad \text{at } z = 0.$$

Solution

With $g(z) = \sinh z$ and $h(z) = \sin^5 z$ we find that $g(0) = 0$, and $h(0) = 0$.

Because $\sinh z = z + z^3/3! + z^5/5! + \cdots$, we see that $g(z)$ has a zero of order 1 at $z = 0$. Since $(\sin z)^5 = (z - z^3/3! + z^5/5! + \cdots)^5$, we see that the lowest power of z in the Maclaurin series for $\sin^5 z$ is 5. Thus $(\sin z)^5$ has a zero of order 5 at $z = 0$. The order of the pole of $f(z)$ at $z = 0$ is, by Rule II, the order of the zero of $(\sin z)^5$ less the order of the zero of $\sinh z$, that is, $5 - 1 = 4$. ◀

EXERCISES

1. Show, by means of a Laurent expansion, that the following functions possess essential singularities at the points stated.

   a) e^{1/z^2}, at $z = 0$

   b) $(z - 1)e^{1/z}$, at $z = 0$

   c) $e^{1/(z-1)} \sin[1/(z - 1)]$, at $z = 1$

2. Use series expansions or L'Hopital's rule to show that the following functions possess removable singularities at their isolated singular points, that is, show that $\lim_{z \to z_0} f(z)$ exists and is finite at each such point. How, in each case, should $f(z_0)$ be defined in order to remove the singularity?

   a) $\dfrac{z^2 + 2z + 1}{z + 1}$ b) $\dfrac{\sin^2 z}{z^2}$ c) $\dfrac{\cos z}{z - \pi/2}$ d) $\dfrac{\sin z \cos z}{\mathrm{Log}\,(1 + z)}$

3. Let $f(z) = g(z)/h(z)$, where $g(z) = b_M(z - z_0)^M + b_{M+1}(z - z_0)^{M+1} + \cdots$ has a zero of order M at $z = z_0$ and $h(z) = a_N(z - z_0)^N + a_{N+1}(z - z_0)^{N+1} + \cdots$, which has a zero of order N at $z = z_0$.

   a) Show that if $N > M$, $f(z)$ has a pole of order $N - M$ at $z = z_0$.

b) If $N \le M$, how should $f(z_0)$ be defined so that the singularity of $f(z)$ at z_0 is eliminated? Consider the cases $N = M$, $N < M$.

4. Find the residues of the functions in Exercise 1 at each of their singularities.

5. State the locations of all the poles for each of the following functions, and give the order of each pole.

a) $\dfrac{z}{(z^2 + z + 1)^2}$　　d) $\dfrac{z^2 + 1}{(z^4 - 1)^2}$　　g) $\dfrac{e^{1/z}}{(z + 1/z)^2}$

b) $\dfrac{z^2 + 3z + 2}{(z^4 + 2z^2 + 1)}$　　e) $\dfrac{\sinh z}{(e^z - 1)^2}$　　h) $\dfrac{\cosh z}{z \sinh z}$

c) $\dfrac{z}{\sin z \cos z}$　　f) $\dfrac{e^z + 1}{e^z - e}$　　i) $\left(\dfrac{\sin z + \cos z}{z^2}\right)$

6.3 FINDING THE RESIDUE

When a function $f(z)$ is known to possess a pole at z_0, there is a straightforward method for finding its residue at this point that does not entail our obtaining a Laurent expansion about z_0. When the order of the pole is known, the technique involved is even easier.

We have found in the preceding section (see Eq. (6.2–4) that when $f(z)$ has a pole of order N at z_0, that $\psi(z) = (z - z_0)^N f(z)$ has the following Taylor series expansion about z_0:

$$\psi(z) = (z - z_0)^N f(z) = c_{-N} + c_{-(N-1)}(z - z_0) + \cdots + c_0(z - z_0)^N + \cdots .$$

(6.3–1)

Suppose $f(z)$ has a simple pole at $z = z_0$. Then $N = 1$. Hence

$$\psi(z) = c_{-1} + c_0(z - z_0) + c_1(z - z_0)^2 + \cdots . \qquad (6.3–2)$$

The residue c_{-1} is easily seen to be $\lim_{z \to z_0} \psi(z) = \lim_{z \to z_0} [(z - z_0)f(z)]$. Hence, we arrive at Rule I.

Rule I　Residues

If $f(z)$ has a pole of order 1 at $z = z_0$, then

$$\text{Res}[f(z), z_0] = \lim_{z \to z_0} [(z - z_0)f(z)]. \quad \blacksquare \qquad (6.3–3)$$

Suppose $f(z)$ has a pole of order 2 at $z = z_0$. We then have, from Eq. (6.3–1),

$$\psi(z) = c_{-2} + c_{-1}(z - z_0) + c_0(z - z_0)^2 + \cdots .$$

We notice that

$$\frac{d\psi}{dz} = c_{-1} + 2c_0(z - z_0) + \cdots ,$$

so that

$$c_{-1} = \lim_{z \to z_0} \frac{d}{dz}(\psi(z)) = \lim_{z \to z_0} \frac{d}{dz}\left[(z - z_0)^2 f(z)\right],$$

from which we obtain Rule II.

Rule II   Residues

If $f(z)$ has a pole of order 2 at $z = z_0$,

$$\text{Res}[f(z), z_0] = \lim_{z \to z_0} \frac{d}{dz}\left[(z - z_0)^2 f(z)\right]. \quad \blacksquare \qquad (6.3\text{--}4)$$

The method can be generalized, the result of which is Rule III.

Rule III   Residues

If $f(z)$ has a pole of order N at $z = z_0$, then

$$\text{Res}[f(z), z_0] = \lim_{z \to z_0} \frac{1}{(N-1)!} \frac{d^{N-1}}{dz^{N-1}}\left[(z - z_0)^N f(z)\right]. \quad \blacksquare \qquad (6.3\text{--}5)$$

Rule II is contained in Rule III (put $N = 2$), as is Rule I if we take $0! = 1$, and $d^{N-1}/dz^{N-1}|_{N=1} = 1$.

If the order of the pole of $f(z)$ at z_0 is known, the application of Eq. (6.3–5) yields the residue directly. If the order is unknown, we might seek to determine its order by means of the methods suggested in Section 6.2. Another possibility is the following method, which is proved in Exercise 7 of this section.

> Guess the order of the pole and use Eq. (6.3–5), taking N as the conjectured value. If the guessed N is less than the actual order, an infinite result, and not the residue, is obtained. However, if this N equals or exceeds the order of the pole, the residue is correctly obtained.

The problem of finding the residue at a pole of first order for a quotient of the form $f(z) = g(z)/h(z)$ occurs so often that we will derive a special formula for this case.

Let us assume that $f(z)$ has a simple pole at z_0 and that $g(z_0) \neq 0$. Thus $h(z)$ has a zero of first order at z_0. Applying Rule I

$$\text{Res}[f(z), z_0] = \lim_{z \to z_0} (z - z_0) \frac{g(z)}{h(z)},$$

which results in the indeterminate form $0/0$. Using L'Hopital's rule, we obtain

$$\lim_{z \to z_0} \frac{(z - z_0)g(z)}{h(z)} = \lim_{z \to z_0} \frac{(z - z_0)g'(z) + g(z)}{h'(z)} = \frac{g(z_0)}{h'(z_0)}.$$

Since $h(z)$ has a zero of order 1 at z_0, $h'(z_0) \neq 0$. We summarize the preceding steps as follows:

> **Rule IV** Residues
>
> The residue of $f(z) = g(z)/h(z)$ at a simple pole, where $g(z_0) \neq 0$, $h(z_0) = 0$, is given by
>
> $$\text{Res}[f(z), z_0] = \frac{g(z_0)}{h'(z_0)}. \quad \blacksquare \qquad (6.3\text{-}6)$$

Example 1

Find the residue of

$$f(z) = \frac{e^z}{(z^2 + 1)z^2}$$

at all poles.

Solution

We rewrite $f(z)$ with a factored denominator:

$$f(z) = \frac{e^z}{(z + i)(z - i)z^2},$$

which shows that there are simple poles at $z = \pm i$ and a pole of order 2 at $z = 0$.
From Rule I we obtain the residue at i:

$$\text{Res}[f(z), i] = \lim_{z \to i} \frac{(z - i)e^z}{(z + i)(z - i)z^2} = \frac{e^i}{(2i)(-1)}.$$

The residue at $-i$ could be similarly calculated. Instead, for variety, let's use Rule IV. Taking $g(z) = e^z/z^2$ (which is nonzero at $-i$) and $h(z) = z^2 + 1$, so that $h'(z) = 2z$, we have

$$\text{Res}[f(z), -i] = \left. \frac{e^z/z^2}{2z} \right|_{z = -i} = \frac{e^{-i}}{2i}.$$

Notice that we could also have taken $g(z) = e^z$, $h(z) = z^2(z^2 + 1)$ and the same result would ultimately be obtained.
The residue at $z = 0$ is computed from Rule II as follows.

$$\text{Res}[f(z), 0] = \lim_{z \to 0} \frac{d}{dz} \frac{z^2 e^z}{(z^2 + 1)(z^2)} = \lim_{z \to 0} \frac{e^z(z^2 + 1) - 2ze^z}{(z^2 + 1)^2} = 1 \quad \blacktriangleleft$$

Example 2

Find the residue of

$$f(z) = \frac{\tan z}{z^2 + z + 1}$$

at all singularities of $\tan z$.

Solution

Rewriting $f(z)$ as $\sin z/[(\cos z)(z^2 + z + 1)]$, we see that there are poles of $f(z)$ for z satisfying $\cos z = 0$, that is, $z = \pi/2 + k\pi$, $k = 0, \pm 1, \pm 2, \ldots$. We can show that these are simple poles by expanding $\cos z$ in a Taylor series about the point $z_0 = \pi/2 + k\pi$. We obtain

$$\cos z = a_1(z - z_0) + a_2(z - z_0)^2 + \cdots, \quad \text{where } a_1 = -\cos k\pi \neq 0.$$

Since $\cos z$ has a zero of order 1 at z_0, $f(z)$ must have a pole of order 1 there.

Let us apply Rule IV, taking $g(z) = \sin z/(z^2 + z + 1)$ and $h(z) = \cos z$, so that $h'(z) = -\sin z$. Thus

$$\text{Res}\big[f(z), \pi/2 + k\pi \big] = - \frac{1}{z^2 + z + 1}\bigg|_{\pi/2+k\pi}$$

$$= \frac{-1}{(\pi/2 + k\pi)^2 + (\pi/2 + k\pi) + 1}, \quad k = 0, \pm 1, \pm 2, \ldots.$$

Instead of first determining the order of the poles at $z = k\pi + \pi/2$, we might have just assumed that they were of first order and then applied Rule I or Rule IV. The finite result thus obtained would justify our guess.

Comment: There are also poles of $f(z)$, for z satisfying $z^2 + z + 1 = 0$. The roots of this quadratic, z_1 and z_2, are $-1/2 \pm i\sqrt{3}/2$. Because the roots are distinct, the quadratic expression is a product of *nonrepeated* factors $(z - z_1)(z - z_2)$, and the poles of $f(z)$ at z_1 and z_2 are thus of first order. The residues at these poles can be found from Rule IV. We take $g(z) = \tan z$, and $h(z) = z^2 + z + 1$. ◄

Example 3

Find the residue of

$$f(z) = \frac{(z)^{1/2}}{z^3 - 4z^2 + 4z}$$

at all poles. Use the principal branch of $(z)^{1/2}$.

Solution

We factor the denominator and obtain

$$f(z) = \frac{z^{1/2}}{(z)(z - 2)^2}.$$

It appears that there is a simple pole at $z = 0$. This is wrong. A pole is an isolated singularity, and $f(z)$ does not have an isolated singularity at $z = 0$. The factor $(z)^{1/2}$ has a branch point at this value of z that in turn causes $f(z)$ to have a branch point there.

However, $f(z)$ does have a pole of order 2 at $z = 2$. Applying Rule II, we find

$$\text{Res}[f(z), 2] = \lim_{z \to 2} \frac{d}{dz} \left[\frac{(z-2)^2 z^{1/2}}{(z)(z-2)^2} \right] = \frac{-1}{4(2)^{1/2}},$$

where, because we are using the principal branch of the square root, $2^{1/2}$ is chosen positive. ◀

Example 4

Find the residue of

$$f(z) = \frac{e^{1/z}}{1 - z}$$

at all singularities.

Solution

Obviously, there is a simple pole at $z = 1$. The residue there, from Rule I, is found to be $-e$. Since

$$e^{1/z} = 1 + z^{-1} + \frac{z^{-2}}{2!} + \cdots$$

has an essential singularity at $z = 0$, this will also be true of $f(z) = e^{1/z}/(1 - z)$.

The residue of $f(z)$ at $z = 0$ is calculable only if we find the Laurent expansion about this point and extract the appropriate coefficient. Since

$$\frac{1}{1 - z} = 1 + z + z^2 + \cdots, \qquad |z| < 1,$$

we have

$$\frac{e^{1/z}}{1 - z} = \left(1 + z + z^2 + z^3 + \cdots\right)\left(1 + z^{-1} + \frac{z^{-2}}{2!} + \cdots\right)$$

$$= \cdots + c_{-2} z^{-2} + c_{-1} z^{-1} + c_0 + \cdots.$$

Our interest is in c_{-1}. If we multiply together the two series and confine our attention to products resulting in z^{-1}, we have

$$c_{-1} z^{-1} = \left[1 + \frac{1}{2!} + \frac{1}{3!} + \cdots\right] z^{-1}.$$

Recalling the definition $e = 1 + 1 + 1/2! + 1/3! + \cdots$, we see that $c_{-1} = e - 1$ $= \text{Res}[f(z), 0]$. ◀

Example 5

Find the residue of

$$f(z) = \frac{e^z - 1}{\sin^3 z}, \quad \text{at } z = 0.$$

Solution

Both numerator and denominator of the given function vanish at $z = 0$. To establish the order of the pole, we will expand both these expressions in Maclaurin series by the usual means.

$$e^z - 1 = z + \frac{z^2}{2!} + \frac{z^3}{3!} + \cdots, \qquad \sin^3 z = z^3 - \frac{z^5}{2} + \cdots.$$

Thus

$$\frac{e^z - 1}{\sin^3 z} = \frac{z + \dfrac{z^2}{2!} + \dfrac{z^3}{3!} + \cdots}{z^3 - \dfrac{z^5}{2} + \cdots}.$$

Since the numerator has a zero of order 1 and the denominator has a zero of order 3, the quotient has a pole of order 2.

To find the residue of $f(z)$ at $z = 0$, we could apply Rule II. However, it will be found, after performing the repeated differentiations, that the expression obtained is indeterminate at $z = 0$. The required limit as $z \to 0$ is found only after successive applications of L'Hopital's rule—a tedious procedure.

Instead, the quotient of the two series appearing above is expanded in a Laurent series by means of long division. We need proceed only far enough to obtain the term containing z^{-1}. Thus

$$z^3 - \frac{z^5}{2} + \cdots \overline{\smash{\big)}\, z + \frac{z^2}{2!} + \frac{z^3}{3!} + \cdots,} \qquad \overset{\displaystyle z^{-2} + \frac{z^{-1}}{2} + \cdots}{}$$

from which we see that the residue is $1/2$. ◀

EXERCISES

1. For each of the following functions state the location, order, and residue for every pole. (In part (d) use principal branch of $z^{1/3}$.)

   a) $\dfrac{z}{z^2 + i}$ b) $\dfrac{1}{z(z^3 + 1)}$ c) $\dfrac{\cos z}{(z - 3)(z^2 - 5z + 6)}$ d) $\dfrac{z^{1/3}}{z(z - 1)}$

2. For each of the functions in Exercise 5, Section 6.2, find the residue at every pole.

3. Compute the residue of the following functions at the indicated singularity.

   a) $\mathrm{Res}[z^2 e^{1/z}, 0]$ c) $\mathrm{Res}[e^z \sin 1/z, 0]$

       Give answer as an infinite sum.

   b) $\mathrm{Res}[z e^{1/(z-1)}, 1]$ d) $\mathrm{Res}\left[\dfrac{e^z}{\mathrm{Log}\, 1/z}, 1\right]$

4. Find the following residues, where k is any integer.

a) $\text{Res}\left[z \tan z, k\pi + \dfrac{\pi}{2}\right]$ e) $\text{Res}\left[\dfrac{1}{(z-1)^{10}}, 1\right]$

b) $\text{Res}\left[\dfrac{\tan z}{\cos z}, k\pi + \dfrac{\pi}{2}\right]$ f) $\text{Res}\left[\dfrac{\cos z}{(z-1)^{10}}, 1\right]$

c) $\text{Res}\left[\dfrac{e^z - 1}{\sin^3 z}, \pi\right]$ g) $\text{Res}\left[\dfrac{1}{\sin(\cos z)}, \dfrac{\pi}{2}\right]$

d) $\text{Res}\left[\dfrac{\text{Log } z}{(z^2 + 1)^2}, i\right]$ h) $\text{Res}\left[\dfrac{1}{(\text{Log } z) - 1}, e\right]$

5. Can a function have a residue of zero at a simple pole? Can a function have a residue of zero at a higher-order pole? Explain.

6. Use residue calculus to evaluate the following integrals:

a) $\oint \dfrac{\cosh z}{z^2 + 1}\,dz$, around $|z - 2i| = 2$;

b) $\oint z e^{1/z}\,dz$, around $\dfrac{x^2}{9} + \dfrac{y^2}{2} = 1$;

c) $\oint \dfrac{z e^{1/z}}{1 - z}\,dz$, around $\dfrac{x^2}{9} + \dfrac{y^2}{2} = 1$;

d) $\oint \dfrac{e^z}{(z^2 + z + 1)}\,dz$, around the square with sides $x = \pm 1, y = \pm 1$;

e) $\oint_{|z|=4} \dfrac{\sinh z}{\sin^2 z}\,dz$;

f) $\oint_{|z|=4} \dfrac{z}{\sin z}\,dz$;

g) $\oint \dfrac{z^{1/2}}{z^2 - 4z + 4}\,dz$, principal branch of $z^{1/2}$, integrate on $|z - 3| = 2$.

7. Let $f(z)$ have a pole of order m at $z = z_0$, so that, about the point z_0, we have the Laurent expansion

$$f(z) = c_{-m}(z - z_0)^{-m} + c_{-(m-1)}(z - z_0)^{-(m-1)} + \cdots .$$

a) Consider $\psi(z) = (z - z_0)^N f(z)$. Suppose $N \geq m$. What is the Taylor expansion for $\psi(z)$ about z_0? Show that

$$\lim_{z \to z_0} \dfrac{1}{(N-1)!} \dfrac{d^{N-1}}{dz^{N-1}}\left[(z - z_0)^N f(z)\right] = c_{-1} = \text{Res}\left[f(z), z_0\right].$$

b) Suppose $1 \leq N < m$. Show that $\psi(z)$ has a Laurent expansion about z_0. Show that

$$\lim_{z \to z_0} \dfrac{1}{(N-1)!} \dfrac{d^{N-1}}{dz^{N-1}}\left[(z - z_0)^N f(z)\right] = \infty .$$

8. a) Consider the analytic function $f(z) = g(z)/h(z)$, having a pole at z_0. Let $g(z_0) \neq 0$, $h(z_0) = h'(z_0) = 0$, $h''(z_0) \neq 0$. Thus $f(z)$ has a pole of second order at $z = z_0$. Show that

$$\text{Res}[f(z), z_0] = \frac{2g'(z_0)}{h''(z_0)} - \frac{2}{3} \frac{g(z_0)h'''(z_0)}{[h''(z_0)]^2}.$$

Hint: Write down the Taylor series expansion, about z_0, for $g(z)$ and $h(z)$, taking note of which coefficients are zero. Divide the two series using long division and so obtain the Laurent expansion of $f(z)$ about z_0.

b) Use the formula of part (a) to obtain

$$\text{Res}\left[\frac{e^z}{\sin^2 z}, 0\right].$$

6.4 EVALUATION OF REAL INTEGRALS WITH RESIDUE CALCULUS, I

Real definite integrals of the type $\int_0^{2\pi} R(\sin\theta, \cos\theta)\, d\theta$, where R is a *rational function* of $\sin\theta$ and or $\cos\theta$ are frequently very difficult to evaluate with the methods of elementary calculus.[†] However, the calculus of residues can give the result in a straightforward manner.

An example of such an integral is $\int_0^{2\pi} 1/(2 + \sin\theta)\, d\theta$. Integrals like this occur in Dirichlet problems for the circle solved by the Poisson integral formula (see Section 4.5).

To evaluate all integrals of the form $\int_0^{2\pi} R(\sin\theta, \cos\theta)\, d\theta$, the approach is the same. The given expression is converted into a line integration in the complex z-plane by the following change of variables:

$$z = e^{i\theta}, \qquad dz = e^{i\theta}i\, d\theta$$

so that

$$d\theta = \frac{dz}{iz},$$

$$\sin\theta = \frac{e^{i\theta} - e^{-i\theta}}{2i} = \frac{z - z^{-1}}{2i}, \qquad (6.4\text{–}1)$$

$$\cos\theta = \frac{e^{i\theta} + e^{-i\theta}}{2} = \frac{z + z^{-1}}{2}.$$

As θ ranges from 0 to 2π, or over any interval of 2π, the variable $z = \cos\theta + i\sin\theta$ proceeds in the counterclockwise direction around the *unit circle* in the complex z-plane. This contour integral is evaluated with residue theory.

The method fails if the integrand for the contour integration has pole singularities *on* the unit circle. However, this can occur only if $\int_0^{2\pi} R(\sin\theta, \cos\theta)\, d\theta$ is an improper integral, that is, if $R(\sin\theta, \cos\theta)$ has a discontinuity on the interval $0 \leq \theta \leq 2\pi$.

[†] Recall that these functions R are quotients of polynomials in $\sin\theta$ and $\cos\theta$.

Example 1

Find

$$I = \int_0^{2\pi} \frac{d\theta}{2 + \sin\theta}$$

by using residues.

Solution

With the change of variables in Eq. (6.4–1) we have

$$I = \oint_{|z|=1} \frac{\dfrac{dz}{iz}}{2 + \dfrac{z - z^{-1}}{2i}} = \oint_{|z|=1} \frac{2\,dz}{z^2 + 4iz - 1}.$$

We now examine the integrand on the right for poles. From the quadratic formula we find that $z^2 + 4iz - 1 = 0$ has roots $z = i\left[-2 \pm \sqrt{3}\right]$. Thus

$$I = \oint_{|z|=1} \frac{2\,dz}{\left[z - i(-2 + \sqrt{3})\right]\left[z - i(-2 - \sqrt{3})\right]}.$$

There is a simple pole at $z_1 = i(-2 + \sqrt{3}) \doteq -0.27i$ and also at $z_2 = i(-2 - \sqrt{3}) \doteq -3.73i$, but only z_1 is inside the unit circle. We thus need only the residue at z_1. Using Rule IV of the previous section, we have

$$\oint \frac{2\,dz}{z^2 + 4iz - 1} = \left. \frac{4\pi i}{2z + 4i} \right|_{z=i(-2+\sqrt{3})} = \frac{2\pi}{\sqrt{3}}. \qquad \blacktriangleleft$$

Functions of the form $\cos n\theta$ and $\sin n\theta$, where n is an integer, are expressible in terms of sums and differences of integral powers of $\cos\theta$ and $\sin\theta$ and are therefore rational functions of $\cos\theta$ and $\sin\theta$. Integrals containing rational expressions in $\cos n\theta$ and $\sin n\theta$ are readily evaluated by the method just discussed. We still take $z = e^{i\theta}$ and use the substitution

$$\cos n\theta = \frac{e^{in\theta} + e^{-in\theta}}{2} = \frac{z^n + z^{-n}}{2}, \qquad \sin n\theta = \frac{e^{in\theta} - e^{-in\theta}}{2i} = \frac{z^n - z^{-n}}{2i}.$$

Example 2

Find

$$I = \int_0^{2\pi} \frac{\cos 2\theta}{5 - 4\sin\theta}\,d\theta.$$

Solution

With the substitutions

$$\cos 2\theta = \frac{z^2 + z^{-2}}{2}, \qquad \sin\theta = \frac{z - z^{-1}}{2i}, \qquad d\theta = \frac{dz}{iz}$$

we have

$$I = \oint_{|z|=1} \frac{\dfrac{z^2 + z^{-2}}{2}}{5 - \dfrac{2}{i}(z - z^{-1})} \left(\frac{dz}{iz}\right) = \oint_{|z|=1} \frac{(z^4 + 1)\, dz}{2iz^2[2iz^2 + 5z - 2i]}.$$

There is a second-order pole at $z = 0$. Solving $2iz^2 + 5z - 2i = 0$, we find simple poles at $i/2$ and $2i$. The pole at $2i$ is outside the circle $|z| = 1$ and can be ignored. Thus

$$I = 2\pi i \sum_{\text{res}} \frac{z^4 + 1}{2iz^2[2iz^2 + 5z - 2i]}, \quad \text{at } z = 0 \text{ and } i/2.$$

From Eq. (6.3–4) we find the residue at $z = 0$:

$$\frac{1}{2i} \frac{d}{dz} \left[\frac{z^4 + 1}{2iz^2 + 5z - 2i} \right]\bigg|_{z=0} = \frac{-5}{8} i,$$

and from Eq. (6.3–6), the residue at $i/2$:

$$\frac{1}{(2i)\left(\dfrac{i}{2}\right)^2} \frac{\left(\dfrac{i}{2}\right)^4 + 1}{\dfrac{d}{dz}[2iz^2 + 5z - 2i]}\bigg|_{z=i/2} = \frac{17i}{24}.$$

Thus

$$I = 2\pi i \left(\frac{-5i}{8} + \frac{17i}{24} \right) = \frac{-\pi}{6}. \qquad \blacktriangleleft$$

EXERCISES

1. Evaluate the following integrals by residue calculus.

a) $\displaystyle\int_0^{2\pi} \frac{d\theta}{2 + \cos\theta}$

b) $\displaystyle\int_0^{2\pi} \frac{d\theta}{a + b\sin\theta}, \quad a > b$

c) $\displaystyle\int_0^{2\pi} \frac{\cos\theta}{a + b\cos\theta}\, d\theta, \quad a > b$

d) $\displaystyle\int_0^{2\pi} \frac{\cos 3\theta}{5 + 4\cos\theta}\, d\theta$

e) $\displaystyle\int_0^{2\pi} \frac{d\theta}{(a + b\sin\theta)^2}, \quad a > b$

f) $\displaystyle\int_0^{2\pi} \frac{d\theta}{9 + \sin^2\theta}$

g) $\displaystyle\int_0^{2\pi} \cos^4\theta\, d\theta$

h) $\displaystyle\int_0^{2\pi} \frac{\cos n\theta}{\cosh a + \cos\theta}\, d\theta,$
where $a \neq 0$ is a real number,
n an integer

2. Use the periodic or symmetric properties of the integrand to convert the following expressions, where necessary, to integrals over an interval of 2π. In part (a), for example, $\int_0^\pi = 1/2 \int_0^{2\pi}$. Why? Evaluate the resulting expression

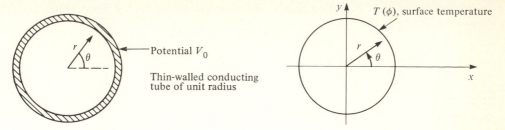

Figure 6.4–1   Figure 6.4–2

with residue calculus. Find

a) $\displaystyle\int_0^{\pi} \frac{\cos\theta}{5 + 4\cos\theta}\,d\theta,$

d) $\displaystyle\int_{-\pi}^{+\pi} \frac{\sin\theta}{5 + 4\cos\theta}\,d\theta,$

b) $\displaystyle\int_{-\pi/2}^{+\pi/2} \frac{\sin\theta}{5 + 4\sin\theta}\,d\theta,$

e) $\displaystyle\int_0^{\pi} \frac{1 + \cos\theta}{1 + \cos^2\theta}\,d\theta,$

c) $\displaystyle\int_0^{\pi/2} \frac{d\theta}{(4 + \sin^2\theta)^2},$

f) $\displaystyle\int_0^{\pi} \frac{\cos 2\theta}{1 + a^2 - 2a\cos\theta}\,d\theta, \qquad -1 < a < 1.$

3. A long metal tube of radius 1 is maintained at a constant voltage V_0 (see Fig. 6.4–1).

   a) Use the Poisson integral formula for the circle (see Section 4.5) to obtain an integral expression for the electric potential at an arbitrary point r, θ inside the tube.

   b) Evaluate this integral by residue calculus and show that the potential inside the tube is constant, that is, independent of r and θ.

4. A cylinder of radius 1 contains a homogeneous heat-conducting material. The surface of the cylinder is maintained at the steady-state temperature $T(\phi) = \cos\phi$, $\phi = \tan^{-1}(y/x)$ (see Fig. 6.4–2).

   a) Use the Poisson integral formula for a circle to obtain an integral expression for $T(r, \theta)$, the temperature at an arbitrary point inside the cylinder.

   b) Evaluate the integral by using residue calculus. Show that $T(r, \theta) = r\cos\theta$ inside the cylinder.

   c) Verify that your result satisfies the given boundary condition.

6.5 EVALUATION OF INTEGRALS, II

In previous work in mathematics and physics the reader has probably encountered "improper" integrals in which one or both limits are infinite, that is, expressions of the form

$$\int_k^{\infty} f(x)\,dx, \qquad \int_{-\infty}^{k} f(x)\,dx, \qquad \int_{-\infty}^{+\infty} f(x)\,dx,$$

where $f(x)$ is a real function of x, and k is a real constant.

Integrals of the first two types are defined in terms of proper integrals (Riemann sums) as follows:

$$\int_k^\infty f(x)\,dx = \lim_{R\to\infty}\int_k^R f(x)\,dx, \tag{6.5-1}$$

$$\int_{-\infty}^k f(x)\,dx = \lim_{R\to\infty}\int_{-R}^k f(x)\,dx, \tag{6.5-2}$$

provided the indicated limits exist.

Such improper integrals do not always exist. Thus, for example,

a) $\displaystyle\int_1^\infty \frac{1}{1+x^2}\,dx = \lim_{R\to\infty}\int_1^R \frac{dx}{1+x^2} = \lim_{R\to\infty}(\arctan R - \arctan 1)$

$$= \frac{\pi}{2} - \frac{\pi}{4}$$

exists; however,

b) $$\int_1^\infty \frac{1}{x}\,dx = \lim_{R\to\infty}(\log R - \log 1)$$

fails to exist, as does

c) $$\int_0^\infty \cos x\,dx = \lim_{R\to\infty}\sin R.$$

In case (b) as x increases, the curve $y = 1/x$ does not fall to zero fast enough for the area under the curve to approach a finite limit. In case (c) a sketch of $y = \cos x$ shows that, along the positive x-axis, the total area under this curve has no meaning.

We define an improper integral with two infinite limits by the following equation.

CAUCHY PRINCIPAL
VALUE
$$\int_{-\infty}^{+\infty} f(x)\,dx = \lim_{R\to\infty}\int_{-R}^{+R} f(x)\,dx \tag{6.5-3}$$

Integrals between $-\infty$ and $+\infty$ are often defined in another, more restrictive way. The definition given in Eq. (6.5–3) is known as the Cauchy principal value of the improper integral. A different definition of an integral between these limits, the standard or ordinary value, is considered in Exercise 3 of this section. It is shown there that if the ordinary value exists, it agrees with the Cauchy principal value, and that there are instances where the Cauchy principal value exists and the ordinary value does not. Unless otherwise stated, we will be using Cauchy principal values of integrals having infinite limits.

Now, if $f(x)$ is an odd function, that is, $f(x) = -f(-x)$, we have $\int_{-R}^{+R} f(x)\,dx = 0$ since the area under the curve $y = f(x)$ to the left of $x = 0$ cancels the area to the right of $x = 0$. Thus, from Eq. (6.5–3),

$$\int_{-\infty}^{+\infty} f(x)\,dx = 0, \quad \text{if } f(x) \text{ is odd} \tag{6.5-4}$$

for the Cauchy principal value of this integral. To illustrate:

$$\int_{-\infty}^{+\infty} \frac{x^3}{x^4+1}\,dx = 0, \qquad \int_{-\infty}^{+\infty} \frac{x}{x^2+1}\,dx = 0, \qquad \int_{-\infty}^{+\infty} x\,dx = 0.$$

When $f(x)$ is an even function of x, we have $f(x) = f(-x)$. Because of the symmetry of $y = f(x)$ about $x = 0$,

$$\int_{-R}^{+R} f(x)\, dx = 2\int_0^R f(x)\, dx.$$

From Eqs. (6.5–1) and (6.5–3) we thus obtain

$$2\int_0^\infty f(x)\, dx = \int_{-\infty}^\infty f(x)\, dx, \quad \text{if } f(x) \text{ is even.} \tag{6.5–5}$$

To illustrate:

$$2\int_0^\infty \frac{1}{x^2 + 1}\, dx = \int_{-\infty}^{+\infty} \frac{dx}{x^2 + 1}.$$

Let us see, with an example, how residue calculus enables us to find the Cauchy principal value of a real integral taken between $-\infty$ and $+\infty$.

Example 1

Find $\int_{-\infty}^{+\infty} x^2/(x^4 + 1)\, dx$ using residues.

Solution

We first consider $\oint_C z^2/(z^4 + 1)\, dz$ taken around the closed contour C (see Fig. 6.5–1) consisting of the line segment $y = 0$, $-R \leq x \leq R$ and the semicircle $|z| = R$, $0 \leq \arg z \leq \pi$. Let us take $R > 1$, which means that C encloses all the poles of $z^2/(z^4 + 1)$ in the upper half plane (abbreviated u.h.p.).

Hence,

$$\oint_C \frac{z^2\, dz}{z^4 + 1} = 2\pi i \sum_{\text{res}} \frac{z^2}{z^4 + 1}, \quad \text{at all poles in u.h.p.}$$

The integral along C is now broken into two parts: an integral along the real axis (here $z = x$) and an integral along the semicircular arc C_1 in the upper half plane.

$$\int_{-R}^{+R} \frac{x^2}{x^4 + 1}\, dx + \int_{C_1} \frac{z^2}{z^4 + 1}\, dz = 2\pi i \sum_{\text{res}} \frac{z^2}{z^4 + 1}, \quad \text{at all poles in u.h.p.}$$

$$\tag{6.5–6}$$

If we let $R \to \infty$, the integral on the extreme left becomes the Cauchy principal value of the real integral being evaluated. For $R \to \infty$ we can show that the second integral on the left becomes zero. To establish this we use the ML inequality (see

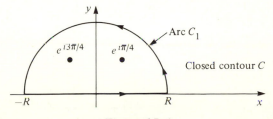

Figure 6.5–1

Section 4.1) and arrive at

$$\left| \int_{C_1} \frac{z^2}{z^4 + 1} \, dz \right| \le ML = M\pi R, \tag{6.5-7}$$

where $L = \pi R$ is the length of the semicircle C_1.

We require $|z^2/(z^4 + 1)| \le M$ on C_1. Since $|z| = R$ on this contour, we can instead require that $R^2/|z^4 + 1| \le M$. By a triangle inequality (see Eq. 1.3–20) $|z^4 + 1| \ge |z^4| - 1 = R^4 - 1$. Hence, $R^2/|z^4 + 1| \le R^2/(R^4 - 1)$. Thus we can put $M = R^2/(R^4 - 1)$ and use it in Eq. (6.5–7) with the result that

$$\left| \int_{C_2} \frac{z^2}{z^4 + 1} \, dz \right| \le \frac{\pi R^3}{R^4 - 1}.$$

As $R \to \infty$, the right side of this equation goes to zero, which means that the integral on the left must also become zero.

Armed with this fact, we put $R \to \infty$ in Eq. (6.5–6). The first integral on the left is the desired Cauchy principal value, the second disappears, and the right side remains unchanged. Thus

$$\int_{-\infty}^{+\infty} \frac{x^2}{x^4 + 1} \, dx = 2\pi i \sum_{\text{res}} \frac{z^2}{z^4 + 1}, \quad \text{at all poles in u.h.p.}$$

The equation $z^4 = -1$ has solutions $e^{i\pi/4}$, $e^{i3\pi/4}$, $e^{-i\pi/4}$, $e^{-i3\pi/4}$, of which only the first two lie in the upper half plane. The residues at the simple poles $e^{i\pi/4}$ and $e^{i3\pi/4}$ are easily found from Eq. (6.3–6) to be $(1/4)e^{-i\pi/4}$ and $(1/4)e^{-i3\pi/4}$, respectively. Thus

$$\int_{-\infty}^{+\infty} \frac{x^2}{x^4 + 1} \, dx = \frac{2\pi i}{4} \left[e^{-i\pi/4} + e^{-i3\pi/4} \right] = \frac{\pi}{\sqrt{2}}.$$

Because $x^2/(x^4 + 1)$ is an even function, we have, as a bonus,

$$\int_0^\infty \frac{x^2}{x^4 + 1} \, dx = \frac{1}{2} \frac{\pi}{\sqrt{2}}. \quad \blacktriangleleft$$

We can solve the problem just considered by using a contour of integration containing a semicircular arc in the lower half plane (abbreviated l.h.p.). Referring to Fig. 6.5–2, we have

$$\int_{-R}^{+R} \frac{x^2}{x^4 + 1} \, dx + \int_{C_2} \frac{z^2}{z^4 + 1} \, dz = -2\pi i \sum_{\text{res}} \frac{z^2}{z^4 + 1}, \quad \text{at all poles in l.h.p.}$$

Note the minus sign on the right. It arises because the closed contour in Fig. 6.5–2 is being negotiated in the negative (clockwise) sense. We again let $R \to \infty$ and apply the arguments of Example 1 to eliminate the second integral on the left. The reader should sum the residues on the right and verify that the same value is obtained for the integral evaluated in that example.

The technique involved in Example 1 is not restricted to the problem just presented but has wide application in the evaluation of other integrals taken between infinite limits. In all cases we must be able to argue that the integral taken over the arc becomes zero as $R \to \infty$. Theorem 3 is of use in asserting that this is so.

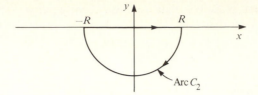

Figure 6.5–2

Theorem 3

Let $f(z)$ have the following property in the half plane Im $z \geq 0$. There exist constants $k > 1$, R_0, and μ such that

$$|f(z)| \leq \frac{\mu}{|z|^k}, \quad \text{for all } |z| \geq R_0 \text{ in this half plane.}$$

Then, if C_1 is the semicircular arc $Re^{i\theta}$, $0 \leq \theta \leq \pi$, and $R > R_0$, we have

$$\lim_{R \to \infty} \int_{C_1} f(z)\, dz = 0. \quad \blacksquare \tag{6.5-8}$$

The preceding merely says that if $|f(z)|$ falls off more rapidly than the reciprocal of the radius of C_1, then the integral of $f(z)$ around C_1 will vanish as the radius of C_1 becomes infinite. A corresponding theorem can be stated for a contour in the lower half plane.

The proof of Eq. (6.5–8) is simple. Assuming $R > R_0$, we apply the ML inequality as follows:

$$\left| \int_{C_1} f(z)\, dz \right| \leq ML = M\pi R,$$

where $L = \pi R$ is the length of C_1. We require $|f(z)| \leq M$ on C_1. By hypothesis $|f(z)| \leq \mu/|z|^k = \mu/R^k$ on C_1. Thus taking $M = \mu/R^k$ in the ML inequality, we have

$$\left| \int_{C_1} f(z)\, dz \right| \leq \frac{\pi R \mu}{R^k}, \quad \text{where } k > 1.$$

As $R \to \infty$, the right side of the preceding inequality goes to zero; thus, since the integral over C_1 must also go to zero, the theorem is proved.

Consider the rational function

$$\frac{P(z)}{Q(z)} = \frac{a_n z^n + a_{n-1} z^{n-1} + \cdots + a_0}{b_m z^m + b_{m-1} z^{m-1} + \cdots + b_0},$$

where m, the degree of the denominator Q, is assumed to exceed n, the degree of the numerator P. As $|z|$ grows without limit, the leading terms in the numerator and denominator become dominant. We therefore see intuitively that

$$\frac{P(z)}{Q(z)} \approx \frac{a_n}{b_m} \frac{z^n}{z^m} = \frac{a_n}{b_m} \frac{1}{z^d}, \quad \text{where } d = m - n.$$

Thus it should seem plausible that for sufficiently large $|z|$ there must exist constants μ and R_0 such that

$$\left| \frac{P(z)}{Q(z)} \right| \le \frac{\mu}{|z|^d}, \quad \text{for } |z| \ge R_0. \tag{6.5-9}$$

The proof can be found in many texts.[†] Thus when $d \ge 2$, the function $f(z) = P(z)/Q(z)$ will satisfy Eq. (6.5-8), and we can assert that

$$\lim_{R \to \infty} \int_{C_1} \frac{P(z)}{Q(z)} \, dz = 0, \tag{6.5-10}$$

where P, Q are polynomials and degree Q − degree $P \ge 2$.

Now, integrating $P(z)/Q(z)$ around contour C of Fig. 6.5-1, and taking R sufficiently large, we have from residue theory

$$\int_{-R}^{+R} \frac{P(x)}{Q(x)} \, dx + \int_{C_1} \frac{P(z)}{Q(z)} \, dz = 2\pi i \sum_{\text{res}} \frac{P(z)}{Q(z)}, \quad \text{at all poles in u.h.p.}$$

On the left z has been set equal to x on the straight portion of the path. Passing to the limit $R \to \infty$, we use Eq. (6.5-10) to eliminate the integral over the arc C_1 and then obtain the following theorem:

Theorem 4

Let $P(x)$ and $Q(x)$ be polynomials in x, and let the degree of $Q(x)$ exceed that of $P(x)$ by two or more. Let $Q(x)$ be nonzero for any real value of x. Then

$$\int_{-\infty}^{+\infty} \frac{P(x)}{Q(x)} \, dx = 2\pi i \sum_{\text{res}} \frac{P(z)}{Q(z)} \quad \text{at all poles in u.h.p.} \quad \blacksquare \tag{6.5-11}$$

The requirement $Q(x) \ne 0$ assures us that the integrand in Eq. (6.5-11) is finite for all x. The question of how to evaluate integrals in which $Q(x) = 0$, for some x, is dealt with later in this chapter.

Example 2

Find $\int_{-\infty}^{+\infty} x^2/(x^4 + x^2 + 1) \, dx$.

Solution

Equation (6.5-11) can be used directly since the degree of the denominator, which is 4, differs from that of the numerator by 2. A difference of at least two is required. Thus

$$\int_{-\infty}^{+\infty} \frac{x^2 \, dx}{x^4 + x^2 + 1} = 2\pi i \sum_{\text{res}} \frac{z^2}{z^4 + z^2 + 1}, \quad \text{at all poles in u.h.p.}$$

[†]See, for example, E. B. Saff and A. D. Snider, *Fundamentals of Complex Analysis* (Englewood Cliffs, N.J.: Prentice-Hall, 1976), p. 252.

Using the quadratic formula, we can solve $z^4 + z^2 + 1 = 0$ for z^2 and obtain

$$z^2 = \frac{-1 \pm i\sqrt{3}}{2} = e^{i2\pi/3}, e^{-i2\pi/3}.$$

Taking square roots yields $z = e^{i\pi/3}, e^{-i2\pi/3}, e^{-i\pi/3}, e^{i2\pi/3}$. Thus $z^2/(z^4 + z^2 + 1)$ has simple poles in the u.h.p. at $e^{i\pi/3}$ and $e^{i2\pi/3}$. Evaluating the residues at these two poles in the usual way (see, for example, Eq. 6.3–6), we find that the value of the given integral is

$$2\pi i \sum_{\text{res}} \frac{z^2}{z^4 + z^2 + 1} = \frac{\pi}{\sqrt{3}}, \quad \text{in u.h.p.} \qquad \blacktriangleleft$$

EXERCISES

1. Which of the following integrals exist?

   a) $\displaystyle\int_0^\infty e^{-x}\, dx$ b) $\displaystyle\int_\infty^0 e^{-x}\, dx$ c) $\displaystyle\int_{-\infty}^0 xe^x\, dx$

2. For which of the following integrals does the Cauchy principal value exist?

   a) $\displaystyle\int_{-\infty}^{+\infty} e^{-x}\, dx$ c) $\displaystyle\int_{-\infty}^{+\infty} \frac{x+1}{x^2+1}\, dx$

   b) $\displaystyle\int_{-\infty}^{+\infty} e^{-x^2}\, dx^\dagger$ d) $\displaystyle\int_{-\infty}^{+\infty} \frac{x^2}{x^2+1}\, dx$

3. The standard or ordinary definition of $\int_{-\infty}^{+\infty} f(x)\, dx$ is given by

   $$\int_{-\infty}^{+\infty} f(x)\, dx = \lim_{b\to\infty} \int_0^b f(x)\, dx + \lim_{a\to\infty} \int_{-a}^0 f(x)\, dx,$$

   where the two limits must exist independently of one another. Work the following without using complex variables.

   a) Show that $\int_{-\infty}^{+\infty} \sin x\, dx$ fails to exist according to the standard definition.

   b) Show that the Cauchy principal value of the preceding integral does exist and is zero.

   c) Show that $\int_{-\infty}^{+\infty} dx/(1+x^2) = \pi$ for both the standard definition and the Cauchy principal value.

   d) Show that if the ordinary value of $\int_{-\infty}^{+\infty} f(x)\, dx$ exists, then the Cauchy principal value must also exist and that the two results agree.

   e) Let $f(x)$ be an even function $[f(x) = f(-x)]$. Show that if the Cauchy principal value of $\int_{-\infty}^{+\infty} f(x)\, dx$ exists, then the ordinary value also exists and agrees with the Cauchy principal value.

---

†*Hint:* Look up the comparison test for improper integrals. See, for example, W. Kaplan, *Advanced Calculus* (Reading, Mass.: Addison-Wesley, 1957), sec. 4.5.

4. Evaluate the following integrals by means of residue calculus. Use the Cauchy principal value where appropriate.

a) $\int_{-\infty}^{+\infty} \dfrac{x^2 + 1}{x^4 + 1} \, dx$

f) $\int_{-\infty}^{+\infty} \dfrac{x^2 \, dx}{(x^2 + 1)(x^2 + 9)(x^2 + 16)}$

b) $\int_{-\infty}^{+\infty} \dfrac{dx}{(x - 3)^2 + 9}$

g) $\int_{-\infty}^{+\infty} \dfrac{x + x^2 + x^3}{(x^2 + 1)(x^2 + 9)} \, dx$

c) $\int_{0}^{\infty} \dfrac{dx}{x^6 + 1}$

Caution: Does Theorem 4 apply here? Express this integral as the sum of two Cauchy principal values. Evaluate each separately.

d) $\int_{-\infty}^{+\infty} \dfrac{dx}{x^2 + x + 1}$

e) $\int_{-\infty}^{0} \dfrac{x^2 \, dx}{x^4 + x^2 + 1}$

5. Restate Eq. (6.5–11) so as to employ only residues in the lower half plane. State the conditions on P and Q.

6. a) Solve Example 1 by using only residues in the lower half plane.

   b) Solve Example 2 by using residues in the lower half plane.

7. What is wrong with this assertion:

$$\int_{-\infty}^{+\infty} \frac{dx}{(x^2 + 1)(x^2 - 2x + 1)} = 2\pi i \sum_{\text{res}} \frac{1}{(z^2 + 1)(z^2 - 2z + 1)}, \quad \text{in u.h.p.}$$

8. Evaluate $\int_{-\infty}^{+\infty} du/\cosh u$ by making the change of variable $x = e^u$ and then applying residues.

9. Let $f(z) = P(z)/Q(z)$, where P and Q are polynomials in z with the property that degree Q − degree $P \geq 2$.

   a) Show that

$$\sum_{\text{res}} \frac{P(z)}{Q(z)} = 0, \quad \text{all poles.}$$

   Hint: Consider $\lim_{R \to \infty} \oint_{|z| = R} f(z) \, dz$. Use Eq. (6.5–10) and its counterpart in the lower half plane.

   b) Verify the result of part (a) by summing the residues of $f(z) = 1/(z^2 - i)$.

10. a) Explain why $\int_{0}^{\infty} x/(x^4 + 1) \, dx$ cannot be evaluated through the use of a closed semicircular contour in the upper or lower half plane (see Fig. 6.5–1 or Fig. 6.5–2).

    b) Consider the quarter circle contour shown in Fig. 6.5–3. C_1 is the arc of radius $R > 1$. Show that

$$\int_{0}^{R} \frac{x}{x^4 + 1} \, dx - \int_{R}^{0} \frac{y}{y^4 + 1} \, dy + \int_{C_1} \frac{z \, dz}{z^4 + 1} = 2\pi i \sum_{\text{res}} \frac{z}{z^4 + 1}, \quad \begin{array}{l} \text{at all poles in} \\ \text{first quadrant.} \end{array}$$

    c) Let $R \to \infty$ and show that $\int_{0}^{\infty} x/(x^4 + 1) \, dx = \pi/4$.

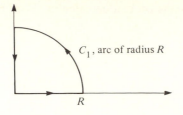

C_1, arc of radius R

R

Figure 6.5–3

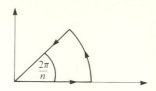

$\dfrac{2\pi}{n}$

Figure 6.5–4

11. Show that

$$\int_0^\infty \frac{x^m}{x^n + 1}\, dx = \frac{\pi}{n \sin\left[\pi(m + 1)/n\right]},$$

where n and m are nonnegative integers and $n - m \geq 2$. (This result is also found in Exercise 12, Section 6.8 for less-restrictive conditions.) *Hint:* Use the method employed in Exercise 10 above, but change to the contour of integration in Fig. 6.5–4.

12. a) Show that

$$\int_0^\infty \frac{u^{1/l}}{u^k + 1}\, du = \frac{\pi}{k \sin\left[\pi(l + 1)/(lk)\right]},$$

where k and l are integers, $l > 0$, which satisfy $l(k - 1) \geq 2$. Take $u^{1/l}$ as a nonnegative real function in the interval of integration. *Hint:* First work Exercise 11 above. Then, in the present problem, make the change of variable $x = u^{1/l}$ and use the result of Exercise 11.

b) What is $\int_0^\infty u^{1/4}/(u^5 + 1)\, du$?

6.6   EVALUATION OF INTEGRALS, III

Integrals of the type $\int_{-\infty}^{+\infty} f(x) \cos px\, dx$ and $\int_{-\infty}^{+\infty} f(x) \sin px\, dx$, where $f(x)$ is a rational function of x, and p is a real constant are often evaluated by methods similar to that just presented. These integrals appear in the theory of Fourier transforms, which is discussed in Section 6.9. Generally, we will determine the Cauchy principal value of such integrals and ignore the question of whether the ordinary values exist (see Exercise 3, Section 6.5).

To give some insight into the method discussed in this section, we try to evaluate $\int_{-\infty}^{+\infty} \cos(3x)/((x - 1)^2 + 1)\, dx$ using the technique of the preceding section. We integrate $\cos 3z/((z - 1)^2 + 1)$ around the closed semicircular contour of Fig. 6.5–1 and evaluate the result with residues. Thus

$$\int_{-R}^{+R} \frac{\cos 3x}{(x - 1)^2 + 1}\, dx + \int_{C_1} \frac{\cos 3z\, dz}{(z - 1)^2 + 1} = 2\pi i \sum_{\text{res}} \frac{\cos 3z}{(z - 1)^2 + 1}, \quad \text{in u.h.p.}$$

As before, C_1 is an arc of radius R in the upper half plane. Although the preceding equation is valid for sufficiently large R it is of no use to us. We would like to show that as $R \to \infty$, the integral over C_1 goes to zero. However,

$$\cos 3z = \frac{e^{3iz} + e^{-3iz}}{2} = \frac{e^{i3x-3y} + e^{-i3x+3y}}{2}.$$

As $R \to \infty$, the y-coordinates of points on C_1 become infinite and the term $e^{-i3x+3y}$, whose magnitude is e^{3y}, becomes unbounded. The integral over C_1 thus does not vanish with increasing R.

The correct approach in solving the given problem is to begin by finding $\int_{-\infty}^{+\infty} e^{3ix}/((x-1)^2 + 1) \, dx$. Its value *can* be determined if we use the technique of the previous section, that is, we integrate $\int e^{3iz}/((z-1)^2 + 1) \, dz$ around the closed contour of Fig. 6.5–1 and obtain

$$\int_{-R}^{+R} \frac{e^{3ix}}{(x-1)^2 + 1} \, dx + \int_{C_1} \frac{e^{3iz}}{(z-1)^2 + 1} \, dz = 2\pi i \sum_{\text{res}} \frac{e^{3iz}}{(z-1)^2 + 1}, \quad \text{in u.h.p.}$$

$$(6.6-1)$$

Assuming we can argue that the integral over arc C_1 vanishes as $R \to \infty$ (the troublesome e^{-3iz} no longer appears), we have, in this limit,

$$\int_{-\infty}^{+\infty} \frac{e^{3ix}}{(x-1)^2 + 1} \, dx = 2\pi i \sum_{\text{res}} \frac{e^{3iz}}{(z-1)^2 + 1}, \quad \text{in u.h.p.}$$

Putting $e^{3ix} = \cos 3x + i \sin 3x$ and rewriting the integral on the left as two separate expressions, we have

$$\int_{-\infty}^{+\infty} \frac{\cos 3x}{(x-1)^2 + 1} \, dx + i \int_{-\infty}^{+\infty} \frac{\sin 3x}{(x-1)^2 + 1} \, dx = 2\pi i \sum_{\text{res}} \frac{e^{3iz}}{(z-1)^2 + 1}, \quad \text{in u.h.p.}$$

When we equate corresponding parts (reals and imaginaries) in this equation, the values of two real integrals are obtained:

$$\int_{-\infty}^{+\infty} \frac{\cos 3x}{(x-1)^2 + 1} \, dx = \text{Re}\left[2\pi i \sum_{\text{res}} \frac{e^{3iz}}{(z-1)^2 + 1} \right], \quad \text{at all poles in u.h.p.}$$

$$(6.6-2)$$

$$\int_{-\infty}^{+\infty} \frac{\sin 3x}{(x-1)^2 + 1} \, dx = \text{Im}\left[2\pi i \sum_{\text{res}} \frac{e^{3iz}}{(z-1)^2 + 1} \right], \quad \text{at all poles in u.h.p.}$$

$$(6.6-3)$$

Equation (6.6–2) contains the result being sought while the integral in Eq. (6.6–3) has been evaluated as a bonus.

Solving the equation $(z-1)^2 = -1$ and finding that $z = 1 \pm i$, we see that, on the right sides of Eqs. (6.6–2) and (6.6–3), we must evaluate a residue at the simple pole $z = 1 + i$. From Eq. (6.3–6) we obtain

$$2\pi i \operatorname{Res}\left(\frac{e^{3iz}}{(z-1)^2 + 1}, 1 + i\right) = 2\pi i \lim_{z \to (1+i)} \frac{e^{3iz}}{2(z-1)}$$

$$= \pi e^{-3+3i} = \pi e^{-3}[\cos 3 + i \sin 3].$$

Using the result in Eqs. (6.6–2) and (6.6–3), we have finally

$$\int_{-\infty}^{+\infty} \frac{\cos 3x}{(x-1)^2 + 1}\, dx = \pi e^{-3} \cos 3 \quad \text{and} \quad \int_{-\infty}^{+\infty} \frac{\sin 3x}{(x-1)^2 + 1}\, dx = \pi e^{-3} \sin 3.$$

Recall now that we still have the task of showing that the second integral on the left in Eq. (6.6–1) becomes zero as $R \to \infty$. Rather than supply the details, we instead prove the following theorem and lemma. These not only perform our required task but many similar ones that we will encounter.

Theorem 5

Let $f(z)$ have the following property in the half plane $\operatorname{Im} z \geq 0$. There exist constants, $k > 0$, R_0, and μ such that

$$|f(z)| \leq \frac{\mu}{|z|^k}, \quad \text{for all } |z| \geq R_0 \text{ in this half plane.}$$

Then, if C_1 is the semicircular arc $Re^{i\theta}$, $0 \leq \theta \leq \pi$, and $R > R_0$, we have

$$\lim_{R \to \infty} \int_{C_1} f(z) e^{i\nu z}\, dz = 0, \quad \text{when } \nu > 0. \quad \blacksquare \qquad (6.6\text{–}4)$$

When $\nu < 0$, there is a corresponding theorem that applies in the lower half plane.

Equation (6.6–4) should be compared with Eq. (6.5–8). Notice that when the factor $e^{i\nu z}$ is not present, as happens in Eq. (6.5–8), we require $k > 1$, whereas the validity of Eq. (6.6–4) requires the less-restrictive condition $k > 0$.

To prove Eq. (6.6–4) we rewrite the integral on the left, which we call I, in terms of polar coordinates; taking $z = Re^{i\theta}$, $dz = Re^{i\theta} i\, d\theta$, we have

$$I = \int_{C_1} f(z) e^{i\nu z}\, dz = \int_0^{\pi} f(Re^{i\theta}) e^{i\nu Re^{i\theta}} i Re^{i\theta}\, d\theta. \qquad (6.6\text{–}5)$$

Recall now the inequality

$$\left|\int_a^b g(\theta)\, d\theta\right| \leq \int_a^b |g(\theta)|\, d\theta$$

derived in Exercise 11 of Section 4.1. Applying this to Eq. (6.6–5) and recalling that $|e^{i\theta}| = 1$, we have

$$|I| \leq R \int_0^{\pi} |f(Re^{i\theta})| |e^{i\nu Re^{i\theta}}|\, d\theta. \qquad (6.6\text{–}6)$$

We see that

$$\left|e^{i\nu Re^{i\theta}}\right| = \left|e^{i\nu R(\cos\theta + i\sin\theta)}\right| = \left|e^{-\nu R\sin\theta}\right|\left|e^{i\nu R\cos\theta}\right|.$$

Now

$$\left|e^{i\nu R\cos\theta}\right| = 1,$$

and since $e^{-\nu R\sin\theta} > 0$, we find that

$$\left|e^{i\nu Re^{i\theta}}\right| = e^{-\nu R\sin\theta}.$$

Rewriting Eq. (6.6–6) with the aid of the previous equation we have

$$|I| \leq R\int_0^\pi |f(Re^{i\theta})|e^{-\nu R\sin\theta}\, d\theta.$$

With the assumption $|f(z)| = |f(Re^{i\theta})| \leq \mu/R^k$ it should be clear that

$$|I| \leq R\int_0^\pi \frac{\mu}{R^k}e^{-\nu R\sin\theta}\, d\theta = \frac{\mu}{R^{k-1}}\int_0^\pi e^{-\nu R\sin\theta}\, d\theta. \tag{6.6–7}$$

Since $\sin\theta$ is symmetric about $\theta = \pi/2$ (see Fig. 6.6–1), we can perform the integration on the right in Eq. (6.6–7) from 0 to $\pi/2$ and then double the result. Hence,

$$|I| \leq \frac{2\mu}{R^{k-1}}\int_0^{\pi/2} e^{-\nu R\sin\theta}\, d\theta. \tag{6.6–8}$$

Figure 6.6–1 also shows that over the interval $0 \leq \theta \leq \pi/2$ we have $\sin\theta \geq 2\theta/\pi$. Thus when $\nu \geq 0$ we find $e^{-\nu R\sin\theta} \leq e^{-\nu R\theta 2/\pi}$ for $0 \leq \theta \leq \pi/2$.

Making use of this inequality in Eq. (6.6–8), we have

$$|I| \leq \frac{2\mu}{R^{k-1}}\int_0^{\pi/2} e^{-\nu R\theta 2/\pi}\, d\theta = \frac{\pi\mu}{\nu R^k}\left[1 - e^{-\nu R}\right].$$

With $R \to \infty$ the right-hand side of this equation becomes zero, which implies $I \to 0$ in the same limit. Thus

$$\lim_{R\to\infty} \int_{C_1} f(z)e^{i\nu z}\, dz = 0.$$

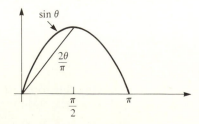

Figure 6.6–1

Any rational function $f(z) = P(z)/Q(z)$, where the degree of the polynomial $Q(z)$ exceeds that of the polynomial $P(z)$ by one or more, will fulfill the requirements of the theorem just presented (see Eq. 6.5–9) and leads us to the following lemma.

JORDAN'S
LEMMA
$$\lim_{R \to \infty} \int_{C_1} \frac{P(z)}{Q(z)} e^{ivz}\, dz = 0, \quad \text{if } v > 0, \text{ degree } Q - \text{degree } P \geq 1 \quad (6.6\text{–}9)$$

Jordan's lemma can be used to assert that the integral over C_1 in Eq. (6.6–1) becomes zero as $R \to \infty$. This was a required step in the derivation of Eqs. (6.6–2) and (6.6–3). We can use this lemma to develop a general formula for evaluating many other integrals involving polynomials and trigonometric functions.

Let us evaluate $\int e^{ivz} P(z)/Q(z)\, dz$ around the closed contour of Fig. 6.5–1 by using residues. All zeros of $Q(z)$ in the u.h.p. are assumed enclosed by the contour, and we also assume $Q(x) \neq 0$ for any real value of x. Therefore,

$$\int_{-R}^{+R} \frac{P(x)}{Q(x)} e^{ivx}\, dx + \int_{C_1} \frac{P(z)}{Q(z)} e^{ivz}\, dz = 2\pi i \sum_{\text{res}} \frac{P(z)}{Q(z)} e^{ivz}, \quad \text{at all poles in u.h.p.}$$

$$(6.6\text{–}10)$$

Now, provided the degrees of Q and P are as described in Eq. (6.6–9), we put $R \to \infty$ in Eq. (6.6–10) and discard the integral over C_1 in this equation by invoking Jordan's lemma. We obtain the following

$$\int_{-\infty}^{-\infty} \frac{P(x)}{Q(x)} e^{ivx}\, dx = 2\pi i \sum_{\text{res}} \frac{P(z)}{Q(z)} e^{ivz}, \quad \text{in u.h.p.} \quad (6.6\text{–}11)$$

The derivation of Eq. (6.6–11) requires that $v > 0$, $Q(x) \neq 0$ for $-\infty < x < \infty$, and the degree of Q exceed the degree of P by at least one.

We now apply Euler's identity on the left in Eq. (6.6–11) and obtain

$$\int_{-\infty}^{+\infty} (\cos vx + i \sin vx) \frac{P(x)}{Q(x)}\, dx = 2\pi i \sum_{\text{res}} e^{ivz} \frac{P(z)}{Q(z)}, \quad \text{in u.h.p.}$$

Now, assume that $P(x)$ and $Q(x)$ are *real* functions of x, that is, the coefficients of x in these polynomials are real numbers. We can then equate corresponding parts (reals and imaginaries) on each side of the preceding equation with the result that

$$\int_{-\infty}^{+\infty} \cos vx\, \frac{P(x)}{Q(x)}\, dx = \text{Re}\left[2\pi i \sum_{\text{res}} e^{ivz} \frac{P(z)}{Q(z)} \right], \quad \text{in u.h.p.} \quad (6.6\text{–}12a)$$

$$\int_{-\infty}^{+\infty} \sin vx\, \frac{P(x)}{Q(x)}\, dx = \text{Im}\left[2\pi i \sum_{\text{res}} e^{ivz} \frac{P(z)}{Q(z)} \right], \quad \text{in u.h.p.,} \quad (6.6\text{–}12b)$$

where degree Q – degree $P \geq 1$, $Q(x) \neq 0$, $-\infty < x < \infty$, $v > 0$.

These equations are useful in the rapid evaluation of integrals like those appearing in Exercise 1 below and in Eqs. (6.6–2) and (6.6–3). When $v = 0$, the integral on the left in Eq. (6.6–12b) is zero while that on the left in Eq. (6.6–12a) is evaluated from Eq. (6.5–11) if the degree of Q exceeds that of P by two or more.

When v is negative, we do not use Eqs. (6.6–11), (6.6–12a) or (6.6–12b). It is instructive and useful for later work to have formulas valid for $v < 0$. The reader should refer to Exercise 2 below where these are stated and derived.

EXERCISES

1. Evaluate the following integrals by means of residue calculus. Use the Cauchy principal value where appropriate.

   a) $\displaystyle\int_{-\infty}^{+\infty} \frac{\cos 2x}{x^2 + 9} \, dx$

   b) $\displaystyle\int_{0}^{\infty} \frac{x^2 \cos 2x}{(x^2 + 9)^2} \, dx$

   c) $\displaystyle\int_{-\infty}^{\infty} \frac{\cos ax \, dx}{(x^2 + b^2)(x^2 + c^2)},$

   where $b \neq c \neq 0$, and a, b, c are real numbers

   d) $\displaystyle\int_{-\infty}^{+\infty} \frac{\sin 3x}{(x^2 + 9)(x^2 + 1)} \, dx$

   e) $\displaystyle\int_{-\infty}^{+\infty} \frac{x \sin x \, dx}{[(x - 1)^2 + 1](x^2 + 1)}$

   f) $\displaystyle\int_{-\infty}^{+\infty} \frac{(x - 1) \cos x}{[(x - 1)^2 + 1]^2} \, dx$

   g) $\displaystyle\int_{0}^{\infty} \frac{x^2 \cos x}{x^4 + x^2 + 1} \, dx$

   h) $\displaystyle\int_{-\infty}^{+\infty} \frac{\cos^2 x}{(x^2 + b^2)(x^2 + c^2)} \, dx,$

   where b and c are real numbers $\neq 0$

   Hint: $\cos^2 x = \dfrac{1 + \cos 2x}{2}$.

   i) $\displaystyle\int_{-\infty}^{+\infty} \frac{\sin ax \cos bx}{(x - c)^2 + d^2} \, dx;$

   where a, b, c, d are real numbers, and $d \neq 0$

   Hint: $\sin ax \cos bx = (1/2) \sin (a+b)x + (1/2) \sin (a - b)x$.

   j) $\displaystyle\int_{-\infty}^{\infty} \frac{\cos (x - 1)}{x^2 + 1} \, dx$

2. a) Refer to the contour shown in Fig. 6.6–2. Let C_2 be the arc of radius R in the lower half plane (l.h.p.). Show that

$$\lim_{R \to \infty} \int_{C_2} \frac{P(z)}{Q(z)} e^{ivz} \, dz = 0$$

   if $v < 0$, where Q and P are polynomials such that degree $Q -$ degree $P \geq 1$. This is Jordan's lemma in the l.h.p. *Hint:* Begin by finding a formula analogous to Eq. (6.6–4) that applies when $v < 0$ and the contour is a semicircular arc in the l.h.p.

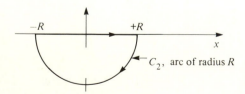

Figure 6.6–2

b) Perform an integration of $P(z)/Q(z)e^{ivz}$ around the closed contour in Fig. 6.6–2, allow $R \to \infty$, and use the result of part (a) to show that

$$\int_{-\infty}^{\infty} \frac{P(x)}{Q(x)} e^{ivx} \, dx = -2\pi i \sum_{\text{res}} \frac{P(z)}{Q(z)} e^{ivz}, \quad \text{in l.h.p.} \quad (6.6-13)$$

if $v < 0$ and $Q(x) \neq 0$ for $-\infty < x < \infty$. Why is there a minus sign in Eq. (6.6–13) that does not appear in Eq. (6.6–11)?

c) Assume that $P(x)$ and $Q(x)$ are real functions. Use Eq. (6.6–13) to show that

$$\int_{-\infty}^{\infty} \cos vx \frac{P(x)}{Q(x)} \, dx = -\text{Re}\left[2\pi i \sum_{\text{res}} \frac{P(z)}{Q(z)} e^{ivz} \right], \quad \text{in l.h.p. for } v < 0,$$

$$(6.6-14a)$$

$$\int_{-\infty}^{\infty} \sin vx \frac{P(x)}{Q(x)} \, dx = -\text{Im}\left[2\pi i \sum_{\text{res}} \frac{P(z)}{Q(z)} e^{ivz} \right], \quad \text{in l.h.p. for } v < 0.$$

$$(6.6-14b)$$

3. Evaluate the following integrals by means of residues. In cases (b) and (c) it is helpful to use Eq. (6.6–13).

a) $\displaystyle\int_{-\infty}^{+\infty} \frac{e^{i3x}}{(x^2 + 1)(x^2 + 2)} \, dx$

b) $\displaystyle\int_{-\infty}^{+\infty} \frac{e^{-i3x}}{(x^2 + 1)(x^2 + 2)} \, dx$

c) $\displaystyle\int_{-\infty}^{+\infty} \frac{e^{-ix}}{(x - 1)^2 + 1} \, dx$

4. a) Explain why $\int_{-\infty}^{+\infty} (\sin 2x)/(x - i) \, dx$ cannot be evaluated by means of Eq. (6.6–12b).

b) Evaluate this integral through the use of Eqs. (6.6–11) and (6.6–13). *Hint:* Express $\sin 2x$ in terms of e^{i2x} and e^{-i2x}. Write the given integral as the sum of two integrals and evaluate each by using residues.

5. Use a method suggested by the previous exercise and find

a) $\displaystyle\int_{-\infty}^{+\infty} \frac{\cos x}{x^2 + i} \, dx;$

b) $\displaystyle\int_{-\infty}^{+\infty} \frac{e^{i\omega x} \cos x}{x^2 + 1} \, dx,$ consider the cases $\omega \leq 1, -1 \leq \omega \leq 1, \omega \geq 1;$

c) $\displaystyle\int_{\infty}^{+\infty} \frac{e^{i\omega x} \sin x}{x^2 + 1} \, dx,$ choose ω as in part (b).

6. Explain why even though $\int_0^\infty (\cos x)/(x^2 + 1) \, dx$ can be evaluated with the aid of Eq. (6.6–12a), $\int_0^\infty (\sin x)/(x^2 + 1) \, dx$ cannot be evaluated with the help of Eq. (6.6–12b). This latter integral must be evaluated with a numerical table or a computer program.

6.7 INTEGRALS INVOLVING INDENTED CONTOURS

In previous sections we gave attention to the evaluation of integrals of the form $\int_{-\infty}^{+\infty} P(x)/Q(x)\, dx$ and $\int_{-\infty}^{+\infty} P(x)/Q(X)(\cos \nu x$ or $\sin \nu x)\, dx$. We required $Q(x) \neq 0$ for all x. In this section we dispense with this requirement, which means that the integrand can become infinite somewhere in the interval of integration. We will find that when the limiting process involved in the integration is appropriately defined, the integral can sometimes be evaluated.

We begin with a brief and ultimately useful digression. Let the function $f(z)$ possess a simple pole at the point z_0, and let C be a circle of radius r centered at z_0. Suppose a_{-1} is the residue of $f(z)$ at z_0. Assuming that $f(z)$ has no other singularities on and inside C, we know immediately that

$$\oint_C f(z)\, dz = 2\pi i a_{-1}. \tag{6.7-1}$$

The reader may wonder if an integration taken only halfway around C would yield $(1/2)(2\pi i a_{-1})$ and if an integration performed $1/4$ of the way around C would yield $(1/4)(2\pi i a_{-1})$. A specific example (see Exercise 1 of this section) shows this is a naive expectation. What is true, however, is that an integral taken around a fraction of C can be evaluated in the *limit* as the radius of C shrinks to zero by our using only the corresponding fraction of the residue of $f(z)$ at z_0. To be more specific, consider Theorem 6.

Theorem 6

Let $f(z)$ have a simple pole at z_0. An arc C_0 of radius r is constructed using z_0 as its center. The arc subtends an angle α at z_0 (see Fig. 6.7–1). Then

$$\lim_{r \to 0} \int_{C_0} f(z)\, dz = 2\pi i \left[\frac{\alpha}{2\pi} \operatorname{Res}[f(z), z_0] \right], \tag{6.7-2}$$

where the integration is done in the counterclockwise direction. (For a clockwise integration, a factor of -1 is placed on the right in Eq. 6.7–2). ∎

Note that when the integration in Theorem 6 is performed around an entire circle, α equals 2π and Eq. (6.7–2) yields a familiar result.[†]

To prove this theorem, we first expand $f(z)$ in a Laurent series about z_0. Because of the simple pole at z_0, the series assumes the form

$$f(z) = \frac{a_{-1}}{(z - z_0)} + \sum_{n=0}^{\infty} a_n (z - z_0)^n = \frac{a_{-1}}{(z - z_0)} + g(z),$$

where $g(z) = \sum_{n=0}^{\infty} a_n (z - z_0)^n$, which is the sum of a Taylor series, is analytic at z_0, and where $a_{-1} = \operatorname{Res}[f(z), z_0]$. We now integrate the series expansion of $f(z)$ along

---

[†] In Eq. (6.7–2) we should strictly write the limit using the special notation $\lim_{r \to 0+}$ to signify that the limit is evaluated as r shrinks to zero through *positive* values. This kind of limit will often be used without special notation throughout the remainder of this book.

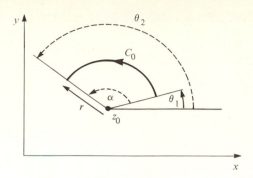

Figure 6.7–1

C_0 in Fig. 6.7–1. Thus

$$\int_{C_0} f(z)\,dz = \int_{C_0} \frac{a_{-1}}{(z - z_0)}\,dz + \int_{C_0} g(z)\,dz. \qquad (6.7\text{–}3)$$

Because $g(z)$ is continuous at z_0, we assert that $|g(z)|$ is bounded in a neighborhood of z_0; that is, there is a real constant M such that $|g(z)| \le M$ in this neighborhood. The radius r of C_0 is taken sufficiently small so that C_0 lies entirely in the neighborhood in question. Applying the ML inequality to the second integral on the right in Eq. (6.7–3), we have

$$\left| \int_{C_0} g(z)\,dz \right| \le Mr\alpha, \qquad (6.7\text{–}4)$$

where $r\alpha$ is the length of contour C_0. From Eq. (6.7–4) we see that

$$\lim_{r \to 0} \int_{C_0} g(z)\,dz = 0. \qquad (6.7\text{–}5)$$

The first integral on the right in Eq. (6.7–3) can be rewritten with a switch to polar variables. With $z = re^{i\theta}$, $dz = ire^{i\theta}\,d\theta$, and with the limits on θ indicated in Fig. 6.7–1, we have

$$\int_C \frac{a_{-1}}{(z - z_0)}\,dz = \int_{\theta_1}^{\theta_1 + \alpha} \frac{a_{-1} ire^{i\theta}\,d\theta}{re^{i\theta}} = a_{-1}\alpha i = 2\pi i \frac{\alpha}{2\pi} a_{-1}. \qquad (6.7\text{–}6)$$

Thus, passing to the limit $r \to 0$ in Eq. (6.7–3) and using Eqs. (6.7–5) and (6.7–6), we prove the theorem at hand. Applications of this theorem will now be discussed.

For integrals of the form $\int_a^b f(x)\,dx$, we sometimes find that $f(x)$ becomes infinite at some point, let us say p, that lies between a and b. Let us assume that $f(x)$ is continuous at all other points in the interval $a \le x \le b$. We thus have an improper integral. Previously we have encountered improper integrals involving infinite limits. The term "improper" is used in both cases because neither integral is expressible as the usual limit of a sum. Some improper integrals of the type we are considering here are

$$\int_{-1}^{+1} \frac{1}{x} dx, \qquad \int_{1}^{3} \frac{1}{(x-2)} dx, \quad \text{and} \quad \int_{-\pi/2}^{\pi/2} \frac{1}{\sin x} dx.$$

In colloquial language each of the integrands "blows up" somewhere between the limits of integration.

To evaluate such expressions we require a suitable definition. The term "Cauchy principal value," which we have used before (with another meaning) is applied to the following definition of this kind of improper integral.

Definition   Cauchy principal value

$$\int_{a}^{b} f(x) \, dx = \lim_{\varepsilon \to 0} \left[\int_{a}^{p-\varepsilon} f(x) \, dx + \int_{p+\varepsilon}^{b} f(x) \, dx \right], \qquad (6.7-7)$$

where $f(x)$ is continuous for $a \le x < p$ and $p < x \le b$ and ε shrinks to zero through *positive* values.   ∎

In both integrals appearing in the brackets the troublesome point $x = p$ is excluded from the interval of integration. In the first integral on the right the point p is approached from the left on the x-axis while in the second integral this same point is approached from the right.

Example 1

Find the Cauchy principal value of $\int_{-1}^{2} 1/x \, dx$.

Solution

Applying Eq. (6.7–7), we take $a = -1$, $b = 2$, and $p = 0$, (since $1/x$ has a discontinuity at $x = 0$). Thus for the Cauchy principal value we have

$$\int_{-1}^{2} \frac{1}{x} dx = \lim_{\varepsilon \to 0} \left[\int_{-1}^{-\varepsilon} \frac{1}{x} dx + \int_{\varepsilon}^{2} \frac{1}{x} dx \right]. \qquad (6.7-8)$$

Using the indefinite integral $\int 1/x \, dx = \text{Log}|x|$, we obtain

$$\int_{-1}^{2} \frac{1}{x} dx = \lim_{\varepsilon \to 0} \left[\text{Log} \frac{|-\varepsilon|}{|-1|} + \text{Log} \frac{2}{\varepsilon} \right] = \lim_{\varepsilon \to 0} \left[\text{Log} \, \varepsilon + \text{Log} \, 2 - \text{Log} \, \varepsilon \right] = \text{Log} \, 2.$$

Notice that in the limit $\varepsilon \to 0$, neither of the integrals in Eq. (6.7–8) exists separately. However, because of the cancellation of negative and positive areas about $x = 0$ (see Fig. 6.7–2), the sum of these integrals does possess a finite limit as $\varepsilon \to 0$. In the exercises we will see that there are integrals whose Cauchy principal value does not exist since the limit in Eq. (6.7–7) does not exist.   ◀

Presently, we will be dealing with integrals that are improper not only because of discontinuities in the integrand but also because of infinite limits of integration. Thus both definitions of the Cauchy principal value are required; for example,

$$\int_{-\infty}^{+\infty} \frac{\cos x}{(x-1)} dx = \lim_{\substack{\varepsilon \to 0 \\ R \to \infty}} \left[\int_{-R}^{1-\varepsilon} \frac{\cos x}{x-1} dx + \int_{1+\varepsilon}^{R} \frac{\cos x}{(x-1)} dx \right].$$

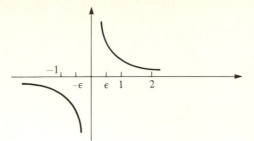

Figure 6.7–2

The point $x = 1$ is approached in a symmetric fashion, as are the limits at infinity. The concept of the Cauchy principal value can readily be extended to cover cases where the integrand has two or more points of discontinuity, as, for example, in this integral:

$$\int_{-\infty}^{\infty} \frac{\cos x}{(x+2)(x-3)} dx = \lim_{\substack{\varepsilon \to 0 \\ \delta \to 0 \\ R \to \infty}} \left[\int_{-R}^{-2-\delta} \frac{\cos x}{(x+2)(x-3)} dx \right.$$

$$\left. + \int_{-2+\delta}^{3-\varepsilon} \frac{\cos x}{(x+2)(x-3)} dx + \int_{3+\varepsilon}^{R} \frac{\cos x}{(x+2)(x-3)} dx \right].$$

The previous ideas, as well as the theorem just presented, can be combined in order to evaluate integrals of the form $\int_{-\infty}^{+\infty} f(x)\, dx$, where the discontinuities of $f(x)$ at real values of x coincide with simple poles of the analytic function $f(z)$. The method to be presented is known as *indentation of contours*, a term whose meaning will become clear with an example.

Example 2

Find the Cauchy principal value of $\int_{-\infty}^{+\infty} (\cos 3x)/(x - 1)\, dx$.

Solution

If we were to proceed according to the methods of Section 6.6, we would consider $\int e^{i3z}/(z-1)\, dz$ integrated along a closed contour like that of Fig. 6.5–1. We use here a contour like that one but with a modification. Because $e^{i3z}/(z-1)$ has a pole at $z = 1$, this point must be avoided by means of a semicircular indentation of radius ε in the contour. The closed contour actually used is shown in Fig. 6.7–3. Notice that $e^{i3z}/(z-1)$ is analytic at all points lying on, and interior to, this contour. Integrating this function around the path shown and putting $z = x$ where appropriate, we have

$$\int_{-R}^{1-\varepsilon} \frac{e^{i3x}}{(x-1)} dx + \int_{|z-1|=\varepsilon} \frac{e^{i3z}}{z-1} dz + \int_{1+\varepsilon}^{R} \frac{e^{i3x}}{(x-1)} dx + \int_{|z|=R} \frac{e^{i3z}}{(z-1)} dz = 0.$$

$$(6.7-9)$$

Allowing $R \to \infty$ and invoking Jordan's lemma (see Eq. 6.6–9) we can easily argue that the integral around the semicircle of radius R goes to zero.

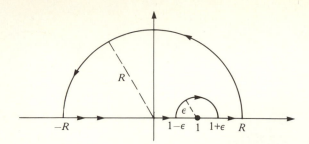

Figure 6.7–3

Taking the limit $\varepsilon \to 0$, we can evaluate the integral over the semicircular indentation at $z = 1$. Using Eq. (6.7–2) with $\varepsilon = r$, $f(z) = e^{i3z}/(z - 1)$, $z_0 = 1$, and $\alpha = \pi$ (for a semicircle), we find that the second integral on the left becomes in the limit $-i\pi e^{3i}$. The minus sign appears because of the clockwise direction of integration. With $R \to \infty$ and $\varepsilon \to 0$ we rewrite Eq. (6.7–9) as

$$\lim_{\substack{R \to \infty \\ \varepsilon \to 0}} \left[\int_{-R}^{-\varepsilon} \frac{e^{i3x}}{x - 1} \, dx + \int_{\varepsilon}^{R} \frac{e^{i3x}}{x - 1} \, dx \right] - i\pi e^{i3} = 0.$$

The sum of the two integrals in the bracket becomes, with the limits indicated, the Cauchy principal value of

$$\int_{-\infty}^{+\infty} \frac{e^{i3x}}{(x - 1)} \, dx = \int_{-\infty}^{+\infty} \frac{\cos 3x + i \sin 3x}{(x - 1)} \, dx.$$

Thus

$$\int_{-\infty}^{+\infty} \frac{\cos 3x}{(x - 1)} \, dx + i \int_{-\infty}^{+\infty} \frac{\sin 3x}{(x - 1)} \, dx = i\pi e^{i3} = i\pi [\cos 3 + i \sin 3].$$

Equating real and imaginary parts on either side, we have

$$\int_{-\infty}^{\infty} \frac{\cos 3x}{(x - 1)} \, dx = -\pi \sin 3 \quad \text{and} \quad \int_{-\infty}^{+\infty} \frac{\sin 3x}{(x - 1)} \, dx = \pi \cos 3. \quad \blacktriangleleft$$

Comment: In this example the real integral evaluated had one discontinuity in its integrand, at $x = 1$. The integral was evaluated by means of a contour integration with the contour indented about the value of z corresponding to this point. In the exercises (see Exercise 4 below) we consider integrals in which the integrand possesses more than one point of discontinuity. Here it becomes necessary to employ contours of integration indented around *each* such point.

EXERCISES

1. a) Find $\oint_{|z|=1} (z + 1)/z \, dz$.

   b) Find $\int_C (z + 1)/z \, dz$, where C is the semicircle $|z| = 1$, $0 \le \arg z \le \pi$. Integrate in the counterclockwise sense.

   c) In part (b) you integrated halfway around the circle used in part (a). Is the answer to part (b) half that of part (a)? Explain.

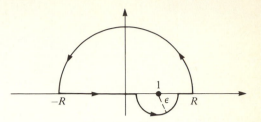

Figure 6.7–4   Figure 6.7–5

2. a) Evaluate $\int_{1-\varepsilon}^{1+\varepsilon}(z + 1)/(z - 1)\,dz$ around the semicircular arc of radius ε shown in Fig. 6.7–4.

b) In the answer to (a) let $\varepsilon \to 0$. Obtain $-2\pi i(1/2)\,\mathrm{Res}[(z + 1)/(z - 1), 1]$.

3. Obtain the Cauchy principal value required in Example 2 by using, instead of Fig. 6.7–3, the contour shown in Fig. 6.7–5. Notice that a pole singularity is now enclosed.

4. Find the Cauchy principal value of each of the following integrals:

a) $\displaystyle\int_{-\infty}^{\infty} \frac{\cos 3x}{(x + 1)}\,dx;$

b) $\displaystyle\int_{-\infty}^{+\infty} \frac{\cos x}{x^2 - 1}\,dx;$

c) $\displaystyle\int_{-\infty}^{+\infty} \frac{\cos 2x}{(x^2 - 1)(x^2 + 1)}\,dx;$

d) $\displaystyle\int_{-\infty}^{+\infty} \frac{\cos(x/2)}{(x^2 - \pi^2)}\,dx;$

e) $\displaystyle\int_{-\infty}^{+\infty} \frac{\sin x\,dx}{(x - \pi)(x + 1)(x^2 + 1)}\,dx;$

f) $\displaystyle\int_{-\infty}^{\infty} \frac{\sin x}{ax^2 + bx + c}\,dx,$

where a, b, c are real numbers, and $b^2 > 4ac$. Explain what difficulty is encountered if $b^2 = 4ac$. Can the Cauchy principal value still be found?

5. Find the Cauchy principal values of the following integrals:

a) $\displaystyle\int_{-\infty}^{+\infty} \frac{\sin x}{x}\,dx$

Explain how this yields $\int_0^{\infty} \sin x/x\,dx.$

b) $\displaystyle\int_{-\infty}^{+\infty} \frac{\sin^2 x}{x^2}\,dx$

Hint: $\sin^2 x = \dfrac{1}{2} - \dfrac{1}{2}\cos 2x = \mathrm{Re}\left[\dfrac{1 - e^{2ix}}{2}\right].$

6. Show that

$$\int_0^{\infty} \left(\frac{\cos t - e^{-t}}{t}\right) dt = 0.$$

Hint: Integrate e^{iz}/z around the contour shown in Fig. 6.7–6 and allow $\varepsilon \to 0$, $R \to \infty$.

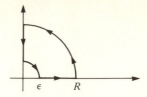

Figure 6.7–6

7. Show that

$$\int_{-\infty}^{+\infty} \frac{\cos ax - \cos bx}{x^2} dx = \pi(b - a), \quad \text{where } b > 0, a > 0.$$

Hint: The integrand equals $\text{Re}\left[\dfrac{e^{iax} - e^{ibx}}{x^2}\right]$.

6.8   CONTOUR INTEGRATIONS INVOLVING BRANCH POINTS AND BRANCH CUTS

Up to now we have been using contour integrations to evaluate integrals whose integrands are either rational functions or products of rational functions and trigonometric functions. In this section we will show how contour integration can often be used to evaluate integrals whose integrands are single-valued branches of multivalued functions; for example, $\int_0^\infty 1/[\sqrt{x}(x - 1)]\, dx$, where we have chosen the positive square root of x along the interval of integration.

Unlike the previous integrals considered, the present ones cannot be evaluated through a prescribed list of rules. What makes these problems more difficult is that their solution involves contour integrations taken along branch cuts and around branch points. Some of the techniques involved are illustrated in the following examples.

Example 1

Find $\int_0^\infty (\text{Log } x)/(x^2 + 4)\, dx$. Notice that Log x has a discontinuity at $x = 0$. This integral is thus defined as

$$\lim_{\substack{\varepsilon \to 0 \\ R \to \infty}} \int_\varepsilon^R \frac{\text{Log } x\, dx}{x^2 + 4}.$$

Solution

Following earlier reasoning, we try to evaluate $\int_C \log z/(z^2 + 4)\, dz$ around a closed contour C, a portion of which in some limit will coincide with the positive x-axis. We will use the principal branch of log z since it agrees with Log x on the positive x-axis, and the integral along this portion of C becomes identical to that of the given problem. Other branches providing such agreement can also be used. The contour C is shown in Fig. 6.8–1.

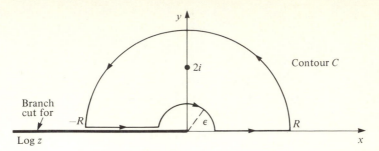

Figure 6.8–1

The integral along the negative real axis is taken along the "upper side" of the branch cut.[†] Since we want the integrand to be analytic on and inside C, the branch point $z = 0$ is avoided by means of a semicircle of radius ε.

We now express $\oint_C \text{Log } z/(z^2 + 4) \, dz$ in terms of integrals taken around the various parts of C. Note that the integrand has a simple pole at $z = 2i$. Thus

$$\int_{-R}^{-\varepsilon} \frac{\text{Log } z}{z^2 + 4} dz + \int_{-\varepsilon}^{+\varepsilon} \frac{\text{Log } z}{z^2 + 4} dz + \int_{\varepsilon}^{R} \frac{\text{Log } x}{x^2 + 4} dx + \int \frac{\text{Log } z}{z^2 + 4} \, dz$$

$$\underset{\overset{|z|=\varepsilon}{\frown}}{} \qquad\qquad \underset{\overset{|z|=R}{\frown}}{}$$

$$= 2\pi i \, \text{Res} \left[\frac{\text{Log } z}{z^2 + 4}, 2i \right]. \quad (6.8-1)$$

Consider the first integral on the left. For the principal branch of the logarithm $\text{Log } z = \text{Log}|z| + i \arg z, \; -\pi < \arg z \leq \pi$. Since we are integrating along the upper side of the branch cut in this integral, we see that $\arg z$ is π while $|z| = |x|$. Thus

$$\int_{-R}^{-\varepsilon} \frac{\text{Log } z \, dz}{z^2 + 4} = \int_{-R}^{-\varepsilon} \frac{\text{Log}|x|}{x^2 + 4} dx + \int_{-R}^{-\varepsilon} \frac{i\pi}{x^2 + 4} dx. \quad (6.8-2)$$

The first integration on the right can be taken between the limits ε and R (the integrand is an even function) and $\text{Log}|x|$ can be replaced by $\text{Log } x$. Therefore

$$\int_{-R}^{-\varepsilon} \frac{\text{Log } z \, dz}{z^2 + 4} = \int_{\varepsilon}^{R} \frac{\text{Log } x \, dx}{x^2 + 4} + \int_{-R}^{-\varepsilon} \frac{i\pi}{x^2 + 4} dx. \quad (6.8-3)$$

Using Eq. (6.8–3) or the far left in Eq. (6.8–1) and combining two identical integrals we have

$$i\pi \int_{-R}^{-\varepsilon} \frac{dx}{x^2 + 4} + 2\int_{\varepsilon}^{R} \frac{\text{Log } x}{x^2 + 4} dx + \int \frac{\text{Log } z}{z^2 + 4} \, dz + \int \frac{\text{Log } z}{z^2 + 4} \, dz$$

$$\underset{\overset{|z|=\varepsilon}{\frown}}{} \qquad\qquad \underset{\overset{|z|=R}{\frown}}{}$$

$$= 2\pi i \, \text{Res} \left[\frac{\text{Log } z}{z^2 + 4}, 2i \right]. \quad (6.8-4)$$

[†]Strictly speaking, we should use a contour of the shape shown in Fig. 6.8–2 and allow $\alpha \to 0 +$. In this way we can define what is meant by the "upper side" of the branch cut.

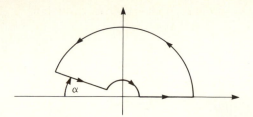

Figure 6.8–2

We need to show that the integrals over the semicircles of radii ε and R go to zero in the limits $\varepsilon \rightarrow 0$ and $R \rightarrow \infty$, respectively. We will present the first result; the derivation of the second is similar. The third integral on the left in Eq. (6.8–4) is rewritten with the polar substitution $z = \varepsilon e^{i\theta}$. Thus

$$\int_{|z| = \varepsilon} \frac{\text{Log } z}{z^2 + 4} \, dz = \int_{\pi}^{0} \frac{\varepsilon e^{i\theta} i \, \text{Log} \, (\varepsilon e^{i\theta}) \, d\theta}{\varepsilon^2 e^{2i\theta} + 4}. \tag{6.8–5}$$

We apply the ML inequality to Eq. (6.8–5), where $L = \pi$ (the interval of integration) and M is a constant such that

$$\left| \frac{\varepsilon e^{i\theta} i \, \text{Log} \, (\varepsilon e^{i\theta})}{\varepsilon^2 e^{i2\theta} + 4} \right| \leq M, \quad 0 \leq \theta \leq \pi. \tag{6.8–6}$$

Thus

$$\left| \int_{\pi}^{0} \frac{\varepsilon e^{i\theta} i \, \text{Log} \, (\varepsilon e^{i\theta}) \, d\theta}{\varepsilon^2 e^{i2\theta} + 4} \right| \leq M\pi. \tag{6.8–7}$$

If $0 \leq \varepsilon \leq 1$, then $|\varepsilon^2 e^{2i\theta} + 4| \geq 3$, and

$$\left| \frac{1}{\varepsilon^2 e^{i2\theta} + 4} \right| \leq \frac{1}{3}. \tag{6.8–8}$$

Also, notice that

$$|\varepsilon e^{i\theta} \, \text{Log} \, (\varepsilon e^{i\theta})| = \varepsilon |\text{Log} \, (\varepsilon) + i\theta| \leq \varepsilon [|\text{Log } \varepsilon| + \pi], \tag{6.8–9}$$

where we have recalled that $0 \leq \theta \leq \pi$. Combining Eqs. (6.8–9) and (6.8–8), we see that

$$\left| \frac{\text{Log} \, (\varepsilon e^{i\theta}) \varepsilon e^{i\theta} i}{\varepsilon^2 e^{i2\theta} + 4} \right| \leq \frac{\varepsilon [|\text{Log } \varepsilon| + \pi]}{3}. \tag{6.8–10}$$

A glance at Eq. (6.8–6) shows that the right side of Eq. (6.8–10) can be identified as M. The right side of Eq. (6.8–7) becomes $(\pi\varepsilon/3)[|\text{Log } \varepsilon| + \pi]$. As $\varepsilon \rightarrow 0$, this expression goes to zero (L'Hopital's rule can be used to resolve the indeterminate form $\varepsilon|\text{Log } \varepsilon|$ in this limit). The integral in Eq. (6.8–5) and the integral over the semicircle of radius ε in Eq. (6.8–4) thus become zero as $\varepsilon \rightarrow 0$.

The residue on the right in Eq. (6.8–4) is easily found to be $(\text{Log}\,2 + i\pi/2)/(4i)$. Passing to the limits $\varepsilon \to 0$ and $R \to \infty$ in Eq. (6.8–4) and using the computed residue, we have

$$i\pi \int_{-\infty}^{0} \frac{dx}{x^2 + 4} + 2\int_{0}^{\infty} \frac{\text{Log}\,x}{x^2 + 4}\,dx = \frac{\pi}{2}\left[\text{Log}\,2 + \frac{i\pi}{2}\right].$$

Identifying real and imaginary parts in the above we have

$$\int_{0}^{\infty} \frac{\text{Log}\,x}{x^2 + 4}\,dx = \frac{\pi}{4}\,\text{Log}\,2 \quad \text{and} \quad \int_{-\infty}^{0} \frac{dx}{x^2 + 4}\,dx = \frac{\pi}{4}.$$

The right-hand result is of course more easily found with the method shown in Section 6.5. ◀

Example 2

Evaluate $\int_{0}^{\infty} dx/[x^{1/\alpha}(x + 1)]$, where $\alpha > 1$ and $x^{1/\alpha} = \sqrt[\alpha]{x}$ for $0 \le x \le \infty$. Notice that as in the previous problem the integrand has a discontinuity at $x = 0$.

Solution

We will consider $\oint_C dz/[z^{1/\alpha}(z + 1)]$, where the contour of integration C lies partly along the x-axis. When we pass to appropriate limits, the integration along this part of the contour reduces to the given integral. In this calculation we must use a specific branch of $z^{1/\alpha}$. With the polar representation $z = re^{i\theta}$ we choose $z^{1/\alpha} = \sqrt[\alpha]{r}\,e^{i\theta/\alpha}$, where $0 \le \theta < 2\pi$. This branch is analytic in a domain defined by a branch cut along $y = 0$, $x \ge 0$. The contour of integration C, which lies in this domain, is shown in Fig. 6.8–3. The circular path of radius ε is necessary in order to exclude the branch point of $z^{1/\alpha}$ from the path of integration.

We express $\oint_C dz/[z^{1/\alpha}(z + 1)]$ as integrals along the four paths shown in the figure. Along path I, $z = r$, $dz = dr$, $z^{1/\alpha} = \sqrt[\alpha]{r}\,e^{i\theta/\alpha}$, where $\theta = 0$. Since $r = x$ on I,

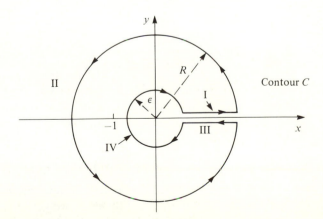

Figure 6.8–3

we have $z = x$, $dz = dx$, $z^{1/\alpha} = \sqrt[\alpha]{x}$. Along path III, $z = re^{i2\pi} = r$, $z^{1/\alpha} = \sqrt[\alpha]{r}\, e^{i\theta/\alpha}$, where $\theta = 2\pi$. Since $r = x$ on III, this becomes $z = x$, $dz = dx$, $z^{1/\alpha} = \sqrt[\alpha]{x}\, e^{i2\pi/\alpha}$. Along path II, $z = Re^{i\theta}$, $z^{1/\alpha} = \sqrt[\alpha]{R}\, e^{i\theta/\alpha}$, $dz = iRe^{i\theta}\, d\theta$. And along path IV, $z = \varepsilon e^{i\theta}$, $z^{1/\alpha} = \sqrt[\alpha]{\varepsilon}\, e^{i\theta/\alpha}$, and $dz = i\varepsilon e^{i\theta}\, d\theta$.

The contour C encloses the simple pole of $1/[z^{1/\alpha}(z + 1)]$ at $z = -1$. We have

$$\text{Res}\left[\frac{1}{z^{1/\alpha}(z + 1)}, -1\right] = \lim_{z \to -1} \frac{z + 1}{z^{1/\alpha}(z + 1)} = \left(\frac{1}{z^{1/\alpha}}\right)_{z = -1}$$

$$= \left[\frac{1}{\sqrt[\alpha]{r}\, e^{i\theta/\alpha}}\right]_{r = 1,\, \theta = \pi} = \frac{1}{e^{i\pi/\alpha}}. \qquad (6.8\text{--}11)$$

We have been careful to use the particular value of $(-1)^{1/\alpha}$ belonging to the chosen branch of $z^{1/\alpha}$. The integral around C, expressed in terms of integrals along the four paths mentioned and evaluated with Eq. (6.8–11), yields the equation

$$\text{I}\int_\varepsilon^R \frac{dx}{\sqrt[\alpha]{x}\,(x + 1)} + \text{II}\int_0^{2\pi} \frac{Re^{i\theta} i\, d\theta}{\sqrt[\alpha]{R}\, e^{i\theta/\alpha}(Re^{i\theta} + 1)} + \text{III}\int_R^\varepsilon \frac{dx}{\sqrt[\alpha]{x}\, e^{i2\pi/\alpha}(x + 1)}$$

$$+ \text{IV}\int_{2\pi}^0 \frac{\varepsilon e^{i\theta} i\, d\theta}{\sqrt[\alpha]{\varepsilon}\, e^{i\theta/\alpha}[\varepsilon e^{i\theta} + 1]} = \frac{2\pi i}{e^{i\pi/\alpha}}. \qquad (6.8\text{--}12)$$

To show that as $R \to \infty$ the integral over path II goes to zero, we employ

$$\left|\int_0^{2\pi} \frac{Re^{i\theta} i\, d\theta}{\sqrt[\alpha]{R}\, e^{i\theta/\alpha}(Re^{i\alpha} + 1)}\right| \le ML,$$

where

$$\left|\frac{Re^{i\theta}}{\sqrt[\alpha]{R}\, e^{i\theta/\alpha}(Re^{i\theta} + 1)}\right| = \frac{R}{\sqrt[\alpha]{R}|Re^{i\theta} + 1|} \le \frac{R}{\sqrt[\alpha]{R}\,(R - 1)} = \frac{1}{\sqrt[\alpha]{R}\,(1 - 1/R)} = M,$$

and $L = 2\pi$. We observe that

$$\lim_{R \to \infty} (ML) = \lim_{R \to \infty} \frac{2\pi}{\sqrt[\alpha]{R}\,(1 - 1/R)} = 0.$$

Thus the integral over path II in Eq. (6.8–12) goes to zero as $R \to \infty$. A similar discussion demonstrates that the integral over path IV in the same equation becomes zero as $\varepsilon \to 0$. Taking the limits $R \to \infty$, $\varepsilon \to 0$ in Eq. (6.8–12), we now have

$$\int_0^\infty \frac{dx}{\sqrt[\alpha]{x}\,(x + 1)} + e^{-i2\pi/\alpha}\int_\infty^0 \frac{dx}{\sqrt[\alpha]{x}\,(x + 1)} = 2\pi i e^{-i\pi/\alpha}. \qquad (6.8\text{--}13)$$

Reversing the limits on the second integral on the left, compensating with a minus sign, and multiplying both sides of Eq. (6.8–13) by $e^{i\pi/\alpha}$, we get

$$\int_0^\infty \frac{dx}{\sqrt[\alpha]{x}\,(x + 1)} (e^{i\pi/\alpha} - e^{-i\pi/\alpha}) = 2\pi i. \qquad (6.8\text{--}14)$$

The exponentials inside the parentheses sum to $2i \sin(\pi/\alpha)$. Dividing by this factor we have

$$\int_0^\infty \frac{dx}{x^{1/\alpha}(x+1)} = \frac{\pi}{\sin(\pi/\alpha)}, \qquad \alpha > 1.$$

Comment: We might try to solve this problem by means of a closed semicircular contour like that used in Example 1. However, because the pole of $1/[z^{1/\alpha}(z+1)]$ at $z = -1$ lies along this contour, an indentation must be made around this point. Such an approach is investigated in Exercise 7 below. ◀

EXERCISES

1. Show that $\int_0^\infty (\text{Log } x)/(x^2+1)\, dx = 0$ by employing a contour like that in Fig. 6.8–1.

2. a) Find $\int_0^\infty (\text{Log } x)/(x^4+16)\, dx$.

   b) Find $\int_0^\infty x^2 (\text{Log } x)/(x^4+16)\, dx$.

3. a) Evaluate $\oint_C e^{iz}/z^{1/2}\, dz$ around the contour shown in Fig. 6.8–4, and allow $\varepsilon \to 0$ and $R \to \infty$ to show

   $$\int_0^\infty \frac{\sin x}{\sqrt{x}}\, dx = \int_0^\infty \frac{\cos x}{\sqrt{x}}\, dx = \int_0^\infty \frac{e^{-y}}{\sqrt{2y}}\, dy.$$

   b) Make the change of variable $u^2 = y$ in the integral on y. Evaluate the resulting integral on u by means of a table and show that

   $$\int_0^\infty \frac{\cos x}{\sqrt{x}}\, dx = \int_0^\infty \frac{\sin x}{\sqrt{x}}\, dx = \sqrt{\frac{\pi}{2}}.$$

4. Show that

$$\int_0^\infty \frac{dx}{x^{1/e}(x+1)^2} = \frac{\pi}{e \sin(\pi/e)}$$

by using a circular contour like that of Example 2. Take $x^{1/e} \geq 0$.

5. Use the contour of the previous problem to show that

$$\int_0^\infty \frac{x^\alpha}{x^2 + 3x + 2}\, dx = \frac{\pi}{\sin(\alpha\pi)}(2^\alpha - 1), \qquad -1 < \alpha < 1,$$

where x^α and 2^α are ≥ 0.

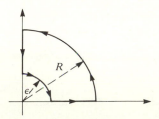

Figure 6.8–4

6. Show that

$$\int_{-\infty}^{+\infty} \frac{e^{\alpha u}}{1 + e^u} du = \frac{\pi}{\sin(\alpha\pi)}, \qquad 0 < \alpha < 1.$$

Hint: Let $x = e^u$.

7. a) Evaluate the integral of Example 2 by using the indented semicircular contour C shown in Fig. 6.8–5 and passing to appropriate limits. *Hint:* Consider $\int_C dz/[z^{1/\alpha}(z + 1)]$. Take the required limits for R and ε. Evaluate the integral by employing a residue. Equate real and imaginary parts on both sides of the resulting equation.

   Solve the resulting pair of equations simultaneously for an unknown integral. Check your result using Example 2.

b) Use a method similar to that of part (a) to show that

$$\int_0^\infty \frac{du}{u^{1/\alpha}(u - 1)} = \pi \cot \frac{\pi}{\alpha}, \qquad \text{(Cauchy principal value)}$$

where $\alpha > 1$ and $u^{1/\alpha} \geq 0$.

8. Show that

$$\int_0^\infty \frac{x^{1/\alpha} \, dx}{x^2 - a^2} = \frac{\pi}{2a} \frac{a^{1/\alpha}}{\sin\left(\dfrac{\pi}{\alpha}\right)} \left[1 - \cos \frac{\pi}{\alpha}\right], \qquad \text{(Cauchy principal value)}$$

where $a > 0$, $a^{1/\alpha} > 0$, $x^{1/\alpha} \geq 0$, and $-1 < 1/\alpha < 1$.

9. a) Show that

$$\int_0^\infty \frac{\sqrt{x} \, \text{Log} \, x}{x^2 + 1} dx = \frac{\pi^2}{2\sqrt{2}}$$

by using an indented semicircular contour in the u.h.p.

b) Show that, as part of the same computation, we find that

$$\int_0^\infty \frac{\sqrt{x}}{x^2 + 1} dx = \frac{\pi}{\sqrt{2}}.$$

10. Show that

$$\int_0^\infty \frac{\text{Log}(1 + x)}{x^\alpha} dx = \frac{\pi}{(1 - \alpha)\sin(\alpha\pi)},$$

where $x^\alpha \geq 0$ and $1 < \alpha < 2$.

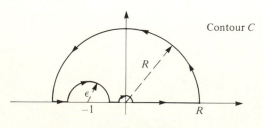

Figure 6.8–5

11. Show that

$$\int_0^\infty \frac{(\text{Log } x)^2}{x^2 + 1} \, dx = \frac{\pi^3}{8}.$$

Hint: You may need the result of Exercise 1.

12. In Exercise 11, Section 6.5 assume that m and n are any real numbers such that $n > m + 1 > 0$. Use the methods of Section 6.8 to show that the result given in that exercise is still valid.

6.9 RESIDUE CALCULUS APPLIED TO FOURIER TRANSFORMS

The theory of Fourier transforms is a branch of mathematics with wide physical application.[†] We do not have the space here to delve into this theory. However, we will see how residue calculus is useful in the evaluation of integrals that arise when one is using Fourier transforms.

A few definitions are first required.

Definition   Absolute integrability

A function $f(t)$ of a real variable is absolutely integrable if

$$\int_{-\infty}^\infty |f(t)| \, dt \quad \text{exists.} \quad \blacksquare \tag{6.9--1}$$

Next, we require the notion of piecewise continuity:

Definition   Piecewise continuity

The function $f(t)$ is piecewise continuous over an interval on the t-axis if this interval can be divided into a finite number of subintervals in which $f(t)$ is continuous. For each subinterval $f(t)$ has a finite limit as the ends are approached from the interior. $\blacksquare$

An example of a real function $y = f(t)$ that is piecewise continuous is shown in Fig. 6.9--1. Note that the only discontinuities experienced by $f(t)$ are "jumps" of finite size. At a typical jump, say t_1, $f(t)$ has finite right- and left-hand limits defined by $\lim_{\delta \to 0+} f(t_1 + \delta) = f(t_1 +)$ and $\lim_{\delta \to 0+} f(t_1 - \delta) = f(t_1 -)$, respectively. Recall that the symbol $\lim_{\delta \to 0+}$ means that δ shrinks to zero only through positive values. Thus in the first case t_1 is approached from the right while in the second case it is approached from the left. A piecewise continuous complex function of t, for example, $\phi(t) + i\psi(t)$, can have jump discontinuities that occur in both $\phi(t)$ and $\psi(t)$.

Suppose we have an absolutely integrable function $f(t)$ that is piecewise continuous over every finite interval along the t-axis. Then, we can define a new function, $F(\omega)$, called the Fourier transform of $f(t)$, which is given by the following definition.

[†]See, for example, R. Bracewell, *The Fourier Transform and its Applications* (New York: McGraw-Hill, 1978).

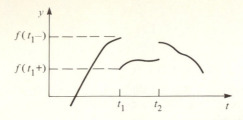

Figure 6.9–1

Definition   Fourier transform

$$F(\omega) = \frac{1}{2\pi} \int_{-\infty}^{+\infty} f(t) e^{-i\omega t} \, dt, \qquad -\infty < \omega < \infty \quad \blacksquare \qquad (6.9\text{–}2)$$

Note that ω is real. A comparison test[†] guarantees the existence of $F(\omega)$. Usually, we will use a lowercase letter (like f) to denote a function of t and the corresponding uppercase letter (here F) to denote its Fourier transform. It is well to note that we have stated *sufficient* conditions for the existence of $F(\omega)$. There are functions that fail to satisfy Eq. (6.9–1) but that do have Fourier transforms (see Exercise 9 of this section).

It can be shown that, within certain limitations, the following formula permits us to recover or find $f(t)$ when its Fourier transform is known.

$$f(t) = \int_{-\infty}^{+\infty} F(\omega) e^{i\omega t} \, d\omega, \qquad (6.9\text{–}3)$$

where the integral is a Cauchy principal value. The limitation on Eq. (6.9–3) is that this formula correctly yields $f(t)$ *except* for values of t where $f(t)$ is discontinuous. Here the formula gives the *average* of the right- and left-hand limits of $f(t)$, that is, $(1/2)f(t+) + (1/2)f(t-)$. The function $f(t)$ and its corresponding function $F(\omega)$ are known as Fourier transform pairs. Equation (6.9–3) is called the Fourier integral representation of $f(t)$. We will often regard the variable t as meaning time.

Example 1

For the function

$$f(t) = \begin{cases} e^{-t}, & t \geq 0, \\ 0, & t < 0, \end{cases} \qquad (6.9\text{–}4)$$

find the Fourier transform and verify the Fourier integral representation shown in Eq. (6.9–3).

Solution

From Eq. (6.9–2) we obtain

$$F(\omega) = \frac{1}{2\pi} \int_0^\infty e^{-t} e^{-i\omega t} \, dt = \frac{1}{2\pi} \int_0^\infty e^{-(1+i\omega)t} \, dt$$

$$= \frac{1}{2\pi} \frac{e^{-(1+i\omega)t}}{-1 - i\omega} \bigg|_0^\infty = \frac{1}{2\pi} \frac{1}{1 + i\omega}.$$

[†]See W. Kaplan, *Advanced Calculus* (Reading, Mass.: Addison-Wesley, 1957), ch. 4.

Substituting this $F(\omega)$ in Eq. (6.9–3), we have

$$f(t) = \frac{1}{2\pi} \int_{-\infty}^{+\infty} \frac{e^{i\omega t}}{1 + i\omega} \, d\omega = \frac{1}{2\pi} \int_{-\infty}^{+\infty} \frac{e^{ixt}}{1 + ix} \, dx. \tag{6.9–5}$$

We have replaced ω by x in order to evaluate our integral with a contour integration in the more familiar z-plane. In example 2, and thereafter, we dispense with this step.

With $t > 0$, Eq. (6.9–5) is readily evaluated from Eq. (6.6–11) and equals $i \, \mathrm{Res}[e^{izt}/(1 + iz), i] = e^{-t}$. With $t < 0$, Eq. (6.9–5) is evaluated from Eq. (6.6–13) and found to be zero since $e^{izt}/(1 + iz)$ has no poles in the lower half of the z-plane.

The case $t = 0$ in Eq. (6.9–5) is considered separately. Evaluating, we find that

$$\frac{1}{2\pi} \int_{-\infty}^{+\infty} \frac{dx}{1 + ix} = \frac{1}{2\pi} \int_{-\infty}^{+\infty} \frac{1 - ix}{x^2 + 1} \, dx = \frac{1}{2\pi} \int_{-\infty}^{+\infty} \frac{dx}{x^2 + 1} - \frac{i}{2\pi} \int_{-\infty}^{+\infty} \frac{x \, dx}{x^2 + 1}.$$

$$\tag{6.9–6}$$

The last integral on the right in Eq. (6.9–6) does not exist in the ordinary sense. However, because the integrand is an odd function, its Cauchy principal value is zero. The remaining integral on the right in Eq. (6.9–6) is readily evaluated with residues as follows:

$$\frac{1}{2\pi} \int_{-\infty}^{+\infty} \frac{dx}{x^2 + 1} = \frac{2\pi i}{2\pi} \, \mathrm{Res}\left[\frac{1}{z^2 + 1}, i\right] = \frac{1}{2}.$$

To summarize:

$$\int_{-\infty}^{+\infty} \frac{1}{2\pi} \frac{e^{-i\omega t}}{1 + i\omega} \, d\omega = \begin{cases} e^{-t}, & t > 0, \\ \frac{1}{2}, & t = 0, \\ 0, & t < 0. \end{cases} \tag{6.9–7}$$

Note that the function of t defined by Eq. (6.9–7) agrees with the given $f(t)$ in Eq. (6.9–4) for all t except $t = 0$. Here, the Fourier integral yields $1/2$, whereas $f(0) = e^{-t}|_0 = 1$. The discrepancy occurs because $f(t)$ in Eq. (6.9–4) is discontinuous at $t = 0$. The right- and left-hand limits of $f(t)$ at $t = 0$ are 1 and 0 (see Fig. 6.9–2). The average of these quantities is $1/2$, which is the value produced by the Fourier integral shown in Eq. (6.9–3). ◀

The Fourier integral representation of $f(t)$ given in Eq. (6.9–3) probably reminds us of the complex phasors discussed in the appendix to Chapter 3. Equation (A3–1),

$$f(t) = \mathrm{Re}[Fe^{st}] = \mathrm{Re}[Fe^{(\sigma + i\omega)t}],$$

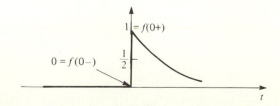

Figure 6.9–2

where F is the complex phasor corresponding to $f(t)$, is in a sense analogous to Eq. (6.9–3). A similarity exists between the Fourier transform $F(\omega)$ and the complex phasor F. Phasor representations are limited to functions of the form $e^{\sigma t} \cos(\omega t + \theta)$, that is, functions exhibiting a single complex frequency $\sigma + i\omega$. The Fourier integral, which represents a function of time by means of an integration over all frequencies, is not limited to the representation of functions possessing a single frequency.

Fourier transforms are used in the solution of differential equations in much the same way as are phasors. A property of Fourier transforms analogous to property (5) of phasors in the appendix to Chapter 3 makes this possible. Thus given the relationship between $f(t)$ and $F(\omega)$ described by Eq. (6.9–2), we want a quick method for finding the Fourier transform of df/dt, that is, for finding $(1/2\pi) \int_{-\infty}^{+\infty} (df/dt) e^{-i\omega t} dt$. Integrating by parts, we have

$$\frac{1}{2\pi} \int_{-\infty}^{+\infty} \frac{df}{dt} e^{-i\omega t} dt = \left. \frac{e^{-i\omega t} f(t)}{2\pi} \right|_{-\infty}^{+\infty} + \frac{1}{2\pi} \int_{-\infty}^{+\infty} f(t) i\omega e^{-i\omega t} dt.$$

Now let us assume that $\lim_{t \to \pm\infty} f(t) = 0$. Thus, in the preceding equation, the term $e^{-i\omega t} f(t)$ evaluated in the limits $t \to \infty$ and $t \to -\infty$ must equal zero. Hence, we obtain

$$\int_{-\infty}^{+\infty} \frac{df}{dt} e^{-i\omega t} dt = \frac{i\omega}{2\pi} \int_{-\infty}^{+\infty} f(t) e^{-i\omega t} dt = i\omega F(\omega), \tag{6.9–8}$$

which shows that if $f(t)$ has Fourier transform $F(\omega)$, then df/dt has Fourier transform $i\omega F(\omega)$. This result is also obtainable from formal differentiation of Eq. (6.9–3); the operator d/dt is placed under the integral sign. The procedure leading to Eq. (6.9–8) can be repeated to yield the Fourier transform of any derivative of $f(t)$. Thus if $f(t)$ has Fourier transform $F(\omega)$, then $d^n f/dt^n$ has Fourier transform $(i\omega)^n F(\omega)$. This is the analogue to property (5) for phasors in Chapter 3.

Let $g(t) = \int^t f(x) \, dx + C$, where, for some choice of the constant C, $g(t)$ is absolutely integrable. Note that $dg/dt = f(t)$, which implies that $i\omega G(\omega) = F(\omega)$ or $G(\omega) = F(\omega)/i\omega$. This is the counterpart to property (6) for phasors in Chapter 3.

The utility of the Fourier transform in the solution of physical problems is demonstrated in the following example.

Example 2

Consider the series electric circuit in Fig. 6.9–3 containing a resistor r and inductance L. The voltage $v(t)$ supplied by the generator is a sine function that is turned on for only two cycles. What is the current $\iota(t)$?

Solution

Applying the Kirchhoff voltage law around the circuit we have

$$v(t) = L \frac{d\iota}{dt} + \iota r. \tag{6.9–9}$$

Unlike Example 1 in the appendix to Chapter 3, the voltage in this problem is not

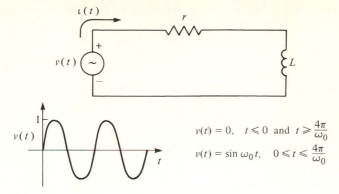

Figure 6.9–3

harmonic for *all* time and therefore does not possess a phasor. Nonetheless, $v(t)$ does have a Fourier transform. Applying Eq. (6.9–2) to $v(t)$, using the exponential form of $\sin \omega_0 t$, and integrating we have

$$V(\omega) = \frac{1}{2\pi} \int_0^{4\pi/\omega_0} \frac{\left[e^{i\omega_0 t} - e^{-i\omega_0 t} \right]}{2i} e^{-i\omega t}\, dt = \frac{1}{2\pi} \left[1 - e^{-i4\pi\omega/\omega_0} \right] \frac{\omega_0}{\omega_0^2 - \omega^2}.$$

$$(6.9\text{--}10)$$

Transforming both sides of Eq. (6.9–9) according to the rules just described we get

$$V(\omega) = i\omega L I(\omega) + I(\omega) r. \qquad (6.9\text{--}11)$$

Using Eq. (6.9–10) in the preceding equation, we solve for $I(\omega)$ and obtain

$$I(\omega) = \frac{1}{2\pi} \left[1 - e^{-i4\pi\omega/\omega_0} \right] \frac{\omega_0}{(i\omega L + r)(\omega_0^2 - \omega^2)}. \qquad (6.9\text{--}12)$$

The desired time function $\iota(t)$ is now produced from Eqs. (6.9–3) and (6.9–12). Thus

$$\iota(t) = \iota_1(t) - \iota_2(t), \qquad (6.9\text{--}13)$$

where

$$\iota_1(t) = \frac{\omega_0}{2\pi} \int_{-\infty}^{+\infty} \frac{e^{i\omega t}}{(\omega_0^2 - \omega^2)(i\omega L + r)}\, d\omega, \qquad (6.9\text{--}14)$$

and

$$\iota_2(t) = \frac{\omega_0}{2\pi} \int_{-\infty}^{+\infty} \frac{e^{i\omega(t - 4\pi/\omega_0)}}{(\omega_0^2 - \omega^2)(i\omega L + r)}\, d\omega. \qquad (6.9\text{--}15)$$

We first evaluate $\iota_1(t)$ for $t > 0$ by means of a contour integration in the complex ω-plane. Because of singularities at $\omega = \pm\omega_0$, we determine the Cauchy principal value of the integral. The contour used is shown below in Fig. 6.9–4. Notice that

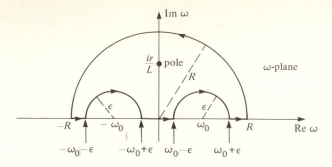

Figure 6.9–4

indentations of radius ε are used around $-\omega_0$ and ω_0. We have

$$
\int_{-R}^{-\omega_0-\varepsilon} \cdots + \underset{|\omega+\omega_0|=\varepsilon}{\int} \cdots + \int_{-\omega_0+\varepsilon}^{\omega_0-\varepsilon} \cdots + \underset{|\omega-\omega_0|=\varepsilon}{\int} \cdots + \int_{\omega_0+\varepsilon}^{R} \cdots
$$

$$
+ \underset{|\omega|=R}{\int} \frac{\omega_0}{2\pi} \frac{e^{i\omega t}}{(\omega_0^2 - \omega^2)(i\omega L + r)} d\omega = 2\pi i \operatorname{Res}\left[\frac{\omega_0}{2\pi} \frac{e^{i\omega t}}{(\omega_0^2 - \omega^2)(i\omega L + rR)}, \frac{ir}{L}\right].
$$

$$
(6.9–16)
$$

Only one singularity of the integrand is enclosed by the contour of integration. This is the pole where $i\omega L + r = 0$ or $\omega = ir/L$. We let $R \to \infty$ in Eq. (6.9–16) and invoke Jordan's lemma (see Eq. 6.6–9) to set the integral over the large semicircle to zero. We allow $\varepsilon \to 0$ and evaluate the integrals over the semicircular indentations of radius ε, in this limit, by using Eq. (6.7–2) with $\alpha = \pi$. The result is

$$
\iota_1(t) = \left[\frac{r \sin(\omega_0 t) - \omega_0 L \cos(\omega_0 t)}{2(r^2 + \omega_0^2 L^2)} + \frac{\omega_0 L e^{-(r/L)t}}{r^2 + \omega_0^2 L^2}\right], \qquad t \geq 0. \quad (6.9–17)
$$

Although our derivation of Eq. (6.9–17) presupposed $t > 0$, we have indicated $t \geq 0$ in Eq. (6.9–17). The case $t = 0$ in Eq. (6.9–14) can be treated with the contour of Fig. 6.9–4. We use Eq. (6.5–10) to argue that the integral over the large semicircle vanishes. It is found that Eq. (6.9–17), with $t = 0$, gives the correct result.

For $t < 0$ we must evaluate $\iota_1(t)$ by means of an integration over a semicircular contour lying in the lower half of the ω-plane. The contour used is obtained by reflecting the one in Fig. 6.9–4 about the real axis. The integrand employed is the same as in Eq. (6.9–16). The new contour does not encircle the pole at ir/L. We can use Eq. (6.6–13) to argue that the integral over the semicircle of radius R vanishes as $R \to \infty$. We ultimately obtain (see Exercise 2 of this section)

$$
\iota_1(t) = \frac{\omega_0 L \cos(\omega_0 t) - r \sin(\omega_0 t)}{2(r^2 + \omega_0^2 L^2)}, \qquad t < 0. \quad (6.9–18)
$$

To evaluate $\iota_2(t)$ we first consider the case $t \geq 4\pi/\omega_0$. The contour of integration used is identical to Fig. 6.9–4, and the integrand is the same as in Eq. (6.9–16), except $(t - 4\pi/\omega_0)$ is substituted for t. We also make this substitution in

Eq. (6.9–17) to obtain $\iota_2(t)$. Note that $\sin[(\omega_0)(t - 4\pi/\omega_0)] = \sin\omega_0 t$ and $\cos[(\omega_0)(t - 4\pi/\omega_0)] = \cos\omega_0 t$. Thus

$$\iota_2(t) = \left[\frac{r\sin(\omega_0 t) - \omega_0 L\cos(\omega_0 t)}{2(r^2 + \omega_0^2 L^2)} + \frac{\omega_0 L e^{(-r/L)(t - 4\pi/\omega_0)}}{r^2 + \omega_0^2 L^2}\right], \qquad t \geq \frac{4\pi}{\omega_0}.$$

To obtain $\iota_2(t)$ for $t < 4\pi/\omega_0$, we follow a procedure analogous to the derivation of Eq. (6.9–18). The contour of integration used is the same, and t in the integrand is replaced by $t - 4\pi/\omega_0$. With this change in Eq. (6.9–18) we get

$$\iota_2(t) = \frac{\omega_0 L\cos(\omega_0 t) - r\sin(\omega_0 t)}{2(r^2 + \omega_0^2 L^2)}, \qquad t < \frac{4\pi}{\omega_0}. \qquad (6.9\text{–}20)$$

Combining the last four equations according to Eq. (6.9–13), we have

$$\iota(t) = \frac{\omega_0 L e^{-rt/L}}{r^2 + \omega_0^2 L^2}[1 - e^{4\pi r/(\omega_0 L)}], \qquad t \geq \frac{4\pi}{\omega_0}; \qquad (6.9\text{–}21)$$

$$\iota(t) = \left[\frac{r\sin(\omega_0 t) - \omega_0 L\cos(\omega_0 t)}{(r^2 + \omega_0^2 L^2)} + \frac{\omega_0 L e^{-rt/L}}{(r^2 + \omega_0^2 L^2)}\right], \qquad 0 \leq t \leq \frac{4\pi}{\omega_0}; \qquad (6.9\text{–}22)$$

$$\iota(t) = 0, \qquad t \leq 0. \blacktriangleleft \qquad (6.9\text{–}23)$$

Although there are physical problems, like the preceding one, that are unsolvable with phasors and solvable with Fourier transforms, these transforms are inconvenient or useless in many situations. The Fourier transform technique is incapable of accounting for the response of a system due to any initial conditions that exist at the instant of excitation. Suppose, in the problem just given, a known current is already flowing in the circuit at $t = 0$. The Fourier transformation solution given takes no account of how the response of the system is affected by this current. The student of differential equations will perhaps realize that to obtain this portion of the response one must make a separate solution of the homogeneous differential equation describing the network.

The method of Laplace transforms, which is probably familiar to the reader, takes direct account of the initial conditions imposed on a system. In the next chapter we discuss Laplace transforms in their relationship to complex variable theory.

EXERCISES

1. If $f(t)$ has the Fourier transform $F(\omega)$ given below, find $f(t)$ for *all* t. Use Cauchy principal values where appropriate.

a) $\dfrac{1}{1 + \omega^2}$

b) $\dfrac{i}{\omega - i} + \dfrac{i}{\omega - 2i}$

c) $\dfrac{1}{(\omega - i)^2}$

d) $\dfrac{\omega}{i(1 + \omega^2)}$

e) $\dfrac{1}{\omega^4 + 2\omega^2 + 1}$

f) $\dfrac{e^{-2i\omega}}{\omega^2 + 1}$

g) $\dfrac{\cos\omega}{\omega^2 + 4}$

h) $\dfrac{i\sin\omega}{\omega^2 + 4}$

i) $\dfrac{1 - e^{-i2\pi\omega}}{\omega^2 - 1}$

j) $\dfrac{\sin\omega}{\omega}$

2. Supply the necessary details for the derivation of Eqs. (6.9–17) and (6.9–18).

3. If $f(t)$ has Fourier transform $F(\omega)$, what is the Fourier transform of $f(t - \tau)$?
Answer: $e^{-i\omega\tau}F(\omega)$.

4. Let

$$f(t) = \begin{cases} 1, & 0 \le t \le T, \\ 0, & t < 0, \\ 0, & t > T. \end{cases}$$

Find the Fourier transform $F(\omega)$. Verify the Fourier integral theorem by obtaining $f(t)$ from $F(\omega)$. Consider all possible real values of t.

5. Repeat the previous exercise using $f(t) = e^{-|t|} \cos \omega_0 t$, $-\infty < t < \infty$.

6. Use the results of Exercises 3 and 4 but no new integrations to obtain the Fourier transform of the function $f(t)$ given by

$$f(t) = \begin{cases} 1, & 5 \le t \le 6, \\ 0, & t < 5, \\ 0, & t > 6. \end{cases}$$

7. In Fig. 6.9–5 let $v(t)$ be the voltage across this parallel r, C circuit, and let $\iota(t)$ be the current supplied by the generator. Then, from Kirchhoff's current law,

$$\frac{v(t)}{r} + C\frac{dv}{dt} = \iota(t).$$

Assume that for $t < 0$, $v(t) = 0$. Let $\iota(t)$ be the function of time defined in Exercise 4. Use the method of Fourier transforms to find $V(\omega)$ and $v(t)$. What is the "system function" $V(\omega)/I(\omega)$?

8. a) Use residues to obtain the Fourier transform of

$$f(t) = \frac{\sin\left(\dfrac{2\pi t}{T}\right)}{t}, \qquad t \ne 0, \qquad f(0) = \frac{2\pi}{T}.$$

b) Sketch $F(\omega)$ and $f(t)$. What are the effects on $f(t)$ and $F(\omega)$ of increasing the period of oscillation T?

9. In Exercise 8 the Fourier transform of $f(t) = (\sin(2\pi t/T))/t$ was found. Show that $\int_{-\infty}^{+\infty} |f(t)| dt$ fails to exist. *Hint:* When A is any nonzero constant, $\sum_{n=1}^{\infty} A/n$

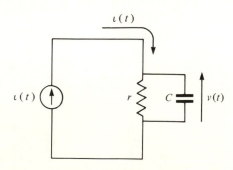

Figure 6.9–5

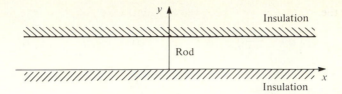

Figure 6.9–6

is known to diverge. Make a comparison test[†] in which you show that, for an appropriate choice of A, the terms in this series are smaller than corresponding terms in the sequence of integrals

$$\int_0^T \frac{\left|\sin \frac{2\pi t}{T}\right|}{t} \, dt, \int_T^{2T} \frac{\left|\sin \frac{2\pi t}{T}\right|}{t} \, dt, \ldots .$$

Thus we see that Eq. (6.9–1) provides a sufficient but not a necessary condition for the existence of a Fourier transform.

10. In Chapter 2 we discussed the equation governing the two-dimensional steady-state flow of heat. Under transient (nonsteady-state) conditions Eq. (2.6–5) must be altered. The temperature $\phi(x, y)$ now satisfies

$$\frac{\partial^2 \phi}{\partial x^2} + \frac{\partial^2 \phi}{\partial y^2} = \frac{1}{a^2} \frac{\partial \phi}{\partial t}, \tag{6.9–24}$$

where t = time and a is a constant characteristic of the heat-conducting material.

Consider an infinite rod extending along the x-axis as shown in Fig. 6.9–6. The rod is assumed to be insulated along all its faces, so that the temperature varies only with x and t. Assume that at $t = 0$ the temperature $\phi_0(x)$ along the rod is known. In this exercise we obtain $\phi(x, t)$ for $t > 0$. We use the Fourier transformation

$$\tilde{\phi}(\omega, t) = \frac{1}{2\pi} \int_{-\infty}^{+\infty} \phi(x, t) e^{-i\omega x} \, dx.$$

a) Take the Fourier transformation of both sides of Eq. (6.9–24) (ϕ is independent of y) to show that

$$-\omega^2 \tilde{\phi}(\omega, t) = \frac{1}{a^2} \frac{d\tilde{\phi}}{dt}(\omega, t).$$

You may assume that the order of a differentiation and an integration can be reversed.

b) Show that

$$\tilde{\phi}(\omega, t) = A(\omega) e^{-\omega^2 a^2 t}, \qquad t \geq 0,$$

[†]See W. Kaplan, *Advanced Calculus* (2d ed.) (Reading, Mass.: Addison-Wesley, 1973), secs. 6.6, 6.22.

where

$$A(\omega) = \frac{1}{2\pi} \int_{-\infty}^{+\infty} \phi_0(x) e^{-i\omega x} \, dx = \frac{1}{2\pi} \int_{-\infty}^{+\infty} \phi_0(\zeta) e^{-i\omega \zeta} \, d\zeta,$$

and a change of variable is made from x to ζ for convenience.

c) Show that

$$\phi(x, t) = \int_{-\infty}^{+\infty} \left[\frac{1}{2\pi} \int_{-\infty}^{+\infty} \phi_0(\zeta) e^{-i\omega \zeta} \, d\zeta e^{-\omega^2 a^2 t} \right] e^{i\omega x} \, d\omega.$$

d) Show that

$$\phi(x, t) = \frac{1}{2a\sqrt{\pi t}} \int_{-\infty}^{+\infty} \phi_0(\zeta) e^{-(x-\zeta)^2/(4a^2 t)} \, d\zeta. \qquad (6.9\text{–}25)$$

Hint: One can interchange the order of integration of the double integral in part (c). To do the integration on ω, one uses the formula[†]

$$\int_{-\infty}^{+\infty} e^{-m^2 u^2} \cos (bu) \, du = \frac{\sqrt{\pi}}{m} e^{-b^2/(4m^2)}, \qquad \text{Re } m^2 > 0.$$

e) Assume $a^2 = 1/2$; $\phi_0(\zeta) = 1$, $-1 \le \zeta \le 1$; and $\phi_0(\zeta) = 0$, $|\zeta| > 1$. Plot $\phi(x, t)$ as a function of x for $t = 0.1$ and also $t = 1$. To evaluate Eq. (6.9–25) either use a table of the error function[‡] or write a computer program.

11. Suppose a taut vibrating string lies along the x-axis except for small displacements $y(x, t)$ parallel to the y-axis (see Fig. 6.9–7). Let the deviation from the axis at any time be described by $y(x, t)$, where t is time. Assuming that the amplitude of vibrations is small enough so that each part of the string moves only in the y-direction and that gravitational forces are negligible, one can show that

$$\frac{\partial^2 y}{\partial x^2} = \frac{1}{c^2} \frac{\partial^2 y}{\partial t^2}, \qquad (6.9\text{–}26)$$

where c is the velocity of propagation of waves along the string. We will use Fourier transforms to solve Eq. (6.9–26) for an infinitely long string that, at $t = 0$, is subjected to the displacement $y_0(x) = y(x, 0)$ and velocity $v_0(x) = \partial y(x, 0)/\partial t$. The behavior of the string for $t > 0$ is sought. We use the transformation

$$Y(\omega, t) = \frac{1}{2\pi} \int_{-\infty}^{+\infty} y(x, t) e^{-i\omega x} \, dx.$$

a) Transform both sides of Eq. (6.9–26) and show that

$$\frac{d^2 Y}{dt^2}(\omega, t) + \omega^2 c^2 Y(\omega, t) = 0.$$

[†]See any good table of integrals.
[‡]See B. Peirce and R. Foster, *A Short Table of Integrals* (New York: Ginn, 1956), p. 128.

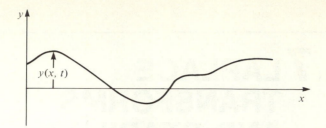

Figure 6.9–7

Hint: Assume that the operation $\partial/\partial t$ and the Fourier transformation can be performed in any order.

b) Show that $Y(\omega, t)$ must be of the form

$$Y(\omega, t) = A(\omega)\cos(\omega ct) + B(\omega)\sin(\omega ct). \qquad (6.9\text{–}27)$$

c) By putting $t = 0$ in the preceding show that

$$A(\omega) = \frac{1}{2\pi}\int_{-\infty}^{+\infty} y_0(x)e^{-i\omega x}\,dx.$$

d) Differentiate Eq. (6.9–27) with respect to time; put $t = 0$, and show that

$$B(\omega) = \frac{1}{\omega c 2\pi}\int_{-\infty}^{+\infty} v_0(x)e^{-i\omega x}\,dx.$$

e) Show that

$$y(x, t) = \int_{-\infty}^{+\infty}\left[A(\omega)\cos(\omega ct) + B(\omega)\sin(\omega ct)\right]e^{i\omega x}\,d\omega, \qquad (6.9\text{–}28)$$

where $A(\omega)$ and $B(\omega)$ are given in parts (c) and (d).

f) Assume that $v_0(x) = 0$ and that $y_0(x) = \Delta e^{-|x|}$, where $\Delta > 0$ is a constant. Using residue calculus to evaluate the integral in Eq. (6.9–28), find $y(x, t)$ for $t > 0$. Consider the three cases: $x > ct$, $-ct < x < ct$, $x < -ct$. Check your answer by considering $\lim_{t \to 0} y(x, t)$.

CHAPTER 7 LAPLACE TRANSFORMS AND STABILITY OF SYSTEMS

7.1 INTRODUCTION AND INVERSION OF LAPLACE TRANSFORMS

In this chapter we presuppose that the reader has some familiarity with the method of Laplace transforms in the solution of differential equations but is unfamiliar with the way in which complex variable theory can be of help when Laplace transforms are used. We will show here how residue calculus can be used in the "inversion" of Laplace transforms. We will also use our knowledge of analytic functions to determine whether the behavior of a physical system, analyzed with Laplace transforms, is stable or unstable.

We begin by briefly reviewing and listing the basic properties of Laplace transforms.

Let $f(t)$ be a real or complex function that is defined for all positive t, and let $s = \sigma + i\omega$ be a complex variable. Then the Laplace transform of $f(t)$, designated $F(s)$, is defined as follows.

Definition Laplace transform

$$F(s) = \int_0^\infty f(t) e^{-st} \, dt \quad \blacksquare \tag{7.1-1}$$

In general, we use lowercase letters to mean functions of t, for example, $f(t)$ and $g(t)$, and uppercase letters to denote the corresponding Laplace transform, in this case $F(s)$ and $G(s)$.

The operation on $f(t)$ described by Eq. (7.1–1) is also written $F(s) = \mathcal{L}f(t)$. The function of t whose Laplace transform is $F(s)$ is written $\mathcal{L}^{-1}F(s)$. Thus $f(t) = \mathcal{L}^{-1}F(s)$. We say that $f(t)$ is the *inverse* Laplace transform of $F(s)$.

Recall that

$$\mathcal{L}e^{-at} = \frac{1}{s+a}, \quad \text{if } \operatorname{Re} s > -\operatorname{Re} a \tag{7.1-2}$$

which is derived from

$$\mathscr{L} e^{-at} = \int_0^\infty e^{-st} e^{-at}\, dt = \int_0^\infty e^{-(s+a)t}\, dt = \frac{e^{-(s+a)t}}{-(s+a)}\bigg|_0^\infty$$

$$= \lim_{t \to \infty}\left[\frac{e^{-(s+a)t}}{-(s+a)}\right] + \frac{1}{(s+a)}. \tag{7.1-3}$$

Taking $s = \sigma + i\omega$, $a = \alpha + i\beta$, we obtain

$$\frac{e^{-(s+a)t}}{s+a} = \frac{e^{-(\sigma+\alpha)t} e^{-i(\beta+\omega)t}}{s+a}.$$

For $\sigma + \alpha > 0$, the preceding expression $\to 0$ as $t \to \infty$. Putting this limit in Eq. (7.1–3) establishes Eq. (7.1–2). The condition $\sigma + \alpha > 0$ is equivalent to Re $s > -$ Re a. The inverse of Eq. (7.1–2) is

$$\mathscr{L}^{-1}\frac{1}{s+a} = e^{-at}, \qquad t > 0. \tag{7.1-4}$$

If necessary, the reader should consult a table to again become familiar with some of the common transforms and their inverses.

Both of the operations $\mathscr{L}$ and $\mathscr{L}^{-1}$ satisfy the *linearity property*. Thus

$$\mathscr{L}\left[c_1 f_1(t) + c_2 f_2(t)\right] = c_1 \mathscr{L} f_1(t) + c_2 \mathscr{L} f_2(t) = c_1 F_1(s) + c_2 F_2(s), \tag{7.1-5}$$

where c_1 and c_2 are constants, and

$$\mathscr{L}^{-1}\left[c_1 F_1(s) + c_2 F_2(s)\right] = c_1 f_1(t) + c_2 f_2(t). \tag{7.1-6}$$

If a function $f(t)$ is piecewise continuous[†] over every finite interval on the line $t \geq 0$ and if there exist real constants k and p and T such that

$$|f(t)| < k e^{pt}, \tag{7.1-7}$$

then $f(t)$ will have a Laplace transform $F(s)$ for all s satisfying Re $s > p$. This transform not only exists in the half plane Re $s > p$, it is an *analytic function* in this half plane.[‡] Functions satisfying Eq. (7.1–7) for some choice of k, p, and T are said to be of *order e^{pt}*.

In any transformation procedure one wonders about uniqueness. According to Eq. (7.1–1), $f(t)$ has only one Laplace transform $F(s)$. It can be shown that if $f(t)$ and $g(t)$ have the same transform $F(s)$, then for $t > 0$, $f(t) = g(t)$, over any finite, nonzero interval; except that $f(t)$ and $g(t)$ can differ at specific isolated points in the interval.

The fundamental property of Laplace transforms relates to the ease with which we may obtain $\mathscr{L}\, df/dt$ in terms of $F(s) = \mathscr{L} f(t)$. Taking $\mathscr{L}\, df/dt = \int_0^\infty df/dt\, e^{-st}\, dt$,

---

[†] Piecewise continuity is discussed in Section 6.9.

[‡] See R. V. Churchill, *Operational Mathematics* (New York: McGraw-Hill, 1958), p. 171.

one can integrate by parts and obtain

$$\mathcal{L}f'(t) = sF(s) - f(0).\tag{7.1-8}$$

The derivation of the preceding equation is valid if $f(t)$ is of order e^{pt}, $f(t)$ is continuous for $t \geq 0$, and $f'(t)$ is piecewise continuous in every finite interval along the line $t \geq 0$. Actually, $f(t)$ can have a jump discontinuity at $t = 0$. We then replace $f(0)$ in Eq. (7.1–8) by its right-hand limit $f(0 +)$, and the equation will still hold true.

The Laplace transform of d^2f/dt^2 can be found in a similar way and is given by

$$\mathcal{L}f''(t) = s^2F(s) - sf(0) - f'(0),$$

and, in general,

$$\mathcal{L}f^{(n)}(t) = s^nF(s) - s^{n-1}f(0) - s^{n-2}f'(0) - s^{n-3}f''(0) - \cdots - f^{(n-1)}(0),\tag{7.1-9}$$

provided $f(t)$ and its 1st, 2nd,...,$(n - 1)$ derivatives are of order e^{pt}, $f(t)$ and these same derivatives are continuous for $t \geq 0$, and $f^{(n)}(t)$ is piecewise continuous over every finite interval on the line $t \geq 0$. This Laplace transform of $f^{(n)}(t)$ will exist in the s-plane for Re $s > p$. The requirements of continuity at $t = 0$ can be relaxed provided we use right-hand limits $f(0 +)$, $f'(0 +)$, etc. in Eq. (7.1–9).

The Laplace transform of the integral of $f(t)$ is easily stated in terms of the Laplace transform $f(t)$. Thus if $f(t)$ is of order e^{pt} and piecewise continuous for every finite interval on $t \geq 0$, then

$$\mathcal{L}\int_0^t f(x)\,dx = \frac{1}{s}\mathcal{L}f(t) = \frac{F(s)}{s}.\tag{7.1-10}$$

Laplace transforms are of use in solving linear differential equations with constant coefficients and prescribed initial conditions. Such equations are converted to algebraic equations involving the Laplace transform of the unknown function. The following example should serve as a reminder of the method. We solve

$$\frac{df}{dt} + 2f(t) = e^{-3t}, \quad \text{for } t \geq 0\tag{7.1-11}$$

with the initial condition $f(0) = 4$.

From Eq. (7.1–8) we have $\mathcal{L}\,df/dt = sF(s) - 4$, and from Eq. (7.1–2), $\mathcal{L}e^{-3t} = 1/(s + 3)$. Employing the linearity property in Eq. (7.1–5) we transform both sides of Eq. (7.1–11) and obtain $sF(s) - 4 + 2F(s) = 1/(s + 3)$. We solve the preceding equation and obtain $F(s) = 1/[(s + 3)(s + 2)] + 4/(s + 2)$. To obtain $f(t) = \mathcal{L}^{-1}F(s)$, we could consult a table of transforms and their inverses and find that $f(t) = 5e^{-2t} - e^{-3t}$. We easily verify that this satisfies the differential equation and its initial condition.

The preceding illustrates one potentially difficult step for the Laplace transform user—performing an inverse transformation to convert $F(s)$ to the actual solution $f(t)$. Often we are lucky enough to find $F(s)$ in a table; we then read off the corresponding $f(t)$. If $F(s)$ is not listed, we must, if possible, rearrange our expression into a sum of simpler terms that do appear in our table. The reader is perhaps familiar with a set of rules for finding $\mathcal{L}^{-1}F(s)$ when $F(s) = P(s)/Q(s)$

and $P(s)$ and $Q(s)$ are polynomials in s. These rules, called the Heaviside expansion formulas, are based implicitly on the fact that rational expressions like $P(s)/Q(s)$ can be written as a sum of partial fractions, each of whose inverse transform is readily found.[†] The technique that we introduce here for finding $f(t)$ is rooted directly in complex variable theory. It is more succinct than the Heaviside method, does not involve the memorization of a set of rules, and is not limited to rational expressions.

Typically, $F(s)$ defined by Eq. (7.1–1) exists when s is confined to a half plane $\operatorname{Re} s > p$; we observed earlier that $F(s)$ is analytic in the same half plane. For example (see Eq. 7.1–2) $\mathscr{L}e^{-2t} = 1/(s + 2)$ exists and is analytic for $\operatorname{Re} s > -2$. The analytic properties of $F(s)$ are important as they enable us to use the tools of complex variable theory.

For the moment, instead of dealing with $F(s)$, let us use $F(z)$ which is $F(s)$ with s replaced by z, that is,

$$F(z) = \int_0^\infty f(t)e^{-zt}\,dt. \qquad (7.1\text{–}12)$$

Suppose that $F(z)$ is analytic in the z-plane everywhere along the line $x = a$ and to the right of this line. We also make an assumption about $F(z)$ tending to 0 as $|z| \to \infty$ along any path in the half plane $\operatorname{Re} z \geq a$. More precisely, there exist positive numbers m, k, and R_0 so that, for $|z| > R_0$ and $\operatorname{Re} z \geq a$ we have

$$|F(z)| \leq \frac{m}{|z|^k}. \qquad (7.1\text{–}13)$$

Now let us apply the Cauchy integral formula to $F(z)$ and use the closed semicircular contour C shown in Fig. 7.1–1. The radius of the arc is b. For simplicity, we take $a \geq 0$, although an easy modification makes the discussion valid for $a < 0$ as well. We take s as some arbitrary point within C and take C_1 as the curved portion of C.

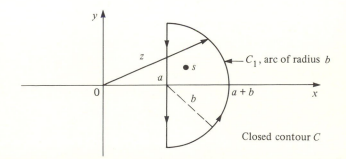

Figure 7.1–1

[†]See C. R. Wylie, *Advanced Engineering Mathematics* (New York: McGraw-Hill, 1975), sec. 7.5.

Integrating in the direction of the arrows we have

$$F(s) = \frac{1}{2\pi i} \oint \frac{F(z)}{z-s} dz = \frac{1}{2\pi i} \left[\int_{a+ib}^{a-ib} \frac{F(z)}{z-s} dz + \int_{C_1} \frac{F(z)}{z-s} dz \right]. \quad (7.1\text{--}14)$$

$$x = a$$

Our plan is to argue that the integral over C_1 tends to zero in the limit $b \to \infty$.

Let us consider an upper bound for $|F(z)|/|(z-s)|$ on C_1. We begin with the numerator. Provided b is sufficiently large, Eq. (7.1–13) provides a bound on the numerator $|F(z)|$. As is shown in Figure 7.1–1, on C_1 we have $|z| \geq b$ or $1/|z| \leq 1/b$. Combining this with Eq. (7.1–13), we have

$$|F(z)| \leq \frac{m}{b^k}, \quad \text{with } z \text{ on } C_1. \quad (7.1\text{--}15)$$

Now we examine $|z - s|$ on C_1.

Some careful study of Fig. 7.1–2 reveals that on C_1 the minimum possible value of $|z - s|$ occurs when the point z lies on the line connecting points s and a. The minimum value of $|z - s|$ is indicated and is equal to $b - |s - a|$. Thus on C_1 we have

$$|z - s| \geq b - |s - a|. \quad (7.1\text{--}16)$$

A triangle inequality $|s - a| \leq |s| + a$ combined with Eq. (7.1–16) yields

$$|z - s| \geq b - (|s| + a). \quad (7.1\text{--}17)$$

We will assume that b is large enough so that the right side of Eq. (7.1–17) is positive. Taking the reciprocal of both sides of Eq. (7.1–17), we get

$$\frac{1}{|z - s|} \leq \frac{1}{b - |s| - a}; \quad (7.1\text{--}18)$$

and multiplying Eq. (7.1–18) by Eq. (7.1–15), we obtain

$$\frac{|F(z)|}{|z - s|} \leq \frac{m}{b^k(b - |s| - a)}. \quad (7.1\text{--}19)$$

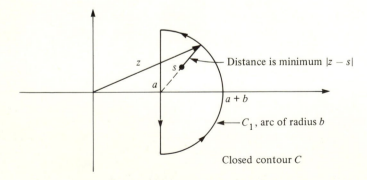

Figure 7.1–2

Now we apply the ML inequality to our integral over C_1, and notice that L, the path length, is πb. Hence

$$\left| \int_{C_1} \frac{F(z)}{z - s} dz \right| \le M\pi b, \tag{7.1-20}$$

where we require that $|F(z)|/|(z - s)| \le M$.

We see that M can be taken as the right side of Eq. (7.1–19). Thus Eq. (7.1–20) becomes

$$\left| \int_{C_1} \frac{F(z)}{z - s} dz \right| \le \frac{m\pi b}{b^k(b - |s| - a)} = \frac{m\pi}{b^k \left(1 - \dfrac{|s| + a}{b} \right)}.$$

Clearly, as $b \to \infty$, the right side of the equation $\to 0$. Therefore the integral contained on the left also goes to zero. Finally, passing to the limit $b \to \infty$ in Eq. (7.1–14) and using the result just derived for the integral on C_1, we have

$$F(s) = \frac{1}{2\pi i} \int_{a+i\infty}^{a-i\infty} \frac{F(z)}{z - s} dz.$$

With a reversal of limits this becomes

$$F(s) = \frac{1}{2\pi i} \int_{a-i\infty}^{a+i\infty} \frac{F(z)}{s - z} dz. \tag{7.1-21}$$

The preceding formula is not limited to functions $F(z)$ that are Laplace transforms but is applicable to any function analytic in the half space $\mathrm{Re}\, z \ge a$ and satisfying Eq. (7.1–13).

Now, consider the Laplace transform of the function e^{zt}. We have, by the usual method of transformation,

$$\mathcal{L} e^{zt} = \frac{1}{s - z}, \qquad \mathrm{Re}\, s > \mathrm{Re}\, z, \tag{7.1-22}$$

which can be confirmed by studying Eq. (7.1–2), where we now have z instead of $-a$. The inverse of Eq. (7.1–22) is

$$\mathcal{L}^{-1} \frac{1}{s - z} = e^{zt}, \qquad \mathrm{Re}\, s > \mathrm{Re}\, z. \tag{7.1-23}$$

Let us apply the $\mathcal{L}^{-1}$ operator to Eq. (7.1–21). Thus

$$\mathcal{L}^{-1} F(s) = \frac{\mathcal{L}^{-1}}{2\pi i} \int_{a-i\infty}^{a+i\infty} \frac{F(z)}{s - z} dz. \tag{7.1-24}$$

Now, we will assume that on the right in Eq. (7.1–24) the order of the $\mathcal{L}^{-1}$ and integration operations can be interchanged. Justification for this is not an obvious matter, and the reader is referred to more specialized texts for a rigorous proof.[†]

[†]See, for example, R. V. Churchill, *Operational Mathematics* (New York: McGraw-Hill, 1958), ch. 6.

We have

$$\mathcal{L}^{-1}F(s) = \frac{1}{2\pi i}\int_{a-i\infty}^{a+i\infty}\mathcal{L}^{-1}\frac{F(z)}{s-z}\,dz = \frac{1}{2\pi i}\int_{a-i\infty}^{a+i\infty}F(z)\mathcal{L}^{-1}\left(\frac{1}{s-z}\right)\,dz.$$

$$(7.1\text{--}25)$$

Since $\mathcal{L}^{-1}$ acts to convert a function of s into a function of t, it regards the variable z as constant; thus we have brought $F(z)$ outside this operator. Now with the result in Eq. (7.1–23) used on the right in Eq. (7.1–25), we have

$$\mathcal{L}^{-1}F(s) = f(t) = \frac{1}{2\pi i}\int_{a-i\infty}^{a+i\infty}F(z)e^{zt}\,dz. \qquad (7.1\text{--}26)$$

The constraint $\operatorname{Re} s > \operatorname{Re} z$ required in Eq. (7.1–23) is satisfied in Eq. (7.1–25) because the point s lies to the right of the straight line contour along which z varies (see Fig. 7.1–1, with $b \to \infty$).

The result in Eq. (7.1–26) is known as the *Bromwich integral formula* or simply as the inversion integral for Laplace transforms. Its path is a vertical line taken to the right of all singularities of the function $F(z)$.

Some remarks are required. The dummy variable z in Eq. (7.1–26) can be eliminated if we carry out our integration in the complex s-plane, where $s = \sigma + i\omega$. We now integrate along the vertical line $\sigma = a$. The principle of path independence (see Section 4.3) will permit us to deviate from this contour to a new path, provided this second path of integration is obtainable by a deformation of the vertical line that does not require our moving it through any singularities of $F(s)$. A valid alternative path is shown in Fig. 7.1–3. The definition of $F(s)$ in Eq. (7.1–1) employs only those values of $f(t)$ defined for $t > 0$. The condition $t > 0$ also applies to Eq. (7.1–26).

Lastly, a careful derivation shows that if $f(t)$ has a jump discontinuity at $t_0 > 0$, then the Bromwich integral, evaluated for $t = t_0$, will yield $1/2[f(t_0 +) + f(t_0 -)]$, that is, the average of the values assumed by $f(t)$ at each side of the jump.

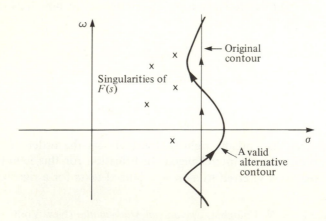

Figure 7.1–3

Summarizing our results:

BROMWICH INTEGRAL FORMULA
(LAPLACE INVERSION FORMULA)

$$\mathcal{L}^{-1}F(s) = f(t) = \frac{1}{2\pi i}\int_{a-i\infty}^{a+i\infty}F(s)e^{st}\,ds,$$

for $t > 0$,   (7.1–27)

where the contour of integration is either a vertical line Re $s = a$ lying to the right of all singularities of $F(s)$ or any other contour into which this line can legally be deformed. The function $F(s)$ is typically defined and analytic throughout some right half space of the complex s-plane, and the analytic continuation[†] of $F(s)$ into the remainder of this plane is often such that the Bromwich integral is evaluated by residues. For example, suppose we must find $\mathcal{L}^{-1}1/(s+1)^2$ without a table of transforms. We have, from Eq. (7.1–27),

$$f(t) = \frac{1}{2\pi i}\int_{a-i\infty}^{a+i\infty}\frac{e^{st}}{(s+1)^2}\,ds,$$

where, because of the pole of $1/(s+1)^2$ at -1, we require $a > -1$. Let us take $a = 0$.

To evaluate our integral, we consider the contour C in Fig. 7.1–4, which consists of the straight line extending from $\omega = -R$ to $\omega = R$ and the semicircular arc C_1 on which $|s| = R$. We have

$$\frac{1}{2\pi i}\oint_C\frac{e^{st}}{(s+1)^2}\,ds = \frac{1}{2\pi i}\int_{-iR}^{iR}\frac{e^{st}}{(s+1)^2}\,ds + \frac{1}{2\pi i}\int_{C_1}\frac{e^{st}}{(s+1)^2}\,ds. \quad (7.1–28)$$

The integral on the left, taken around the closed contour C, is readily evaluated with residues as follows:

$$\frac{2\pi i}{2\pi i}\,\mathrm{Res}\,\frac{e^{st}}{(s+1)^2} = \lim_{s\to -1}\frac{d}{ds}e^{st} = te^{-t}. \quad (7.1–29)$$

Passing to the limit $R \to \infty$ in Eq. (7.1–28), we see that the first integral on the right is now taken along the entire imaginary axis. One can easily show that as $R \to \infty$ the integral over the arc C_1 tends to zero whenever $t \geq 0$. The details, which involve the ML inequality, can be supplied by the reader.

[†]Analytic continuation is discussed briefly in Exercise 15, Section 5.4. To expand a bit on this subject: Let R and R_1 be two regions in the z-plane, with R contained inside the larger region R_1. Suppose a given $f(z)$ is analytic throughout R. Suppose we find a new function $f_1(z)$ that is analytic throughout R_1 and equals $f(z)$ everywhere in R. Then there is a unique function $f_1(z)$ with this property. It is called the analytic continuation of $f(z)$ from R into the larger region R_1. Exercise 15, Section 5.4 illustrates the analytic continuation of a function defined by a series. As another example, consider the function defined by the integral $\int_0^\infty e^{-zt}\,dt$ (see Eqs. 7.1–2 and 7.1–3). This function is equal to $1/z$ in the half plane Re $z > 0$. If Re $z \leq 0$, the integral does not converge; no function is defined. The *analytic continuation* of the function defined by this integral, into a region containing points from the half plane Re $z \leq 0$, is, of course, $1/z$.

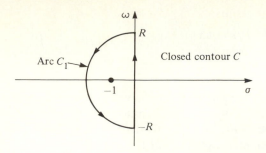

Figure 7.1–4

Thus, letting $R \to \infty$ in Eq. (7.1–28) and using Eq. (7.1–29), we have

$$te^{-t} = \frac{1}{2\pi i} \int_{-i\infty}^{i\infty} \frac{e^{st} \, ds}{(s+1)^2}.$$

On the right we have a Bromwich integral for the evaluation of $\mathcal{L}^{-1}1/(s+1)^2$. Thus $te^{-t} = \mathcal{L}^{-1}1/(s+1)^2$, which tables of transforms confirm as being correct.

The procedure just used should be generalized to permit inversion of a variety of transforms. We will therefore prove the following theorem.

Theorem 1

Let $F(s)$ be analytic in the s-plane except for a finite number of poles that lie to the left of some vertical line $\text{Re } s = a$. Suppose there exist positive constants m, R_0, and k such that for all s lying in the half plane $\text{Re } s \le a$, and satisfying $|s| > R_0$, we have $F(s) \le m/|s|^k$. Then for $t > 0$,

$$\mathcal{L}^{-1}F(s) = \sum_{\text{res}} F(s)e^{st}, \quad \text{at all poles of } F(s). \quad \blacksquare \qquad (7.1\text{–}30)$$

Theorem 1 requires that ultimately $F(s)$ falls off at least as rapidly as $m/|s|^k$ when s lies in a certain half space.

The proof proceeds as follows: Consider $(1/2\pi i) \oint_C F(s)e^{st} \, ds$ taken around the contour C shown in Fig. 7.1–5. We choose R greater than R_0 of the theorem, and a is chosen so that all poles of $F(s)$ are inside C. Recall that e^{st} is an entire function. From residue calculus we have

$$\frac{1}{2\pi i} \int_C F(s)e^{st} \, ds = \sum_{\text{res}} F(s)e^{st}, \quad \text{at all poles of } F(s). \qquad (7.1\text{–}31)$$

We now rewrite the integral around C in terms of integrals along the straight segment and the various arcs. Thus

$$\frac{1}{2\pi i} \oint_C F(s)e^{st} \, ds = \frac{1}{2\pi i} \int_{a-ib}^{a+ib} F(s)e^{st} \, ds + \frac{1}{2\pi i} \int_{C_1} F(s)e^{st} \, ds$$

$$+ \frac{1}{2\pi i} \int_{C_2} F(s)e^{st} \, ds + \frac{1}{2\pi i} \int_{C_3} F(s)e^{st} \, ds + \frac{1}{2\pi i} \int_{C_4} F(s)e^{st} \, ds,$$

$$(7.1\text{–}32)$$

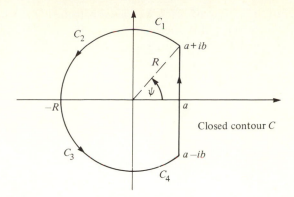

Figure 7.1–5

where, as shown in Fig. 7.1–5, C_1 extends from $a + ib$ to iR, C_2 goes from iR to $-R$, etc. Our goal is to let $R \to \infty$ and argue that the integrals taken over C_1, C_2, C_3, C_4 become zero. The first integral on the right in Eq. (7.1–32) becomes the Bromwich integral in this limit while the left-hand side of Eq. (7.1–32) is found from residues.

Let us consider I_1, the integral over C_1. We make a switch to polar variables so that $s = Re^{i\theta}$, $ds = Re^{i\theta} i \, d\theta$, and obtain

$$I_1 = \frac{1}{2\pi i} \int_\psi^{\pi/2} F(Re^{i\theta}) e^{tRe^{i\theta}} Re^{i\theta} i \, d\theta$$

$$= \frac{1}{2\pi i} \int_\psi^{\pi/2} F(Re^{i\theta}) e^{tR\cos\theta} e^{iRt\sin\theta} Re^{i\theta} i \, d\theta,$$

where we have put $e^{Rte^{i\theta}} = e^{Rt(\cos\theta + i\sin\theta)}$. We now use the inequality, shown in Eq. (4.1–26); it is rewritten here with different variables:

$$\left| \int_{\theta_1}^{\theta_2} u(\theta) \, d\theta \right| \leq \int_{\theta_1}^{\theta_2} |u(\theta)| \, d\theta, \quad \text{if } \theta_2 \geq \theta_1,$$

and $u(\theta)$ is any integrable function. We can thus assert that

$$|I_1| \leq \left| \frac{1}{2\pi i} \right| \left| \int_\psi^{\pi/2} |F(Re^{i\theta})| \, |e^{tR\cos\theta}| \, |e^{iRt\sin\theta}| R |e^{i\theta}| \, |i| \, d\theta.$$

Now, $|i| = 1$ and $|e^{iRt\sin\theta}| = 1$. Further, $e^{tR\cos\theta}$ is positive, and its magnitude signs can be dropped from inside the integral. Thus

$$|I_1| \leq \frac{1}{2\pi} \int_\psi^{\pi/2} |F(Re^{i\theta})| e^{Rt\cos\theta} R \, d\theta.$$

Since, by hypotheses, $|F(Re^{i\theta})| \leq m/|s|^k = m/R^k$, then

$$|I_1| \leq \frac{1}{2\pi} \int_\psi^{\pi/2} \frac{m}{R^k} e^{Rt\cos\theta} R \, d\theta = \frac{1}{2\pi} \frac{m}{R^{k-1}} \int_\psi^{\pi/2} e^{Rt\cos\theta} \, d\theta. \quad (7.1\text{–}33)$$

As θ varies from ψ to $\pi/2$, $e^{Rt\cos\theta}$ becomes progressively smaller. Thus, over the interval of integration, $e^{Rt\cos\theta} \le e^{Rt\cos\psi}$. We can substitute $e^{Rt\cos\psi}$ for $e^{Rt\cos\theta}$ in Eq. (7.1–33) and preserve the inequality there. Note that $R\cos\psi = a$ (see Fig. 7.1–5) or $\psi = \cos^{-1}(a/R)$. Rewriting the far right side of Eq. (7.1–33), we have

$$|I_1| \le \frac{1}{2\pi} \frac{m}{R^{k-1}} \int_{\cos^{-1}(a/R)}^{\pi/2} e^{at}\, d\theta = \frac{1}{2\pi} \frac{me^{at}}{R^{k-1}} \left[\frac{\pi}{2} - \cos^{-1} \frac{a}{R} \right],$$

and, because

$$\frac{\pi}{2} - \cos^{-1} \frac{a}{R} = \sin^{-1} \frac{a}{R},$$

we have

$$|I_1| \le \frac{1}{2\pi} \frac{m}{R^{k-1}} e^{at} \sin^{-1} \frac{a}{R}. \tag{7.1–34}$$

Now, we will use the inequality

$$\sin^{-1} p \le \frac{\pi}{2} p, \quad \text{if } 0 \le p \le 1 \tag{7.1–35}$$

The validity of this is demonstrated if we sketch $\sin^{-1} p$ and $(\pi/2)p$ over $0 \le p \le 1$. Thus, with $p = a/R$, we have, by combining Eqs. (7.1–34) and (7.1–35),

$$|I_1| \le \frac{1}{2\pi} \frac{m}{R^{k-1}} e^{at} \frac{\pi}{2} \frac{a}{R} = \frac{m}{4} \frac{e^{at}}{R^k}.$$

As $R \to \infty$, the expression on the right $\to 0$. Thus I_1 must approach the same limit.

Now I_2, the integral over C_2, will be treated in a similar fashion.

$$I_2 = \frac{1}{2\pi i} \int_{\pi/2}^{\pi} F(Re^{i\theta}) e^{tRe^{i\theta}} iRe^{i\theta}\, d\theta \quad \text{and} \quad |I_2| \le \frac{1}{2\pi} \int_{\pi/2}^{\pi} e^{Rt\cos\theta} |F(Re^{i\theta})| R\, d\theta.$$

Since $|F(re^{i\theta})| \le m/R^k$, we have

$$|I_2| \le \frac{1}{2\pi} \frac{m}{R^{k-1}} \int_{\pi/2}^{\pi} e^{Rt\cos\theta}\, d\theta. \tag{7.1–36}$$

A sketch of $\cos\theta$ and $1 - (2/\pi)\theta$ shows that $\cos\theta \le 1 - (2/\pi)\theta$ for $\pi/2 \le \theta \le \pi$. Thus

$$e^{Rt\cos\theta} \le e^{Rt(1-(2/\pi)\theta)}, \quad \text{for } \frac{\pi}{2} \le \theta \le \pi. \tag{7.1–37}$$

Combining the inequalities in Eqs. (7.1–37) and (7.1–36), we get

$$|I_2| \le \frac{1}{2\pi} \frac{m}{R^{k-1}} \int_{\pi/2}^{\pi} e^{Rt(1-(2/\pi)\theta)}\, d\theta.$$

Evaluating the above, we obtain $|I_2| \le [m/(2\pi R^{k-1})][\pi/(2Rt)](1 - e^{-Rt})$. This shows that as $R \to \infty$, we have $I_2 \to 0$.

An argument much like the one just presented shows that as $R \to \infty$, the integral over arc C_3 in Eq. (7.1–32) (see Fig. 7.1–5) goes to zero. Finally, a discussion much like the one given for I_1 (the integral over C_1) can be used to show that the integral over C_4 in Eq. (7.1–32) becomes zero as $R \to \infty$.

If $R \rightarrow \infty$ in Eq. (7.1–32) with a kept constant, then b must also become infinite. Passing to this limit and using the limiting values of all integrals, we have

$$\lim_{R \rightarrow \infty} \frac{1}{2\pi i} \int_C F(s) e^{st}\, ds = \frac{1}{2\pi i} \int_{a-i\infty}^{a+i\infty} F(s) e^{st}\, ds. \tag{7.1–38}$$

Since Eq. (7.1–31) is still valid as $R \rightarrow \infty$, it can be used on the left side of Eq. (7.1–38) with the result that

$$\sum_{res} F(s) e^{st} = \frac{1}{2\pi i} \int_{a-i\infty}^{a+i\infty} F(s) e^{st}\, ds, \quad \text{at poles of } F(s).$$

Since the right-hand side of this equation is $\mathcal{L}^{-1}F(s)$, we have proved the theorem under discussion.

Let $F(s) = P/Q$, where P and Q are polynomials in s whose degrees are n and l, respectively, with $l > n$. Then $|F(s)| \approx c/|s|^{l-n}$ for large $|s|$, where c is a constant. The conditions of the theorem are satisfied,[†] and $f(t)$ can be found.

Example 1

Find

$$\mathcal{L}^{-1} \frac{1}{(s-2)(s+1)^2} = f(t).$$

Solution

With $F(s) e^{st} = e^{st}/[(s-2)(s+1)^2]$ we use Eq. (7.1–30) recognizing that this function has poles at $s = 2$ and $s = -1$. Thus

$$f(t) = \text{Res}\left[\frac{e^{st}}{(s-2)(s+1)^2}, 2\right] + \text{Res}\left[\frac{e^{st}}{(s-2)(s+1)^2}, -1\right].$$

The first residue is easily found to be $e^{2t}/9$ while the second, which involves a pole of second order, is

$$\lim_{s \rightarrow -1} \frac{d}{ds} \frac{(s+1)^2 e^{st}}{(s-2)(s+1)^2} = \lim_{s \rightarrow -1} \frac{(s-2)\frac{d}{ds}(e^{st}) - e^{st}}{(s-2)^2} = \frac{-3te^{-t} - e^{-t}}{9}.$$

Notice that the expression te^{-t} arises when we differentiate e^{st} *with respect to s*. Thus, summing residues, we obtain

$$\mathcal{L}^{-1} \frac{1}{(s-2)(s+1)^2} = \frac{e^{2t}}{9} - \frac{te^{-t}}{3} - \frac{e^{-t}}{9}. \qquad \blacktriangleleft$$

The example just presented involves an $F(s)$ whose only singularities are a finite number of poles. The inversion was easily performed with a summation of residues,

---

[†]See M. Spiegel, *Laplace Transforms*, Schaum's Outline Series (New York: McGraw-Hill, 1965), p. 212, for a detailed proof.

but these inversions can also be performed, as noted earlier, with the Heaviside expansion formulas. Suppose we are given an $F(s)$ whose singularities are an infinite number of poles. Then, as outlined in Exercise 12 of this section, we are often justified in using Eq. (7.1–30) to obtain $f(t)$ provided that we evaluate the *infinite* summation of all the residues of $F(s)e^{st}$.

If, however, $F(s)$ is defined by means of a branch cut, then a summation of residues or a Heaviside formula will not yield $f(t)$. Instead we must return to the Bromwich integral (see Eq. 7.1–27) and deform the path of integration into some other valid contour in the s-plane along which the integration is more easily performed. An example follows.

Example 2

Find $\mathcal{L}^{-1}(1/s^{1/2})$ by means of the Bromwich integral.

Solution

From Eq. (7.1–27) we find that

$$\mathcal{L}^{-1}\frac{1}{s^{1/2}} = \frac{1}{2\pi i}\int_{a-i\infty}^{a+i\infty}\frac{1}{s^{1/2}}e^{st}\,ds. \tag{7.1–39}$$

The vertical line $s = a$ must be chosen to lie to the right of all singularities of $1/s^{1/2}$. Now $1/s^{1/2}$ is a multivalued function. A single-valued branch can be established by means of a branch cut extending from the origin to infinity. The branch specified by the cut along $\mathrm{Re}\,s = \sigma \leq 0$ will be used. When s assumes positive real values, we will take $s^{1/2} = \sqrt{s} > 0$. The path of integration can now be chosen as the vertical line $\mathrm{Re}\,s = a > 0$ shown in Fig. 7.1–6. To evaluate Eq. (7.1–39) along this line we first consider $(1/2\pi i)\oint_C e^{st}/s^{1/2}\,ds$ taken around the closed contour C shown in Fig. 7.1–7. As $e^{st}/s^{1/2}$ is analytic on and inside C, we have $(1/2\pi i)\oint_C e^{st}/s^{1/2}\,ds = 0$. This integral is rewritten in terms of integrations taken along the various portions of C. We have

$$\frac{1}{2\pi i}\int_{a-ib}^{a+ib}\frac{e^{st}}{s^{1/2}}\,ds + \frac{1}{2\pi i}\int_{C_1}\frac{e^{st}}{s^{1/2}}\,ds + \frac{1}{2\pi i}\underset{\text{above cut}}{\int_{-R}^{-\varepsilon}}\frac{e^{\sigma t}}{\sigma^{1/2}}\,d\sigma$$

$$+ \frac{1}{2\pi i}\oint_{|s|=\varepsilon}\frac{e^{st}}{s^{1/2}}\,ds + \frac{1}{2\pi i}\underset{\text{below cut}}{\int_{-\varepsilon}^{-R}}\frac{e^{\sigma t}}{\sigma^{1/2}}\,d\sigma + \frac{1}{2\pi i}\int_{C_2}\frac{e^{st}}{s^{1/2}}\,ds = 0, \tag{7.1–40}$$

where C_1 is the circular arc extending from $a + ib$ to $-R$, while C_2 is the circular arc extending from $-R$ to $a - ib$. For the integrals along the straight lines that are above and below the branch cuts we recognize that $s = \sigma$. Because $1/s^{1/2}$ has a branch point singularity at $s = 0$, we make a circular detour of radius ε around this point.

We now consider the limiting values of the integrals along arcs C_1 and C_2 as the radius $R \to \infty$. Referring to the derivation of Eq. (7.1–30) and to Fig. 7.1–5, we find that the discussion used there to justify setting the integrals over C_1, C_2, C_3, and C_4 to zero as $R \to \infty$ can be applied directly to the present problem. Thus, as $R \to \infty$, the integrals over C_1 and C_2 in Eq. (7.1–40) become zero.

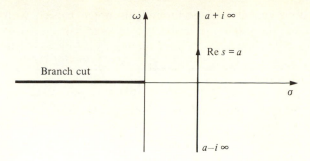

Figure 7.1–6

We study now the integral in Eq. (7.1–40) that is taken around $|s| = \varepsilon$. We make our usual switch to polar coordinates $s = \varepsilon e^{i\theta}$, $ds = \varepsilon e^{i\theta} i \, d\theta$, $s^{1/2} = \sqrt{\varepsilon} \, e^{i\theta/2}$ and so

$$\frac{1}{2\pi i} \oint_{|s| = \varepsilon} \frac{e^{st}}{s^{1/2}} ds = \frac{1}{2\pi} \int_{\pi}^{-\pi} \frac{e^{\varepsilon t e^{i\theta}} \varepsilon e^{i\theta} i \, d\theta}{\sqrt{\varepsilon} \, e^{i\theta/2}} = \frac{\sqrt{\varepsilon}}{2\pi} i \int_{\pi}^{-\pi} e^{\varepsilon t e^{i\theta}} e^{i\theta/2} \, d\theta.$$

As $\varepsilon \to 0$, the integral on the far right is bounded and its coefficient $\sqrt{\varepsilon}/2\pi \to 0$. As $\varepsilon \to 0$, the right side of the preceding equation $\to 0$. We thus have

$$\lim_{\varepsilon \to 0} \frac{1}{2\pi i} \oint_{|s| = \varepsilon} \frac{e^{st}}{s^{1/2}} ds = 0.$$

Along the top edge of the branch cut in Fig. 7.1–7 $s^{1/2} = \sigma^{1/2}$ is the square root of a negative real variable. The correct value of this multivalued expression must be established. By assumption, $s^{1/2}$ is a positive real number for any s lying on the positive real axis. Here $\arg s^{1/2} = 0$. As we proceed to the top side of the branch cut, $\arg s$ increases by π (see Fig. 7.1–8). Thus $\arg s^{1/2}$ increases from 0 to $\pi/2$. Hence, when s is a negative real number and lies on the top of the cut, we have $\arg s^{1/2} = \pi/2$, which implies that $\sigma^{1/2} = i\sqrt{|\sigma|} = i\sqrt{-\sigma}$.

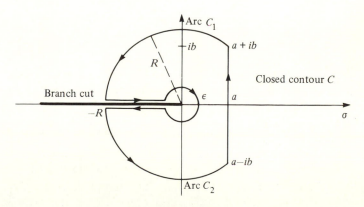

Figure 7.1–7

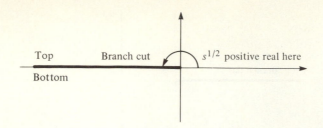

Figure 7.1–8

A similar discussion shows that along the bottom of the branch cut $\sigma^{1/2} = -i\sqrt{-\sigma}$. Note from Fig. 7.1–7 that as $R \to \infty$, with a fixed, we have $b \to \infty$. Passing to the limits $R \to \infty$, $\varepsilon \to 0$ in Eq. (7.1–40) and using the limiting values of all integrals, we obtain

$$\frac{1}{2\pi i}\int_{a-i\infty}^{a+i\infty}\frac{e^{st}}{s^{1/2}}\,ds + \frac{1}{2\pi i}\int_{-\infty}^{0}\frac{e^{\sigma t}}{i\sqrt{-\sigma}}\,d\sigma + \frac{1}{2\pi i}\int_{0}^{-\infty}\frac{e^{\sigma t}}{-i\sqrt{-\sigma}}\,d\sigma = 0. \quad (7.1\text{–}41)$$

The second and third integrals on the left are equal and can be combined into $-(1/\pi)\int_{-\infty}^{0}(e^{\sigma t}/\sqrt{-\sigma})\,d\sigma$. Thus Eq. (7.1–41) becomes

$$\frac{1}{2\pi i}\int_{a-i\infty}^{a+i\infty}\frac{e^{st}}{s^{1/2}}\,ds = \frac{1}{\pi}\int_{-\infty}^{0}\frac{e^{\sigma t}}{\sqrt{-\sigma}}\,d\sigma.$$

The left side of this equation is the desired $\mathcal{L}^{-1}(1/s^{1/2})$. The integral on the right can be somewhat simplified with the change of variables $x = \sqrt{-\sigma}$, $x^2 = -\sigma$, $2x\,dx = -d\sigma$. Thus

$$\mathcal{L}^{-1}\frac{1}{s^{1/2}} = \frac{2}{\pi}\int_{0}^{\infty}e^{-x^2 t}\,dx.$$

The right-hand side here contains a well-known definite integral. From tables we find that $\int_{0}^{\infty}e^{-x^2 t}\,dx = (1/2)\sqrt{\pi/t}$, which means

$$\mathcal{L}^{-1}\frac{1}{s^{1/2}} = \frac{1}{\sqrt{\pi t}}. \qquad \blacktriangleleft$$

EXERCISES

1. Use residues to find the inverse Laplace transforms of the following functions.

a) $\dfrac{1}{(s+1)(s-2)(s)}$

b) $\dfrac{s+3}{(s^2+1)(s^2+3)}$

c) $\dfrac{s}{s^2+s+1}$

d) $\dfrac{s}{(s+1)^2}$

e) $\dfrac{1}{(s^2+1)^2}$

f) $\dfrac{s+3}{(s^2+s+1)^2}$

g) $\dfrac{1}{(s^3)(s+1)}$

2. Let $F(s) = P(s)/Q(s)$, where P and Q are polynomials in s, the degree of Q exceeding that of P.

a) Suppose that $Q(s) = C(s - a_1)(s - a_2), \ldots, (s - a_n)$, where $a_j \neq a_k$ if $j \neq k$. C is a constant. Thus $Q(s)$ has only first-order zeros. Show, using residues, that

$$f(t) = \mathcal{L}^{-1}F(s) = \sum_{j=1}^{n} \frac{P(a_j)}{Q'(a_j)} e^{a_j t}.$$

This is the most elementary of the Heaviside formulas and is usually derived from a partial fraction expansion of $F(s)$.

b) Use the formula derived in part (a) to solve Exercise 1(a) above.

3. For the series L, C, R circuit shown in Fig. 7.1–9 the charge on the capacitor for time $t \geq 0$ is $q(t)$ coulombs. The switch is open for $t < 0$ and closed for $t \geq 0$. When the switch is closed the capacitor C contains an initial charge q_0. After the switch is closed the charge on the capacitor is

$$q(t) = q_0 + \int_0^t \iota(t') \, dt', \qquad t \geq 0,$$

where $\iota(t')$ is the current in the circuit. From Kirchhoff's voltage law, when $t \geq 0$, the sum of the voltage drops around the three elements must be zero. The voltages across the inductor L, capacitor C, and resistor R are, respectively, $L \, d\iota/dt$, $q(t)/C$, $\iota(t)R$. Thus

$$0 = L \frac{d\iota}{dt} + \frac{1}{C} \left[q_0 + \int_0^t \iota(t') \, dt' \right] + \iota(t)R.$$

From physical considerations we also require that $\iota(0) = 0$ and that $\iota(t)$ be a continuous function.

a) Show that

$$I(s) = \frac{-q_0}{C\left(Ls^2 + Rs + \dfrac{1}{C}\right)}.$$

Hint: Transform the preceding integrodifferential equation. Recall that $\mathcal{L}1 = 1/s$ (see Eq. 7.1–2, with $a = 0$; also use Eqs. 7.1–10 and 7.1–8).

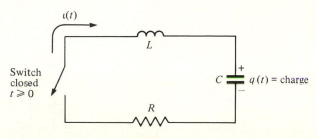

Figure 7.1–9

b) Use residues to find $\iota(t)$ for $t > 0$. Consider three separate cases:

$$\text{(i)}\ R > 2\sqrt{\frac{L}{C}}\,, \qquad \text{(ii)}\ R < 2\sqrt{\frac{L}{C}}\,, \qquad \text{(iii)}\ R = 2\sqrt{\frac{L}{C}}\,.$$

Describe the qualitative differences in your results. These cases are known as *overdamped*, *underdamped*, and *critically damped*. How, in each case, is the location and order of the poles of $I(s)$ related to the type of damping?

4. Figure 7.1–10 illustrates a mechanical problem whose solution requires coupled differential equations. A pair of masses m_1 and m_2 lie on a perfectly smooth surface and are separated by the three identical springs having elastic constant k. Mass m_1 is located by the coordinate x_1, and mass m_2 is located by the coordinate x_2. These coordinates are measured from the equilibrium configuration of the system, that is, with m_1 at $x_1 = 0$ and m_2 at $x_2 = 0$ neither mass experiences any *net* force. From Newton's second law the motion of these masses, as a function of time, is governed by the following pair of coupled differential equations:

$$m_1\frac{d^2x_1}{dt^2} = -2kx_1 + kx_2, \qquad m_2\frac{d^2x_2}{dt^2} = kx_1 - 2kx_2,$$

where $x_1(t)$ and $x_2(t)$ are continuous functions.

a) Suppose $k = 1$, $m_1 = 1$, $m_2 = 2$, $dx_1/dt = dx_2/dt = 0$ at $t = 0$, $x_1(0) = 1$, $x_2(0) = 0$. Take the Laplace transform of the above differential equations and obtain the simultaneous algebraic equations

$$s^2X_1(s) - s = -2X_1(s) + X_2(s),$$
$$2s^2X_2(s) = X_1(s) - 2X_2(s).$$

b) Solve the equations derived in part (a) for $X_1(s)$ and $X_2(s)$.

c) Use the method of residues to obtain $x_1(t)$ and $x_2(t)$ for $t > 0$.

5. A pair of electrical circuits are coupled by means of a transformer having mutual inductance M and self-inductances L_1 and L_2 (see Fig. 7.1–11). If these terms are unfamiliar, see any standard textbook on electric circuits. One can show that the time-varying currents $\iota_1(t)$ and $\iota_2(t)$ circulating around the left- and right-hand circuits in the directions shown satisfy the differential equations

$$V_1(t) = R_1\iota_1(t) + L_1\frac{d\iota_1}{dt} + M\frac{d\iota_2}{dt}, \qquad 0 = M\frac{d\iota_1}{dt} + R_2\iota_2 + L_2\frac{d\iota_2}{dt}.$$

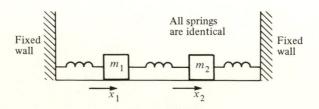

Figure 7.1–10

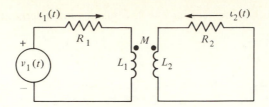

Figure 7.1–11

a) Perform a Laplace transformation on these equations and obtain a pair of simultaneous algebraic equations for $I_1(s)$ and $I_2(s)$. Assume the initial conditions $\iota_1(0) = \iota_2(0) = 0$. Take $V_1(t) = e^{-\alpha t}$, $t \geq 0$, $\alpha > 0$.

b) Assume $L_1 = L_2 = 1$, $M = 0.5$, $R_1 = 1$, $R_2 = 1$, $\alpha = 1$. Solve the equations obtained in part (a) for $I_1(s)$ and $I_2(s)$.

c) Use residues to obtain $\iota_1(t)$ and $\iota_2(t)$.

6. This exercise deals with connection between Fourier and Laplace transforms. In Chapter 6 we presented the Fourier transform pair

$$F(\omega) = \frac{1}{2\pi} \int_{-\infty}^{+\infty} f(t')e^{-i\omega t'}\,dt', \qquad f(t) = \int_{-\infty}^{+\infty} F(\omega)e^{i\omega t}\,d\omega,$$

which implies

$$f(t) = \int_{-\infty}^{+\infty} \frac{1}{2\pi}\left(\int_{-\infty}^{+\infty} f(t')e^{-i\omega t'}\,dt'\right)e^{i\omega t}\,d\omega. \qquad (7.1\text{–}42)$$

a) Suppose in the preceding equation we take $f(t) = g(t)e^{-at}$, where $g(t) = 0$ for $t < 0$. We must, of course, also replace $f(t')$ on the right in Eq. (7.1–42) by $g(t')e^{-at'}$. Using Eq. (7.1–42), show that

$$g(t) = \frac{1}{2\pi} \int_{-\infty}^{+\infty}\left[\int_0^\infty g(t')e^{-(a+i\omega)t'}\,dt'\right]e^{(a+i\omega)t}\,d\omega.$$

b) Make the change of variables $s = a + i\omega$, where a is a real constant. Show that the equation derived in part (a) can be written

$$g(t) = \frac{1}{2\pi i} \int_{a-i\infty}^{a+i\infty}\left[\int_0^\infty g(t')e^{-st'}\,dt'\right]e^{st}\,ds.$$

c) Show that the equation derived in part (b) can be written

$$g(t) = \frac{1}{2\pi i} \int_{a-i\infty}^{a+i\infty} G(s)e^{st}\,ds,$$

where $G(s)$ is the Laplace transform of $g(t)$. Thus we have derived the Bromwich integral (see Eq. 7.1–27) for the inversion of Laplace transforms.

7. a) By starting with the contour shown in Fig. 7.1–12 and passing to the appropriate limits, show that

$$\mathcal{L}^{-1}\frac{1}{s^{1/2}(s-k)} = \frac{e^{kt}}{\sqrt{k}} - \frac{1}{\pi}\int_0^\infty \frac{e^{-ut}}{\sqrt{u}\,(u+k)}\,du,$$

where $k > 0$ and $s^{1/2}$ is the principal branch of this function.

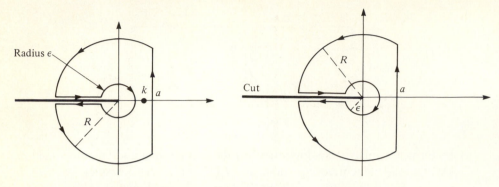

Figure 7.1–12   Figure 7.1–13

b) Show from formula 7.4.9 in M. Abramowitz and I. Stegun, *Handbook of Mathematical Functions* (New York: Dover Books, 1968) that the preceding result can be rewritten

$$\mathcal{L}^{-1}\frac{1}{s^{1/2}(s-k)} = \frac{e^{kt}\operatorname{erf}\sqrt{kt}}{\sqrt{k}},$$

where erf $p = (2/\sqrt{\pi})\int_0^p e^{-s^2}\,ds$ is the error function (see also formulas 7.1.1 and 7.1.2 in the same reference).

8. Show that

$$\mathcal{L}^{-1}\frac{e^{-b(s^{1/2})}}{s} = 1 - \frac{1}{\pi}\int_0^{\infty}\frac{e^{-xt}\sin\left(b\sqrt{x}\right)dx}{x},$$

where $b \geq 0$, and $s^{1/2}$ is the principal branch. *Hint:* Take the branch cut of s as shown in Fig. 7.1–13 using $s^{1/2} > 0$ on the positive real axis. Do an integration around the contour indicated and then allow $R \to \infty$, $\varepsilon \to 0$. Since Re $s^{1/2} \geq 0$ on the arcs of radius R, we have, for $b \geq 0$,

$$|e^{-b(s^{1/2})}| \leq 1 \quad \text{and} \quad \left|\frac{e^{-b(s^{1/2})}}{s}\right| \leq \frac{1}{s}.$$

Thus an argument much like that leading to Eq. (7.1–30) can be used to assert that the integrals over the two curves of radius R become zero as $R \to \infty$. Note that as $\varepsilon \to 0$, the integral around $|s| = \varepsilon$ approaches a nonzero value. Incidentally (this is not a hint) one can show that

$$\frac{1}{\pi}\int_0^{\infty}\frac{e^{-xt}\sin\left(b\sqrt{x}\right)}{x}dx = \operatorname{erf}\left(\frac{b}{2\sqrt{t}}\right),$$

where erf p is defined in Exercise 7.

9. Show that

$$\mathcal{L}^{-1}\frac{1}{(s^2+1)^{1/2}} = \frac{1}{\pi}\int_{-1}^{+1}\frac{e^{i\sigma t}}{\sqrt{1-\sigma^2}}\,d\sigma.$$

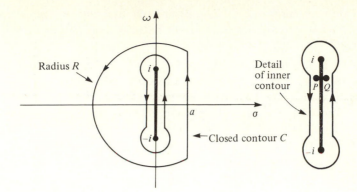

Figure 7.1–14

The integral on the right is $J_0(t)$ the Bessel function of zero order. A branch cut connecting i with $-i$ defining $(s^2 + 1)^{1/2}$ is shown in Fig. 7.1–14. We take $(s^2 + 1)^{1/2} > 0$ on the positive real axis. *Hint:* Recall the principle of deformation of contours (see Chapter 4). Use this concept in order to show that $(1/2\pi i) \int F(s)e^{st}\, ds$ taken around the inner contour in Figure 7.1–14 equals this same integration performed around the closed contour C. Allow $R \to \infty$. Notice that at points such as P and Q the values assumed by $1/\sqrt{s^2 + 1}$ are identical but opposite in sign.

10. Suppose we wish to determine $\mathcal{L}^{-1}F(s) = \mathcal{L}^{-1}[e^{-s}/s^2]$ by means of a contour integration. Since $F(s)$ has a pole at $s = 0$, we perform the Bromwich integration (see Eq. 7.1–27) along a line for which $a > 0$.

a) Explain why, for $t > 1$, we evaluate the Bromwich integral by using residue calculus to evaluate an integral taken around a contour consisting of a line and an arc that closes the contour in the left half plane (see Fig. 7.1–15a). The radius R is allowed to become infinite. Find $f(t)$ for $t > 1$.

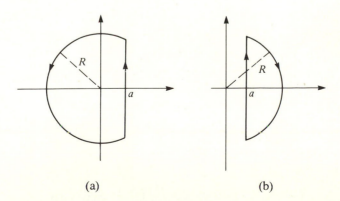

(a)   (b)

Figure 7.1–15

b) Explain why, for $0 < t < 1$, the Bromwich integral is evaluated by our using a contour whose arc lies in the right half plane as shown in Fig. 7.1–15(b). Find $f(t)$ for $0 < t < 1$.

11. Let $f(t) = 0$ for $t < 1$ and $f(t) = 1$ for $t \geq 1$. This function has a jump discontinuity at $t = 1$. Use Eq. (7.1–1) to verify that $\mathcal{L}f(t) = e^{-s} = F(s)$. Show for this $F(s)$ that the Bromwich integral (see Eq. 7.1–27) has the numerical value $1/2$ when $t = 1$. Thus the inverse of $F(s)$ yields, at the jump $t = 1$, the average $(1/2)[f(1 +) + f(1 -)]$. *Hint:* Consider $(1/2\pi i)\int_{a-ib}^{a+ib} 1/s \, ds$, $a > 0$, taken along $\text{Re}(s) = a$. Evaluate, and let $b \to \infty$ in your result.

12. The derivation of Eq. (7.1–30) was based on the assumption that $F(s)$ has a finite number of poles. If $F(s)$ contains an infinite number of poles, the proof given is invalid since, as R goes continuously to infinity (see Fig. 7.1–5), the contour of integration will pass through singularities of $F(s)$.

   However, Eq. (7.1–30) is often still valid provided we use on the right the *infinite sum* of all the residues of $F(s)e^{st}$. The justification for this procedure involves replacing the contour C in Fig. 7.1–5 by an infinite sequence of expanding contours $C_1, C_2, \ldots, C_n$ (see Fig. 7.1–16) that are chosen in such a way that no contour passes through any pole of $F(s)$. If $\oint_{C_n} F(s)e^{st} \, ds$ tends to zero along the curved portion of C_n as $n \to \infty$, then one can show that the required Bromwich integral along the line $s = a$ equals the sum of all the residues of $F(s)e^{st}$. Convince yourself that the functions $F(s)$ below satisfy these conditions provided $0 < x < b$ and prove:

a)

$$\mathcal{L}^{-1} \frac{\sinh sx}{s \sinh sb} = \frac{x}{b} + \frac{2}{\pi} \sum_{k=1}^{\infty} \frac{(-1)^k}{k} \sin\left(\frac{k\pi x}{b}\right) \cos\left(\frac{k\pi t}{b}\right)$$

   Hint: Show there are simple poles at $s = 0$ and at $s = ik\pi/b$, $k = \pm 1, \pm 2, \ldots$.

b)

$$\mathcal{L}^{-1} \frac{\cosh sx}{s^2 \cosh sb} = t + \frac{8b}{\pi^2} \sum_{k=1}^{\infty} \frac{(-1)^k}{(2k-1)^2} \cos\left[\frac{(2k-1)\pi x}{2b}\right] \sin\left[\frac{(2k-1)\pi t}{2b}\right]$$

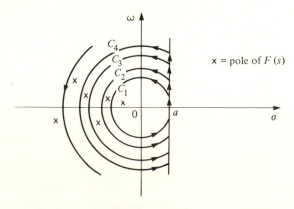

Figure 7.1–16

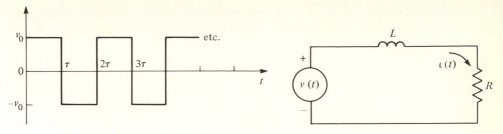

Figure 7.1–17   Figure 7.1–18

13. a) Show for $f(t)$, as defined in Fig. 7.1–17, that

$$F(s) = \frac{v_0}{s} \tanh\left(\frac{s\tau}{2}\right).$$

b) For the electric circuit in Fig. 7.1–18 we can use the Kirchhoff voltage law to show that the current $\iota(t)$ satisfies

$$v(t) = L\frac{d\iota}{dt} + \iota(t)R.$$

Assume that $v(t)$ is given by $f(t)$ in part (a), and that $\iota(0) = 0$. Show that

$$I(s) = \frac{v_0}{s} \frac{\tanh\left(\dfrac{s\tau}{2}\right)}{Ls + R}.$$

c) Show that $I(s)$ has simple poles at $s = -R/L$ and $s = \pm i\pi(2m - 1)/\tau$, $m = 1, 2, 3, \ldots$. Show that $I(s)$ has a finite limit as $s \to 0$.

d) Notice that $I(s)$ satisfies the conditions described in Exercise 12 with regard to its behavior on a sequence of expanding arcs. Show that

$$i(t) = \frac{v_0}{R} e^{-Rt/L} \tanh\left[\frac{R\tau}{2L}\right]$$

$$+ \frac{4\tau}{\pi} v_0 \sum_{m=1}^{\infty} \frac{\tau R \sin\left[(2m - 1)\dfrac{\pi t}{\tau}\right] - (2m - 1)\pi L \cos\left[(2m - 1)\dfrac{\pi t}{\tau}\right]}{(2m - 1)\left[\tau^2 R^2 + (2m - 1)^2 \pi^2 L^2\right]}.$$

7.2 STABILITY—AN INTRODUCTION

One of the needs of the design engineer is to distinguish between two different kinds of functions of time—those that "blow up," that is, become unbounded, and those that do not. We will be a bit more precise. Let us consider a function $f(t)$ defined for $t > 0$.

Definition

A function $f(t)$ is bounded for positive t if there exists a constant M such that

$$|f(t)| < M, \quad \text{for all } t > 0. \quad \blacksquare \qquad (7.2-1)$$

We will usually just use the word "bounded" to describe an $f(t)$ satisfying Eq. (7.2–1). If no constant M can be found that remains larger than $|f(t)|$, we will say that $f(t)$ becomes unbounded. Such a function "blows up."

Although $f(t) = 1/(t - 1)^2$ is unbounded because of a singularity at $t = 1$, our concern here is primarily with functions that fail to satisfy Eq. (7.2–1) because they grow without limit as $t \to \infty$. Thus $f(t) = e^{-t}$ is bounded because this function is less than 1, but $f(t) = e^t$ is unbounded since, for sufficiently large t, it will exceed any preassigned constant. The same is true of the functions $t \sin t$ and $e^t \cos t$, which exhibit oscillations of steadily growing size as t increases. The main subject of this and the following section is the relationship between bounded and unbounded functions and their Laplace transforms. We will also be concerned with knowing whether the response of a system (typically electrical or mechanical) is bounded or unbounded.

Example 1

It is easy to devise an innocent looking physical problem whose response or solution is unbounded. In Fig. 7.2–1 a mass of size m is attached to a spring whose elasticity constant is k. The mass is subjected to a harmonically varying external force $F_0 \cos \omega_0 t$ for $t \geq 0$. Let y be the displacement of the mass from that position in which the spring exerts no force. Then, if there are no frictional losses, Newton's second law asserts that

$$m\frac{d^2y}{dt^2} + ky = F_0 \cos(\omega_0 t). \qquad (7.2–2)$$

We will assume that at $t = 0$, $dy/dt = 0$, and $y(t) = 0$. Taking the Laplace transform of Eq. (7.2–2), subject to the initial conditions, we have

$$ms^2 Y(s) + kY(s) = F_0 \frac{s}{s^2 + \omega_0^2},$$

or

$$Y(s) = \left(\frac{1}{ms^2 + k}\right)\left(\frac{F_0 s}{s^2 + \omega_0^2}\right) = \left(\frac{1}{s^2 + \dfrac{k}{m}}\right)\left(\frac{F_0 s}{m(s^2 + \omega_0^2)}\right).$$

Studying this result, we see that for $\omega_0 \neq \sqrt{k/m}$, $Y(s)$ has simple poles at $s = \pm i\omega_0$ and $s = \pm i\sqrt{k/m}$. If $\omega_0 = \sqrt{k/m}$, then $Y(s)$ has a pole of order 2 at $s = \pm i\omega_0$. In

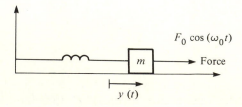

Figure 7.2–1

either case we find $y(t)$ from Eq. (7.1–30) with the result that

a) $\qquad \omega_0 \neq \sqrt{\dfrac{k}{m}}$, $\qquad y(t) = \dfrac{F_0}{\left(\omega_0^2 - \dfrac{k}{m}\right)m}\left[-\cos(\omega_0 t) + \cos\left(\sqrt{\dfrac{k}{m}}\, t\right)\right]$,

b) $\qquad\qquad \omega_0 = \sqrt{\dfrac{k}{m}}$, $\qquad y(t) = \dfrac{F_0}{2m\omega_0} t\sin(\omega_0 t)$.

Expression (a) consists of 2 cosine waves, while (b) is a sine wave whose amplitude grows with increasing t. The first result is bounded. The second is unbounded. ◀

The qualitatively different results found in cases (a) and (b) of Example 1 have something to do with the kinds of poles possessed by $Y(s)$. The response $y(t)$ of many physical systems, including the one just studied, has a Laplace transform of the form

$$Y(s) = \frac{P(s)}{Q(s)}, \qquad (7.2\text{–}3)$$

where P and Q are polynomials in s having real coefficients, and the degree of Q exceeds that of P. We will assume that $Y(s)$ is an irreducible expression, that is, any identical factors of the form $(s - s_0)$ belonging to both P and Q have been divided out.

From Eq. (7.1–30) we have

$$\mathcal{L}^{-1}\frac{P(s)}{Q(s)} = f(t) = \sum_{\text{res}}\left(\frac{P(s)}{Q(s)}e^{st}\right), \qquad \text{at all poles.} \qquad (7.2\text{–}4)$$

The poles of P/Q occur at the values of s for which $Q(s) = 0$. Those roots, designated $s_1, s_2, \ldots$, are also called the zeros of $Q(s)$. We can write $Q(s)$ in the factored form

$$Q(s) = k(s - s_1)^{N_1}(s - s_2)^{N_2}, \ldots, (s - s_n)^{N_n}, \qquad (7.2\text{–}5)$$

where k is a constant. The number N_j tells the multiplicity of the root s_j; it indicates the number of times this root is repeated. From Section 6.2 we see that N_j is also the order of the zero of $Q(s)$ at s_j.

If $Q(s) = 0$ has a root at $s = a + ib$ and if this root occurs with multiplicity N, then $e^{st}P/Q$ has a pole of order N at $s = a + ib$. The residue of $e^{st}P/Q$ at $a + ib$ contributes functions of time to $f(t)$ in Eq. (7.2–4) that can vary as

$$t^{N-1}e^{at}\cos(bt),\ t^{N-1}e^{at}\sin(bt),\ t^{N-2}e^{at}\cos(bt),$$
$$t^{N-2}e^{at}\sin(bt),\ldots,\ e^{at}\cos(bt),\ e^{at}\sin(bt). \qquad (7.2\text{–}6)$$

The reader can verify this fact by direct calculation or consultation with a table. Let us now consider three possibilities:

1. The root is in the right half of the s-plane. Thus $a > 0$. Each of the terms in Eq. (7.2–6) represents an unbounded function of time of either an oscillatory ($b \neq 0$) or nonoscillatory ($b = 0$) nature.

2. The root is in the left half of the s-plane, which means $a < 0$. With $a < 0$ the decay of e^{at} with increasing t causes each term in Eq. (7.2–6) to become zero as $t \to \infty$. Each term is bounded.

3. The root of $Q(s) = 0$ is on the imaginary axis. This means $a = 0$. The terms contained in Eq. (7.2–6) are now of the form

$$t^{N-1} \cos (bt), \, t^{N-1} \sin (bt), \, t^{N-2} \cos (bt), \, t^{N-2} \sin (bt), \ldots, \cos (bt), \sin (bt).$$

$$(7.2-7)$$

If the root is repeated, that is, if $N > 1$, then the amplitude of all these oscillatory terms except the last two, $\cos(bt)$ and $\sin(bt)$, grow without limit and are therefore unbounded. If $N = 1$, only the bounded terms $\cos(bt)$ and $\sin(bt)$ are present.

A root at the origin is a special case of a root on the imaginary axis. We put $b = 0$ in Eq. (7.2–7) and again find that for a repeated root there is an unbounded contribution to $f(t)$ while for a nonrepeated root the contribution is bounded.

The presence of one or more unbounded contributions to $f(t)$ on the right in Eq. (7.2–4) causes $f(t)$ to be unbounded. Our findings for possibilities 1–3 can be summarized in the following theorem.

Theorem 2   Condition for bounded $f(t)$

Let $f(t) = \mathcal{L}^{-1}F(s) = \mathcal{L}^{-1}P(s)/Q(s)$, where P and Q are polynomials in s having no common factors, and the degree of Q exceeds that of P. Then, $f(t)$ is bounded if and only if $Q(s) = 0$ has no roots to the right of the imaginary axis and any roots on the imaginary axis are nonrepeated; in other words, if and only if poles of $F(s)$ do not occur in the right half of the s-plane and any poles on the imaginary axis are simple. ∎

In case (b) of the oscillating mass problem just considered (see Example 1) the unbounded result was caused by the second-order poles of $Y(s)$ lying on the imaginary axis at ω_0 and also at $-\omega_0$. In case (a) the result was bounded. The poles of $Y(s)$ were simple and lay on the imaginary axis at $\pm i\omega_0$ and $\pm i\sqrt{k/m}$.

Example 2

Consider

$$f(t) = \mathcal{L}^{-1}\frac{s}{s^2 + \beta s + 1}, \quad \text{where } \beta \text{ is real.}$$

Discuss the boundedness of $f(t)$ for the cases $\beta = 1$, $\beta = -1$, $\beta = 0$.

Solution

We take

$$F(s) = \frac{P(s)}{Q(s)} = \frac{s}{s^2 + \beta s + 1}.$$

The roots s_1 and s_2 of $Q(s) = 0$ are found from the quadratic formula. Thus for

$s^2 + \beta s + 1 = 0$ we have

$$s_{1,2} = \frac{-\beta + (\beta^2 - 4)^{1/2}}{2}.$$

a) With $\beta = 1$,

$$s_{1,2} = \frac{-1 \pm i\sqrt{3}}{2}.$$

Here $F(s)$ has poles only in the left plane. Thus $f(t)$ is bounded.

b) With $\beta = -1$,

$$s_{1,2} = \frac{1 \pm i\sqrt{3}}{2}.$$

Now $F(s)$ has poles in the right half plane. Thus $f(t)$ is unbounded.

c) With $\beta = 0$,

$$s_{1,2} = \pm i.$$

These roots are on the imaginary axis. Now $Q(s) = (s + i)(s - i)$. The poles of $F(s)$ lie only on the imaginary axis and are simple. Thus $f(t)$ is bounded. ◀

In control theory, electronics, and often in biology and medicine one deals with systems (electrical, mechanical, or animal) subjected to an input or excitation that produces some kind of output or response. The systems analyst must answer this question: If the input $x(t)$, defined for $t > 0$, is bounded for $t > 0$, will the output $y(t)$ also be bounded?

To answer this question we will employ transforms. The Laplace transforms $Y(s)$ and $X(s)$, of $y(t)$ and $x(t)$, for the systems we will be studying are related through an expression of the form

$$Y(s) = G(s)X(s). \tag{7.2-8}$$

Here $G(s)$ is known as the transfer function of the system. We make the following definition.

Definition Transfer function

The *transfer function* of a system is that function that must be multiplied by the Laplace transform of the input to yield the Laplace transform of the output. ∎

To see how a relationship like Eq. (7.2–8) can arise, let us consider a system in which the variables $x(t)$ and $y(t)$ satisfy a linear differential with constant coefficients, that is,

$$a_n \frac{d^n y}{dt^n} + a_{n-1} \frac{d^{n-1} y}{dt^{n-1}} + \cdots + a_1 \frac{dy}{dt} + a_0 y = x(t). \tag{7.2-9}$$

An elementary example of such a system is the spring and oscillating mass considered earlier in this section. The input is the force $F_0 \cos \omega_0 t$ while the output is the displacement of the mass $y(t)$.

Besides Eq. (7.2–9) we are typically given initial conditions at $t = 0$ for the function $y(t)$ and its time derivatives. We assume here, as we do in the remainder of this section, that all such values are zero. This is equivalent to requiring that there be no energy stored in the system at $t = 0$.

Taking the Laplace transform of Eq. (7.2–9) in the usual way we obtain the algebraic equation

$$a_n s^n Y(s) + a_{n-1} s^{n-1} Y(s) + \cdots + a_0 Y(s) = X(s),$$

which yields

$$Y(s) = \frac{X(s)}{a_n s^n + a_{n-1} s^{n-1} + \cdots + a_0}. \tag{7.2–10}$$

Comparing Eq. (7.2–10) with Eq. (7.2–8) we see that, for the systems described by the differential equation (7.2–9), the transfer function is given by

$$G(s) = \frac{1}{a_n s^n + a_{n-1} s^{n-1} + \cdots + a_0}. \tag{7.2–11}$$

Some common systems are characterized not by differential equations but by integrodifferential equations (see, for example, Exercise 7 of this section). Here the transfer function relating output to input is not simply the reciprocal of a polynomial expression in s, as in Eq. (7.2–11), but is the ratio of two polynomials in s. This complication also occurs in the feedback systems considered in Exercise 6, Section 7.3. To allow for such systems we will assume a transfer function of the form

$$G(s) = \frac{A(s)}{B(s)}, \tag{7.2–12}$$

where A and B are polynomials in s. The coefficients in these polynomials are invariably real numbers. Note that Eq. (7.2–11) is a special case of Eq. (7.2–12).

Now we return to our original question. If it is given that the input $x(t)$ is a bounded function, what conditions must be imposed on the transfer function $G(s)$ so that the output $y(t)$ is a bounded function? This leads naturally to the following definition.

Definition   Stable system

A *stable system* is one that produces a bounded output for *every* bounded input.   ∎

A system that is not stable will, of course, be called *unstable*.

Let us study Eq. (7.2–8) for a moment: $Y(s) = G(s)X(s)$. If $x(t)$ is bounded, then $X(s)$ has no poles to the right of the imaginary axis, and any poles on the axis are simple. Now, if $G(s)$ has all its poles lying to the left of the imaginary axis, then the product $G(s)X(s) = Y(s)$ has no poles to the right of the imaginary axis. Whatever poles the product GX possesses on the imaginary axis are the poles of $X(s)$ and are simple. Thus the poles of $Y(s)$ are such that $y(t)$ is bounded. This leads to the following theorem.

Theorem 3 Poles of stable and unstable systems

The transfer function $G(s)$ of a stable system has all its poles lying to the left of the imaginary s-axis. A system whose $G(s)$ has one or more poles on, or to the right of, the imaginary axis is unstable. ■

A pole of $G(s)$ lying to the right of the imaginary s-axis results, in general, in $Y(s)$ having such a pole. Thus $y(t)$ will be unbounded. Now, comparing Theorems 2 and 3 we see that the conditions required for a bounded function are not quite the same as those required for a stable system. The transform of a bounded function can have simple poles on the imaginary axis, while the transfer function of a stable system cannot. To see why this is so, consider Eq. (7.2–8). If $G(s)$ has a simple pole on the imaginary axis and if $X(s)$ also has a simple pole at the same location, then the product $G(s)X(s)$ would have a second-order pole at this point. With $Y(s)$ now having a second-order pole on the imaginary axis the output $y(t)$ would be unbounded. For the $G(s)$ just described a bounded output is obtained for all bounded inputs except those whose transforms $X(s)$ have a simple pole coinciding with the simple pole of $G(s)$ on the imaginary axis. To describe this situation, the following definition is useful.

Definition Marginal instability

An unstable system with a transfer function whose poles on the imaginary axis are simple and with no poles to the right of this axis is called *marginally unstable*. ■

Marginally unstable systems are thus special kinds of unstable systems. The term "marginally unstable" is not a universal one. Some authors use the term "marginally stable" to mean the same thing.

Example 3

The mass–spring system considered in Example 1 is marginally unstable. We had

$$Y(s) = \left(\frac{1}{ms^2 + k} \right)\left(\frac{F_0 s}{s^2 + \omega_0^2} \right).$$

The first term on the right is $G(s)$, the second, the transformed input $X(s)$. Now $G(s)$ has poles on the imaginary axis at $s = \pm i\sqrt{k/m}$. The poles of $X(s)$ are at $s = \pm i\omega_0$. When $\omega_0 = \sqrt{k/m}$, the poles of $X(s)$ are identical to those of $G(s)$, and an unbounded output varying as $t \sin(\omega_0 t)$ occurs. For $\omega_0 \neq \sqrt{k/m}$ the output is bounded and varies with both $\cos(\omega_0 t)$ and $\cos(\sqrt{k/m}\, t)$. ◄

Example 4

The input $x(t)$ and the output $y(t)$ of a certain system are related by

$$\frac{d^3 y}{dt^3} - a\frac{d^2 y}{dt^2} + b^2\frac{dy}{dt} - ab^2 = x(t).$$

For what real values of a and b is this a stable system?

Solution

With all initial values taken as zero we transform this equation and obtain

$$(s^3 - as^2 + b^2 s - ab^2)Y(s) = X(s),$$

which we can rewrite as

$$(s - a)(s^2 + b^2)Y(s) = X(s) \quad \text{or} \quad Y(s) = \frac{X(s)}{(s - a)(s^2 + b^2)}.$$

We see that

$$G(s) = \frac{1}{(s - a)(s^2 + b^2)}$$

and that $G(s)$ has simple poles at $s = a$ and at $s = \pm ib$. If $a > 0$, $G(s)$ has a pole to the right of the imaginary axis. The system is unstable. If $a < 0$, $G(s)$ has no poles to the right of the imaginary axis. Now, with $a < 0$, assume $b \neq 0$. $G(s)$ has simple poles on the imaginary axis at $\pm ib$. Thus the system is marginally unstable.

If $a = 0$ and $b \neq 0$, the poles of $G(s)$ are simple and lie on the imaginary axis at $s = 0$ and $\pm ib$. The system is marginally unstable.

If $b = 0$ and $a \neq 0$, $G(s)$ has a second-order pole on the imaginary axis at $s = 0$. The system is unstable.

If $b = 0$ and $a = 0$, there is a third-order pole at $s = 0$. Thus the system is unstable. ◀

EXERCISES

1. Which of the functions $F(s)$ given below have inverse Laplace transforms $f(t)$ that are bounded functions of t?

a) $\dfrac{1}{s(s^2 + 1)}$

b) $\dfrac{1}{(s)(s^2 + 1)^2}$

c) $\dfrac{s + 1}{s^2 + s + 1}$

d) $\dfrac{s}{s^2 + 1}$

e) $\dfrac{s - 1}{s^2 - s - 2}$

f) $\dfrac{s + 1}{s^3 + 2s^2 + 2s}$

g) $\dfrac{1}{s^4 + 2s^2 + 1}$

2. Let

$$F(s) = \frac{s + 1}{s^2 + \beta s + 1}, \quad \text{where } \beta \text{ is a real number.}$$

a) For $\beta \geq 0$ show that $f(t)$ is bounded, and for $\beta < 0$ show that $f(t)$ is unbounded.

b) For $-2 < \beta < 2$ show that $f(t)$ oscillates with t. For which values of β do the oscillations grow with t and for which values do they decay?

3. Assume that the expressions given in Exercise 1 are transfer functions $G(s)$. In each case is the system stable or unstable? If the system is unstable, is it marginally unstable?

4. For the marginally unstable systems characterized by the following transfer functions find a bounded input $x(t)$ that will produce an unbounded output $y(t)$.

a) $\dfrac{1}{s^2 + 1}$ b) $\dfrac{1}{(s)(s + 1)}$

5. For a certain system the output y lags behind the input x by 1 time unit and is $1/2$ the input, that is,

$$y(t) = \frac{1}{2}x(t - 1), \qquad t > 1.$$

Now assume that $x(t - 1) = 0$, $t < 1$. Find the transfer function $G(s)$ of the system. *Hint:* If $\mathcal{L}x(t) = X(s)$ find $\mathcal{L}x(t - 1)$ in terms of $X(s)$. This is an example of a system whose transfer function is not a rational function. Is this a stable system?

6. In Example 1 (see Fig. 7.2–1) assume that the mass is moving through a fluid that exerts a retarding force on the object proportional to the velocity of motion dy/dt. Thus the differential equation describing the motion is now given by

$$m\frac{d^2y}{dt^2} + ky + \alpha\frac{dy}{dt} = x(t),$$

where $\alpha \geq 0$ is a constant, and $x(t)$ is the external force applied to the mass. Assume that $y = 0$ and $dy/dt = 0$ at $t = 0$.

a) Show that the transfer function relating the input $x(t)$ and the response $y(t)$ is

$$G(s) = \frac{1}{ms^2 + \alpha s + k}.$$

b) Show that for $\alpha > 0$ the system is stable and that for $\alpha = 0$ the system is marginally unstable.

7. For the R, L circuit in Fig. 7.2–2 the input is the voltage $v_i(t)$ and the output is the voltage $v_0(t)$. The current is $\iota(t)$. From Kirchhoff's voltage law we have

$$v_i(t) = v_0(t) + \iota(t)R.$$

If $\iota(0) = 0$, then by Faraday's law,

$$\iota(t) = \frac{1}{L}\int_0^t v_0(t')\,dt'.$$

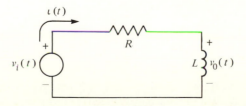

Figure 7.2–2

Thus v_i and v_0 are related by the integrodifferential equation

$$v_i(t) = v_0(t) + \frac{R}{L} \int_0^t v_0(t') \, dt'.$$

Show that the transfer function is

$$\frac{V_0(s)}{V_i(s)} = \frac{Ls}{Ls + R} = G(s).$$

Is the system stable?

8. Certain electrical devices, for example, Gunn diodes, exhibit a negative electrical resistance. A certain negative resistance diode is characterized by the equivalent circuit shown in the broken box (see Fig. 7.2–3). The indicated resistance R has a negative numerical value $-R_d$, where $R_d > 0$. The applied voltage $v(t)$ is a bounded excitation, while the supplied current $\iota_1(t)$ is the response. Writing Kirchhoff's voltage law around the two electrical meshes shown, we obtain the coupled equations

$$v(t) = L\frac{d\iota_1}{dt} + \frac{1}{C}\int_0^t [\iota_1(t) - \iota_2(t)] \, dt,$$

$$0 = \frac{1}{C}\int_0^t (\iota_2 - \iota_1) \, dt - \iota_2 R_d.$$

a) Assume all initial conditions are zero, and take the Laplace transform of this pair of equations. Obtain a pair of algebraic equations involving $I_1(s)$ and $I_2(s)$. Show that $G(s) = (1 - sCR_d)/(Ls(1 - sCR_d) - R_d) = I_1(s)/V(s)$.

b) Examine the poles of the transfer function $G(s)$, and show that the system is unstable.

c) Show that if $R_d > (\sqrt{L/C})/2$, the current ι_1 exhibits oscillations that grow exponentially in time, and that if $R_d < (\sqrt{L/C})/2$, the current exhibits a nonoscillatory exponential growth.

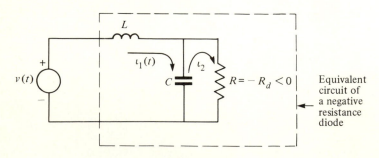

Figure 7.2–3

7.3   PRINCIPLE OF THE ARGUMENT AND INTRODUCTION TO THE NYQUIST STABILITY CRITERION

We have just seen that a stable system has a transfer function $G(s)$ whose poles lie entirely to the left of the imaginary axis. If $G(s)$ is a rational function, for example,

$$G(s) = \frac{A(s)}{B(s)}, \tag{7.3-1}$$

the task of determining whether $G(s)$ describes an unstable system is equivalent to seeing whether the polynomial $B(s)$ satisfies

$$B(s) = 0 \tag{7.3-2}$$

for any s on or to the right of the imaginary axis.

The exercises given in Section 7.2 were carefully selected so that the roots of Eq. (7.3–2) would be easily found. However, it is comparatively easy to construct a problem where the location of roots is not so obvious; for example, consider

$$B(s) = s^4 + s^3 - 2s + 1.$$

The problem of determining the presence of the roots of a polynomial in a half plane or, equivalently, the problem of determining the presence of poles of an expression like Eq. (7.3–1) in a half plane is important in control theory. The engineer uses a variety of mathematical tools to deal with this question. One convenient method, which uses complex variable theory, will be discussed here. The reader is referred to more specialized texts for other techniques.[†]

We will be using the Nyquist method. It is based on a theorem called the *principle of the argument*, which we now derive.

Consider a function $f(z)$ that is analytic on a simple closed contour C, and, except at a finite number of pole singularities, is analytic inside C as well. The preceding guarantees that, as we make one complete circuit around C, the initial and final numerical values assumed by $f(z)$ are identical. As C is completely negotiated, there is no reason, however, why the initial and final values of the argument of $f(z)$ must be identical.

Suppose we write

$$f(z) = |f(z)| e^{i(\arg f(z))}. \tag{7.3-3}$$

We will use the notation $\Delta_C \arg f(z)$ to mean the *increase in argument* of $f(z)$ (final minus initial value) as the contour C is negotiated once in the positive direction.

Let us consider an elementary example. We take $f(z) = z$, and, as a contour C, the circle $|z| = 1$. On C we have $z = |z| e^{i \arg z} = e^{i \arg z}$. As we proceed around C once in the clockwise direction, we see from Fig. 7.3–1 that $\arg z$ progresses from 0 to 2π. Thus in this case $\Delta_C \arg f(z) = 2\pi$. Note that $|f(z)|$ returns to its original numerical value as C is negotiated.

[†]See, for example, J. DiStefano, A. Stubberud, I. Williams, *Feedback and Control Systems*, Schaum's Outline Series (New York: McGraw-Hill, 1967).

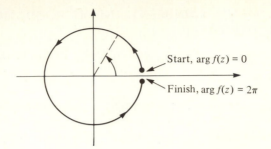

Start, arg $f(z) = 0$

Finish, arg $f(z) = 2\pi$

Figure 7.3–1

To choose another example, if $f(z) = 1/z^2$ and C is any closed contour encircling the origin, then the reader should verify that $\Delta_C \arg f(z) = -4\pi$.

Now consider

$$I = \frac{1}{2\pi i} \oint \frac{f'(z)}{f(z)} \, dz. \tag{7.3–4}$$

We observe that

$$\frac{d}{dz} \log f(z) = \frac{f'(z)}{f(z)}$$

if we use some analytic branch of $\log(f(z))$. If we require that C not pass through any zero of $f(z)$, we see that our integral I can be evaluated by a standard procedure (see Eq. 4.3–4), and thus

$$\frac{1}{2\pi i} \oint_C \frac{f'(z)}{f(z)} dz = \frac{1}{2\pi i} \oint \frac{d}{dz} \log f(z) \, dz = \frac{1}{2\pi i} \oint d(\log f(z))$$

$$= \frac{1}{2\pi i} \left[\text{increase in } \log f(z) \text{ in going around } C \right]$$

$$= \frac{1}{2\pi i} \left[\text{increase in } \left[\log |f(z)| + i \arg f(z) \right] \text{ in going around } C \right].$$

Now $|f(z)|$ necessarily returns to its original numerical value as C is negotiated. However $\arg f(z)$ need not. Thus

$$\frac{1}{2\pi i} \oint_C \frac{f'(z)}{f(z)} dz = \frac{1}{2\pi} \Delta_C \arg f(z). \tag{7.3–5}$$

One can also evaluate Eq. (7.3–4) by residues. If $f(z)$ is analytic on C and at all points interior to C except at poles, then $f'(z)$ will be analytic on and interior to C except at these same poles. (Recall from Section 4.4 that the derivative of an analytic function is analytic.) The quotient $f'(z)/f(z)$ is thus analytic on C and interior to C except where $f'(z)$ has a pole or when $f(z) = 0$. Thus to evaluate Eq. (7.3–4) with residues we must determine the residue of $f'(z)/f(z)$ at all zeros and poles of $f(z)$ lying interior to C.

Suppose $f(z)$ has a zero of order n at ζ. Recall (see Section 6.2) that this means $f(z)$ has a Taylor expansion about ζ of the form

$$f(z) = a_n(z - \zeta)^n + a_{n+1}(z - \zeta)^{n+1} + \cdots, \qquad a_n \neq 0.$$

Thus factoring out $(z - \zeta)^n$, we have

$$f(z) = (z - \zeta)^n \phi(z), \tag{7.3-6}$$

where $\phi(z)$ is a function that is analytic at ζ and has the series expansion

$$\phi(z) = a_n + a_{n+1}(z - \zeta) + a_{n+2}(z - \zeta)^2 + \cdots .$$

Note that $\phi(\zeta) = a_n \neq 0$. Differentiating Eq. (7.3–6), we arrive at

$$f'(z) = n(z - \zeta)^{n-1}\phi(z) + (z - \zeta)^n \phi'(z). \tag{7.3-7}$$

Dividing Eq. (7.3–7) by Eq. (7.3–6), we obtain

$$\frac{f'(z)}{f(z)} = \frac{n}{z - \zeta} + \frac{\phi'(z)}{\phi(z)}. \tag{7.3-8}$$

The first term on the right in Eq. (7.3–8) has a simple pole at ζ in the z-plane. The residue is n. Recalling that $\phi(\zeta) \neq 0$, we see that the second term on the right has no singularity at ζ. Thus, at ζ, the residue of $f'(z)/f(z)$ is identical to the residue of $n/(z - \zeta)$ and equals n. In other words, the residue of $f'(z)/f(z)$ at ζ is equal to the order (multiplicity) of the zero of $f(z)$ at ζ.

Suppose that $f(z)$ has a pole of order p at a point α inside C. Proceeding much as before, (see Exercise 3 of this section) we find that the residue of $f'(z)/f(z)$ at α is equal to (-1) times the order p of the pole of $f(z)$ at α.

We can use the information just derived to sum the residues of $f'(z)/f(z)$ at all the singularities that this expression possesses inside C and thus evaluate the integral I in Eq. (7.3–4) (see Fig. 7.3–2). If $f(z)$ has zeros of order n_1 at ζ_1, n_2 at $\zeta_2, \ldots$ and poles of order p_1 at α_1, p_2 at $\alpha_2, \ldots$, we have

$$\frac{1}{2\pi i} \oint_C \frac{f'(z)}{f(z)} dz = N - P, \tag{7.3-9}$$

where

$$N = n_1 + n_2, \ldots \tag{7.3-10}$$

is the total number of zeros of $f(z)$ inside C and

$$P = p_1 + p_2, \ldots \tag{7.3-11}$$

is the total number of poles of $f(z)$ inside C. In both Eqs. (7.3–10) and (7.3–11) zeros and poles are counted, according to their multiplicities; for example, a zero of order 2 at some point contributes the number 2 to the sum on the right in Eq. (7.3–10) and a pole of order 3 results in a contribution of 3 in Eq. (7.3–11). We will assume that $f(z)$ has a finite number of poles inside C. One can show that $f(z)$ has a finite number of zeros inside C.[†]

Equations (7.3–9) and (7.3–5) provide two different ways of evaluating the same integral. We dispense with the integral and utilize the right side of each equation. This provides the following theorem.

[†]It can be shown that a function that is analytic in a bounded region except at pole singularities has a finite number of zeros in that region.

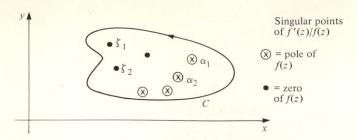

Figure 7.3–2

Theorem 4 Principle of the argument

Let $f(z)$ be analytic on a simple closed contour C and analytic inside C except possibly at a finite number of poles. Also, assume $f(z)$ has no zeros on C. Then

$$\frac{1}{2\pi}\Delta_C \arg f(z) = N - P, \qquad (7.3\text{--}12)$$

where N is the total number of zeros of $f(z)$ inside C, and P is the total number of poles of $f(z)$ inside C. In each case the number of poles and zeros are counted according to their multiplicities. ■

The preceding theorem is called the *principle of the argument*. We should also recall that $\Delta_C \arg f(z)$ in Eq. (7.3–12) is computed when C is traversed in the positive sense. This quantity can be positive, zero, or negative depending on the relative sizes of N and P.

Example 1

Let $f(z) = z^2 - 1$, and let C be the circle $|z - 1| = 1$. Verify the correctness of Eq. (7.3–12) in this case.

Solution

We will use two planes, the usual z-plane and the w-plane, the latter showing values assumed by $w = f(z)$ as z travels around C. We write $w = u + iv$.

   A few points $a, b, c, \ldots$ (see Fig. 7.3–3a) lying on C are considered. We determine the corresponding image points $a', b', \ldots$ under the transformation $w = f(z)$ (see Table 2) and plot these points in the w-plane (see Fig. 7.3–3b). Using the image points we can quickly sketch the locus C' of all the values that $f(z)$ assumes on C. Notice that because $f(z) = z^2 - 1$ is a polynomial in z with real coefficients, we have

$$f(\bar{z}) = \overline{f(z)}.$$

For example, the values assumed by $f(z)$ at the conjugate points b and f are complex conjugates of each other. Notice the relationship of the points b, b', f, and f' in Fig. 7.3–3. Since the curve C in the xy-plane is symmetric about the x-axis, the curve C' in the uv-plane must be symmetric about the u-axis.

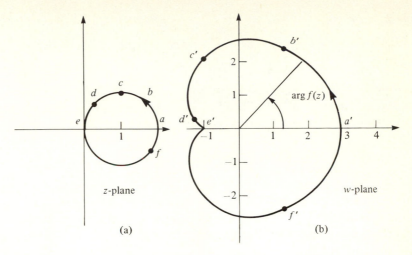

Figure 7.3–3

We see from Fig. 7.3–3(a) and (b) that the argument of $f(z)$ increases by 2π as C is negotiated in the counterclockwise direction from a to b to c... and back to a again; that is, one complete counterclockwise encirclement of the origin has been made in the w-plane. Thus on the left in Eq. (7.3–12) we have

$$\frac{\Delta_C \arg f(z)}{2\pi} = 1.$$

To evaluate the right side of Eq. (7.3–12) we see that $f(z) = z^2 - 1$ has no pole singularities. We have $P = 0$. Since $(z^2 - 1) = (z - 1)(z + 1)$, we see (compare with Eq. 7.3–6) that $f(z)$ possesses two zeros, each of order (or multiplicity) 1. Only the zero at $z = 1$ lies within C. Thus $N - P = 1$. The correctness of Eq. (7.3–12) has been verified in this case. ◀

TABLE 2

point in z-plane	z	$z^2 - 1 = w$	point in w-plane
a	2	3	a'
b	$1 + \dfrac{1}{\sqrt{2}} + \dfrac{i}{\sqrt{2}}$	$\sqrt{2} + i(1 + \sqrt{2})$	b'
c	$1 + i$	$2i - 1$	c'
d	$1 - \dfrac{1}{\sqrt{2}} + \dfrac{i}{\sqrt{2}}$	$-\sqrt{2} + i(\sqrt{2} - 1)$	d'
e	0	-1	e'
f	$1 + \dfrac{1}{\sqrt{2}} - \dfrac{i}{\sqrt{2}}$	$\sqrt{2} - i(1 + \sqrt{2})$	f'

Example 2

Verify Eq. (7.3–12), where $f(z) = z/(z + 1)^2$ and C is the circular contour $|z| = 20$.

Solution

A typical point on C is described by $z = 20e^{i\theta}$ (see Fig. 7.3–4). The corresponding point in the w-plane is $20e^{i\theta}/(1 + 20e^{i\theta})^2$. For our purposes we make an excellent approximation by ignoring 1 in the denominator. Thus

$$f(z) \approx \frac{20e^{i\theta}}{400e^{2i\theta}} = \frac{1}{20}e^{-i\theta}.$$

All the values of $f(z)$ that we will encounter on C therefore lie approximately on a circle in the w-plane of radius $1/20$. As we move around C in the positive direction the angle θ increases by 2π. However, the argument of $f(z)$, which is $-\theta$, *decreases* by 2π, that is, it increases by -2π. Thus $\Delta_C \arg f(z) = -2\pi$. The left side of Eq. (7.3–12) is -1.

Now $f(z) = z/(z + 1)^2$ contains a zero of multiplicity 1 at the origin of the z-plane. A vanishing denominator causes $f(z)$ to have a pole of order 2 at $z = -1$. Both the zero and the pole are inside C. Thus the right side of Eq. (7.3–12) is $1 - 2 = -1$. The formula is verified. ◀

Comment: In many texts the left side of Eq. (7.3–12) is written in a different form. If, as C is traversed, the locus of $f(z)$ makes one complete encirclement of the origin in the w-plane, then the argument of $f(z)$ increases by 2π. Every such additional encirclement results in an additional contribution of 2π to the expression $\Delta_C \arg f(z)$. Thus, on the left side of Eq. (7.3–12), $\Delta_C \arg f(z)/2\pi$ tells the net number of counterclockwise encirclements that $f(z)$ makes about the point $w = 0$. Letting

$$E = \frac{\Delta_C \arg f(z)}{2\pi}$$

be this number, we have, from Eq. (7.3–12),

$$E = N - P. \tag{7.3–13}$$

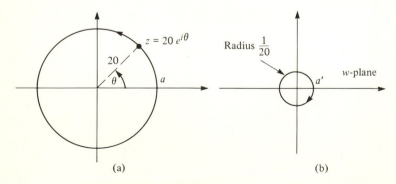

(a)   (b)

Figure 7.3–4

In Example 2, $E = -1$ since the origin in Fig. 7.3–4(b) was encircled once in the *clockwise* direction, while in Example 1 we had $E = 1$.

We will now see how the principle of the argument can be used to settle some questions about stability. We know that if the transfer function of a system is an irreducible rational function of the form $G(s) = A(s)/B(s)$, then the system is unstable if $B(s)$ has any zeros in the right half plane (abbreviated r.h.p) Re $s > 0$ or on the imaginary axis of this plane. Our procedure involves, as in the preceding examples, the use of two planes—the s-plane with real and imaginary axes σ and ω and the w-plane, having axes u and v, in which values of $B(s)$ are plotted.

We first determine whether $B(s)$ has any zeros in the r.h.p. We consider $\Delta_C \arg B(s)$, where C, depicted in Fig. 7.3–5, is the closed semicircular contour of "large" radius R. The diameter of C lies on the axis. R is taken large enough so that C encloses all possible zeros of $B(s)$ lying in the r.h.p. For the moment we postpone consideration of what happens if $B(s) = 0$ on the imaginary axis.

Now, beginning high on the ω-axis in Fig. 7.3–5 at $\omega = R$ (see "start"), we allow ω to become less positive, shrink to zero, become increasingly negative, and stop at $\omega = -R$. While negotiating this straight line in the s-plane, we compute values of $B(s) = u(s) + iv(s)$ and plot these quantities in the w-plane (that is, the uv-plane) to trace the locus of $B(s)$. Since, typically, $B(s)$ is a polynomial with real coefficients, this locus is symmetric about the u-axis (see Example 1).

The next step involves our moving, in the direction of the arrow, along the dotted semicircular arc shown in Fig. 7.3–5. As s proceeds along here and returns to the plane marked "start," we continue to trace the locus of $B(s)$ in the w-plane. Our task here is easy. If $B(s)$ is a polynomial of degree n,

$$B(s) = a_n s^n + a_{n-1} s^{n-1} + \cdots + a_0.$$

On the arc $s = Re^{i\theta}$ so that

$$B(s) = a_n R^n e^{in\theta} + a_{n-1} R^{n-1} e^{i(n-1)\theta} + \cdots + a_0,$$

or

$$B(s) = a_n R^n e^{in\theta} \left[1 + \frac{a_{n-1} e^{-i\theta}}{a_n R} + \frac{a_{n-2} e^{-i2\theta}}{a_n R^2} + \cdots + \frac{a_0}{a_n R^n} e^{-in\theta} \right].$$

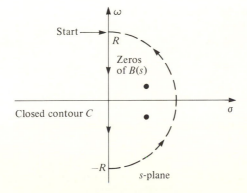

Figure 7.3–5

For arbitrarily large R the expression in the brackets can be made arbitrarily close to 1, and so

$$B(s) \approx a_n R^n e^{in\theta},$$

which means

$$|B(s)| \approx a_n R^n \quad \text{and} \quad \arg B(s) \approx n\theta.$$

Thus as s moves along the arc of radius R, $B(s)$ is closely confined in the w-plane to an arc of radius $a_n R^n$. As the argument of s changes from $-\pi/2$ to $\pi/2$ along the arc in Fig. 7.3–5, $\arg B(s)$ increases by $\approx [n\pi/2 - - n\pi/2] = n\pi$. Since we are letting $R \to \infty$, we will replace $\approx$ here by $=$.

The quantity $\Delta_C \arg B(s)$, which is the *total* increase in the argument of $B(s)$ as the semicircle C is traversed, is the sum of two parts: The increase in the argument of $B(s)$ as the diameter of C is negotiated plus $n\pi$, which arises from the contribution along the curved path.

The function $B(s)$ has no singularities. Therefore Eq. (7.3–12) becomes for our contour C

$$\frac{\Delta_C \arg B(s)}{2\pi} = N, \qquad (7.3\text{–}14)$$

where N is the total number of zeros of $B(s)$ in the r.h.p. Thus if $\Delta_C \arg B(s)$ is found to be nonzero, then $G(s) = A(s)/B(s)$ describes an unstable system. This determination is called the Nyquist criterion applied to polynomials, and the locus of $B(s)$ employed is called a *Nyquist diagram*. Other kinds of Nyquist criteria and diagrams are used when we deal with feedback systems. This subject is briefly treated in Exercise 6 of this section.

Regarding $\Delta_C \arg B(s)/2\pi$ in terms of encirclements, we can state the *Nyquist criterion for polynomials:*

Suppose as s traverses the closed semicircle of Fig. 7.3–5, the locus of $w = B(s)$ makes, in total, a nonzero number of encirclements of $w = 0$ (for $R \to \infty$). Then $B(s) = 0$ has at least one root in the right half of the s-plane.

If, as s moves along the diameter of C in Fig. 7.3–5, $B(s)$ passes *through* the origin in the w-plane, then $\arg B(s)$ becomes undefined. We cannot then compute $\Delta_C \arg B(s)$. However, such an occurrence indicates that $B(s)$ has a zero on the imaginary axis of the s-plane. Since $G(s) = A(s)/B(s)$ has a pole on the imaginary axis, the system in question is unstable. To see whether it could be classified as marginally unstable, we can use a technique illustrated in Example 5.

Example 3

Discuss the stability of the system whose transfer function is

$$G(s) = \frac{s + 1}{s^3 + s^2 + 9s + 4}.$$

Solution

We must see whether $B(s) = s^3 + s^2 + 9s + 4$ has any zeros in the r.h.p. If s lies on the ω-axis, we have $s = i\omega$. Thus

$$B(s) = u + iv = -i\omega^3 - \omega^2 + 9i\omega + 4,$$

which implies

$$u = -\omega^2 + 4, \qquad u = 0, \quad \text{when } \omega = \pm 2;$$
$$v = -\omega^3 + 9\omega, \qquad v = 0, \quad \text{when } \omega = \pm 3 \text{ and } 0.$$

At the point "start" in Fig. 7.3–5 ω is very large and positive. Thus u and v are large negative numbers with v (having a higher power of ω) dominating u. As ω becomes less positive, both u and v diminish in magnitude. Ultimately, when $\omega = 3$, we have $v = 0$ and u is still negative. When ω diminishes to 2, $u = 0$ while v here is positive. Finally, when $\omega = 0$, we have $v = 0$ while u is positive. The locus of $B(s)$ just described is shown in Fig. 7.3–6.

Because $B(s)$ has real coefficients, the locus generated by $B(s)$ as s moves along the negative imaginary axis is the mirror image of that just obtained for the positive axis. The result is also shown in Fig. 7.3–6.

The entire path of $B(s)$ when $s = i\omega$, $-\infty < \omega < \infty$, is shown by the solid line in Fig. 7.3–6. For $\omega \to \infty$, the argument of $B(i\omega)$ is $3\pi/2$. As ω falls to 3, $\arg B(\omega)$ becomes π. At $\omega = 0$, $\arg B(i\omega) = 0$, etc. When $\omega \to -\infty$, $\arg B(i\omega) = -3\pi/2$. Thus the net increase of $\arg B(s)$ as s moves over the imaginary axis in Fig. 7.3–5 is the final value less the initial value:

$$-\frac{3\pi}{2} - \frac{3\pi}{2} = -3\pi.$$

Now, along the large semicircular arc in Fig. 7.3–5 we have $B(s) \approx s^3$, which, as noted previously, means that the increase in argument of $B(s)$ as s moves along this arc is 3π. The *total* increase in argument of $B(s)$ as s ranges over the contour C in Fig. 7.3–5 is $\Delta_C \arg B(s) = 3\pi - 3\pi = 0$. Thus $B(s)$ has *no* roots in the r.h.p.

Since the solid curve in Fig. 7.3–6 does not pass through the origin, $B(s)$ has no zeros on the imaginary axis. The system described by $G(s)$ is stable.

The arc indicated by the broken line in Fig. 7.3–6 is the locus taken by $B(s) \approx s^3$ as s moves along the semicircular arc of Fig. 7.3–5. The net change in

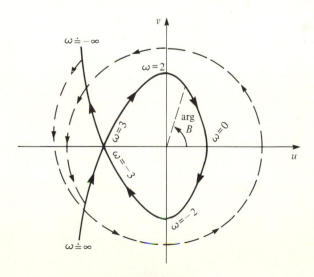

Figure 7.3–6

arg $B(s)$ over the entire path (broken and solid) in Fig. 7.3–6 is zero—a fact we have already noted. Equivalently, observe that this path makes zero net encirclements of the origin. ◀

Example 4

Discuss the stability of the system with transfer function

$$G(s) = \frac{s + 2}{s^3 + s^2 + 3s + 16}.$$

Solution

Letting $B(s) = s^3 + s^2 + 3s + 16$, we proceed as in Example 3. With $s = i\omega$ we have

$$B(s) = u + iv = -i\omega^3 - \omega^2 + 3i\omega + 16,$$

so that

$$u = 16 - \omega^2, \qquad u = 0, \quad \text{when } \omega = \pm 4;$$
$$v = 3\omega - \omega^3, \qquad v = 0, \quad \text{when } \omega = \pm\sqrt{3} \text{ and } 0.$$

A sketch in Fig. 7.3–7, indicated by the solid line, shows the locus taken by $B(s)$ as s ranges downward along the ω-axis is the s-plane. The argument of B ranges from $-\pi/2$ (for $\omega \to \infty$) to $\pi/2$ (for $\omega \to -\infty$). Thus the argument of B increases by π. Notice that this computation is unrelated to the fact that $B(s)$ executes a loop over the range $3 > \omega > -3$. We can ignore the loop since it fails to encircle the origin of the w-plane (compare this with Example 3). As s moves along the semicircular arc in Fig. 7.3–5, the argument of $B(s)$ increases by 3π (see broken line in Fig. 7.3–7). Thus, adding contributions, $\Delta_C \arg B(s) = \pi + 3\pi = 4\pi$. According to Eq. (7.3–14), $B(s)$ has *two* zeros in the r.h.p. The system is unstable. ◀

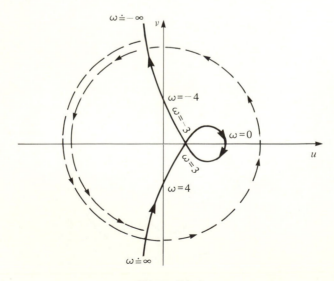

Figure 7.3–7

Example 5

Discuss the stability of the system whose transfer function is

$$G(s) = \frac{s}{s^5 + 6s^4 + 12s^3 + 12s^2 + 11s + 6}.$$

Solution

Proceeding as before, with $B(s) = u + iv$ as the denominator of $G(s)$ and $s = i\omega$, we have on the imaginary s-axis

$$u = 6\omega^4 - 12\omega^2 + 6 = 6(\omega^2 - 1)^2,$$

$$v = \omega^5 - 12\omega^3 + 11\omega = \omega[\omega^4 - 12\omega^2 + 11] = \omega(\omega^2 - 11)(\omega^2 - 1).$$

As s moves downward along the imaginary axis, $B(s)$ traces out the path shown in Fig. 7.3–8. When $s = \pm i$ (or $\omega = \pm 1$), notice that $u = 0$ *and* $v = 0$, that is, $B(s) = 0$. The system is unstable since $B(s)$ has zeros on the imaginary axis.

Perhaps the system is marginally unstable. This would occur if both $s - i$ and $s + i$ are nonrepeated factors of $B(s)$ and, in addition, if $B(s)$ has no zeros in the r.h.p. Writing

$$B(s) = (s - i)(s + i)R(s),$$

we find $R(s)$ by long division:

$$R(s) = \frac{B(s)}{(s - i)(s + i)} = \frac{B(s)}{s^2 + 1}$$

$$= \frac{s^5 + 6s^4 + 12s^3 + 12s^2 + 11s + 6}{s^2 + 1} = s^3 + 6s^2 + 11s + 6.$$

An application of the Nyquist method shows that $R(s)$ has no zeros on the

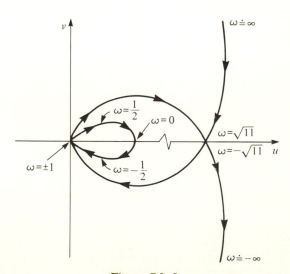

Figure 7.3–8

imaginary axis or in the r.h.p. Thus $B(s)$ has zeros of order 1 on the imaginary axis and no zeros in the r.h.p. The system is marginally unstable. ◄

EXERCISES

1. a) Let $f(z) = (z - 1)/(z + 1)$ and verify Eq. (7.3–12) by taking C as the circle $|z| = 2$ and sketching the locus of $f(z)$ in the w-plane as z moves around this circle.

   b) Repeat part (a) but take as C the circle $|z - 1| = 1$.

2. Use the Nyquist method to investigate the stability of the systems whose transfer functions are given below. State the number of zeros of the denominator that occur in the r.h.p. and on the imaginary axis. State which systems are marginally unstable.

   a) $\dfrac{s^3 + 1}{s^3 + 2s^2 + 2s + 1}$ d) $\dfrac{s + 3}{s^4 + s^3 + 2s^2 + 2s + 1}$

   b) $\dfrac{s - 1}{s^4 + 3s^3 + 3s^2 + 3s + 2}$ e) $\dfrac{s^4 + 1}{s^5 + s^4 + 18s^3 + 18s^2 + 81s + 81}$

   c) $\dfrac{s - 3}{s^4 + 5s^3 + 8s^2 + 7s + 3}$

3. Show that if $f(z)$ has a pole of order p at α, then the residue of $f'(z)/f(z)$ at α is $-p$. *Hint:* $f(z)$ can be expressed as $g(z)/(z - \alpha)^p$, where $g(\alpha) \neq 0$ and $g(z)$ is analytic at α. Why?

4. Let $f(z)$ and $g(z)$ be analytic on and everywhere inside a simple closed contour C. Suppose $|f(z)| > |g(z)|$ on C. We will prove that $f(z)$ and $(f(z) + g(z))$ have the same number of zeros inside C. This is known as *Rouché's theorem*.

   a) Explain why

   $$\frac{\Delta_C \arg f(z)}{2\pi} = N_f,$$

   and

   $$\frac{\Delta_C \arg (f(z) + g(z))}{2\pi} = N_{f+g},$$

   where N_f is the number of zeros of $f(z)$ inside C, and N_{f+g} is the number of zeros of $f(z) + g(z)$ inside C.

   b) Show that

   $$N_{f+g} = \frac{1}{2\pi}\Delta_C \arg f(z) + \frac{1}{2\pi}\Delta_C \arg\left(1 + \frac{g(z)}{f(z)}\right).$$

   Hint: $f + g = f[1 + g/f]$.

   c) If $|g|/|f| < 1$ on C, explain why $\Delta_C \arg[1 + g/f] = 0$. *Hint:* Let $w(z) = 1 + g/f$. As z goes along C, suppose that $w(z)$ encircles the origin of the w-plane. This implies that $w(z)$ assumes a negative real value for some z. Why does this contradict our assumption $|g|/|f| < 1$ on C?

   d) Combine the results of parts (a), (b), (c) to show that $N_f = N_{f+g}$.

5. Let $h(z) = a_n z^n + a_{n-1} z^{n-1} + \cdots + a_0 z^0$ be a polynomial of degree n. We will prove that $h(z)$ has exactly n zeros (counted according to multiplicities) in the z-plane. This is called the *Fundamental Theorem of Algebra*.

   a) Let

   $$f(z) = a_n z^n,$$

   $$g(z) = a_{n-1} z^{n-1} + a_{n-2} z^{n-2} + \cdots + a_1 z + a_0 z^0.$$

   Note that $h = f + g$. Consider a circle C of radius $r > 1$ centered at $z = 0$. Show that on C

   $$\left| \frac{g(z)}{f(z)} \right| < \frac{|a_0| + |a_1| + \cdots + |a_{n-1}|}{|a_n| r}.$$

   How does this inequality indicate that for sufficiently large r we have $|g(z)| < |f(z)|$ on C?

   b) Use Rouché's theorem (see Exercise 4) to argue that, for C chosen with a radius as just described, the number of zeros of $h(z) = f(z) + g(z)$ inside C is identical to the number of zeros of $f(z)$ inside C. How many zeros (counting multiplicities) does $f(z)$ have?

6. The kinds of systems discussed in this and the previous section can be represented schematically as shown in Fig. 7.3–9. Here, $G(s)$ is the transfer function of the system, $x(t)$ and $y(t)$ the input and output, $X(s)$ and $Y(s)$ their Laplace transforms. Note that $G(s) = Y(s)/X(s)$.

   A more complicated system employs the principle of *feedback*. Such systems are often used to control a physical process requiring continuous monitoring and adjustment, for example, the regulation of a furnace so as to maintain a house within a comfortable range of temperature. A block diagram of a feedback system is shown in Fig. 7.3–10. We see that an additional path, called a *feedback path* or *loop*, has been added to the original system of Fig. 7.3–9. The original system function $G(s)$ is now here called the *forward transfer function*. The input to the total system is $x_i(t)$ and the output is $y(t)$. The output $y(t)$ is monitored and sent down the feedback path into the system whose transfer function is $H(s)$. The output $y_f(t)$ of this subsystem is called the *feedback signal*. This feedback signal is fed into the device designated c, a comparator. The comparator provides an input signal $x(t)$ for the subsystem described by $G(s)$. Here $x(t) = x_i(t) - y_f(t)$ is the difference between the overall input signal and the feedback signal. Note that $X(s) = X_i(s) - Y_f(s)$ and $H(s) = Y_f(s)/Y(s)$ and, as before, $G(s) = Y(s)/X(s)$. The transfer function of the whole feedback system is defined as $T(s) = Y(s)/X_i(s)$.

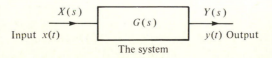

The system

Figure 7.3–9

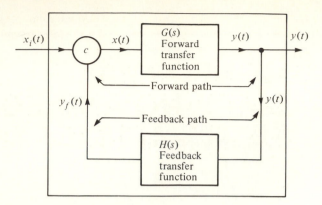

Figure 7.3–10

a) Show that

$$T(s) = \frac{G(s)}{1 + G(s)H(s)}.$$

b) Typically, both $G(s)$ and $H(s)$ are ratios of polynomials in the variable s and describe stable systems. Thus the poles of $G(s)$ and $H(s)$ lie in the left half of the s-plane. A plot of the locus of $1 + G(s)H(s)$ as s negotiates the semicircle of Fig. 7.3–5 (with $R \to \infty$) can tell us whether $T(s)$ has any poles in the plane $\operatorname{Re} s \geq 0$ and thus whether the feedback system is unstable. Explain why we can instead plot $w = G(s)H(s)$ as s negotiates the same semicircle; the feedback system is unstable if this locus encircles the point $w = -1$, in the positive sense, a total of one or more times or if this locus passes through $w = -1$. Otherwise the system is stable. This is a form of the Nyquist test for feedback systems.

c) Let

$$G(s)H(s) = \frac{-1}{(s + 1)(s + 2)(s + 3)}.$$

Determine whether the feedback system is stable or unstable by investigating encirclements of -1 as described in part (b). Note that the locus of GH as s negotiates the *arc* in Fig. 7.3–5 degenerates to a point as $R \to \infty$.

d) Repeat part (c) but take

$$GH = \frac{8}{(s)(s + 1)(s + 2)}.$$

e) Repeat part (c) but take

$$GH = \frac{-1}{(s + 1)^2(s + 1/2)}.$$

8 CONFORMAL MAPPING AND SOME OF ITS APPLICATIONS

8.1 INTRODUCTION

When we first began our discussion of functions of a complex variable in Section 2.1, we learned that a functional relationship $w = f(z)$ cannot be studied by the conventional graphing procedure of elementary algebra. Instead, two planes were used, the z-plane (with axes x and y) and the w-plane (with axes u and v). We found that $w = f(z)$ sets up a correspondence between points in the z-plane and points in the w-plane. Corresponding points in the two planes are called *images* of each other.

We can also take the view that $w = f(z)$ *maps* or *transforms* points from the z-plane into points in the w-plane. Sometimes we will superimpose the w-plane on top of the z-plane so that their axes and origins coincide. We can imagine that the vector representing a point, say A, in the z-plane has been rotated, stretched (or some combination of the two) by $w = f(z)$ in order to create the vector for A' (the image of A). A typical case is shown in Fig. 8.1–1, where we see that counterclockwise rotation *and* stretching are required to obtain the vector representing A' from that for A.

If we use $w = f(z)$ to map all the points lying in a domain D_1 of the z-plane, they may form a domain D_2 in the w-plane. If this is the case, we say that D_1 is

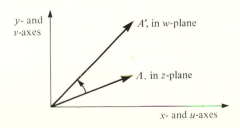

y- and
v-axes

A', in w-plane

A, in z-plane

x- and u-axes

Figure 8.1–1

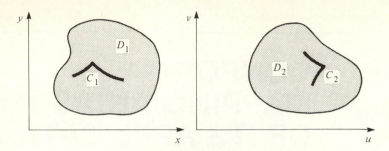

Figure 8.1–2

mapped *onto* D_2 by the transformation $w = f(z)$ (see Fig. 8.1–2). (Similarly, we can also speak of one region being mapped onto another.) If a curve C_1 is constructed in D_1 and all the points on this curve are mapped into the w-plane, we typically find that the image points form a curve C_2. Thus one curve is transformed into another. In Section 7.3 we had some experience in transforming large semicircular curves from the s-plane into the w-plane by way of $w = B(s)$.

In this chapter we will study, in some detail, how points, domains, and especially curves are mapped from the z-plane into the w-plane when the transformation $w = f(z)$ is an analytic function of z. Later, we will take a real function $\phi(x, y)$ and by a change of variables convert $\phi(x, y)$ to $\phi(u, v)$ defined in the w-plane. If $\phi(x, y)$ is a harmonic function in the z-plane, which implies

$$\frac{\partial^2 \phi}{\partial x^2} + \frac{\partial^2 \phi}{\partial y^2} = 0, \tag{8.1–1}$$

and if

$$w = u(x, y) + iv(x, y) = f(z),$$

which defines the change of variables is analytic, we can show that

$$\frac{\partial^2 \phi}{\partial u^2} + \frac{\partial^2 \phi}{\partial v^2} = 0, \tag{8.1–2}$$

and $\phi(u, v)$ is harmonic in the w-plane. We will find that the preservation of the harmonic property when ϕ is "transferred" from one plane to another, together with a knowledge of how contours are transformed from one plane to another by $w = f(z)$ will enable us to solve a greater variety of physical problems in electrostatics, fluid flow, etc. than those treated in Section 4.5.

8.2   THE CONFORMAL PROPERTY

To see how curves can be transformed by an analytic function, let us consider the specific case

$$w = \text{Log } z \tag{8.2–1}$$

applied to the arc $|z| = 1$, $\pi/6 \le \arg z \le \pi/4$ and also to the line segment $\arg z = \pi/6$, $1 \le |z| \le 2$. Both the arc and the line are shown in Fig. 8.2–1(a). Each point

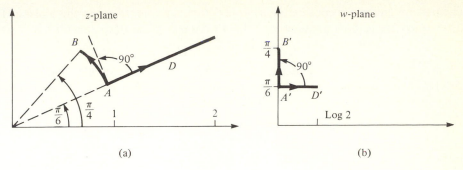

Figure 8.2–1

on the arc is described by $e^{i\theta}$, where $\pi/6 \leq \theta \leq \pi/4$ and the corresponding image point is Log $e^{i\theta} = i\theta$. As θ advances from $\pi/6$ to $\pi/4$ along the circle, w traces out the vertical line segment $A'B'$ shown in Fig. 8.2–1(b).

Under the transformation in Eq. (8.2–1) each point on the line arg $z = \pi/6, 1 \leq |z| \leq 2$ has an image

$$w = \text{Log}|z| + i \arg z = \text{Log}|z| + i\frac{\pi}{6}.$$

As z advances from 1 to 2, the locus of w is the horizontal line $A'D'$ shown in Fig. 8.2–1(b). The two original curves in Fig. 8.2–1(a) intersect at point A with a 90° angle (that is, their tangents, at A, have this angle of intersection). In Fig. 8.2–1(b) the image curves intersect at A' with a 90° angle. Moreover, the *sense* of the angle of intersection is preserved, that is, the tangents to the two curves at their intersection in Fig. 8.2–1(a) have an angular displacement from each other that is in the same direction as the corresponding tangents in Fig. 8.2–1(b). The preservation of both the magnitude and sense of the angle of intersection of these curves under this transformation is not an accident and will occur extensively throughout this chapter. The following definition will be useful in our discussion.

Definition   Conformal mapping

A mapping $w = f(z)$ that preserves the size and sense of the angle of intersection between any two curves intersecting at z_0 is said to be *conformal* at z_0. A mapping that is conformal at every point in a domain D is called conformal in D. ∎

Occasionally one speaks of *isogonal mappings*. In this case the magnitudes of angles of intersection are preserved but not necessarily their sense.

In a moment we will be able to show why $w = \text{Log } z$ is conformal at the point A and also decide when functions $f(z)$ are conformal in general. The following theorem will be proved and used:

Theorem 1   Condition for conformal mapping

Let $f(z)$ be analytic in a domain D. Then $f(z)$ is conformal at every point in D where $f'(z) \neq 0$. ∎

The proof requires our considering a curve C that is a smooth arc in the z-plane. The curve is generated by a parameter t, which we might think of as time. Thus

$$z(t) = x(t) + iy(t)$$

traces out the curve C as t increases (see Fig. 8.2–2a). We assume $x(t)$ and $y(t)$ to be differentiable functions of t. The curve C can be transformed into an image curve C' (see Fig. 8.2–2b) by means of the analytic function

$$w = f(z) = u(x, y) + iv(x, y).$$

The arrows on C and C' indicate the sense in which these contours are generated as t increases. At any point on C or C' we can define a directed tangent. This is a vector that is tangent to the curve and points in the direction in which the curve is being generated.

At time t_0 we are at $z(t_0) = z_0$ on C, and at the later time $t_0 + \Delta t$ we are at $z(t_0 + \Delta t) = z_0 + \Delta z$. The vector Δz connecting $z(t_0)$ with $z(t_0 + \Delta t)$ is shown in Fig. 8.2–2(a).

Now refer to Fig. 8.2–2(b). The point z_0 is mapped into the image $w_0 = f(z_0)$ on C' and $z_0 + \Delta z$ has the image point $f(z_0 + \Delta z) = w_0 + \Delta w$ on C'. If $\Delta t \to 0$, then Δz, and consequently Δw, both shrink to zero. In Fig. 8.2–2(a) we see that, as the vector Δz shortens, its direction approaches that of the directed tangent to the curve C at z_0. Similarly, as the vector Δw shortens, its direction approaches the tangent to the image curve C' at w_0. Since Δt is real, both the vectors $\Delta z/\Delta t$ and $\Delta w/\Delta t$ have the same direction as the vectors Δz and Δw, respectively. Thus $\lim_{\Delta t \to 0} \Delta z/\Delta t = dz/dt$ and $\lim_{\Delta t \to 0} \Delta w/\Delta t = dw/dt$ are tangent to C and C' at z_0 and w_0, respectively. Note that

$$\left.\frac{dz}{dt}\right|_{z_0} = \left.\frac{dx}{dt}\right|_{z_0} + i \left.\frac{dy}{dt}\right|_{z_0}$$

and that the slope of this vector is $(dy/dx)|_{z_0}$, the slope of the curve C at z_0. Similarly $(dw/dt)|_{w_0}$ is the slope of the curve C' at w_0.

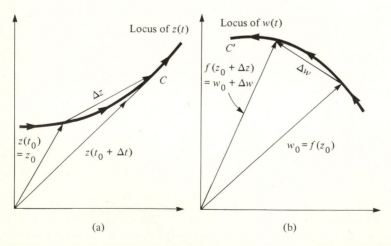

(a)   (b)

Figure 8.2–2

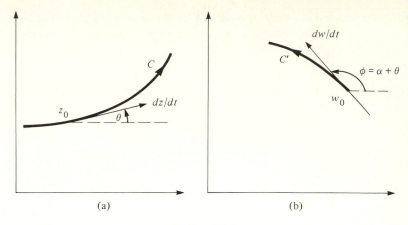

Figure 8.2–3

From the chain rule for differentiation

$$\frac{dw}{dt} = \frac{dw}{dz}\frac{dz}{dt} = f'(z)\frac{dz}{dt}.$$

Setting $t = t_0$ so that $z = z_0$ and $w = w_0$ in the preceding we have

$$\left.\frac{dw}{dt}\right|_{w_0} = f'(z_0)\left.\frac{dz}{dt}\right|_{z_0}.$$

Equating the arguments of each side of the above we obtain

$$\arg\left.\frac{dw}{dt}\right|_{w_0} = \arg f'(z_0) + \arg\left.\frac{dz}{dt}\right|_{z_0}. \tag{8.2–2}$$

Let

$$\phi = \arg\left.\frac{dw}{dt}\right|_{w_0}, \qquad \alpha = \arg f'(z_0), \qquad \theta = \arg\left.\frac{dz}{dt}\right|_{z_0}.$$

Thus Eq. (8.2–2) becomes

$$\phi = \alpha + \theta. \tag{8.2–3}$$

We should recall that θ and ϕ specify the directions of the tangents to the curves C and C' at z_0 and w_0, respectively (see Fig. 8.2–3). Using Eq. (8.2–3), we realize that under the mapping $w = f(z)$ the directed tangent to the curve C, at z_0, is rotated through an angle $\alpha = \arg f'(z_0)$. The rotation of the tangent is shown in Fig. 8.2–3(b).

Another smooth arc, say C_1, intersecting C at the point z_0 with angle ψ (the angle between the tangents to the curves) can be mapped by $w = f(z)$ into the image curve C_1'. The tangent to C_1 at z_0 is also rotated through the angle $f'(z_0) = \alpha$ by the mapping.

The mapping $w = f(z)$ rotates the tangents to C and C_1 by identical amounts in the same direction. Thus the image curves C' and C_1' have the same angle of intersection ψ as do C and C_1. The sense (direction) of the intersection is also preserved, as shown in Fig. 8.2–4.

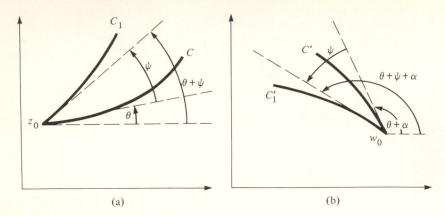

Figure 8.2–4

If $f'(z_0) = 0$, the preceding discussion will break down since the angle $\alpha = \arg(f'(z_0))$, through which tangents are rotated, is undefined. There is no guarantee of a conformal mapping where $f'(z_0) = 0$. One can show that if $f'(z_0) = 0$ the mapping cannot be conformal at z_0. A value of z for which $f'(z) = 0$ is known as a *critical point* of the transformation.

Example 1

Consider the contour C defined by $x = y$, $x \geq 0$ and the contour C_1 defined by $x = 1$, $y \geq 1$. Map these two curves using $w = 1/z$ and verify that their angle of intersection is preserved in size and direction.

Solution

Our transformation is

$$w = \frac{1}{z} = u + iv = \frac{1}{x + iy} = \frac{x}{x^2 + y^2} - \frac{iy}{x^2 + y^2},$$

so that

$$u = \frac{x}{x^2 + y^2}, \tag{8.2–4}$$

$$v = \frac{-y}{x^2 + y^2}. \tag{8.2–5}$$

On C, $y = x$, which, when substituted in Eqs. (8.2–4) and (8.2–5), yields

$$u = \frac{1}{2x} = -v. \tag{8.2–6}$$

Since $x \geq 0$, we have $u \geq 0$ and $v \leq 0$. The line defined by Eq. (8.2–6) is shown as C' in Fig. 8.2–5(b). As we move outward from the origin along C, the corresponding image point moves toward the origin on C' since, according to Eq. (8.2–6), both u and v tend to zero with increasing x.

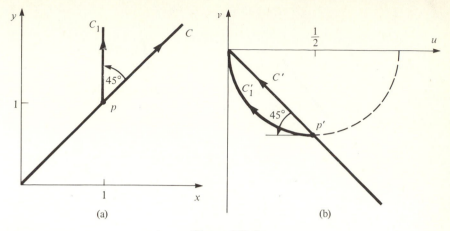

Figure 8.2–5

On C_1, $x = 1$ which, when used in Eqs. (8.2–4) and (8.2–5), yields

$$u = \frac{1}{1 + y^2}, \qquad (8.2\text{–}7)$$

$$v = -\frac{y}{1 + y^2}. \qquad (8.2\text{–}8)$$

This implies that

$$v = -uy. \qquad (8.2\text{–}9)$$

From Eq. (8.2–7) we easily obtain $y = \sqrt{1/u - 1}$ which, combined with Eq. (8.2–9), yields $v = -\sqrt{u - u^2}$. We can square both sides of this equation and make some algebraic rearrangements to show that $(u - 1/2)^2 + v^2 = (1/2)^2$.

Thus points on C_1 have their images on a circle of radius $1/2$, centered at $(1/2, 0)$ in the w-plane. As y increases from 1 to ∞ along C_1, then, according to Eq. (8.2–7), the u-coordinate of the image point varies from $1/2$ to 0 along the circle. Since v remains negative (see Eq. 8.2–8), the image of C_1 is the arc C_1' shown in Fig. 8.2–5(b).

From plane geometry we recall that the angle between a tangent and a chord of a circle is $1/2$ the angle of the intercepted arc. Thus the angle of intersection between C_1' and C' in Fig. 8.2–5(b) is $45°$, the same angle existing between C_1 and C. Observe in Fig. 8.2–5(a) and (b) that the sense of the angular displacement between the tangents to C and C_1 is the same as for C' and C_1'. ◀

Suppose a small line segment, not necessarily straight, connecting the points z_0 and $z_0 + \Delta z$ is mapped by means of the analytic transformation $w = f(z)$ (see Fig. 8.2–6). The image line segment connects the point $w_0 = f(z_0)$ with the point $w = f(z_0 + \Delta z)$.

Now consider

$$|f'(z_0)| = \lim_{\Delta z \to 0} \left| \frac{f(z_0 + \Delta z) - f(z_0)}{\Delta z} \right|. \qquad (8.2\text{–}10)$$

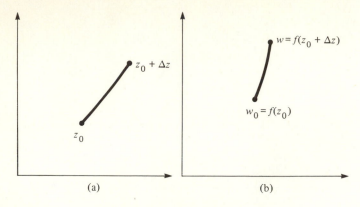

(a)   (b)

Figure 8.2–6

Equation (8.2–10) follows from the definition of the derivative, Eq. (2.3–3), and this easily proved fact: $|\lim_{z \to z_0} g(z)| = \lim_{z \to z_0} |g(z)|$ when $\lim_{z \to z_0} g(z)$ exists. The expression $|(f(z_0 + \Delta z) - f(z))/\Delta z|$ is the approximate ratio of the lengths of the line segments in Fig. 8.2–6(b) and (a). Thus a small line segment starting at z_0 is magnified in length by approximately $|f'(z_0)|$ under the transformation $w = f(z)$. As the length of this segment approaches zero, the amount of magnification tends to the limit $|f'(z_0)|$.

We see that if $f'(z_0) \neq 0$, all small line segments passing through z_0 are approximately magnified under the mapping by the same nonzero factor $R = |f'(z_0)|$. A "small" figure composed of line segments and constructed near z_0 will, when mapped into the w-plane, have each of its sides approximately magnified by the same factor $|f'(z_0)|$. The shape of the new figure will conform to the shape of the old one although its size and orientation will typically have been altered. Because of the magnification in lengths, the image figure in the w-plane will have an area approximately $|f'(z_0)|^2$ times as large as that of the original figure. The conformal mapping of a small figure is shown in Fig. 8.2–7. The similarity in shapes and the magnification of areas need not hold if we map a "large" figure since $f'(z)$ may deviate significantly from $f'(z_0)$ over the figure.

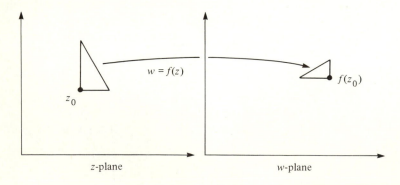

z-plane   w-plane

Figure 8.2–7

Example 2

Discuss the way in which $w = z^2$ maps the grid $x = x_1, x = x_2, \ldots; y = y_1, y = y_2, \ldots$ (see Fig. 8.2–8a) into the w-plane. Verify that the angles of intersection are preserved and that a small square is approximately preserved in shape.

Solution

With $w = u + iv$, $z = x + iy$, the transformation is $u + iv = (x + iy)^2 = x^2 - y^2 + i2xy$ so that

$$u = x^2 - y^2, \tag{8.2–11}$$

$$v = 2xy. \tag{8.2–12}$$

On the line $x = x_1$, $-\infty \leq y \leq \infty$ we have

$$u = x_1^2 - y^2, \tag{8.2–13}$$

$$v = 2x_1 y. \tag{8.2–14}$$

We can use Eq. (8.12–14) to eliminate y from Eq. (8.2–13) with the result that

$$u = x_1^2 - \frac{v^2}{4x_1^2}. \tag{8.2–15}$$

As the y-coordinate of a point on $x = x_1$ increases from $-\infty$ to ∞, Eq. (8.2–14) indicates that v progresses from $-\infty$ to ∞ (if $x_1 > 0$). A parabola described by Eq. (8.2–15) is generated. This curve, which passes through $u = x_1^2$, $v = 0$, is shown by the solid line in Fig. 8.2–8. This parabola is the image of $x = x_1$. Also illustrated is the image of $x = x_2$, where $x_2 > x_1$.

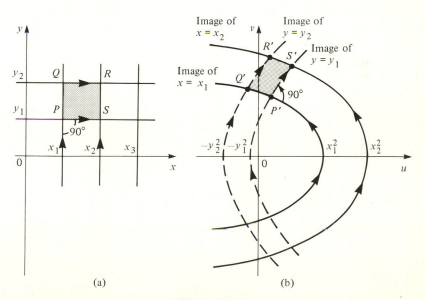

(a)   (b)

Figure 8.2–8

Mapping a horizontal line $y = y_1$, $-\infty \le x \le \infty$, we have from Eqs. (8.2–11) and (8.2–12) that

$$u = x^2 - y_1^2, \tag{8.2–16}$$

$$v = 2xy_1. \tag{8.2–17}$$

Using Eq. (8.2–17) to eliminate x from Eq. (8.2–16), we have

$$u = \frac{v^2}{4y_1^2} - y_1^2. \tag{8.2–18}$$

This is also the equation of a parabola—one opening to the right. One can easily show that, as the x-coordinate of a point moving along $y = y_1$ increases from $-\infty$ to ∞, its image traces out a parabola shown by the broken line in Fig. 8.2–8(b). The direction of progress is indicated by the arrow. Also shown in Fig. 8.2–8(b) is the image of the line $y = y_2$. The point P at (x_1, y_1) is mapped by $w = z^2$ into the image $u_1 = x_1^2 - y_1^2$, $v_1 = 2x_1y_1$ shown as P' in Fig. 8.2–8(b). P' lies at the intersection of the images of $x = x_1$ and $y = y_1$. Although these curves have two intersections, only the upper one corresponds to P' since Eq. (8.2–12) indicates that $v > 0$ when $x > 0$ and $y > 0$.

The slope of the image of $x = x_1$ is found from Eq. (8.2–15). Differentiating implicitly, we have

$$du = -\frac{2v\,dv}{4x_1^2},$$

or

$$\frac{du}{dv} = \frac{-v}{2x_1^2}. \tag{8.2–19}$$

Similarly, from Eq. (8.2–18), the slope of the image of $y = y_1$ is

$$\frac{du}{dv} = \frac{v}{2y_1^2}. \tag{8.2–20}$$

Substituting $v_1 = 2x_1y_1$, which is valid at the point of intersection, into Eqs. (8.2–19) and (8.2–20), we find that the respective slopes are $-y_1/x_1$ and x_1/y_1. As these values are negative reciprocals of each other, we have established the orthogonality of the intersection of the two parabolas at P'. Since $x = x_1$ intersects $y = y_1$ at a right angle, the transformation has preserved the angle of intersection. Notice that the rectangular region with corners at P, Q, R, and S shown shaded in Fig. 8.2–8(a) is mapped onto the nearly rectangular region having corners at P', Q', R', and S' shown shaded in Fig. 8.2–8(b).

With $f(z) = z^2$, we have $f'(z) = 0$ at $z = 0$. Our theorem on conformal mapping no longer guarantees a conformal transformation at $z = 0$. Lines intersecting here require special attention. The vertical line $x = 0$, $-\infty < y < \infty$ is transformed (see Eqs. 8.2–11 and 8.2–12) into $u = -y^2$, $v = 0$, the negative real axis. The horizontal line $y = 0$, $-\infty < x < \infty$ is, by the same equations, mapped into $u = x^2$, $v = 0$, the positive u-axis. The lines $x = 0$ and $y = 0$, which intersect at the origin at $90°$, have images in the uv-plane intersecting at $180°$ (see Fig. 8.2–9b). Notice that the small

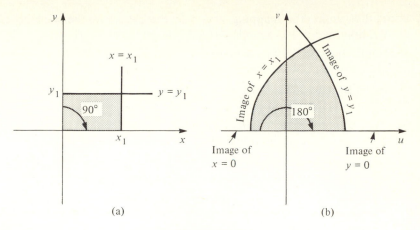

Figure 8.2–9

rectangle $0 \le x \le x_1$, $0 \le y \le y_1$ in Fig. 8.2–9(a) is mapped onto the nonrectangular shape in Fig. 8.2–9(b). The breakdown of the conformal property is again evident. ◀

EXERCISES

1. We can readily show that $w = z^2$ establishes a conformal mapping for all $z \ne 0$. Is $w = (\bar{z})^2$ conformal for $z \ne 0$? If not, why isn't it?

2. Two semiinfinite lines (see Fig. 8.2–10) $y = ax$, $x \ge 0$ and $y = bx$, $x \ge 0$ are mapped by the transformation $w = u + iv = z^2$. Find the equation of each image curve in the form $v = g(u)$. If the two given lines intersect at angle α as shown, what is the angle of intersection of their images?

3. What are the critical points of the following transformations?

   a) $w = e^{z^2}$ b) $w = \text{Log } z - z$

4. a) What is the image of the semicircular arc $|z| = 1$, $0 \le \arg z \le \pi$ under the transformation $w = z + 1/z$? *Hint:* Put $z = e^{i\theta}$.

   b) What is the image of the line $y = 0$, $x \ge 1$ under this same transformation?

   c) Do the image curves found in parts (a) and (b) have the same angle of intersection in the w-plane as do the original curves in the xy-plane? Explain.

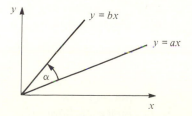

Figure 8.2–10

5. a) Show that under the mapping $w = 1/z$ the image in the w-plane of the infinite line Im $z = 1$ is a circle. What is its center and radius?

   b) Find the equation in the w-plane of the image of $x + y = 1$ under the mapping $w = 1/z$. What kind of curve is obtained?

6. Consider the straight line segment directed from $(2, 2)$ to $(2.1, 2.1)$ in the z-plane. The segment is mapped into the w-plane by $w = \text{Log } z$.

   a) Obtain the approximate length of the image of this segment in the w-plane by using the derivative of the transformation at $(2, 2)$ (see Eq. 8.2–10).

   b) Obtain the exact value of the length of this image. Use a numerical table or pocket calculator to convert this to a decimal, and compare your result with part (a).

   c) Use the derivative of the transformation to find the angle through which the given segment is rotated when mapped into the w-plane.

7. The square boundary of the region $1 \le x \le 1.1$, $1 \le y \le 1.1$ is transformed by means of $w = e^z$.

   a) Use the derivative of the transformation at $(1, 1)$ to obtain a numerical approximation to the area of the image of the square in the w-plane.

   b) Obtain the exact value of the area of the image, and compare your result with part (a).

8.3   ONE-TO-ONE MAPPINGS AND MAPPINGS OF REGIONS

It is now necessary to study with some care the correspondence that the analytic transformation $w = f(z)$ creates between points in the z-plane and points in the w-plane. Let all the points in a region R be mapped into the w-plane so as to form an image region R'. Let z_1 be any point in R. Since $f(z)$ is single valued in R, z_1 is mapped into a unique point $w_1 = f(z_1)$. Given the point w_1, can we assert that it is the image of a unique point, that is, if $w_1 = f(z_1)$ and $w_1 = f(z_2)$, where z_1 and z_2 are points in R, does it follow that $z_1 = z_2$? The following definition is useful in dealing with this question.

Definition   One-to-one mapping

If the equation $f(z_1) = f(z_2)$ implies for any points z_1 and z_2 in a region R of the z-plane that $z_1 = z_2$, we say that the mapping of the region R provided by $w = f(z)$ is *one to one*.

A one-to-one correspondence is said to exist between the points in R and the points in an image region R' of the w-plane, that is, R is mapped one to one onto R'.   ∎

A hypothetical mapping that fails to be one to one is shown in Fig. 8.3–1. Some specific, obvious cases of failure are not hard to find. For example, let R be the disc $|z| \le 2$, and let $w = z^2$. Without considering how the entire disc is mapped into the w-plane, observe that $w = 1$ is the image of both $z = -1$ and $z = +1$, that is, $1 = z^2$ implies $z = \pm 1$. Clearly $w = z^2$ cannot map the region R one to one.

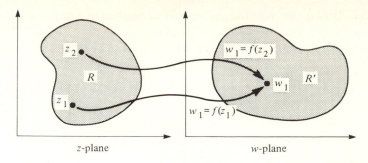

Figure 8.3–1

The failure of $w = z^2$ to establish a one-to-one mapping for R is easily demonstrated if we solve this equation for z and obtain $z = w^{1/2}$. Given w, we see that two values of z are possible whose arguments are 180° apart. Since the given R contains numbers whose arguments differ by 180° a one-to-one transformation is not possible. However, by using an R in which this condition cannot occur, a one-to-one mapping can be obtained. (Example 1 will provide further discussion.)

A solution for the inverse mapping, $z = g(w)$, as in the previous paragraph, allows us to decide if a mapping is one to one. The analytic transformation $w = iz + 2$, for example, can be solved to yield $z = (w - 2)/i$. A point w_1 has a unique inverse point $z_1 = (w_1 - 2)/i$. Thus $w = iz + 2$ can map any region of the z-plane one to one onto a region in the w-plane.

The transformation $w = u + iv = f(z)$ can be regarded as a pair of equations $u = u(x, y)$ and $v = v(x, y)$. Thus x_0, y_0 is mapped into (u_0, v_0), where $u_0 = u(x_0, y_0)$ and $v_0 = v(x_0, y_0)$. In texts in advanced calculus it is shown that if the Jacobian of the mapping, given by the determinant

$$\begin{vmatrix} \dfrac{\partial u}{\partial x} & \dfrac{\partial u}{\partial y} \\[2ex] \dfrac{\partial v}{\partial x} & \dfrac{\partial v}{\partial y} \end{vmatrix}$$

is not zero at (x_0, y_0), then $w = f(z)$ yields a one-to-one mapping of a neighborhood of (x_0, y_0) onto a corresponding neighborhood of (u_0, v_0).

Expanding the above determinant, we have the requirement

$$\left. \frac{\partial u}{\partial x} \frac{\partial v}{\partial y} - \frac{\partial u}{\partial y} \frac{\partial v}{\partial x} \right|_{x_0, y_0} \neq 0 \tag{8.3–1}$$

for a one-to-one mapping. Using the Cauchy–Riemann equations $\partial v / \partial y = \partial u / \partial x$ and $-\partial u / \partial y = \partial v / \partial x$, we can rewrite Eq. (8.3–1) as

$$\left[\left(\frac{\partial u}{\partial x} \right)^2 + \left(\frac{\partial v}{\partial x} \right)^2 \right]_{x_0, y_0} \neq 0. \tag{8.3–2}$$

From Eq. (2.3–6) we observe that the left side of Eq. (8.3–2) is $|f'(z_0)|^2$, where

$z_0 = x_0 + iy_0$. Hence, the requirement for one to oneness in a neighborhood of z_0 is $f'(z_0) \neq 0$.

The preceding is summarized in Theorem 2.

Theorem 2   One-to-one mapping

Let $f(z)$ be analytic at z_0 and $f'(z_0) \neq 0$. Then $w = f(z)$ provides a one-to-one mapping of a neighborhood of z_0.   ■

It can also be shown that if $f'(z) = 0$ in any point of a domain, then $f(z)$ cannot give a one-to-one mapping of that domain.

A corollary to Theorem 2 asserts that if $f'(z_0) \neq 0$, then $w = f(z)$ can be solved for an inverse $z = g(w)$ that is single valued in a neighborhood of $w_0 = f(z_0)$.

One must employ Theorem 2 with some amount of caution since it deals only with the *local* properties of the transformation $w = f(z)$. If we consider the interior of a *sufficiently small* circle centered at z_0, the theorem can guarantee a one-to-one mapping of the interior of this circle. However, if we make the circle too large, the mapping can fail to be one to one even though $f'(z) \neq 0$ throughout the circle.

Example 1

Discuss the possibility of obtaining a one-to-one mapping from the transformation $w = z^2$.

Solution

We make a switch to polar coordinates and take $z = re^{i\theta}$, $w = \rho e^{i\phi}$. Substituting these into the given transformation, we find that $\rho = r^2$ and $\phi = 2\theta$. We observe that the wedge-shaped region in the z-plane bounded by the rays $\theta = \alpha$, $\theta = \beta$, $r \geq 0$, (where $0 \leq \alpha < \beta$) shown in Fig. 8.3–2(a) is mapped onto the wedge bounded by the rays $\phi = 2\alpha$, $\phi = 2\beta$, $\rho \geq 0$ shown in Fig. 8.3–2(b).

The wedge bounded by the rays $\theta = \alpha + \pi$, $\theta = \beta + \pi$, $r \geq 0$ shown in Fig. 8.3–2(a) is mapped onto the wedge bounded by the rays $\phi = 2(\alpha + \pi) = 2\alpha$ and

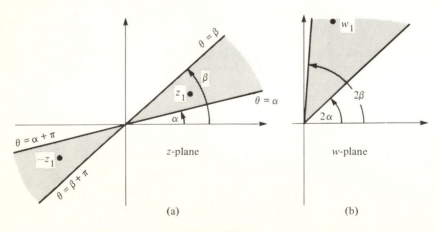

(a)   (b)

Figure 8.3–2

$\phi = 2(\beta + \pi) = 2\beta$, $\rho \geq 0$ shown in Fig. 8.3–2(b). Thus both wedges in Fig. 8.3–2(a) are mapped onto the identical wedge Fig. 8.3–2(b).

The inverse of our transformation is $z = w^{1/2}$. Applying this to w_1 shown in Fig. 8.3–2(b), we obtain $w_1^{1/2}$ whose values z_1 and $-z_1$ lie in the upper and lower wedges of Fig. 8.3–2(a).

Either of the wedges in Fig. 8.3–2(a) can be mapped one to one since $w^{1/2}$ has only one value in each wedge. Similarly, any domain in either wedge in Fig. 8.3–2(a) can be mapped one to one onto a domain in the w-plane. Notice that any domain containing $z = 0$ must necessarily contain points from both wedges in Fig. 8.3–2(a) and cannot be used for a one-to-one mapping. However, we know that such a domain must be avoided since it contains the solution of $f'(z) = 2z = 0$.

The region $0 \leq \theta < \pi, r > 0$, which is the upper half of the z-plane plus the axis $y = 0$, $x \geq 0$, can be mapped one to one since it contains no two points that are negatives of each other (observe the necessity for excluding the negative real axis). The image of this region is the entire w-plane (see Fig. 8.3–3).

An alternative solution to this example, not using polar coordinates, is given in Exercise 1 of this section. ◄

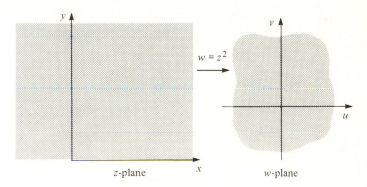

Figure 8.3–3

Example 2

Discuss the way in which the infinite strip $0 \leq \text{Im } z \leq a$, is mapped by the transformation

$$w = u + iv = e^z = e^{x+iy}. \tag{8.3–3}$$

Take $0 \leq a < 2\pi$.

Solution

We first note the desirability of taking $0 \leq a < 2\pi$. It arises from the periodic property $e^z = e^{z+2\pi i}$. By making the width of the strip (see Fig. 8.3–4a) less than 2π, we avoid having two points inside with identical real parts and imaginary parts that differ by 2π. A pair of such points are mapped into identical locations in the w-plane and a one-to-one mapping of the strip becomes impossible.

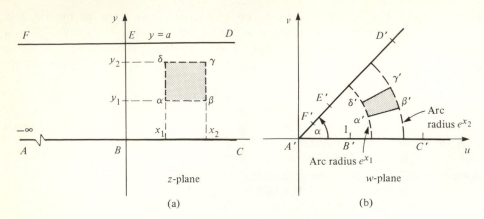

Figure 8.3–4

The bottom boundary of the strip, $y = 0$, $-\infty < x < \infty$ is mapped by our setting $y = 0$ in Eq. (8.3–3) to yield $e^x = u + iv$. As x ranges from $-\infty$ to ∞, the entire line $v = 0$, $0 < u < \infty$ is generated. This line is shown in Fig. 8.3–4(b). The points A', B', and C' are the images of A, B, and C in Fig. 8.3–4(a).

The upper boundary of the strip is mapped by our putting $y = a$ in Eq. (8.3–3) so that

$$u = e^x \cos a, \tag{8.3–4}$$

$$v = e^x \sin a. \tag{8.3–5}$$

Dividing the second equation by the first, we have $v/u = \tan a$ or

$$v = u \tan a, \tag{8.3–6}$$

which is the equation of a straight line through the origin in the uv-plane. If $\sin a$ and $\cos a$ are both positive ($0 < a < \pi/2$), we see from Eqs. (8.3–4) and (8.3–5) that, as x ranges from $-\infty$ to ∞, only that portion of the line lying in the first quadrant of the w-plane is generated. Such a line is shown in Fig. 8.3–4(b). It is labeled with the points D', E', and F', which are the images of D, E, and F in Fig. 8.3–4(a). The slope of the line is $\tan a$, and it makes an angle a with the real axis. If a satisfied the condition $\pi/2 < a < \pi$ or $\pi < a < 3\pi/2$ or $3\pi/2 < a < 2\pi$, lines lying in respectively the 2nd or 3rd or 4th quadrant would have been obtained. The cases $\pi/2 = a$, $3\pi/2 = a$, and $\pi = a$ yield lines along the coordinate axes.

The strip in Fig. 8.3–4(a) is mapped onto the wedge-shaped region shown in Fig. 8.3–4(b). An important mapping occurs if the strip is chosen to have width $a = \pi$. The upper boundary passing through F, E, D in Fig. 8.3–4(a) is transformed into the negative real axis in the w-plane. The wedge shown in Fig. 8.3–4(b) evolves into the half plane $v \geq 0$, which is now the image of the strip.

The inverse transformation of Eq. (8.3–3), that is,

$$z = \log w \tag{8.3–7}$$

can be used to obtain the image in the z-plane of any point in the wedge of Fig. 8.3–4(b). Of course $\log w$ is multivalued, but there is only one value of $\log w$ that lies in the strip of Fig. 8.3–4(a). The shaded rectangular area bounded by the lines $x = x_1$, $x = x_2$, $y = y_1$, $y = y_2$ shown in Fig. 8.3–4(a) is readily mapped onto a region in the w-plane. With $x = x_1$ we have from Eq. (8.3–3) that

$$u = e^{x_1} \cos y,$$

$$v = e^{x_1} \sin y,$$

so that

$$u^2 + v^2 = e^{2x_1},$$

which is the equation of a circle of radius e^{x_1}. The line segment $x = x_1$, $0 \le y \le a$, is transformed into an arc lying on this circle and illustrated in Fig. 8.3–4(b). The line segment $x = x_2$, $0 \le y \le a (x_2 > x_1)$ is transformed into an arc of larger radius, which is also shown.

The images of the lines $y = y_1$ and $y = y_2$ are readily found from Eqs. (8.3–4) and (8.3–5) if we replace a by y_1 or y_2. Rays are obtained with slopes $\tan y_1$ and $\tan y_2$, respectively. These rays (see Fig. 8.3–4b) together with the arcs of radius e^{x_1} and e^{x_2} form the boundary of a nonrectangular shape (shaded in Fig. 8.3–4b) that is the image of the rectangle shown in Fig. 8.3–4(a). Notice that the corners of the nonrectangular shape have right angles as in the original rectangle. ◀

Example 3

Discuss the way in which $w = \sin z$ maps the strip $y \ge 0$, $-\pi/2 \le x \le \pi/2$.

Solution

Because $\sin z$ is periodic, that is, $\sin z = \sin(z + 2\pi)$, any two points in the z-plane having identical imaginary parts and real parts differing by 2π (or its multiple) will be mapped into identical locations in the w-plane. This situation cannot occur for points in the given strip (see Fig. 8.3–5a) because its width is π.

Rewriting the given transformation using Eq. (3.2–9), we have

$$w = (u + iv) = \sin(x + iy) = \sin x \cosh y + i \cos x \sinh y,$$

which means

$$u = \sin x \cosh y, \tag{8.3–8}$$

$$v = \cos x \sinh y. \tag{8.3–9}$$

The bottom boundary of the strip is $y = 0$, $-\pi/2 \le x \le \pi/2$. Here $u = \sin x$ and $v = 0$. As we move from $x = -\pi/2$ to $x = \pi/2$ along this bottom boundary, the image point in the w-plane advances from -1 to $+1$ along the line $v = 0$. The image of the line segment B, C, D of Fig. 8.3–5(a) is the line B', C', D' in Fig. 8.3–5(b).

Along the left boundary of the given strip $x = -\pi/2$, $y \ge 0$. From Eqs. (8.3–8) and (8.3–9) we have

$$u = \sin\left(-\frac{\pi}{2}\right) \cosh y = -\cosh y,$$

$$v = \cos\left(-\frac{\pi}{2}\right) \sinh y = 0.$$

As we move from $y = \infty$ to $y = 0$ along the left boundary, these equations indicate

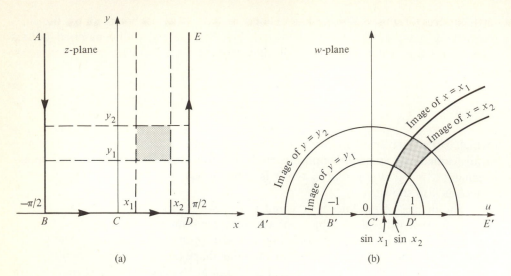

(a)   (b)

Figure 8.3–5

that the u-coordinate of the image goes from $-\infty$ to -1 along $v = 0$. The image of this boundary is thus that portion of the u-axis lying to the left of B' in Fig. 8.3–5(b). Similarly, the image of the right boundary of the strip, $x = \pi/2, 0 \le y \le \infty$, is that portion of the u-axis lying to the right of D' in Fig. 8.3–5(b).

The image of the semiinfinite vertical line $x = x_1$, $0 \le y \le \infty$ is found from Eqs. (8.3–8) and (8.3–9). We have

$$u = \sin x_1 \cosh y, \tag{8.3–10}$$

$$v = \cos x_1 \sinh y. \tag{8.3–11}$$

Recalling that $\cosh^2 y - \sinh^2 y = 1$, we find that

$$\frac{u^2}{\sin^2 x_1} - \frac{v^2}{\cos^2 x_1} = 1,$$

which is the equation of a hyperbola. We will assume that $0 < x_1 < \pi/2$. Because $y \ge 0$, Eqs. (8.3–10) and (8.3–11) reveal that only that portion of the hyperbola lying in the first quadrant of the w-plane is obtained by this mapping. This curve is shown in Fig. 8.3–5(b); also indicated is the image of $x = x_2, y \ge 0$, where $x_2 > x_1$. If x_1 or x_2 had been negative, the portions of the hyperbolas obtained would be in the second quadrant of the w-plane.

The horizontal line segment $y = y_1(y_1 > 0)$, $-\pi/2 \le x \le \pi/2$ in Fig. 8.3–5(a) can be mapped into the w-plane with the aid of Eqs. (8.3–8) and (8.3–9), which yield

$$u = \sin x \cosh y_1, \tag{8.3–12}$$

$$v = \cos x \sinh y_1. \tag{8.3–13}$$

Since $\sin^2 x + \cos^2 x = 1$, we have

$$\frac{u^2}{\cosh^2 y_1} + \frac{v^2}{\sinh^2 y_1} = 1,$$

which describes an ellipse. Because $y_1 > 0$ and $-\pi/2 \le x \le \pi/2$, Eq. (8.3–13) indicates that $v \ge 0$, that is, only the upper half of the ellipse is the image of the given segment. In Fig. 8.3–5(b) we have shown elliptic arcs that are the images of the two horizontal line segments inside the strip in Fig. 8.3–5(a).

The rectangular area $x_1 \le x \le x_2$; $y_1 \le y \le y_2$, in the z-plane is mapped onto the four-sided figure bounded by two ellipses and two hyperbolas, which we see shaded in Fig. 8.3–5(b). The four corners of this figure have right angles.

It should be evident that the interior of our semiinfinite strip, in the z-plane, is mapped by $w = \sin z$ onto the upper half of the w-plane. The transformation of other strips is considered in Exercise 2 of this section.

The transformation $w = \sin z$ fails to be conformal where $d \sin z/dz = \cos z = 0$. This occurs at $z = \pm\pi/2$. The line segments AB and BC in Fig. 8.3–5(a) intersect at $z = -\pi/2$ at right angles. However, their images intersect in the w-plane at a 180° angle. The same phenomenon occurs for segments CD and DE. ◀

EXERCISES

1. a) The transformation $w = u + iv = z^2$ is applied to a certain region R in the first quadrant of the z-plane. The rectangular-shaped image region R' satisfying $0 < u_1 \le u \le u_2$, $0 < v_1 \le v \le v_2$ is obtained as the image of R. Describe R giving the equations of all boundaries. Does $w = z^2$ establish a one-to-one mapping of R?

   b) The same transformation is applied to a region R_1 in the third quadrant of the z-plane. The region R' given in part (a) is still obtained. Describe R_1. Does $w = z^2$ establish a one-to-one mapping of R_1?

2. a) Consider the infinite strip $|\text{Re } z| \le a$, where a is a constant satisfying $0 < a < \pi/2$. Find the image of this strip, under the transformation $w = \sin z$, by mapping its boundaries.

   b) Is the mapping in part (a) one to one?

   c) Suppose $a = \pi/2$. Is the mapping now one to one?

3. How does the transformation $w = \cos z$ map the following regions? Is the mapping one to one in each case?

   a) The infinite strip $a \le \text{Re } z \le b$, where $0 < a < b < \pi$

   b) The infinite strip $0 \le \text{Re } z \le \pi$

   c) The semiinfinite strip $0 \le \text{Re } z \le \pi$, $\text{Im } z \ge 0$

4. Consider the wedge-shaped region $0 \le \arg z \le \alpha$, $|z| < 1$. This region is to be mapped by $w = z^4$. What restriction must be placed on α to make the mapping one to one?

5. a) Refer to Example 3 on page 359. Show that at their point of intersection the images of $x = x_1$ and $y = y_1$ are orthogonal. Work directly with the equation of each image.

   b) In this same example what inverse transformation $z = g(w)$ will map the upper half of the w-plane onto the semiinfinite strip of Fig. 8.3–5(a)? State the branches of any logarithms and square roots in your function, and verify that point D' is mapped into D, that C' is mapped into C, and that $w = i$ has an image lying inside the strip.

6. The semiinfinite strip $0 < \text{Im } z < \pi$, $\text{Re } z > 0$ is mapped by means of $w = \cosh z$. Find the image of this domain.

7. a) Consider the half disc-shaped domain $|z| < 1$, $\text{Im } z > 0$. Find the image of this domain under the transformation

$$w = \left(\frac{z - 1}{z + 1}\right)^2.$$

Hint: Map the semicircular arc bounding the top of the disc by putting $z = e^{i\theta}$ in the above formula. The resulting expression reduces to a simple trigonometric function.

b) What inverse transformation $z = g(w)$ will map the domain found in part (a) back onto the half disc? State the appropriate branches of any square roots.

8. Following Theorem 2 there is a remark asserting that if $f'(z) = 0$ at any point in a domain, then $w = f(z)$ cannot map that domain one to one. However, in Example 1 we found that a wedge containing $z = 0$ can be mapped one to one by $f(z) = z^2$ even though $f'(0) = 0$. Is there a contradiction here? Explain.

8.4 THE BILINEAR TRANSFORMATION

The bilinear transformation defined by

$$w = \frac{az + b}{cz + d}, \quad \text{where } a, b, c, d \text{ are complex constants,} \tag{8.4-1}$$

which is also known as the *linear fractional transformation* and the *Möbius transformation*, is especially useful in the solution of a number of physical problems, some of which are discussed in this chapter. The utility of this transformation arises from the way in which it maps straight lines and circles.

Equation (8.4–1) defines a finite value of w for all $z \neq -d/c$. One generally assumes that

$$ad \neq bc. \tag{8.4-2}$$

If we take $ad = bc$, we can readily show that Eq. (8.4–1) reduces to a constant value of w, that is, $dw/dz = 0$ for all z, and the mapping is not conformal.

In general, from Eq. (8.4–1), we have

$$\frac{dw}{dz} = \frac{a(cz + d) - c(az + b)}{(cz + d)^2} = \frac{ad - bc}{(cz + d)^2}, \tag{8.4-3}$$

which is nonzero if Eq. (8.4–2) is satisfied.

The inverse transformation of Eq. (8.4–1) is obtained by our solving this equation for z. We have

$$z = \frac{-dw + b}{cw - a}, \tag{8.4-4}$$

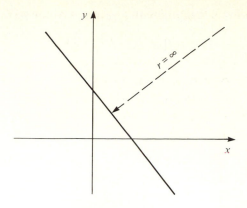

Figure 8.4–1

which is also a bilinear transformation and defines a finite value of z for all $w \neq a/c$.

For reasons that will soon be evident, we now employ the extended w-plane and the extended z-plane (see Section 1.5), that is, planes that include the "points" $z = \infty$ and $w = \infty$.

Consider Eq. (8.4–1) for the case $c = 0$. We have

$$w = \frac{a}{d}z + \frac{b}{d}, \tag{8.4–5}$$

which defines a value of w for every finite value of z. As $|z| \to \infty$, we have $|w| \to \infty$. Thus $z = \infty$ is mapped into $w = \infty$.

Suppose however that $c \neq 0$ in Eq. (8.4–1). As $z \to -d/c$, we have $|w| \to \infty$. Thus $z = -d/c$ is mapped into $w = \infty$. If $|z| \to \infty$, we have $w \to a/c$. Thus $z = \infty$ is mapped into $w = a/c$.

Referring to the inverse transformation (see Eq. 8.4–4), if $c = 0$, we have

$$z = \frac{d}{a}w - \frac{b}{a}, \tag{8.4–6}$$

which also indicates that $z = \infty$ and $w = \infty$ are images (for $c = 0$). If $c \neq 0$, Eq. (8.4–4) shows that $w = \infty$ has image $z = -d/c$, whereas $w = a/c$ has image $z = \infty$. In summary, Eq. (8.4–1) provides a one-to-one mapping of the extended z-plane onto the extended w-plane.

Suppose now we regard infinitely long straight lines in the complex plane as being circles of infinite radius (see Fig. 8.4–1). Thus we will use the word "circle" to mean not only circles in the conventional sense but infinite straight lines as well. Circle (without the quotation marks) will mean a circle in the conventional sense. We will now prove the following theorem.

Theorem 3

The bilinear transformation always transforms "circles" into "circles." ∎

Our proof of Theorem 3 begins with a restatement of Eq. (8.4–1):

$$w = \frac{a}{c} + \frac{bc - ad}{c} \frac{1}{cz + d},$$ (8.4–7)

where we assume $c \neq 0$. If we put Eq. (8.4–7) over a common denominator its equivalence to Eq. (8.4–1) becomes apparent.

The transformation described by Eq. (8.4–7) can be treated as a sequence of mappings. Consider a transformation involving a mapping from the z_1-plane into the w_1-plane, from the w_1-plane into the w_2-plane, and so on according to the following scheme:

$$w_1 = cz$$ (8.4–8a)

$$w_2 = w_1 + d = cz + d$$ (8.4–8b)

$$w_3 = \frac{1}{w_2} = \frac{1}{cz + d}$$ (8.4–8c)

$$w_4 = \frac{bc - ad}{c} w_3 = \frac{bc - ad}{c(cz + d)}$$ (8.4–8d)

$$w = \frac{a}{c} + w_4 = \frac{a}{c} + \frac{bc - ad}{c(cz + d)}$$ (8.4–8e)

Equation (8.4–8e) confirms that these five mappings are together equivalent to Eq. (8.4–7).

There are three distinctly different kinds of operations contained in Eqs. (8.4–8a–e). Let k be a complex constant. There are translations of the form

$$w = z + k,$$ (8.4–9)

as in Eqs. (8.4–8b) and (8.4–8e). There are rotation-magnifications of the form

$$w = kz,$$ (8.4–10)

as in Eqs. (8.4–8a) and (8.4–8d). And there are inversions of the form

$$w = \frac{1}{z},$$ (8.4–11)

as in Eq. (8.4–8c).

If we can show that "circles" are mapped into "circles" under each of these three operations, we will have proved Theorem 3. It should be apparent that under a displacement the geometric character of any shape (circles, triangles, straight lines) is preserved since every point on whatever shape we choose is merely displaced by the complex vector k (see Fig. 8.4–2).

We can rewrite Eq. (8.4–10) as

$$w = |k|e^{i\theta_k}z,$$ (8.4–12)

where $\theta_k = \arg k$. Under this transformation a point from the z-plane is rotated through an angle θ_k and its distance from the origin is magnified by the factor $|k|$. The process of rotation will preserve the shape of any figure as shown in Fig. 8.4–3. We can show that under magnification "circles" are mapped into "circles." For a

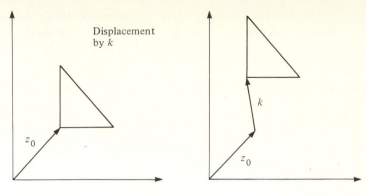

Displacement by k

Figure 8.4–2

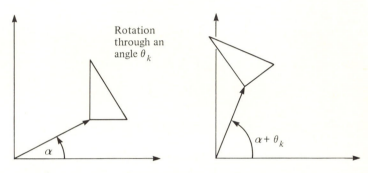

Rotation through an angle θ_k

Figure 8.4–3

magnification

$$w = u + iv = |k|z = |k|(x + iy),$$

and so

$$u = |k|x, \qquad v = |k|y. \tag{8.4–13}$$

A circle in the xy-plane is of the form

$$(x - x_0)^2 + (y - y_0)^2 = r^2. \tag{8.4–14}$$

The center is at (x_0, y_0) and the radius is r. Solving Eq. (8.4–13) for x and y and using these values in Eq. (8.4–14), we can obtain

$$(u - |k|x_0)^2 + (v - |k|y_0)^2 = r^2|k|^2,$$

which is the equation of a circle, in the w-plane, having center at $|k|x_0, |k|y_0$ and radius $r|k|$. A similar argument shows that the straight line $y = mx + b$ is mapped into the straight line $v = mu + b|k|$.

To prove that Eq. (8.4–11) maps "circles" into "circles," consider the algebraic equation

$$A(x^2 + y^2) + Bx + Cy + D = 0, \tag{8.4–15}$$

where A, B, C, and D are all real numbers. If $A = 0$, this is obviously the equation of a straight line. Assuming $A \neq 0$, we divide Eq. (8.4–15) by A to obtain

$$\left(x^2 + y^2\right) + \frac{B}{A}x + \frac{C}{A}y + \frac{D}{A} = 0,$$

which, if we complete two squares, can be rewritten

$$\left(x + \frac{B}{2A}\right)^2 + \left(y + \frac{C}{2A}\right)^2 = -\frac{D}{A} + \left(\frac{B}{2A}\right)^2 + \left(\frac{C}{2A}\right)^2. \quad (8.4\text{–}16)$$

A comparison with Eq. (8.4–14) reveals this to be the equation of a circle provided

$$-\frac{D}{A} + \left(\frac{B}{2A}\right)^2 + \left(\frac{C}{2A}\right)^2 \geq 0$$

(the squared radius cannot be negative). The above condition can be rearranged as

$$B^2 + C^2 \geq 4AD. \quad (8.4\text{–}17)$$

In Eq. (8.4–15) we make the following well-known substitutions:

$$x^2 + y^2 = z\bar{z}, \qquad x = \frac{z + \bar{z}}{2}, \qquad y = \frac{1}{i}\frac{(z - \bar{z})}{2}.$$

Thus

$$Az\bar{z} + \frac{B}{2}(z + \bar{z}) + \frac{C}{2i}(z - \bar{z}) + D = 0. \quad (8.4\text{–}18)$$

With A, B, C, D real numbers Eq. (8.4–18) is the equation of a circle if $A \neq 0$, and in addition Eq. (8.4–17) is satisfied. It is the equation of a straight line if $A = 0$.

We now replace z by $1/w$ in Eq. (8.4–18) in order to find the image of our "circle" under an inversion. The "circle" is transformed into a curve satisfying

$$A\left(\frac{1}{w\bar{w}}\right) + \frac{B}{2}\left(\frac{1}{w} + \frac{1}{\bar{w}}\right) + \frac{C}{2i}\left(\frac{1}{w} - \frac{1}{\bar{w}}\right) + D = 0.$$

Multiplying both sides by $w\bar{w}$ and rearranging terms slightly, we get

$$Dw\bar{w} + \frac{B}{2}(w + \bar{w}) - \frac{C}{2i}(w - \bar{w}) + A = 0. \quad (8.4\text{–}19)$$

Equation (8.4–19) is identical in form to Eq. (8.4–18) with D in Eq. (8.4–19) now playing the role of A, A now playing the role of D, and $-C$ taking the part of C. The meaning of B is unaltered.

With these changes in Eq. (8.4–17) it is found that this inequality remains unaltered. Thus if $D \neq 0$, Eq. (8.4–19) describes a circle as long as Eq. (8.4–18) describes one.

If $D = 0$, Eq. (8.4–19) describes a straight line, that is, a "circle." Thus we have shown that $w = 1/z$ maps "circles" into "circles."

Notice that if $D = 0$, Eq. (8.4–15) is satisfied for $z = 0$, that is, the "circle" described in the z-plane passes through the origin. With $D = 0$ Eq. (8.4–19) yields a straight line. Thus a "circle" passing through the origin of the z-plane is transformed by $w = 1/z$ into a straight line in the w-plane.

Under the assumption $c \neq 0$ we have shown that the bilinear transformation (see Eq. 8.4–1 and, equivalently, Eq. 8.4–7) can be decomposed into a sequence of transformations, each of which transforms "circles" into "circles." If $c = 0$ in Eq. (8.4–1), it is an easy matter to show that the resulting transformation

$$w = \frac{a}{d}z + \frac{b}{d},$$

which involves a rotation-magnification and displacement also has this property. Thus our proof of Theorem 3 is complete.

Because the inverse of a bilinear transformation is also a bilinear transformation (see Eq. 8.4–4), a "circle" in the z-plane is the image of a "circle" in the w-plane (and vice versa). An elementary example of this, involving the simple bilinear transformation $w = 1/z$, was studied in Section 8.2 in Example 1 as well as in Exercise 5.

Example 1

Figure 8.4–4 shows an elementary electric circuit. V_i and V_0 are the phasor input and output potentials for the electric circuit. (See the appendix to Chapter 3 for a discussion of phasors.) The amplification factor $A = V_0/V_i$ for the circuit is given by

$$A(s) = \frac{1 + s}{2 + s}, \tag{8.4–20}$$

where $s = \sigma + i\omega$ is the complex frequency describing the potentials. If we restrict ourselves to potentials that vary in time as sine or cosine functions of fixed amplitude, then $\sigma = 0$ and $s = i\omega$. What is the locus of $A(s)$ in the complex plane as s varies in the complex frequency plane over the line $s = i\omega$, $-\infty \leq \omega \leq \infty$?

Solution

We are mapping the infinite line $\sigma = 0$ (see Fig. 8.4–5a) into the A-plane (see Fig. 8.4–5b) by means of the bilinear transformation in Eq. 8.4–20. The image must be either a circle or a straight line. A straight line must pass through $A = \infty$. From Eq. (8.4–20) we see that $A = \infty$ is the image of $s = -2$. But $s = -2$ does not lie on the line $s = i\omega$ ($-\infty \leq \omega \leq \infty$). Hence the image we are seeking is a circle.

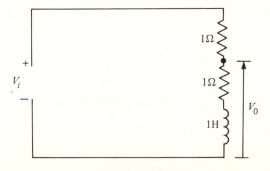

Figure 8.4–4

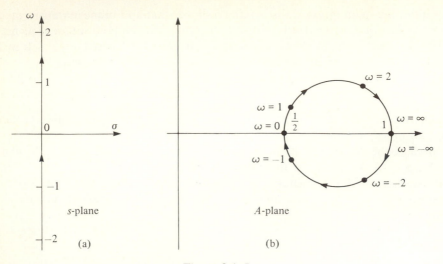

Figure 8.4–5

Observe from Eq. (8.4–20) that the image of $s = (0, 0)$ is $1/2$. As $|s| \to \infty$ along the ω-axis, the same equation shows that $A \to 1$, that is, the image of $s = \infty$ is $A = 1$.

It is easy to show from Eq. (8.4–20) that

$$A(\sigma + i\omega) = \overline{A}(\sigma - i\omega),$$

that is, values of A at conjugate points in the s-plane are conjugates of each other. In particular, a pair of conjugate points on the ω-axis have images that are conjugates. This means the circle that is the image of the entire ω-axis must be symmetric about the real A-axis.

We now have enough information to draw the circle in Fig. 8.4–5(b). The images of a few individual points in the complex A-plane are indicated, for example, if $s = i$,

$$A(i) = \frac{1 + i}{2 + i} \doteq \frac{\sqrt{2}}{\sqrt{5}} \underline{/18.43°} \, . \qquad \blacktriangleleft$$

A common problem is that of finding a specific bilinear transformation that will map certain points in the z-plane into preassigned images in the w-plane. One also seeks transformations capable of mapping a given line or circle into some other specific line or circle. In these problems one must establish the constants a, b, c, d in the bilinear transformation

$$w = \frac{az + b}{cz + d} \, . \qquad (8.4\text{–}21)$$

Let us assume that one of the coefficients, say a, is nonzero. Then we can rewrite Eq. (8.4–21) as

$$w = \frac{z + \dfrac{b}{a}}{\dfrac{c}{a} z + \dfrac{d}{a}} = \frac{z + c_1}{c_2 z + c_3} \, . \qquad (8.4\text{–}22)$$

Thus only three coefficients, c_1, c_2, and c_3 need be found. Given three points z_1, z_2, and z_3, which we must map into w_1, w_2, and w_3, respectively, we replace w and z in Eq. (8.4–22) first of all by w_1 and z_1, respectively, then by w_2 and z_2, and finally by w_3 and z_3. Three simultaneous linear equations are obtained in the unknowns c_1, c_2, and c_3. Solving for these unknowns, we have determined our bilinear transformation (Eq. 8.4–22). If a solution does not exist, it is because $a = 0$ in Eq. (8.4–21). We then could obtain b, c, and d by simultaneously solving three equations obtained by substituting w_1 and z_1, w_2 and z_2, and finally w_3 and z_3 into Eq. (8.4–21) with a set equal to zero.

A more direct way of solving for a bilinear transformation involves the cross ratio.

Definition   Cross ratio

The cross ratio of four distinct complex numbers (or points) z_1, z_2, z_3, z_4 is defined by

$$(z_1, z_2, z_3, z_4) = \frac{(z_1 - z_2)(z_3 - z_4)}{(z_1 - z_4)(z_3 - z_2)}. \tag{8.4–23}$$

If any of these numbers, say z_j, is ∞, the cross ratio in Eq. (8.4–23) is redefined so that the quotient of the two terms on the right containing z_j, that is $(z_j - z_k)/(z_j - z_m)$, is taken as 1.   ∎

The cross ratio of the four image points w_1, w_2, w_3, w_4 is obtained by replacing z_1, z_2, etc. in the preceding definition by w_1, w_2, etc. The order of the points in a cross ratio is important. The reader should verify, that, for example, $(1, 2, 3, 4) = -1/3$, whereas $(3, 1, 2, 4) = 4$.

We will now prove the following theorem.

Theorem 4   Invariance of cross ratio

Under the bilinear transformation (Eq. 8.4–21) the cross ratio of four points is preserved, that is,

$$\frac{(w_1 - w_2)(w_3 - w_4)}{(w_1 - w_4)(w_3 - w_2)} = \frac{(z_1 - z_2)(z_3 - z_4)}{(z_1 - z_4)(z_3 - z_2)}. \quad\blacksquare \tag{8.4–24}$$

The proof of Theorem 4 is straightforward. From Eq. (8.4–21) z_i is mapped into w_i, that is

$$w_i = \frac{az_i + b}{cz_i + d}, \tag{8.4–25}$$

and similarly

$$w_j = \frac{az_j + b}{cz_j + d},$$

so that

$$w_i - w_j = \frac{az_i + b}{cz_i + d} - \frac{az_j + b}{cz_j + d} = \frac{(ad - bc)(z_i - z_j)}{(cz_i + d)(cz_j + d)}. \tag{8.4–26}$$

With $i = 1$, $j = 2$ in Eq. (8.4–26) we obtain $(w_1 - w_2)$ in terms of z_1 and z_2. Similarly, we can express $w_3 - w_4$ in terms of $z_3 - z_4$, etc. In this manner the entire left side of Eq. (8.4–24) can be written in terms of z_1, z_2, z_3, and z_4. After some simple algebra we obtain the right side of Eq. (8.4–24). The reader should supply the details.

If one of the points z_1, z_2,... is at infinity, the invariance of the cross ratio must be proved differently. If, say, $z_1 = \infty$, its image is $w_1 = a/c$ (see, for example, Eq. 8.4–25 as $z_i = z_1 \to \infty$). Thus the left side of Eq. (8.4–24) becomes

$$\frac{\left(\dfrac{a}{c} - w_2\right)(w_3 - w_4)}{\left(\dfrac{a}{c} - w_4\right)(w_3 - w_2)}.$$

If Eqs. (8.4–26) and (8.4–25) are used in this expression, the values w_2, w_3, and w_4 can be rewritten in terms of z_2, z_3, z_4. After some manipulation, the expression $(z_3 - z_4)/(z_3 - z_2)$ is obtained. This is the cross ratio (z_1, z_2, z_3, z_4) when $z_1 = \infty$.

The invariance of the cross ratio is useful when we seek the bilinear transformation capable of mapping three specific points z_1, z_2, z_3 into three specific images w_1, w_2, w_3. The point z_4 in Eq. (8.4–24) is taken as a general point z whose image is w (instead of w_4). Thus our working formula becomes

$$\frac{(w_1 - w_2)(w_3 - w)}{(w_1 - w)(w_3 - w_2)} = \frac{(z_1 - z_2)(z_3 - z)}{(z_1 - z)(z_3 - z_2)}, \tag{8.4–27}$$

which must be suitably modified if any point is at ∞. The solution of w in terms of z yields the required transformation.

Example 2

Find the bilinear transformation that maps $z_1 = 1$, $z_2 = i$, $z_3 = 0$ into $w_1 = 0$, $w_2 = -1$, $w_3 = -i$.

Solution

We substitute these six complex numbers into the appropriate location in Eq. (8.4–27) and obtain

$$\frac{(0 - (-1))(-i - w)}{(0 - w)(-i + 1)} = \frac{(1 - i)(0 - z)}{(1 - z)(0 - i)}.$$

With some minor algebra we get

$$w = \frac{i(z - 1)}{z + 1}. \tag{8.4–28}$$

This result can be checked by our letting z assume the three given values 1, i, and 0. The desired values of w are obtained. ◀

Example 3

For the transformation found in Example 2 what is the image of the circle passing through $z_1 = 1$, $z_2 = i$, $z_3 = 0$, and what is the image of the interior of this circle?

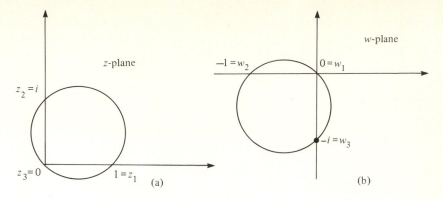

Figure 8.4–6

Solution

The given circle is shown in Fig. 8.4–6(a). From elementary geometry, its center is found to be at $(1 + i)/2$, and its radius is $1/\sqrt{2}$. The circle is described by

$$\left| z - \frac{(1 + i)}{2} \right| = \frac{1}{\sqrt{2}}.$$

The image of the circle under Eq. (8.4–28) must be a straight line or circle in the w-plane. The image is known to pass through $w_1 = 0$, $w_2 = -1$, $w_3 = -i$. The circle determined by these three points is shown in Fig. 8.4–6(b) (no straight line can connect w_1, w_2, and w_3) and is described by

$$\left| w + \frac{(1 + i)}{2} \right| = \frac{1}{\sqrt{2}}. \tag{8.4–29}$$

This disc-shaped domain

$$\left| z - \frac{(1 + i)}{2} \right| < \frac{1}{\sqrt{2}},$$

which is the interior of the circle of Fig. 8.4–6(a) has, under the given transformation (Eq. 8.4–28), an image that is also a domain.[†] The boundary of this image is the circle in Fig. 8.4–6(b). Thus the image domain must be either the disc interior to this latter circle or the annulus exterior to it.

Let us consider some point inside the circle of Fig. 8.4–6(a), say $z = 1/2$. From Eq. (8.4–28) we discover that $w = -i/3$ is its image. This lies inside the circle of Fig. 8.4–6(b) and indicates that the domain

$$\left| z - \frac{(1 + i)}{2} \right| < \frac{1}{\sqrt{2}}$$

[†]A domain is always mapped onto a domain by a nonconstant analytic function. See, for example, E. B. Saff and A. D. Snyder, *Fundamentals of Complex Analysis* (Englewood Cliffs, N.J.: Prentice-Hall, 1976), p. 305.

is mapped onto the domain

$$\left| w + \frac{(1 + i)}{2} \right| < \frac{1}{\sqrt{2}}. \qquad \blacktriangleleft$$

Example 4

Find the bilinear transformation that maps $z_1 = 1$, $z_2 = i$, $z_3 = 0$ into $w_1 = 0$, $w_2 = \infty$, $w_3 = -i$.

Solution

Note that z_1, z_2, and z_3 are the same as in Example 2. We again employ Eq. (8.4–27). However, since $w_2 = \infty$, the ratio $(w_1 - w_2)/(w_3 - w_2)$ on the left must be replaced by 1. Thus

$$\frac{-i - w}{-w} = \frac{(1 - i)(-z)}{(1 - z)(-i)},$$

whose solution is

$$w = \frac{1 - z}{i - z}. \qquad (8.4\text{–}30)$$

Note that the circle in Fig. 8.4–6(a) passing through 1 and i and 0 is transformed into a "circle" passing through 0 and ∞ and $-i$, that is, an infinite straight line lying along the imaginary axis in the w-plane. The half plane to the left of this line is the image of the interior of the circle in Fig. 8.4–6(a), as the reader can readily verify. ◀

Example 5

Find the transformation that will map the domain $0 < \arg z < \pi/2$ from the z-plane onto $|w| < 1$ in the w-plane (see Fig. 8.4–7).

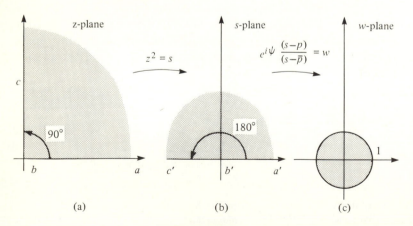

Figure 8.4–7

Solution

The boundary of the given domain in the z-plane, that is, the positive x- and y-axes, must be transformed into the unit circle $|w| = 1$ by the required formula. A bilinear transformation will map an infinite straight line into a circle but cannot transform a line with a 90° bend into a circle. (Why?) Hence, our answer cannot be a bilinear transformation.

Notice however, that the transformation

$$s = z^2 \tag{8.4-31}$$

(see Example 1, Section 8.3) will map our 90° sector onto the *upper* half of the s-plane. If we can find a second transformation that will map the upper half of the s-plane onto the interior of the unit circle in the w-plane, we can combine the two mappings into the required transformation.

Borrowing a result derived in Exercise 15 of this section and changing notation slightly, we observe that

$$w = e^{i\psi} \frac{(s - p)}{(s - \bar{p})}, \quad \text{where } \psi \text{ is a real number} \quad \text{and} \quad \text{Im } p > 0, \tag{8.4-32}$$

will transform the real axis from the s-plane into the circle $|w| = 1$ and map the domain Im $s > 0$ onto the interior of this circle.

Combining Eqs. (8.4–31) and (8.4–32), we have as our result

$$w = e^{i\psi} \frac{(z^2 - p)}{(z^2 - \bar{p})}, \quad \text{where } \psi \text{ is a real number} \quad \text{and} \quad \text{Im } p > 0. \tag{8.4-33}$$

A particular example of Eq. (8.4–33) is

$$w = \frac{z^2 - i}{z^2 + i}.$$

The method just used can be modified so that angular sectors of the form $0 < \arg z < \alpha$, where α is not constrained to be $\pi/2$, can be mapped onto the interior of the unit circle (see Exercise 17 of this section). ◄

EXERCISES

1. a) Derive Eq. (8.4–4) from Eq. (8.4–1).

   b) Verify that Eq. (8.4–7) is equivalent to Eq. (8.4–1).

2. Suppose that the bilinear transformation (see Eq. 8.4–1) has real coefficients a, b, c, d. Show that a curve that is symmetric about the x-axis has an image under this transformation that is symmetric about the u-axis.

3. Derive Eq. (8.4–24) (the invariance of the cross ratio) by following the steps suggested in the text.

4. If a transformation $w = f(z)$ maps z_1 into w_1, where z_1 and w_1 have the same numerical value, we say that z_1 is a *fixed point* of the transformation.

a) For the bilinear transformation (Eq. 8.4–1) show that a fixed point must satisfy

$$cz^2 - (a - d)z - b = 0.$$

b) Show that unless $a = d \neq 0$ and $b = c = 0$ are simultaneously satisfied, there are at most two fixed points for this bilinear transformation.

c) Why are all points fixed points if $a = d \neq 0$ and $b = c = 0$ are simultaneously satisfied? Refer to Eq. (8.4–1).

5. a) A bilinear transformation has fixed points -1 and $+1$. What is the most general form of this transformation?

   b) Repeat part (a), but use the fixed points 1 and i.

6. For the transformation $w = 1/z$, what are the images of the following curves? Give the result as an equation in u and v, where $w = u + iv$.

   a) $x + y = 1$ c) $(x - 1/2)^2 + (y - 1/2)^2 = 1/2$

   b) $(x + 1)^2 + y^2 = 1$ d) $(x + 2)^2 + y^2 = 1$

7. Under the transformation $w = (z - i)/(z + i)$, what are the images of each of the curves in Exercise 6?

8. How is the disc-shaped domain $|z - 1| < 1$ mapped by each of the following transformations?

   a) $w = 1/z$ c) $w = z/(z + 1)$

   b) $w = 1/(z - 1)$ d) $w = e^{i\pi/4}z/(z + 1)$

9. Find the image of the domain $1 < \operatorname{Re} z < 2$ under each of the following transformations.

   a) $w = 1/(z + 1)$ b) $w = i/(z - 1)$

10. Consider the grid of straight lines $\ldots$, $x = -2$, $x = -1$, $x = 0$, $x = 1, \ldots$; $y = -2$, $y = -1$, $y = 0$, $y = 2, \ldots$. Show with a sketch how this grid is transformed by $w = 1/(z + 1)$.

11. The complex impedance at the input of the circuit in Fig. 8.4–8 when driven by a sinusoidal generator of radian frequency ω is $Z(\omega) = R + i\omega L$. When ω increases from 0 to ∞, $Z(\omega)$ progresses in the complex plane from $(R, 0)$ to infinity along the semiinfinite line $\operatorname{Re} Z = R$, $\operatorname{Im} Z \geq 0$. The complex admittance of the circuit is defined by $Y(\omega) = 1/Z(\omega)$. Use the properties of the bilinear transformation to determine the locus of $Y(\omega)$ in the complex plane as ω goes from 0 to ∞. Sketch the locus and indicate $Y(0)$, $Y(R/L)$ and $Y(\infty)$.

Figure 8.4–8

12. a) A circle of radius $\rho > 0$ and center $(x_0, 0)$ is transformed by the inversion $w = 1/z$ into another circle. Locate the intercepts of the image circle on the real w-axis and show that this new circle has center $x_0/(x_0^2 - \rho^2)$ and radius $\rho/|x_0^2 - \rho^2|$.

   b) Is the image of the center of the original circle under the transformation $w = 1/z$ identical to the center of the image circle? Explain.

   c) Does the general bilinear transformation (see Eq. 8.4–1) always map the center of a circle in the z-plane into the center of the image of that circle in the w-plane? Explain.

   d) Consider the special case of Eq. (8.4–1), $w = az + b$. Show that the circle $|z - z_0| = \rho$ is mapped by this transformation into a circle centered at $w_0 = az_0 + b$ with radius $a\rho$. Thus in this special case the original circle and its image have centers that are images of each other under the given transformation.

13. A bilinear transformation

$$w_1 = \frac{a_1 z + b_1}{c_1 z + d_1}$$

defines a mapping from the z-plane to the w_1-plane. Additionally, a second bilinear transformation

$$w = \frac{a_2 w_1 + b_2}{c_2 w_1 + d_2}$$

yields a mapping from the w_1-plane to the w-plane. Show that these two successive bilinear transformations, which together relate z and w, can be combined into a single bilinear transformation

$$w = \frac{az + b}{cz + d}.$$

What are a, b, c, d in terms of a_1, a_2, etc.?

14. For the electric circuit shown in Fig. 8.4–9 the ratio of the phasor output voltage to the phasor input voltage is given by

$$\frac{V_0}{V_i} = A(s) = \frac{1 + s}{1 + 2s},$$

where $s = \sigma + i\omega$ is the complex frequency.

Figure 8.4–9

a) Draw the locus in the complex plane of $A(s)$ as the complex frequency varies in the s-plane along the line $\sigma = 0$, $-\infty < \omega < \infty$. What is the equation of the locus? Indicate on the locus the values of A when $\omega = 0, \pm 1/2, \pm 1, \pm \infty$.

b) Suppose the complex frequency is of the form $s = -1/2 + i\omega$ (which implies a simultaneously oscillating and decaying signal). As ω varies from $-\infty$ to ∞, indicate the locus of A in the complex plane.

15. This exercise establishes the general bilinear transformation that maps the upper half of the z-plane ($\operatorname{Im} z > 0$) onto the unit disc $|w| < 1$.

a) Put Eq. (8.4–1) in the form

$$w = \left(\frac{a}{c}\right)\frac{z + b/a}{z + d/c}.$$

Explain why the desired transformation must transform the x-axis into $|w| = 1$. Let $|z| \to \infty$ on this axis and argue that $|a|/|c| = 1$. Thus $a/c = e^{i\psi}$, where ψ is any real number.

b) From

$$w = e^{i\psi}\frac{z + b/a}{z + d/c}$$

we have with $z = x$ and the use of magnitudes that

$$1 = \left|\frac{x + b/a}{x + d/c}\right|,$$

or

$$\left|x + \frac{d}{c}\right| = \left|x + \frac{b}{a}\right|.$$

Explain why this equation can be satisfied only if

$$\frac{d}{c} = \frac{b}{a},$$

or

$$\frac{d}{c} = \overline{\left(\frac{b}{a}\right)}.$$

Explain why the first choice must be discarded.

c) Taking $-p = b/a$, we have now

$$w = e^{i\psi}\frac{(z - p)}{(z - \bar{p})}. \qquad (8.4–34)$$

Note that $z = p$ has image $w = 0$. Explain why we require $\operatorname{Im} p > 0$ in Eq. (8.4–34) so that $|w| < 1$ will be the image of $\operatorname{Im} z > 0$.

d) Suppose in Eq. (8.4–34) we take $\operatorname{Im} p < 0$. What is the image of $\operatorname{Im} z > 0$ under this transformation?

16. a) Find a bilinear transformation capable of mapping the domain to the right of $x + y = 1$ onto the disc $|w| < 1$.

b) Repeat part (a), but use the domain $|w| > 1$.

17. a) Find a transformation that will map the wedge-shaped domain $0 < \arg z < \pi/6$ onto the disc $|w| < 1$.

b) Find a transformation that will map the wedge $0 < \arg z < \alpha$ onto the same disc. Take $\alpha < 2\pi$.

18. We wish to find a conformal mapping that will map the oval-shaped domain shared by the two discs (see Fig. 8.4–10a) $|z - 1| < 2$ and $|z + 1| < 2$ onto the upper half of the w-plane. We will use the following steps:

a) First find the bilinear transformation that maps the points z_1, z_2, and z_3 into ∞, 0, and 1, respectively. Why does this transform the boundaries of the oval into the pair of lines in Fig. 8.4–10(b) that intersect at an angle α? What is the numerical value of α?

b) By finding the image of $z = 0$ verify that the transformation found in part (a) maps the oval-shaped domain onto the sector of angle α, shown in Fig. 8.4–10(b).

c) Obtain the solution to the exercise by mapping the sector of angle α onto the upper half plane.

19. a) Consider the domain lying between the circles $|z| = 2$ and $|z - i| = 1$. Find a bilinear transformation that will map this domain onto a strip $0 < \operatorname{Im} w < k$, where the line $\operatorname{Im} w = 0$ is the image of $|z| = 2$. The reader may choose k. *Hint:* Transform the outer circle into $\operatorname{Im} w = 0$ by finding the bilinear transformation that maps $-2i$ into 0, 2 into 1, and $2i$ to ∞. Why will the inner boundary $|z - 1| = 1$ be transformed into a line parallel to $\operatorname{Im} w = 0$ by this transformation? Verify that the domain bounded by the circles is mapped onto the strip. What is the value of k for your answer?

b) Use the transformation derived in part (a) and a modification of the transformation in Example 2, Section 8.3 to obtain a transformation that will map the domain bounded by the circles in part (a) onto the upper half plane.

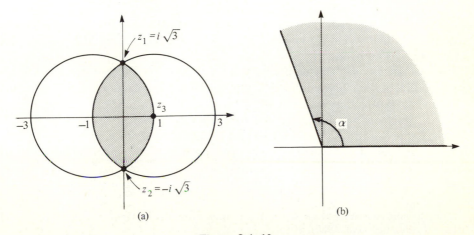

(a)

(b)

Figure 8.4–10

20. The Smith chart is a graphical device used in electrical engineering for the analysis of high frequency transmission lines. It relates two complex variables: the normalized impedance $z = r + ix$ ($r \geq 0$, $-\infty < x < \infty$) and the reflection coefficient $\Gamma = a + ib$. The mapping

$$\Gamma(z) = \frac{z-1}{z+1} \tag{8.4--35}$$

is applied to a grid of infinite vertical and semiinfinite horizontal lines in the right half of the z-plane. The image of this grid in the Γ-plane is the Smith chart.

a) Show that the image of the region $\mathrm{Re}\, z \geq 0$ under the mapping in Eq. (8.4--35) is the disc $|\Gamma| \leq 1$.

b) Sketch and give the equation of the image of the following infinite vertical lines under the transformation (Eq. 8.4--35) $r = 0$, $r = 1/2$, $r = 1$, $r = 2$.

c) Sketch the image and give the equation of each of the following semiinfinite horizontal lines under the transformation (Eq. 8.4--35) $x = 0$, $r \geq 0$; $x = 1/2$, $r \geq 0$; $x = -1/2$, $r \geq 0$; $x = 2$, $r \geq 0$; $x = -2$, $r \geq 0$. The collection of images sketched in parts (b) and (c) form a primitive Smith chart.[†]

d) Solve Eq. (8.4--35) for $z(\Gamma)$. Show that $z(\Gamma) = 1/z(-\Gamma)$. Thus values of z corresponding to values of Γ that are diametrically opposed with respect to the origin of the Smith chart are reciprocals of each other.

8.5 CONFORMAL MAPPING AND BOUNDARY VALUE PROBLEMS[‡]

Earlier in this book (see Section 2.6) we established the close connection that exists between harmonic functions and two-dimensional physical problems involving heat conduction, fluid flow, and electrostatics. Later (see Section 4.5) we returned to physical configurations when we investigated Dirichlet problems. We saw that when the values of a harmonic function (for example, temperature or voltage) are specified on the surface of the cylinder, the values assumed by the harmonic function inside the cylinder can be found. A similar procedure was developed to find a function that is harmonic above a plane surface when the values taken by the function on the plane are specified. What we know now are solutions of the Dirichlet problem for two simple types of boundaries.

In this section we will combine what we know about conformal mapping, harmonic functions, analytic functions, and the complex potential to solve Dirichlet problems whose boundaries are not limited to planes and cylinders. Electrostatic and heat-flow problems will be considered here. In the section after this one we will study heat and fluid-flow problems in which we seek an unknown harmonic function whose normal derivative is specified over some portion of a boundary. Although this is not a Dirichlet problem, we will again find that conformal mapping helps us to

---

[†]For a more elaborate chart see S. Liao, *Microwave Devices and Circuits* (Englewood Cliffs, N.J.: Prentice-Hall, 1980), p. 85.

[‡]The reader should review Sections 2.5, 2.6, and 4.5.

find a solution. The utility of conformal mapping in the solution of a wide variety of physical problems derives from the following theorem:

Theorem 5

Let the analytic function $w = f(z)$ map the domain D from the z-plane onto the domain D_1 of the w-plane. Suppose $\phi_1(u, v)$ is harmonic in D_1, that is, in D_1

$$\frac{\partial^2 \phi_1}{\partial u^2} + \frac{\partial^2 \phi_1}{\partial v^2} = 0. \tag{8.5-1}$$

Then, under the change of variables

$$u(x, y) + iv(x, y) = f(z) = w, \tag{8.5-2}$$

we have that $\phi(x, y) = \phi_1(u(x, y), v(x, y))$ is harmonic in D, that is, in D

$$\frac{\partial^2 \phi}{\partial x^2} + \frac{\partial^2 \phi}{\partial y^2} = 0. \ \blacksquare \tag{8.5-3}$$

Loosely, a solution of Laplace's equation remains a solution of Laplace's equation when transferred from one plane to another by a conformal transformation. Let us verify Theorem 5 in an elementary example. The function $\phi_1(u, v) = e^u \cos v$, which is $\mathrm{Re}\, e^w$, satisfies Eq. (8.5–1) (see Theorem 6, Chapter 2). Let

$$w = z^2 = x^2 - y^2 + i2xy = u + iv,$$

so that $u = x^2 - y^2$ and $v = 2xy$. Now $\phi(x, y) = e^{x^2 - y^2} \cos(2xy)$ is readily found to satisfy Eq. (8.5–3), as the reader should verify.

We can easily prove Theorem 5 when the domain D_1 is simply connected. A more difficult proof, which dispenses with this requirement, is given in many texts. We rely here on Theorem 7, Chapter 2, which guarantees that, with $\phi_1(u, v)$ satisfying Eq. (8.5–1) in D_1, there exists an analytic function in D_1:

$$\Phi_1(w) = \phi_1(u, v) + i\psi_1(u, v), \tag{8.5-4}$$

where $\psi_1(u, v)$ is the harmonic conjugate of $\phi_1(u, v)$. Since $w = f(z)$ is an analytic function in D, we have that $\Phi_1(f(z)) = \Phi(z)$ is an analytic function of an analytic function in D. Thus (see Theorem 5, Chapter 2) $\Phi(z)$ is analytic in D. Now

$$\Phi_1(w) = \Phi_1(f(z)) = \Phi(z) = \phi(x, y) + i\psi(x, y). \tag{8.5-5}$$

Since $f(z) = u(x, y) + iv(x, y)$, we have by comparing Eqs. (8.5–4) and (8.5–5) that $\phi(x, y) = \phi_1(u(x, y), v(x, y))$ and $\psi(x, y) = \psi_1(u(x, y), v(x, y))$.

Because $\phi(x, y)$ is the real part of an analytic function $\Phi(z)$, $\phi(x, y)$ must be harmonic in D. A parallel argument establishes that $\psi(x, y)$, the imaginary part of $\Phi(z)$, must be harmonic in D.

To see the usefulness of Theorem 5 imagine we are given a domain D in the z-plane. We seek a function $\phi(x, y)$ (say temperature or voltage) that is harmonic in D and that assumes certain prescribed values on the boundary of D. Suppose we can

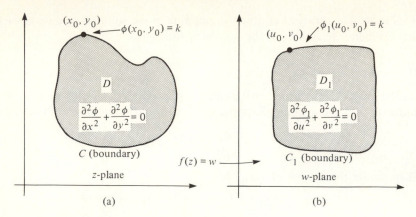

Figure 8.5–1

find an analytic transformation $w = u + iv = f(z)$ that maps D onto a domain D_1 in the w-plane, and D_1 has a simpler or more familiar shape than D. Assume that we can find a function $\phi_1(u, v)$ that is harmonic in D_1 and that assumes values at each boundary point of D_1 exactly equal to the value required of $\phi(x, y)$ at the image of that point on the boundary of D. Then, by Theorem 5, $\phi(x, y) = \phi_1(u(x, y), v(x, y))$ will be harmonic in D and also assume the desired values on the boundary of D. The method is illustrated schematically in Fig. 8.5–1. We have mapped D onto D_1 using $w = f(z)$. The boundaries of the domains are C and C_1; (x_0, y_0) is an arbitrary point on C with image (u_0, v_0) on C_1. The harmonic function $\phi_1(u, v)$ assumes the same value k at (u_0, v_0), as does $\phi(x, y)$ at (x_0, y_0).

We should note that it can be difficult to find an analytic transformation that will map a given domain onto one of some specified simpler shape. We can refer to dictionaries of conformal mappings as an aid.[†] Often experience or trial and error help. The Riemann mapping theorem[‡] guarantees the existence of an analytic transformation that will map any simply connected domain (except the entire z-plane) onto the unit disc $|w| < 1$. The boundary of the domain is transformed into $|w| = 1$. The Poisson integral formula for the circle (see Section 4.5) can then be used to solve the transformed Dirichlet problem. The Riemann mapping theorem does not tell us how to obtain the required mapping, only that it exists.

Example 1

Two cylinders are maintained at temperatures of $0°$ and $100°$, as shown in Fig. 8.5–2(a). An infinitesimal gap separates the cylinders at the origin. Find $\phi(x, y)$, the temperature in the domain between the cylinders.

[†]See, for example, H. Kober, *Dictionary of Conformal Representations*, 2d ed. (New York: Dover, 1957).

[‡]See R. Nevanlinna and V. Paatero, *Introduction to Complex Analysis* (Reading, Mass.: Addison-Wesley, 1969), ch. 17.

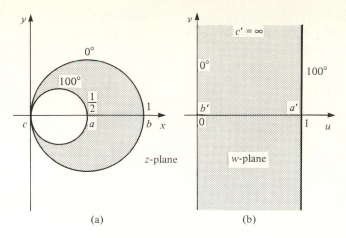

Figure 8.5–2

Solution

The shape of the given domain is complicated. However, because the bilinear transformation will map circles into straight lines, we can transform this domain into the more tractable infinite strip shown in Fig. 8.5–2(b). We follow the method of the previous section and find that the bilinear transformation that maps a, b, and c from Fig. 8.5–2(a) into $a' = 1$, $b' = 0$, $c' = \infty$ is

$$w = \frac{1 - z}{z}. \qquad (8.5\text{--}6)$$

Under the transformation, the cylindrical boundary at $100°$ is transformed into the line $u = 1$, whereas the cylinder at $0°$ becomes the line $u = 0$.

The strip $0 < u < 1$ is the image of the region between the two circles shown in Fig. 8.5–2(a). Our problem now is the simpler one of finding $\phi_1(u, v)$, which is harmonic in the strip. We must also fulfill the boundary conditions $\phi_1(0, v) = 0$ and $\phi_1(1, v) = 100$.

The problem is now easy enough so that we might guess the result, or we can study the similar Example 1 in Section 2.6. From symmetry we expect that $\phi_1(u, v)$ is independent of v, and we notice that

$$\phi_1(u, v) = 100u \qquad (8.5\text{--}7)$$

satisfies both boundary conditions and is harmonic. This is the temperature distribution in the transformed problem.

Now $\phi_1(u, v)$ is the real part of an analytic function. Using the methods of Section 2.5 or employing common sense, we see that $\phi_1(u, v) = \text{Re}(100w)$. Thus the complex temperature (see Section 2.6) in the strip is

$$\Phi_1(w) = 100w, \qquad (8.5\text{--}8)$$

and the corresponding stream function is

$$\psi_1(u, v) = \text{Im}(100w) = 100v. \qquad (8.5\text{--}9)$$

To obtain the temperature $\phi(x, y)$ and stream function $\psi(x, y)$ for Fig. 8.5–2(a) we must transform $\phi_1(u, v)$ and $\psi_1(u, v)$ back into the z-plane by means of Eq. (8.5–6). From Eq. (8.5–6) we have

$$w = u + iv = \frac{1}{z} - 1 = \frac{1}{x + iy} - 1 = \frac{x}{x^2 + y^2} - 1 - \frac{iy}{x^2 + y^2}, \quad (8.5-10)$$

which implies

$$u = \frac{x}{x^2 + y^2} - 1, \quad (8.5-11a)$$

$$v = \frac{-y}{x^2 + y^2}. \quad (8.5-11b)$$

From Eqs. (8.5–11a) and (8.5–7) we have

$$\phi(x, y) = 100\left(\frac{x}{x^2 + y^2} - 1\right) \quad (8.5-12)$$

for the temperature distribution between the given cylinders. Combining Eqs. (8.5–11b) and (8.5–9), we have

$$\psi(x, y) = \frac{-100y}{x^2 + y^2}. \quad (8.5-13)$$

The complex potential $\Phi(z) = \phi(x, y) + i\psi(x, y)$ can be obtained by combining Eqs. (8.5–12) and (8.5–13) or more directly through the use of Eq. (8.5–6) in Eq. (8.5–8). Thus

$$\Phi(z) = 100\frac{1 - z}{z}. \quad (8.5-14)$$

The singularity at $z = 0$ is typical of the behavior of complex potentials at a point where a boundary condition is discontinuous. The shape of the isotherms (surfaces of constant temperature) for Fig. 8.5–2(a) are of interest. If on some surface the temperature is T_0, the locus of this surface must be, from Eq. (8.5–12),

$$T_0 = 100\left(\frac{x}{x^2 + y^2} - 1\right). \quad (8.5-15)$$

From physical considerations we know that T_0 cannot be greater than the temperature of the hottest part of the boundary nor can it be less than the temperature of the coldest part of the boundary, that is, $0 \le T_0 \le 100°$ (see also Exercises 15 and 16, Section 4.4). We can rearrange Eq. (8.5–15) and complete a square to obtain

$$\left(x - \frac{1/2}{\left(1 + \frac{T_0}{100}\right)}\right)^2 + y^2 = \left(\frac{1/2}{1 + \frac{T_0}{100}}\right)^2.$$

Thus an isotherm of temperature T_0 is a cylinder whose axis passes through

$$y_0 = 0, \qquad x_0 = \left(\frac{1/2}{1 + \frac{T_0}{100}}\right).$$

The cross sections of a few such cylinders are shown as circles in Fig. 8.5–3.

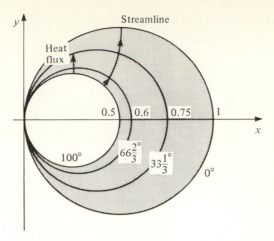

Figure 8.5–3

In Exercise 1 of this section we show that the streamlines describing the heat flow are circles that intersect the isotherms at right angles. A streamline, with an arrow indicating the direction of heat flow, is shown in Fig. 8.5–3.

The complex potential (see Eq. 8.5–14) readily yields the complex heat flux density $q(z)$ between the cylinders. Recall from Eq. (2.6–14) that

$$q(z) = -k \left(\overline{\frac{d\Phi}{dz}} \right), \qquad (8.5–16)$$

where

$$q(z) = Q_x(x, y) + iQ_y(x, y). \qquad (8.5–17)$$

We should remember that Q_x and Q_y give the components of the vector heat flow at a point in a material whose thermal conductivity is k. From Eqs. (8.5–16) and (8.5–14) we have

$$q(z) = Q_x + iQ_y = 100k \left(\overline{\frac{1}{z^2}} \right) = \frac{100k}{(\bar{z})^2}.$$

Thus, for example, at $x = 1/4$, $y = 1/4$ (on top of the inner cylinder) we find

$$Q_x + iQ_y = 100k \frac{1}{\left(\dfrac{1}{4} - \dfrac{1}{4}i \right)^2} = 800ki.$$

Since $Q_x = 0$ and $Q_y = 800k$, the heat flow at $(1/4, 1/4)$ is parallel to the y-axis, as we have indicated schematically in Fig. 8.5–3. ◀

Example 2

Refer to Fig. 8.5–4(a). An electrically conducting strip has a cross section described by $y = 0$, $-1 < x < 1$. It is maintained at 0 volts electrostatic potential. A half cylinder shown in cross section, described by the arc $|z| = 1$, $0 < \arg z < \pi$, is

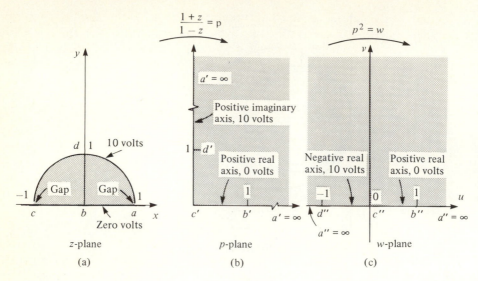

Figure 8.5–4

maintained at 10 volts. Find the potential $\phi(x, y)$ inside the semicircular tube bounded by the two conductors.

Solution

If the boundary of the given configuration (see Fig. 8.5–4a) were transformed into the line Im $w = 0$, in the w-plane, with the half disc in Fig. 8.5–4(a) mapped onto the half space Im $w > 0$, we could use the Poisson integral formula for the half plane (see Section 4.5) to obtain $\phi_1(u, v)$ for the transformed region. A bilinear transformation by itself cannot be used to convert the semicircular boundary in Fig. 8.5–4(a) into a straight line. However, a bilinear transformation *can* map the semicircle into a pair of semiinfinite lines corresponding to the positive real and positive imaginary axes.

One readily verifies that

$$p = \frac{1 + z}{1 - z} \qquad (8.5\text{–}18)$$

performs this transformation (see Fig. 8.5–4b) with the half disc being mapped onto the first quadrant of the p-plane.

Referring now to Fig. 8.4–7(a) and Fig. 8.4–7(b), we see that an additional transformation involving a squaring of p in Eq. (8.5–18) will map the first quadrant of Fig. 8.5–4(b) onto the upper half plane in Fig. 8.5–4(c). Combining both transformations, we find that

$$w = \left(\frac{1 + z}{1 - z}\right)^2 \qquad (8.5\text{–}19)$$

maps the half disc of Fig. 8.5–4(a) onto the half space of Fig. 8.5–4(c). Corresponding boundary points and transformed boundary conditions are indicated in Fig. 8.5–4.

To find the potential in the half space of Fig. 8.5–4(c) that satisfies the transformed boundary conditions $\phi_1(u, 0) = 0$, $u > 0$ and $\phi_1(u, 0) = 10$, $u < 0$ we can use the result of Example 2, Section 4.5, which employed the Poisson integral formula for the half plane. Replacing T_0 of that example by 10 volts and x and y by u and v, we have

$$\phi_1(u, v) = \frac{10}{\pi} \tan^{-1} \frac{v}{u} = \frac{10}{\pi} \arg w, \qquad (8.5\text{--}20)$$

where $0 \leq \arg w \leq \pi$. Notice that

$$\phi_1(u, v) = \text{Re}\left[-i\frac{10}{\pi} \text{Log } w\right],$$

which implies that the complex potential for the configuration of Fig. 8.5–4(c) is

$$\Phi_1(w) = \frac{-10i}{\pi} \text{Log } w. \qquad (8.5\text{--}21)$$

To transform $\phi_1(u, v)$ of Eq. (8.5–20) into $\phi(x, y)$ for the half cylinder we recall the identity $\arg(s^2) = 2 \arg s$. With Eq. (8.5–19) used in Eq. (8.5–20) we have

$$\phi(x, y) = \frac{10}{\pi} \arg\left(\frac{1 + z}{1 - z}\right)^2 = \frac{20}{\pi} \arg \frac{1 + z}{1 - z} = \frac{20}{\pi} \arg \frac{x + 1 + iy}{1 - x - iy}$$

$$= \frac{20}{\pi} \arg\left(\frac{1 - x^2 - y^2}{(x - 1)^2 + y^2} + \frac{i2y}{(x - 1)^2 + y^2}\right),$$

and finally since $\arg s = \tan^{-1}[\text{Im } s/\text{Re } s]$, we find

$$\phi(x, y) = \frac{20}{\pi} \tan^{-1} \frac{2y}{1 - x^2 - y^2}. \qquad (8.5\text{--}22)$$

We require that $0 \leq \tan^{-1}(\cdots) \leq \pi/2$ since $\phi(x, y)$ must satisfy $0 \leq \phi(x, y) \leq 10$. Notice, with this branch of the arctangent, that Eq. (8.5–22) satisfies the required boundary conditions, that is,

$$\lim_{(x^2+y^2)\to 1} \phi(x, y) = 10 \quad \text{(on the curved boundary)},$$

$$\lim_{y\to 0} \phi(x, y) = 0 \quad \text{(on the flat boundary)}.$$

One can easily show that the equipotentials are circular arcs. ◀

A common concern in electrostatics is the amount of capacitance between two conductors. If Q is the electrical charge on either conductor[†] and ΔV is the difference in potential between one conductor and another, then the capacitance c is defined to be

$$c = \frac{|Q|}{|\Delta V|}. \qquad (8.5\text{--}23)$$

[†] The two conductors carry charges that are equal in magnitude and opposite in sign.

In two-dimensional problems we compute the capacitance per unit length of a pair of conductors whose cross section is typically displayed in the complex plane. In Eq. (8.5–23) we take Q as the charge on an amount of one conductor that is 1 unit long in a direction perpendicular to the complex plane. In the appendix to this chapter we establish Theorem 6, which is useful in capacitance calculations.

Theorem 6

The electrical charge per unit length on a conductor that belongs to a charged two-dimensional configuration of conductors is

$$Q = \varepsilon \Delta \psi(z), \qquad (8.5\text{–}24)$$

where ε is a constant (the permittivity of the surrounding material) and $\Delta \psi$ is the decrement (initial value minus final value) of the stream function as we proceed in the positive direction once around the boundary of the cross section of the conductor in the complex plane. ∎

Usually $\psi(z)$ will be a multivalued function defined by means of a branch cut. Thus $\psi(z)$ does not return to its original value when we encircle the conductor, and thus $\Delta \psi \neq 0$. The direction of encirclement is the positive one used in the contour integration, that is, the interior is on the left. Combining Eqs. (8.5–23) and (8.5–24), we have

$$c = \varepsilon \frac{|\Delta \psi|}{|\Delta V|}. \qquad (8.5\text{–}25)$$

Also derived in the appendix to this chapter is this interesting fact:

Theorem 7

The capacitance of a two-dimensional system of conductors is unaffected by a conformal transformation of its cross section. ∎

The usefulness of the two preceding theorems is illustrated by the following example.

Example 3

a) The pair of coaxial electrically conducting tubes in Fig. 8.5–5(a) having radii a and b are maintained at potentials V_a and 0, respectively. Find the electrostatic potential between the tubes, and find their capacitance. This system of conductors is called a *coaxial transmission line*.

b) Use a conformal transformation and the result of part (a) to determine the capacitance of the transmission system consisting of the two conducting tubes shown in Fig. 8.5–5(b). This is called a *two-wire line*.

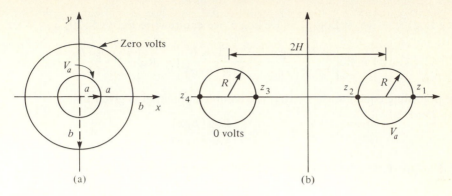

Figure 8.5–5

Solution

Part (a): We seek a function $\phi(x, y)$ harmonic in the domain $a < |z| < b$. The boundary conditions are

$$\lim_{\sqrt{x^2+y^2} \to a} \phi(x, y) = V_a, \qquad (8.5-26)$$

$$\lim_{\sqrt{x^2+y^2} \to b} \phi(x, y) = 0. \qquad (8.5-27)$$

These requirements suggest that the equipotentials are circles concentric with the boundary. We might recall from Example 2, Section 2.5, that $(1/2) \text{Log}(x^2 + y^2) = \text{Log} \sqrt{x^2 + y^2}$ is harmonic and does produce circular equipotentials. However, this function fails to meet the boundary conditions. The more general harmonic function

$$\phi(x, y) = A \,\text{Log}\left(\sqrt{x^2 + y^2}\right) + B, \quad \text{where } A \text{ and } B \text{ are real numbers,} \qquad (8.5-28)$$

also yields circular equipotentials and can be made to meet the boundary conditions. From Eq. (8.5–26) we obtain

$$V_a = A \,\text{Log}\, a + B,$$

and from Eq. (8.5–27) we have

$$0 = A \,\text{Log}\, b + B.$$

Solving these equations simultaneously, we get

$$A = \frac{-V_a}{\text{Log}\,(b/a)}, \qquad B = \frac{V_a \,\text{Log}\, b}{\text{Log}\,(b/a)},$$

which, when used in Eq. (8.5–28), shows that

$$\phi(x, y) = \frac{-V_a \,\text{Log}\, \sqrt{x^2 + y^2}}{\text{Log}\,(b/a)} + \frac{V_a \,\text{Log}\, b}{\text{Log}\,(b/a)}. \qquad (8.5-29)$$

Since $\text{Log}\sqrt{x^2 + y^2} = \text{Log}|z| = \text{Re Log }z$, we can rewrite Eq. (8.5–29) as

$$\phi(x, y) = \text{Re}\left[\frac{-V_a \text{Log }z}{\text{Log}(b/a)} + \frac{V_a \text{Log }b}{\text{Log}(b/a)}\right] = \text{Re}\left[\frac{V_a \text{Log}(b/z)}{\text{Log}(b/a)}\right].$$

The preceding equation shows that the complex potential is given by

$$\Phi(z) = V_a \frac{\text{Log}(b/z)}{\text{Log}(b/a)}. \tag{8.5–30}$$

The stream function, $\psi(x, y) = \text{Im }\Phi(x, y)$ is found from Eq. (8.5–30) to be

$$\psi(x, y) = \text{Im}\left[\frac{V_a(\text{Log }b - \text{Log }z)}{\text{Log}(b/a)}\right] = -\frac{V_a \arg z}{\text{Log}(b/a)}, \tag{8.5–31}$$

where the principal value of $\arg z$, defined by a branch cut on the negative real axis, is used. We now proceed in the counterclockwise direction once around the inner conductor and compute the decrease in ψ (see Fig. 8.5–6).

Just below the branch cut

$$\psi = \frac{-V_a(-\pi)}{\text{Log}(b/a)} = \frac{V_a \pi}{\text{Log}(b/a)},$$

while just above the branch cut

$$\psi = \frac{V_a(-\pi)}{\text{Log}(b/a)}.$$

The decrease in ψ on this circuit is

$$\Delta\psi = \frac{2\pi V_a}{\text{Log}(b/a)}.$$

The magnitude of the potential difference between the two conductors is $|\Delta V| = V_a$ since the outer conductor is at zero potential. Thus, according to Eq. (8.5–25),

$$c = \frac{\varepsilon}{V_a}\frac{2\pi V_a}{\text{Log}(b/a)} = \frac{2\pi\varepsilon}{\text{Log}(b/a)}, \tag{8.5–32}$$

which is a useful result in electrical engineering.

Part (b): Suppose a bilinear transformation with real coefficients can be found that transforms the left-hand circle in Fig. 8.5–5(b) into a circle in the w-plane (see Fig. 8.5–7) in such a way that the points $z_4 = -H - R$ and $z_3 = -H + R$ are mapped into $w_4 = 1$ and $w_3 = -1$.

The real coefficients of the transformation ensure that the left-hand circle in Fig. 8.5–5(b) is transformed into a unit circle centered at the origin in Fig. 8.5–7 (see Exercise 2, Section 8.4). Suppose now the same formula transforms the right-hand circle in Fig. 8.5–5(a) into a circle $|w| = \rho$, with $z_1 = H + R$ having image $w_1 = \rho$, and $z_2 = H - R$ having image $w_2 = -\rho$ (see Fig. 8.5–7). We can solve for ρ by

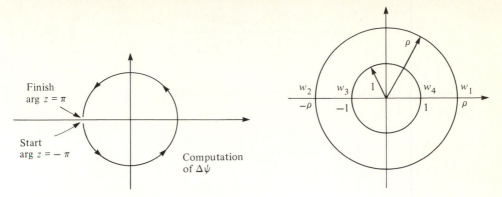

Finish
arg $z = \pi$

Start
arg $z = -\pi$

Computation
of $\Delta\psi$

Figure 8.5–6

Figure 8.5–7

using the equality of the cross ratios (z_1, z_2, z_3, z_4) and (w_1, w_2, w_3, w_4). Thus from Eq. (8.4–24)

$$\frac{2\rho(-2)}{(\rho - 1)(\rho - 1)} = \frac{(2R)(2R)}{(2H + 2R)(-2H + 2R)}. \tag{8.5–33}$$

Some rearrangement yields a quadratic

$$\rho^2 + \left(2 - \frac{4H^2}{R^2}\right)\rho + 1 = 0, \tag{8.5–34}$$

whose solution is

$$\rho = \frac{2H^2}{R^2} - 1 \pm \frac{2H}{R}\sqrt{\frac{H^2}{R^2} - 1}. \tag{8.5–35}$$

Since the two cylinders in Fig. 8.5–5(b) are not touching, we know that $H/R > 1$. The root containing the plus sign in Eq. (8.5–35) therefore exceeds 1. We should recall from our knowledge of quadratic equations that the product of the roots of Eq. (8.5–34) must equal 1. Thus the root containing the minus sign in Eq. (8.5–35) must lie between 0 and 1. Either root can be selected, and the same result will be obtained for the capacitance.

Let us arbitrarily choose the plus sign in Eq. (8.5–35). This corresponds to Fig. 8.5–7, where $\rho > 1$. Notice that with this choice of sign we can rewrite Eq. (8.5–35) as

$$\rho = \left(\frac{H}{R} + \sqrt{\frac{H^2}{R^2} - 1}\right)^2. \tag{8.5–36}$$

We can compute the capacitance of the coaxial system of Fig. 8.5–7 by using Eq. (8.5–32) and taking $a = 1$, $b = \rho$. Thus using Eq. (8.5–36),

$$C = \frac{2\pi\varepsilon}{\text{Log}\left[\left(\frac{H}{R} + \sqrt{\frac{H^2}{R^2} - 1}\right)^2\right]},$$

or

$$C = \frac{\pi \varepsilon}{\mathrm{Log}\left(\dfrac{H}{R} + \sqrt{\dfrac{H^2}{R^2} - 1}\right)}. \qquad (8.5\text{--}37)$$

By Theorem 7 this must be the capacitance of the image of Fig. 8.5–7, that is, the two-wire line of Fig. 8.5–5(b). ◀

EXERCISES

1. a) For Example 1 find the equation of the streamline along which ψ assumes a constant value β. Show that this locus, if drawn in Figure 8.5–2(a), is a circle that is centered on the y-axis and passes through the origin.

   b) Use an argument based on plane geometry to show why such a circle must intersect the isotherms found in this example at right angles.

2. A heat-conducting material occupies the wedge $0 \le \arg z \le \alpha$. The boundaries are maintained at temperatures T_1 and T_2 as shown in Fig. 8.5–8.

   a) Show that $w = u + iv = \mathrm{Log}\, z$ transforms the wedge given above into a strip parallel to the u-axis.

   b) The isotherms in the strip are obviously parallel to the u-axis. Show that the temperature in this region can be described by an expression of the form

   $$\phi(u, v) = Av + B,$$

   and find A and B.

   c) Show that the temperature in the given wedge is

   $$\phi(x, y) = \frac{T_2 - T_1}{\alpha}\, \tan^{-1}\left(\frac{y}{x}\right) + T_1.$$

   d) Show that the complex temperature in the wedge is

   $$\Phi(x, y) = -i\left[\frac{T_2 - T_1}{\alpha}\right] \mathrm{Log}\, z + T_1.$$

   e) Describe the streamlines and isotherms in the wedge.

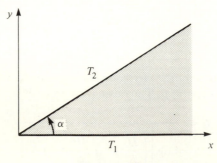

Figure 8.5–8

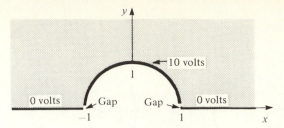

Figure 8.5–9

3. A system of electrical conductors has the cross section shown in Fig. 8.5–9. The potentials of the conductors are maintained as indicated. Determine the complex potential $\Phi(z)$ for the shaded region Im $z > 0$, $|z| > 1$. *Hint:* Consider $w = -1/z$, and try to use the result of Example 2.

4. A cylinder of unit diameter is maintained at a temperature of $100°$. It is tangent to a plane maintained at $0°$ (see Fig. 8.5–10). A material of heat conductivity k exists between the cylinder and the plane, that is for Re $z > 0$, $|z - 1/2| > 1/2$.

a) Find the temperature $\phi(x, y)$ in the material of conductivity k.

b) Find the stream function $\psi(x, y)$.

c) Find the complex heat flux density $q = Q_x(x, y) + iQ_y(x, y)$ in general and at $x = 1$, $y = 1$.

5. An electrically conducting cylinder of radius R has its axis a distance H from an electrically conducting plane (see Fig. 8.5–11a). A bilinear transformation will map the cross section of this configuration into the pair of concentric circles shown in the w-plane (see Fig. 8.5–11b). Image points are indicated with subscripts. Find ρ, the radius of the circle that is the image of the line $x = 0$. Assume $\rho > 1$. Use your result to show that the capacitance, per unit length, between the cylinder and the plane is

$$c = \frac{2\pi\varepsilon}{\mathrm{Log}\left(\dfrac{H}{R} + \sqrt{\dfrac{H^2}{R^2} - 1}\right)}, \qquad H > R.$$

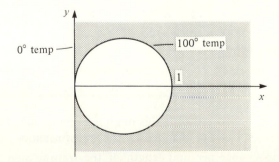

Figure 8.5–10

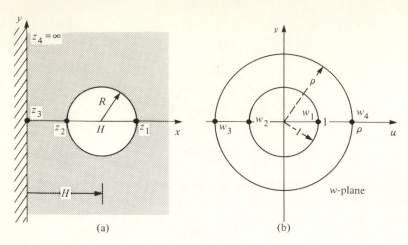

Figure 8.5–11

6. In part (b) of Example 3 show that we cannot obtain a positive value for ρ if the locations of w_3 and w_4 in Fig. 8.5–7 are exchanged, with w_1 and w_2 being kept the same. Assume that z_1, z_2, z_3, and z_4 are unchanged.

7. a) A transmission line consists of two electrically conducting tubes with cross sections as shown in Fig. 8.5–12. Their axes are displaced a distance D. Note that $D + R_1 < R_2$. Show that the capacitance per unit length is given by

$$c = \frac{2\pi\varepsilon}{\text{Log}\,\rho},$$

where

$$\rho = \frac{R_1^2 + R_2^2 - D^2}{2R_1R_2} + \sqrt{\left(\frac{R_1^2 + R_2^2 - D^2}{2R_1R_2}\right)^2 - 1}.$$

Express the capacitance in terms of a hyperbolic function.

b) Take $D = R_1 = 1$, $R_2 = 3$. Let the inner conductor be at 1 volt, the outer at 0. Find the electrostatic potential $\phi(x, y)$ in the domain bounded by the two circles in Fig. 8.5–12.

8. Consider the transformation $z = k\cosh w$, where $k > 0$.

a) Show that the line segment $u = \cosh^{-1}(A/k)$, $-\pi < v \leq \pi$ is transformed into the ellipse $x^2/A^2 + y^2/(A^2 - k^2) = 1$ (see Fig. 8.5–13). Take $A > K$.

b) Show that this transformation takes the infinite line $u = \cosh^{-1}(A/k)$, $-\infty < v < \infty$ into the ellipse of part (a). Is the mapping one to one?

c) Show that the line segment $u = 0$, $-\pi < v \leq \pi$ is transformed into the line segment $y = 0$, $-k \leq x \leq k$. Is this mapping one to one? How is the infinite line $u = 0$, $-\infty < v < \infty$ mapped by the transformation?

d) Find the capacitance per unit length of the transmission line whose cross section is shown in Fig. 8.5–14(a). *Hint:* Find the electrostatic potential

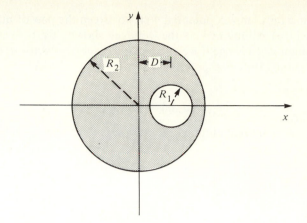

Figure 8.5–12

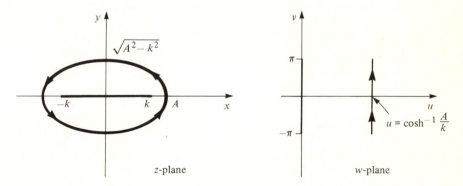

Figure 8.5–13

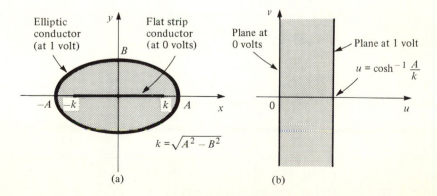

Figure 8.5–14

$\phi(u, v)$ and the complex potential $\Phi(w)$ between the pair of infinite planes in Fig. 8.5–14(b) maintained at the voltages shown. By how much does the stream function ψ change if we encircle the inner conductor in Fig. 8.5–14(a)? *Hint:* Negotiate the corresponding path of Fig. 8.5–14(b).

9. a) The function

$$\phi(u, v) = A \arg w + B, \tag{8.5–38}$$

where A and B are real numbers, and $\arg w$ is the principal value, is harmonic since

$$\phi(u, v) = \text{Re } \Phi(w),$$

where

$$\Phi(w) = -Ai \, \text{Log } w + B.$$

Assume that the line $v = 0$, $u > 0$ is the cross section of an electrical conductor maintained at V_2 volts and that the line $v = 0$, $u < 0$ is similarly a conductor at V_1 volts (see Fig. 8.5–15). Find A and B in Eq. (8.5–38) so that $\phi(u, v)$ will be the electrostatic potential in the space $v > 0$.

b) Obtain $\phi(u, v)$, harmonic for $v > 0$ and satisfying these same boundary conditions along $v = 0$ by using the Poisson integral formula for the upper half plane (see Section 4.5).

c) Find A_1, A_2, and B (all real numbers) so that

$$\phi(u, v) = A_1 \arg (w - u_1) + A_2 \arg (w - u_2) + B$$
$$= \text{Re}\left[-A_1 i \, \text{Log}(w - u_1) - A_2 i \, \text{Log}(w - u_2) + B \right] \tag{8.5–39}$$

is the solution in the space $v \geq 0$ of the electrostatic boundary value problem shown in Fig. 8.5–16, that is, $\phi(u, v)$ is harmonic for $v \geq 0$ and satisfies

$$\phi(u, 0) = V_1, \quad u < u_1; \quad \phi(u, 0) = V_2, \quad u_1 < u < u_2;$$
$$\phi(u, 0) = V_3, \quad u > u_2.$$

d) Let $u_1 = -1$ and $u_2 = 1$, $V_1 = V_3 = 0$ and $V_2 = V$ in the configuration of part (c) (see Fig. 8.5–17). Use the result of part (c) to show that $\phi(u, v)$,

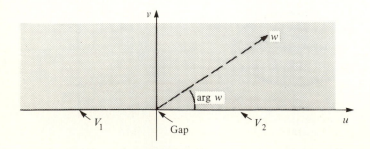

Figure 8.5–15

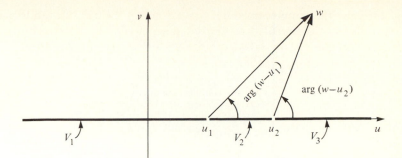

Figure 8.5–16

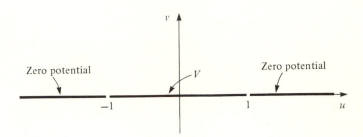

Figure 8.5–17

harmonic for $v > 0$ and meeting these boundary conditions, is given by

$$\phi(u, v) = \text{Re } \Phi(w),$$

where the complex potential is

$$\Phi(w) = -i\frac{V}{\pi} \text{ Log}\left(\frac{w - 1}{w + 1}\right),$$

and

$$\phi(u, v) = \frac{V}{\pi} \tan^{-1}\frac{2v}{u^2 + v^2 - 1}$$

for $0 \leq \tan^{-1}(\cdots) \leq \pi$.

e) Sketch on Fig. 8.5–17 the equipotentials for which $\phi(x, y) = V/2$ and $\phi(x, y) = V/4$. Give the equation of each equipotential.

10. a) A material having heat conductivity k has a cross section occupying the first quadrant of the z-plane. The boundaries are maintained as shown in Fig. 8.5–18. Find the temperature in the material at (x, y). *Hint:* Try to transform the given configuration into one resembling that in part (d) of Exercise 9. Use the result of that exercise.

b) Sketch the variation in temperature with distance along the line $x = y$ from $x = 0$ to $x = 2$.

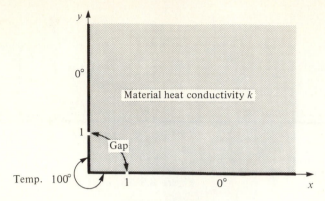

Figure 8.5–18

c) Find the stream function $\psi(x, y)$ associated with $\phi(x, y)$.

d) Find $q(x, y) = Q_x + iQ_y$, the complex heat flux density in the conducting material.

11. a) The boundaries of a heat conducting material are maintained at the temperatures shown in Fig. 8.5–19. Find the temperature $\phi(x, y)$, $-a < x < a$, $y > 0$ inside the material. *Hint:* Begin with the transformation in Example 3, Section 8.3.

    b) Plot $\phi(x, a)$ for $-a \leq x \leq a$.

12. a) Show that under the transformation $w = \cos^{-1}(z/a)$ or $z = a \cos w$, that the lines $y = 0$, $-\infty < x \leq -a$ and $y = 0$, $a \leq x < \infty$ are mapped into the w-plane as shown in Fig. 8.5–20. Assume $a > 0$.

    b) Show that the transformation maps one to one the domain consisting of the z-plane, with the points satisfying $y = 0$, $|x| \geq a$ removed, onto the strip $0 < u < \pi$ shown in Fig. 8.5–20(b). Consider the rectangular region in the strip satisfying $u_1 \leq u \leq u_2$, $v_1 \leq v \leq v_2$. What is its image in the z-plane?

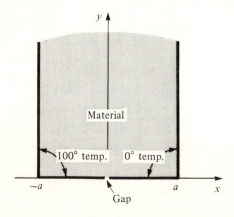

Figure 8.5–19

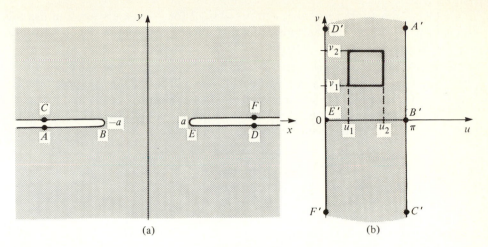

Figure 8.5-20

c) Two semiinfinite electrically conducting sheets, shown in cross section in Fig. 8.5–21, are separated a distance $2a$. The conductors are maintained at voltages V_0 and 0. Show that the complex potential in the surrounding space is given by

$$\Phi(z) = \frac{V_0}{\pi} \cos^{-1}\left(\frac{z}{a}\right).$$

d) For $a = 1$ show that the electrostatic potential is given by

$$\phi(x, y) = \frac{V_0}{\pi} \cos^{-1} \pm \sqrt{\left(\frac{x^2 + y^2 + 1}{2}\right) \pm \sqrt{\left(\frac{x^2 + y^2 + 1}{2}\right)^2 - x^2}},$$

where $0 \le \cos^{-1}(\cdots) \le \pi$. *Hint:* Find $\phi(u)$ in the w-plane. Write u in terms of x and y by observing that $x = \cos u \cosh v$, $y = -\sin u \sinh v$ so that $(x^2/\cos^2 u) - (y^2/\sin^2 u) = 1$. Solve this for $u(x, y)$.

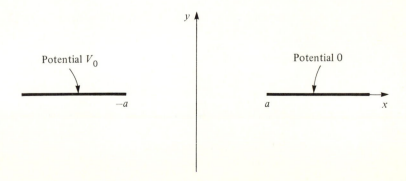

Figure 8.5–21

e) Use symmetry to argue that $\phi(0, y) = V_0/2$ for $-\infty < y < \infty$, and use this fact to establish the sign preceding the inner square root in $\phi(x, y)$ given in part (d).

f) Verify that $\phi(x, y)$ in part (d) satisfies the assigned boundary conditions along the lines $x \geq 1, y = 0$ and $x \leq 1, y = 0$ provided the sign preceding the outer square root in the expression is taken as positive in quadrants 1 and 4 and negative in quadrants 2 and 3. Does this mean that $\phi(x, y)$ is discontinuous as we cross the y-axis? Explain.

g) For $a = 1$ show that the complex electric field is given by $V_0\overline{(1 - z^2)^{-1/2}}/\pi$, where $(1 - z^2)^{-1/2}$ is defined by means of branch cuts along lines corresponding to the conductors in Fig. 8.5–21.

8.6 MORE ON BOUNDARY VALUE PROBLEMS—
STREAMLINES AS BOUNDARIES

In the Dirichlet problems just considered a harmonic function was obtained that assumed certain preassigned values on the boundary of a domain. In this section we will study boundary value problems that are not Dirichlet problems; a function that is harmonic in a domain will be sought, but the values assumed by this function everywhere on the boundary are not necessarily given. Instead, information regarding the derivative of the function on the boundary is supplied. We will see how this can happen in some heat-flow problems and will use conformal mapping in their solution. In the exercises we will also see how this occurs in configurations involving fluid flow.

If a heat-conducting material is surrounded by certain surfaces that provide perfect thermal insulation, then, by definition, there can be no flow of heat into or out of these surfaces. The heat flux density vector Q cannot have a component normal to the surface (see Eq. 2.6–2). We will consider only two-dimensional configurations and employ a complex temperature (see Eq. 2.6–6) of the form

$$\Phi(x, y) = \phi(x, y) + i\psi(x, y), \tag{8.6–1}$$

where $\phi(x, y)$ is the actual temperature, and $\psi(x, y)$ is the stream function. In Section 2.6 we observed that the streamlines, that is, the lines on which $\psi(x, y)$ assumes fixed values, are tangent at each point to the heat flux density vector. Fig. 8.6–1 shows the cross section of a heat-conducting material whose boundary is in part insulated.

The insulated part of the boundary must coincide with a streamline.

Otherwise, the heat flux density vector Q would have a component normal to the insulation. We observed in Section 2.6 that the streamlines and isotherms form a mutually orthogonal set of curves. Suppose, starting at the insulated surface shown in Fig. 8.6–1, we proceed along the line N, which is *normal* to the insulation. At the insulation we must be moving along an isotherm, and ϕ does not change. Mathematically, this is stated as

$$\frac{d\phi}{dn} = 0 \quad \text{(at insulated surface)}, \tag{8.6–2}$$

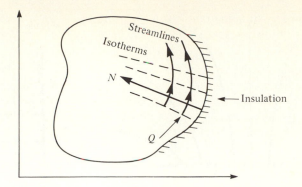

Figure 8.6–1

where n is the distance measured along the normal N. Equation 8.6–2 asserts that the "normal derivative" of the temperature vanishes at an insulated boundary.

When we are given boundary-value problems in heat conduction in which the temperature is specified on some portions of the boundary, while the remaining portions are insulated, we proceed in a manner like that used in the Dirichlet problems of the previous section: We map the cross section of the configuration from, say, the z-plane into a simpler or more familiar shape in the w-plane by means of an analytic transformation $w = f(z)$. As before, at the boundaries of the new domain in the w-plane we assign those temperatures, if known, that exist at the corresponding image points in the z-plane. There will now also be insulated boundaries in the w-plane corresponding to insulated boundaries in the z-plane.

We now seek a complex potential, an analytic function $\Phi_1(w) = \phi_1(u, v) + i\psi_1(u, v)$, such that $\phi_1(u, v)$ assumes the known assigned values at boundary points in the w-plane. We also require that $\psi_1(u, v)$ produce streamlines coinciding with the insulated boundaries in the w-plane. As before, a transformation back into the z-plane yields the analytic function $\Phi(z) = \phi(x, y) + i\psi(x, y)$, where $\phi(x, y)$ is the required temperature and $\psi(x, y)$ is the associated stream function. Now $\psi(x, y)$ will produce streamlines coinciding with the insulated boundaries and $\phi(x, y)$ will assume the prescribed values on the remaining boundaries. An example of the method follows.

Example 1

Refer to Fig. 8.6–2(a). A heat-conducting material fills the space $y > 0$. The boundary $y = 0$, $x > 1$ is maintained at $100°$, the boundary $y = 0$, $x < -1$ is at $0°$ while $y = 0$, $|x| < 1$ is insulated. Find the temperature distribution $\phi(x, y)$ and the complex temperature $\Phi(z)$ in the material.

Solution

We seek a transformation that will take the given region into a more tractable shape. Referring to Example 3, Section 8.3 and to Fig. 8.3–5 (or to a table of transformations), we have a useful clue. Reversing the role of z and w in that example, we find that the transformation

$$z = \sin w \tag{8.6–3}$$

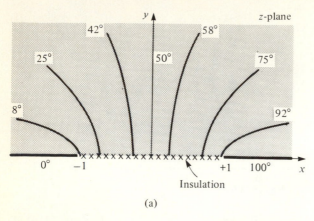

Figure 8.6–2

or

$$w = \sin^{-1}(z)$$

maps the region shown in Fig. 8.6–2(a) onto the region in Fig. 8.6–2(b). The boundary conditions are transformed as indicated.

To solve the transformed problem we note that a temperature $\phi_1(u)$ of the form

$$\phi_1(u) = Au + B, \quad \text{where } A \text{ and } B \text{ are real numbers} \quad \text{and} \quad -\frac{\pi}{2} \leq u \leq \frac{\pi}{2},$$

$$(8.6–4)$$

will produce isotherms coincident with the boundaries along $u = -\pi/2$ and $u = \pi/2$. The associated streamlines will be parallel to the insulated boundary. In particular there will be a streamline coincident with $v = 0$, $-\pi/2 \leq u \leq \pi/2$ as required. To determine A and B, we apply the boundary conditions $\phi_1(-\pi/2) = 0$ and $\phi_1(\pi/2) = 100$ in Eq. (8.6–4). The first condition yields

$$0 = -A\frac{\pi}{2} + B,$$

and the second yields

$$100 = A\frac{\pi}{2} + B.$$

Solving these equations simultaneously, we have $A = 100/\pi$ and $B = 50$ so that Eq. (8.6–4) becomes

$$\phi_1(u) = \frac{100}{\pi}u + 50, \quad -\frac{\pi}{2} \leq u \leq \frac{\pi}{2}. \qquad (8.6–5)$$

Noticing that $\phi_1(u) = \text{Re}[(100/\pi)w + 50]$, we realize that the complex temperature is

$$\Phi_1(w) = \frac{100}{\pi}w + 50, \quad |\text{Re } w| \leq \frac{\pi}{2}, \qquad (8.6–6)$$

and the stream function is

$$\psi_1(v) = \text{Im}\,\Phi_1(w) = \frac{100v}{\pi}. \tag{8.6-7}$$

Since $w = \sin^{-1}(z)$, the complex temperature in the z-plane can be obtained, from a substitution in Eq. (8.6–6):

$$\Phi(z) = \frac{100}{\pi}\sin^{-1}z + 50, \tag{8.6-8}$$

where $-\pi/2 \le \text{Re}\,\sin^{-1}z \le \pi/2$.

To obtain the actual temperature $\phi(x, y)$ we have from Eq. (8.6–3)

$$z = (x + iy) = \sin w = \sin u \cosh v + i \cos u \sinh v,$$

so that

$$x = \sin u \cosh v,$$
$$y = \cos u \sinh v,$$

and since $\cosh^2 v - \sinh^2 v = 1$, we find that

$$\frac{x^2}{\sin^2 u} - \frac{y^2}{\cos^2 u} = 1.$$

We now eliminate $\cos^2 u$ from the above by employing $\cos^2 u = 1 - \sin^2 u$, which yields

$$\frac{x^2}{\sin^2 u} - \frac{y^2}{1 - \sin^2 u} = 1. \tag{8.6-9}$$

We multiply both sides of Eq. (8.6–9) by $\sin^2 u(1 - \sin^2 u)$ and obtain a quadratic equation in $\sin^2 u$ (a quartic in $\sin u$). The quadratic formula yields

$$\sin^2 u = \frac{(x^2 + y^2 + 1)}{2} \pm \sqrt{\left(\frac{x^2 + y^2 + 1}{2}\right)^2 - x^2}.$$

We take the square root of both sides of this expression and then use $u = \sin^{-1}(\sin u)$ to obtain

$$u = \sin^{-1}\left[\pm\sqrt{\frac{x^2 + y^2 + 1}{2} \pm \sqrt{\left(\frac{x^2 + y^2 + 1}{2}\right)^2 - x^2}}\,\right]. \tag{8.6-10}$$

A substitution in Eq. (8.6–5) yields

$$\phi(x, y) = \frac{100}{\pi}\sin^{-1}\left(\pm\sqrt{\frac{x^2 + y^2 + 1}{2} \pm \sqrt{\left(\frac{x^2 + y^2 + 1}{2}\right)^2 - x^2}}\,\right) + 50. \tag{8.6-11}$$

The conditions attached to Eq. (8.6–5) here require $-\pi/2 \le \sin^{-1}(\cdots) \le \pi/2$. From physical arguments we can establish that the temperature in the heat-conducting material can be no less than 0, nor can it exceed 100. To determine the appropriate signs for the square roots note that $x = 0$, $y = 0+$ lies midway between the two conductors and by symmetry will be at a temperature of 50. This condition requires that the inner $\pm$ operator be negative. The boundary conditions $\phi(x, 0) = 0$, $x < -1$ and $\phi(x, 0) = 100$, $x > 1$ demand that the outer $\pm$ operator be positive in the first quadrant and negative in the second quadrant. Note that there is no discontinuity in temperature as we cross the positive y-axis. The isotherms in the w-plane are (from Eq. 8.6–5) those surfaces on which u is constant. According to Eq. (8.6–9) these isotherms become hyperbolas in the xy-plane. Some are sketched in Fig. 8.6–2(a) for various temperatures. ◀

EXERCISES

1. A material of heat conductivity k has a cross section that occupies the first quadrant. The boundaries are maintained at the temperatures indicated in Fig. 8.6–3.

   a) Show that the complex temperature inside the conducting material is given by

$$\Phi(z) = \frac{100}{\pi} \sin^{-1}(z^2) + 50, \qquad -\frac{\pi}{2} \le \text{Re} \sin^{-1}(\cdots) \le \frac{\pi}{2}.$$

   Hint: Map the region of this problem onto that presented in Example 1.

   b) Show that the temperature inside the material is given by

$$\phi(x, y) = 50 + \frac{100}{\pi} \sin^{-1}$$

$$\pm \sqrt{\frac{1 + (x^2 + y^2)^2}{2} \pm \sqrt{\frac{[1 + (x^2 + y^2)^2]^2}{4} - (x^2 - y^2)^2}}$$

   for $-\pi/2 < \sin^{-1}(\cdots) < \pi/2$, and the appropriate signs are used in front of each square root.

   c) Plot a curve showing the variation in temperature with distance along the insulated boundary lying along the x-axis.

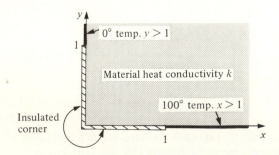

Figure 8.6–3

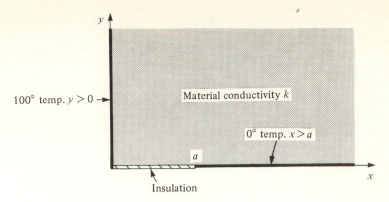

Figure 8.6–4

d) Show that the complex heat flux density is

$$q = -\frac{k200}{\pi} \left(\frac{z}{(1 - z^4)^{1/2}} \right) = Q_x + iQ_y.$$

e) Let $k = 1$. By choosing the appropriate values of the square root, give the numerical values of the components of q, that is, Q_x and Q_y, at the following locations:

$$x = 1/2, \, y = 0+ \, ; \qquad x = 2, \, y = 0+ \, ; \qquad x = 0+ \, , \, y = 1/2;$$
$$x = 0+ \, , \, y = 2.$$

2. a) A material of heat conductivity k has boundaries as shown in Fig. 8.6–4. Show that the complex temperature in the material is given by

$$\Phi = -\frac{200}{\pi} \sin^{-1}\left(\frac{z}{a} \right) + 100, \quad \text{for } 0 \le \operatorname{Re} \sin^{-1}(\cdots) \le \frac{\pi}{2}.$$

Hint: Consider the mapping $z = a \sin w$ applied to the strip $0 \le \operatorname{Re} w \le \pi/2$, $\operatorname{Im} w \ge 0$.

b) Show that the isotherm having temperature T lies on the hyperbola described by

$$\frac{x^2}{a^2 \sin^2 u} - \frac{y^2}{a^2 \cos^2 u} = 1,$$

where $u = (100 - T)\pi/200$.

c) Sketch the isotherm $T = 50$.

3. The outside of a heat-conducting rod of unit radius is maintained at the temperatures shown in Fig. 8.6–5. One half of the boundary is insulated. Show that the complex temperature inside the rod is given by

$$\Phi(z) = 50 - \frac{100}{\pi} \sin^{-1} \frac{i(z + i)}{z - i}, \qquad -\frac{\pi}{2} \le \operatorname{Re} \sin^{-1}(\cdots) \le \frac{\pi}{2}.$$

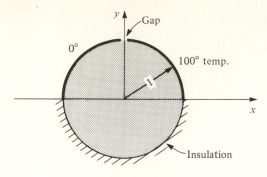

Figure 8.6–5

Hint: A bilinear transformation will map the configuration into that of Example 1 (see Fig. 8.6–2a).

The following problems treat fluid flow in the presence of a rigid, impenetrable boundary.

4. When a rigid, impenetrable obstacle is placed within a moving fluid, no fluid passes through the surface of that object. At each point on the object's surface the component of the fluid velocity vector normal to the surface must vanish; otherwise, there would be penetration by the fluid. Since flow is tangential to the surface of the obstacle, its boundary must be coincident with a streamline.

The simplest type of fluid motion in the presence of a boundary is that of uniform flow parallel to and above an infinite plane (see Fig. 8.6–6). The complex potential describing the fluid flow is $\Phi = Aw$, where $w = u + iv$, and A is a real number. A is positive for flow to the right, negative for flow to the left.

a) Compute the complex fluid velocity from Φ, and verify that the flow is indeed uniform and parallel to the plane, that is, parallel to the u-axis.

b) Find the stream function $\psi(u, v)$ for the flow. What is the value of ψ along the boundary? Plot the loci of $\psi = 0$, $\psi = A$, $\psi = 2A$ on Fig. 8.6–6.

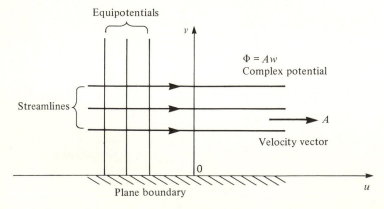

Figure 8.6–6

c) The fluid flow in the space $v \geq 0$ described in Fig. 8.6–6 is transformed into the z-plane by means of $z = w^{1/2}$, where the principal branch of the square root is used. Show that the plane boundary of Fig. 8.6–6 is mapped into the right angle boundary in Fig. 8.6–7. Show that the complex velocity potential $\Phi(w)$ of Fig. 8.6–6 is transformed into $\Phi(z) = Az^2$, which describes flow within the boundary.

d) Show that the complex velocity for flow in the corner is $2Ax - i2Ay$. Show that the speed with which the fluid moves at a point varies directly with the distance of that point from the corner. Show that fluid flow is in the negative y-direction along the wall $x = 0$, $y > 0$ and in the positive x-direction along the wall $y = 0$, $x > 0$.

e) Find the equation for the stream function $\psi(x, y)$ for flow in the corner. Show that the streamlines are hyperbolas and that one coincides with the fluid boundary.

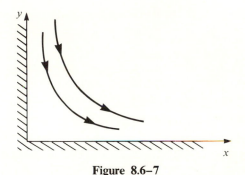

Figure 8.6–7

5. Fluid flows into and out of the 135° corner shown in Fig. 8.6–8.

a) Show that the complex potential describing the flow is of the form

$$\Phi(z) = Az^{4/3}, \quad \text{where } A \text{ is a positive real constant.}$$

Hint: Find a transformation that will map the region $v \geq 0$ from Fig. 8.6–6 onto the region of flow in Fig. 8.6–8. Apply this same transformation to the uniform flow in Fig. 8.6–6.

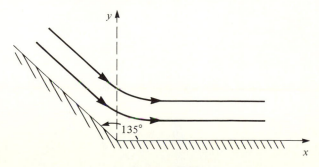

Figure 8.6–8

b) Use $z = re^{i\theta}$ to convert $\Phi(z)$ to polar coordinates, and show that the velocity potential and stream function are given, respectively, by

$$\phi(r, \theta) = Ar^{4/3} \cos \frac{4\theta}{3} \quad \text{and} \quad \psi(r, \theta) = Ar^{4/3} \sin \frac{4\theta}{3}.$$

c) Use $\psi(r, \theta)$ to sketch the streamlines $\psi = 0$ and $\psi = A$.

d) Find the complex fluid velocity. Show that the fluid moves parallel to the wall at $\theta = 135°$ and that the flow is toward the vertex. Show that the fluid moves parallel to the wall at $\theta = 0°$ and that the flow is away from the vertex.

6. In this exercise we study fluid flow into a closed channel by transforming the uniform fluid flow described in Fig. 8.6–6. Using

$$z = \sin^{-1} w, \quad -\frac{\pi}{2} \le \text{Re} \sin^{-1} w \le \frac{\pi}{2}.$$

a) Show that the plane boundary $v = 0$ of Fig. 8.6–6 is transformed into the closed channel shown in Fig. 8.6–9.

b) Show that the complex fluid velocity in the channel is given by $A \cos x \cosh y + iA \sin x \sinh y$.

c) Show that fluid flows in the negative y-direction along the left wall in the channel, in the positive x-direction along the end of the channel, and in the positive y-direction along the right wall of the channel. Assume $A > 0$.

d) Show that the stream function that describes flow in this channel is $\psi = A \cos x \sinh y$.

e) Plot the streamlines $\psi = 0$, $\psi = A/2$, and $\psi = A$ on Fig. 8.6–9.

7. a) A fluid flows with uniform velocity V_0 in the direction shown in Fig. 8.6–10 along a channel of width π. Show that the complex potential $\Phi = iwV_0$ describes the flow and satisfies the requirement that the walls of the channel be streamlines.

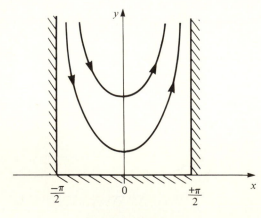

Figure 8.6–9

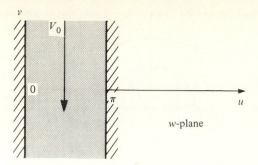

Figure 8.6–10

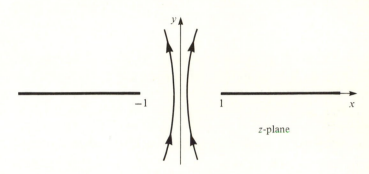

Figure 8.6–11

b) Use the transformation $z = \cos w$ to map this channel and its flow into the z-plane (see Exercise 12, Section 8.5). Show that the flow in the z-plane is through an aperture of width 2 located in the plane $y = 0$. (see Fig. 8.6–11). What is the complex potential describing the flow in the z-plane?

c) With what speed and in what direction is the fluid flowing at the center of the aperture $z = 0$?

d) Find the equation and sketch the locus of the streamline passing through $y = 0$, $x = 1/2$.

e) Show that "far" from the aperture, $|z| \gg 1$, the components of velocity are given approximately by

$$v_x = \frac{V_0 \cos \theta}{r}, \qquad v_y = \frac{V_0 \sin \theta}{r}, \qquad 0 \le \theta \le \pi,$$

and

$$v_x = \frac{-V_0 \cos \theta}{r}, \qquad v_y = \frac{-V_0 \sin \theta}{r}, \qquad \pi \le \theta \le 2\pi,$$

where $z = re^{i\theta}$.

8. In this exercise we study flow around a half cylinder obstruction in a plane. Fluid flow above an infinite plane, as described in Fig. 8.6–6, is transformed by

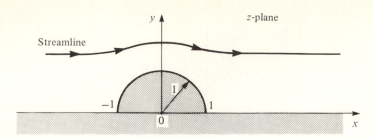

Figure 8.6–12

means of the formula $z = w/2 + (w^2/4 - 1)^{1/2}$, The transformation involves branch cuts extending from $w = \pm 2$ into the lower half plane. The image of $w = 0$ is $z = i$.

a) Show that the image of the axis $v = 0$ in Fig. 8.6–6 is the fluid boundary shown in Fig. 8.6–12 and that the space $v > 0$ is mapped onto the region above this boundary. *Hint:* Show that the inverse of our transformation is $w = z + 1/z$. Use this to transform the boundary in Fig. 8.6–12 into $v = 0$ in Fig. 8.6–6.

b) Show that $\Phi(z) = A(z + z^{-1})$ is the complex potential describing flow in the z-plane.

c) Show that the complex fluid velocity in Fig. 8.6–12 is $A(1 - 1/(\bar{z})^2)$. Why does this indicate a uniform flow of fluid to the right in Fig. 8.6–12 when we are far from the half cylinder obstruction?

d) Let $z = re^{i\theta}$. Show that in polar coordinates the stream function describing the flow is $\psi = A(r - 1/r)\sin\theta$. What is the value of ψ on the streamline that coincides with the fluid boundary? Sketch the streamline $\psi = A$ on Fig. 8.6–12.

8.7   THE SCHWARZ–CHRISTOFFEL TRANSFORMATION

Many physical problems in heat conduction, fluid mechanics, and electrostatics involve boundaries whose cross sections form a polygon. In the domain bounded by the polygon we seek a harmonic function satisfying certain boundary conditions. A one-to-one mapping $w = f(z)$ that would transform this domain from the z-plane onto the upper half of the w-plane, with the polygonal boundary transformed into the real axis, would greatly assist us in solving our problem because of the simplified shape now obtained. We discuss here something close to what is required; the Schwarz–Christoffel transformation is a formula that will transform the real axis of the w-plane (the u-axis) into a polygon in the z-plane. Once this formula is obtained (often a formidable task) an inversion can sometimes be applied that yields the desired $w = f(z)$.

A rigorous derivation of the Schwarz–Christoffel transformation will not be presented. Instead we will first convince the reader of its plausibility and then move on to some examples of its use.

To see how the formula operates, consider the simple transformation

$$z = (w - u_1)^{\alpha_1/\pi}, \qquad 0 \le \alpha_1 \le 2\pi, \tag{8.7-1}$$

where $(u_1, 0)$ is a point on the real axis of the w-plane. Equation (8.7–1) is defined by means of a branch cut originating at $w = u_1$ and going into the lower half of this plane. Equating arguments on both sides of Eq. (8.7–1), we have

$$\arg z = \frac{\alpha_1}{\pi} \arg(w - u_1). \tag{8.7-2}$$

If w is real with $w > u_1$, we take $\arg(w - u_1) = 0$, and from Eq. (8.7–2)

$$\arg z = 0. \tag{8.7-3}$$

Now if w is real with $w < u_1$, we have $\arg(w - u_1) = \pi$, and from Eq. (8.7–2)

$$\arg z = \frac{\alpha_1}{\pi} \pi = \alpha_1. \tag{8.7-4}$$

Equation (8.7–1) indicates that the points $w = u_1$ and $z = 0$ are images of each other. Refer now to Fig. 8.7–1(a) and (b). If we consider a line segment on the u-axis to the right of $w = u_1$, it must, according to Eq. (8.7–3), be transformed into a line segment in the z-plane emanating from the origin and lying along the x-axis. The line segment to the left of $w = u_1$ is, according to Eq. (8.7–4), transformed into a ray making an angle α_1 with the positive x-axis.

To summarize: $z = (w - u_1)^{\alpha_1/\pi}$ bends a straight line segment that lies on the u-axis and passes through $w = u_1$ into a pair of line segments intersecting at the origin of the z-plane with angle α_1. The more complicated transformation

$$z = c_1(w - u_1)^{\alpha_1/\pi} + c_2, \tag{8.7-5}$$

applied to the straight line of Fig. 8.7–1(a), results in the pair of line segments shown in Fig. 8.7–1(c). The angle of intersection is still α_1 but the segments no longer emanate from the origin and, in general, are rotated from their original orientation.

A transformation $z = g(w)$ that simultaneously bends several line segments on the u-axis into straight line segments in the z-plane intersecting at various angles and

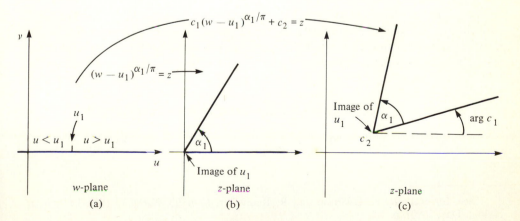

Figure 8.7–1

different locations should, in principle, transform the entire u-axis into a polygon in the z-plane. Notice from Eq. (8.7–5) that

$$\frac{dz}{dw} = c_1 \frac{\alpha_1}{\pi} (w - u_1)^{(\alpha_1/\pi) - 1}.$$

This suggests our considering the following formula in order to transform the u-axis into a polygon.

$$\frac{dz}{dw} = A(w - u_1)^{(\alpha_1/\pi) - 1}(w - u_2)^{(\alpha_2/\pi) - 1} \ldots (w - u_n)^{(\alpha_n/\pi) - 1},$$

where $(u_1, 0)$, $(u_2, 0), \ldots$, etc. are the images in the w-plane of the vertices of the polygon and $\alpha_1, \alpha_2, \ldots, \alpha_n$ are the angles of intersection of the sides of the polygon in the z-plane. In fact, our assumption about this formula is correct and is summarized in the following theorem:

Theorem 8   The Schwarz–Christoffel transformation

The real axis in the w-plane is transformed into a polygon in the z-plane having vertices at $z_1, z_2, \ldots, z_n$ and corresponding interior angles $\alpha_1, \alpha_2, \ldots, \alpha_n$ by the formula

$$\frac{dz}{dw} = A(w - u_1)^{(\alpha_1/\pi) - 1}(w - u_2)^{(\alpha_2/\pi) - 1} \ldots (w - u_n)^{(\alpha_n/\pi) - 1}, \quad (8.7\text{–}6)$$

or

$$z = A \int^w (\zeta - u_1)^{(\alpha_1/\pi) - 1}(\zeta - u_2)^{(\alpha_2/\pi) - 1} \ldots (\zeta - u_n)^{(\alpha_n/\pi) - 1} d\zeta + B,$$

$$(8.7\text{–}7)$$

where $(u_1, 0)$, $(u_2, 0), \ldots, (u_n, 0)$ are mapped into the vertices $z_1, z_2, \ldots, z_n$. If $w = \infty$ is mapped into one vertex, say z_j, then the term containing $(w - u_j)$ is absent in Eq. (8.7–6), and the term containing $(\zeta - u_j)$ is absent in Eq. (8.7–7). The size and orientation of the polygon is determined by A and B. The half plane $\text{Im } w > 0$ is mapped onto the interior of the polygon. ■

A lower limit has not been specified for the integral in Eq. (8.7–7). The reader can choose this quantity arbitrarily. Note, however, that any constant this might produce can be absorbed into B. The integration is performed on the dummy variable ζ. Differentiation of both sides of Eq. (8.7–7) with respect to w yields Eq. (8.7–6) according to the fundamental theorem of integral calculus[†] applied to contour integrals.

To see how the transformation operates, we have from Eq. (8.7–6) that

$$dz = A(w - u_1)^{(\alpha_1/\pi) - 1}(w - u_2)^{(\alpha_2/\pi) - 1} \ldots (w - u_n)^{(\alpha_n/\pi) - 1} dw.$$

[†]See, for example, N. Levinson and R. Redheffer, *Complex Variables* (San Francisco: Holden-Day, 1970), p. 100.

Equating the arguments on both sides, we have

$$\arg dz = \arg A + \left(\frac{\alpha_1}{\pi} - 1\right)\arg(w - u_1) + \left(\frac{\alpha_2}{\pi} - 1\right)\arg(w - u_2) + \cdots$$

$$+ \left(\frac{\alpha_n}{\pi} - 1\right)\arg(w - u_n) + \arg dw. \tag{8.7-8}$$

Imagine now that the point w lies at the location marked P in Fig. 8.7–2(b). We take P to the left of $u_1, u_2, \ldots, u_n$.

As w moves through the increment $dw = du$ along the real axis toward u_1, we have $\arg dw = 0$. Since $(w - u_1), (w - u_2), \ldots, (w - u_n)$ are all negative real numbers when w is to the left of u_1, the arguments of these terms are all π in Eq. (8.7–8), and $\arg dz$ in this equation remains constant as w proceeds toward u_1. The argument of dz can only remain fixed along some locus if that locus is a straight line. Hence, as w moves toward u_1 in Fig. 8.7–2(b), the locus traced out by z, as defined in Eqs. (8.7–6) and (8.7–7), is a line segment.

When w moves through u_1, $\arg(w - u_1)$ in Eq. (8.7–8) abruptly decreases by π. However, all other arguments in this equation remain constant at their original values. According to Eq. (8.7–8), $\arg dz$ will change abruptly in value. It *decreases* by $((\alpha_1/\pi) - 1)\pi = \alpha_1 - \pi$ or *increases* by $\pi - \alpha_1$. If w, which is now to the right of u_1, moves toward u_2 along the u-axis in Fig. 8.7–2(b), $\arg dz$ remains fixed at its new value and a new line segment is traced in the z-plane. The increase in argument $\pi - \alpha_1$ just noted causes the two line segments that have been generated in the z-plane to intersect at z_1 with an angle α_1 (see Fig. 8.7–2a).

As w continues to the right along the u-axis in Fig. 8.7–2(b), we see from Eq. (8.7–8) that $\arg dz$ will abruptly increase by $\pi - \alpha_2$, and as w moves between u_2 and u_3 a new line segment is generated in the z-plane making an angle α_2 with the previous one. In this way the transformation defined by Eq. (8.7–6) or Eq. (8.7–7) generates an entire polygon in the z-plane as w progresses along the whole real axis from $-\infty$ to $+\infty$ in the w-plane.

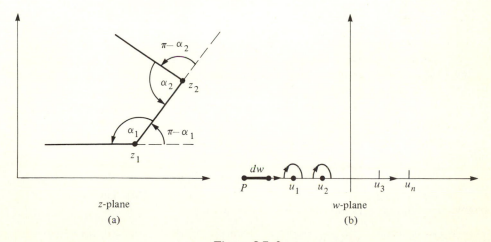

z-plane
(a)

w-plane
(b)

Figure 8.7–2

Recall from plane geometry that the sum of the exterior angles of a *closed* polygon is 2π. The exterior angle at z_1 in Fig. 8.7–2(a) is $\pi - \alpha_1$, at z_2 it is $\pi - \alpha_2$, etc. Thus

$$(\pi - \alpha_1) + (\pi - \alpha_2) + \cdots + (\pi - \alpha_n) = 2\pi.$$

If we divide both sides of this equation by π and then multiply by (-1), we obtain a relationship that the exponents in Eqs. (8.7–6) and (8.7–7) must satisfy if the u-axis is to be transformed into a closed polygon.

$$\frac{\alpha_1}{\pi} - 1 + \frac{\alpha_2}{\pi} - 1 + \cdots + \frac{\alpha_n}{\pi} - 1 = -2. \qquad (8.7–9)$$

This relationship holds in Examples 1 and 2, which follow, but not in Example 3, where an open polygon is considered.

Let us see how Eqs. (8.7–6) and (8.7–7) must be modified if one vertex of a polygon, say z_n, is to have the image $u_n = \infty$. First we divide and multiply the right side of Eq. (8.7–6) by

$$(-u_n)^{(\alpha_n/\pi)-1}.$$

Thus

$$\frac{dz}{dw} = A(-u_n)^{(\alpha_n/\pi)-1}(w - u_1)^{(\alpha_1/\pi)-1}(w - u_2)^{(\alpha_2/\pi)-1} \cdots \left(\frac{w - u_n}{-u_n}\right)^{(\alpha_n/\pi)-1}.$$

As $u_n \to \infty$, the last factor on the right can be taken as 1 while the product of the first two factors $A(-u_n)^{\alpha_n/\pi - 1}$ is maintained finite in the limit and absorbed into a new constant independent of w, which we will again choose to call A. Integration of this new equation yields Eq. (8.7–7) but with the factor $(w - u_n)^{\alpha_n/\pi - 1}$ deleted. This procedure can be applied to any vertex, z_1, z_2, etc.

We will not prove that the Schwarz–Christoffel formula maps the domain $\text{Im } w > 0$ one to one onto the interior of the polygon. The reader is referred to more advanced texts.[†] Note, however, that for the correspondence between the two domains to exist, the images of the consecutive points $u_1, u_2, \ldots, u_n$, which are encountered as we move from left to right along the u-axis, must be $z_1, z_2, \ldots, z_n$, which are encountered in this order as we move around the polygon while keeping its interior on our left.

Another important fact, not proved here, is that given a polygon in the z-plane having vertices at specified locations $z_1, z_2, \ldots, z_n$, we find that the Schwarz–Christoffel formula can be used to create a correspondence between the polygon and the u-axis in such a way that three of the vertices will have images at any three distinct points we choose on the u-axis. The location of the images of the other $(n - 3)$ vertices are then predetermined.

Example 1

Find the Schwarz–Christoffel transformation that will transform the real axis of the w-plane into the right isosceles triangle shown in Fig. 8.7–3(a). The vertices of the triangle have the images indicated in Fig. 8.7–3(b).

[†]See, for example, R. Nevanlinna and V. Paatero, *Introduction to Complex Analysis* (Reading, Mass.: Addison-Wesley, 1969), ch. 17.

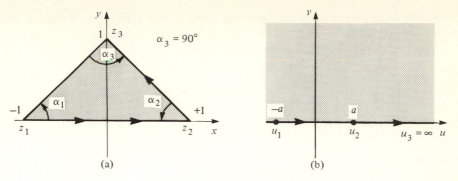

Figure 8.7–3

Solution

For the vertex at z_1, we have $\alpha_1 = \pi/4$ and the image point is $u_1 = -a$. For the vertex at z_2, we have $\alpha_2 = \pi/4$ and the image is $u_2 = a$. Finally, for the vertex at z_3, we have $\alpha_3 = \pi/2$ and the image is $w = \infty$. Since $w = \infty$ is mapped into z_3, a term containing w_3 and α_3 will not appear in Eq. (8.7–7). Using this formula, with the lower limit of integration chosen somewhat arbitrarily as 0, we obtain

$$z = A \int_0^w (\zeta + a)^{-3/4} (\zeta - a)^{-3/4} \, d\zeta + B$$

$$= A \int_0^w (\zeta^2 - a^2)^{-3/4} \, d\zeta + B. \qquad (8.7\text{--}10)$$

This integral cannot be evaluated in terms of conventional functions. It must be found numerically for each value of w of interest. To find A and B we impose these requirements: If $w = a$ on the right side of Eq. (8.7–10), then z must equal 1, whereas if $w = -a$, the corresponding value of z is -1. Thus

$$1 = A \int_0^a (\zeta^2 - a^2)^{-3/4} \, d\zeta + B, \qquad (8.7\text{--}11)$$

$$-1 = A \int_0^{-a} (\zeta^2 - a^2)^{-3/4} \, d\zeta + B. \qquad (8.7\text{--}12)$$

We multiply the preceding equation by (-1) and reverse the limits of integration to obtain

$$1 = A \int_{-a}^0 (\zeta^2 - a^2)^{-3/4} \, d\zeta - B. \qquad (8.7\text{--}13)$$

Adding Eqs. (8.7–11) and (8.7–13), we get

$$2 = A \int_{-a}^{+a} (\zeta^2 - a^2)^{-3/4} \, d\zeta = 2A \int_0^a (\zeta^2 - a^2)^{-3/4} \, d\zeta.$$

The last step follows from the even symmetry of the integrand. We solve the preceding equation for A to obtain

$$A = \frac{1}{\displaystyle\int_0^a (\zeta^2 - a^2)^{-3/4} \, d\zeta}, \qquad (8.7\text{--}14)$$

which also must be evaluated numerically. Substituting this result in Eq. (8.7–11), we see that $B = 0$. With Eq. (8.7–14) used in Eq. (8.7–10) we have

$$z = \frac{\int_0^w (\zeta^2 - a^2)^{-3/4} d\zeta}{\int_0^a (\zeta^2 - a^2)^{-3/4} d\zeta}, \tag{8.7–15}$$

which is the required transformation. ◄

Example 2

Find the Schwarz–Christoffel transformation that will map the half plane $\operatorname{Im} w > 0$ onto the semiinfinite strip $\operatorname{Im} z > 0$, $-1 < \operatorname{Re} z < 1$. The u-axis is to be mapped as indicated.

Solution

From Fig. 8.7–4(a) we see that the strip (a degenerate polygon) can be regarded as the limiting case of the triangle whose top vertex is moved to infinity. In this limit $\alpha_1 = \pi/2$, $\alpha_2 = \pi/2$, and $\alpha_3 = 0$. From Fig. 8.7–4(b) we have $u_1 = -1$, $u_2 = 1$, $u_3 = \infty$. Substituting the preceding values for α and u in Eq. (8.7–7) (without any term involving u_3), we have

$$z = A \int^w (\zeta + 1)^{-1/2} (\zeta - 1)^{-1/2} d\zeta + B = A \int^w \frac{d\zeta}{(\zeta^2 - 1)^{1/2}} + B.$$

$$\tag{8.7–16}$$

To simplify the final answer we shall put $A/i = A_1$ so that

$$\frac{A}{(\zeta^2 - 1)^{1/2}} = \frac{A_1}{(1 - \zeta^2)^{1/2}},$$

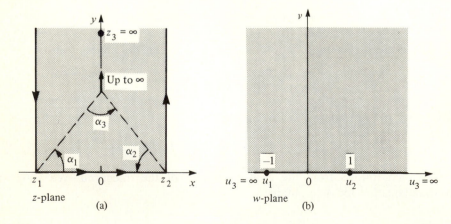

z-plane
(a)

w-plane
(b)

Figure 8.7–4

and Eq. (8.7–16) becomes

$$z = A_1 \int^w \frac{d\zeta}{(1 - \zeta^2)^{1/2}} + B.$$

The indefinite integration is readily performed and the constant absorbed into B. Thus

$$z = A_1 \sin^{-1} w + B. \tag{8.7–17}$$

Since $w = 1$ is mapped into $z = 1$, the preceding implies

$$1 = A_1 \sin^{-1} 1 + B = A_1 \frac{\pi}{2} + B. \tag{8.7–18}$$

Similarly, because $w = -1$ is mapped into $z = -1$, we have from Eq. (8.7–17)

$$-1 = A_1 \sin^{-1}(-1) + B = -A_1 \frac{\pi}{2} + B. \tag{8.7–19}$$

Solving these last two equations simultaneously, we find that $B = 0$ and $A_1 = 2/\pi$. Thus Eq. (8.7–17) becomes

$$z = \frac{2}{\pi} \sin^{-1}(w). \tag{8.7–20}$$

This same transformation (except for a change in scale) has already been studied in Example 3, Section 8.3. ◀

Example 3

Find a transformation that maps the domain $\operatorname{Im} w > 0$ onto the domain *outside* the semiinfinite strip shown in Fig. 8.7–5(a). Boundary points u_1 and u_2 should be mapped into z_1 and z_2 as indicated.

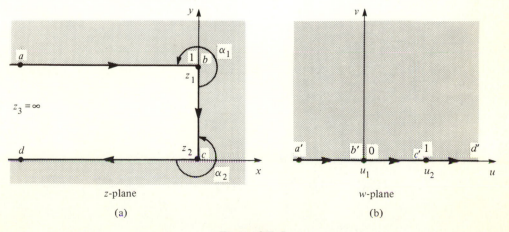

z-plane

(a)

w-plane

(b)

Figure 8.7–5

Solution

As we proceed from left to right along the u-axis in Fig. 8.7–5(b), the corresponding image point advances in the direction indicated by the arrow in Fig. 8.7–5(a). The "interior" of the polygon, which we regard as the domain outside the strip in Fig. 8.7–5(a), should be on our left as we negotiate the path a, b, c, d. In Eq. (8.7–7) we must include terms corresponding to $z_1 = i$ and $z_2 = 0$. The vertex of the polygon at $z_3 = \infty$ is mapped to $w = \infty$ and does not appear in Eq. (8.7–7). Notice that $\alpha_1 = 3\pi/2$ and $\alpha_2 = 3\pi/2$. The angles are measured along arcs passing through the *interior* of the polygon. Thus from Eq. (8.7–7) we have

$$z = A \int^w \zeta^{1/2} (\zeta - 1)^{1/2} \, d\zeta + B.$$

By replacing $(\zeta - 1)^{1/2}$ with $i(1 - \zeta)^{1/2}$ and absorbing i into A, we obtain

$$z = A \int^w \zeta^{1/2} (1 - \zeta)^{1/2} \, d\zeta + B.$$

The integral can be evaluated from tables. Thus

$$z = \frac{A}{4} \left[(2w - 1)(w(w - 1))^{1/2} - \frac{1}{2} \log \left((w(w - 1))^{1/2} + w - \frac{1}{2} \right) \right] + B. \tag{8.7–21}$$

When $w = 0$, we require $z = i$. Thus Eq. (8.7–21) yields

$$i = \frac{A}{4} \left(-\frac{1}{2} \log \left(-\frac{1}{2} \right) \right) + B.$$

Arbitrarily choosing the principal value of the logarithm we get

$$i = \frac{A}{8} [\text{Log} \, 2 - i\pi] + B. \tag{8.7–22}$$

When $w = 1$, we require $z = 0$, which from Eq. (8.7–21) means

$$0 = \frac{A}{4} \left[-\frac{1}{2} \log \left(\frac{1}{2} \right) \right] + B,$$

or, with the principal value

$$0 = \frac{A}{8} \, \text{Log} \, 2 + B. \tag{8.7–23}$$

Solving Eqs. (8.7–22) and (8.7–23) simultaneously, we obtain

$$A = -\frac{8}{\pi}, \qquad B = \frac{\text{Log} \, 2}{\pi}.$$

With these values in Eq. (8.7–21) we have

$$z = -\frac{2}{\pi} \left[(2w - 1)(w(w - 1))^{1/2} - \frac{1}{2} \log \left((w(w - 1))^{1/2} + w - \frac{1}{2} \right) \right] + \frac{\text{Log} \, 2}{\pi}. \tag{8.7–24}$$

Let us verify, by appropriate choices of branches, that the point $w = 1/2$ is mapped into a point on the imaginary z-axis between 0 and 1. With $w = 1/2$ in Eq. (8.7–24) we have

$$z = -\frac{2}{\pi}\left[-\frac{1}{2}\log\left(-\frac{1}{4}\right)^{1/2}\right] + \frac{\text{Log}\,2}{\pi}$$

$$= \frac{1}{\pi}\log\left(\pm\frac{i}{2}\right) + \frac{\text{Log}\,2}{\pi}.$$

Using $+i$ in this expression and the principal value of the logarithm, we obtain $z = i/2$.

The reader should map several points from the u-axis into the z-plane and become convinced that the desired mapping of points in Fig. 8.7–5 can be achieved if we

a) use the principal branch of the log in Eq. (8.7–24) and

b) define $f(w) = (w(w - 1))^{1/2}$ by means of branch cuts extending into the lower half of the w-plane from $w = 0$ and $w = 1$ and take $f(w) > 0$ when $w > 1$. ◀

EXERCISES

1. Is the mapping defined in Eqs. (8.7–6) and (8.7–7) conformal for $w = u_1$, $w = u_2$, etc? Explain.

2. Use the Schwarz–Christoffel formula to find the transformation that will map the sector shown in Fig. 8.7–6 onto the upper half of the w-plane. Map A into $(-1, 0)$, B into $(0, 0)$ and C to ∞.

3. Use the Schwarz–Christoffel formula to find a transformation that will map the region in Fig. 8.7–7 onto the upper half of the w-plane. Map $z = 0-$ to $w = -1$, $z = i$ to $w = 0$, and $z = 0+$ to $w = 1$. *Hint:* Consider the boundary in Fig. 8.7–7 as the limit of the boundary depicted in Fig. 8.7–8 as α_1, α_2, and α_3 achieve appropriate values.

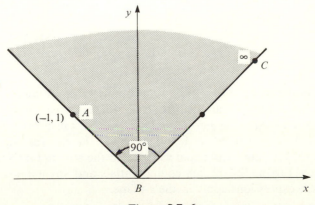

Figure 8.7–6

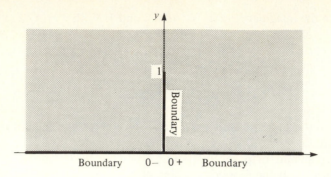

Figure 8.7–7

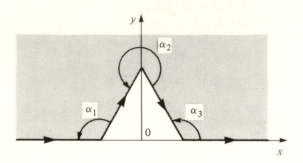

Figure 8.7–8

4. Find a transformation that maps the line $v = 0$ from the w-plane onto the open polygon in the z-plane shown in Fig. 8.7–9(a). The mapping of the boundary should be as shown in the figure.

5. a) Find the transformation that maps the upper half of the w-plane onto the shaded region of the z-plane in Fig. 8.7–10(a). The boundary is mapped as shown in Fig. 8.7–10(a) and (b). *Hint:* Consider the region in Fig. 8.7–10(a) as the limit of the region in Fig. 8.7–10(c) as $\alpha_1 \rightarrow 3\pi/2$ and $\alpha_2 \rightarrow 0$.

   Answer:

   $$z = \frac{2}{\pi}(1 + w)^{1/2} + \frac{1}{\pi} \text{Log}\left(\frac{(1 + w)^{1/2} - 1}{(1 + w)^{1/2} + 1}\right). \qquad (8.7–25)$$

   b) Let the bent line a, b, c in Fig. 8.7–10(a) be the cross section of a conductor maintained at voltage V_0, and let the straight line c, d be the cross section of a conductor maintained at zero potential. Use the transformation derived in part (a) to show that if $|z| \gg 1$ and $\text{Re } z \geq 0$ the electrostatic potential is $\phi(x, y) \approx (2V_0/\pi) \arg z$ (principal value), and the stream function is $\psi(x, y) \approx -(2V_0/\pi) \text{Log}|\pi z/2|$. Sketch equipotentials and streamlines where these approximate expressions apply in the z-plane.

   c) Use Eq. (8.7–25) to show that if $\text{Re } z \ll -1$ and $0 \leq \text{Im } z \leq 1$ the electrostatic potential is $\phi(x, y) \approx V_0 y$, and the stream function is $\psi \approx -V_0 x$.

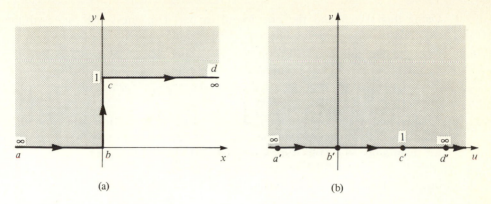

(a)

(b)

Figure 8.7–9

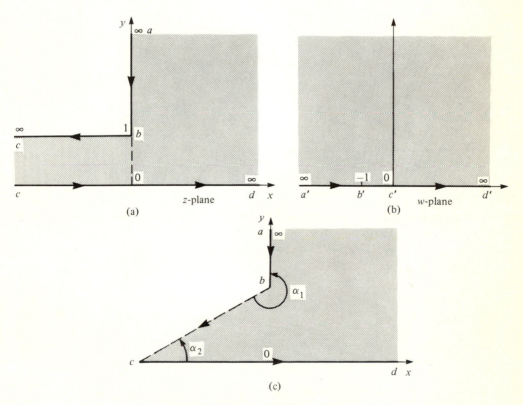

(a)   *z*-plane

(b)   *w*-plane

(c)

Figure 8.7–10

Hint: For $|w| \ll 1$,

$$\frac{(1 + w)^{1/2} - 1}{(1 + w)^{1/2} + 1} \approx \frac{w}{4}$$

from a Maclaurin expansion. Sketch the streamlines and equipotentials in the *z*-plane, where these approximate expressions for $\phi(x, y)$ and $\psi(x, y)$ apply.

Using this sketch and the one found in part (b), guess the shape of the streamlines and equipotentials near the bend in the upper conductor.

6. a) Find the mapping that will transform the upper half of the w-plane onto the strip $0 < \text{Im } z < 1$. The boundary is to be mapped as shown in Fig. 8.7–11. *Hint:* Consider an appropriate limit of the triangle formed by the broken lines in Fig. 8.7–11(a).

   b) Assume that in the z-plane the line $y = 1$ is the cross section of a conductor maintained at an electrostatic potential of 1 volt, whereas $y = 0$ is a conductor maintained at 0 volts. One easily finds the complex potential $\Phi(z)$ in the strip $0 \le y \le 1$. Transform this result into the upper half of the w-plane using the transformation found in part (a), and find $\phi(u, v) = \text{Re } \Phi(w)$.

   c) What boundary condition does $\phi(u, v)$ satisfy on the line $v = 0$?

7. a) Find the transformation that will map the upper half of the w-plane onto the domain indicated in the z-plane in Fig. 8.7–12. The mapping of the boundaries are as shown. Take the inverse of this transformation so as to obtain w as a

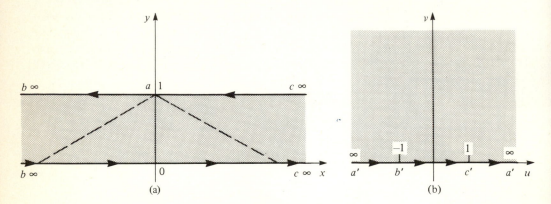

Figure 8.7–11

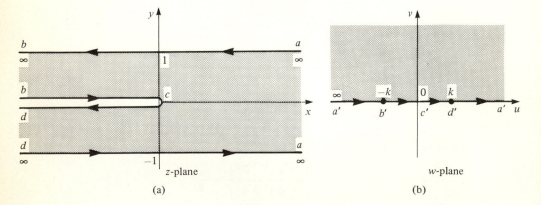

Figure 8.7–12

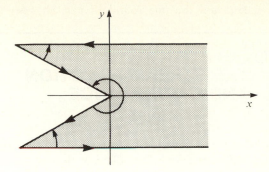

Figure 8.7–13

function of z, and describe the appropriate branch of the function obtained. *Hint:* Consider Fig. 8.7–13. Let the angles shown pass to appropriate limits.

b) Let $k = 1$ in Fig. 8.7–12(b). Suppose the domain shown shaded in Fig. 8.7–12(a) is the cross section of a heat-conducting material. The line $y = 0$, $-\infty < x < 0$ represents a boundary maintained at $0°$ temperature, whereas the lines $y = \pm 1$, $-\infty < x < \infty$ are both maintained at $1°$. Use the transformation found in part (a) as well as the result of Exercise 6(b) or the result of Exercise 9(c) in Section 8.5 to find the complex temperature in the heated material.

c) Plot the actual temperature $\phi(x, y)$ with distance along the line $x = 0$, $0 \le y \le 1$ and along the line $y = 0$, $0 \le x \le 2$.

APPENDIX TO CHAPTER 8
THE STREAM FUNCTION
AND CAPACITANCE

We show here how the stream function $\psi(x, y)$ is useful in capacitance computations and also in fluid-flow problems. Recall from Section 2.6 that the amount of electric flux crossing a differential surface of area dS is

$$df = D_n \, dS, \qquad (A8-1)$$

where D_n is the component of electric flux density normal to the surface. Now consider surfaces whose cross sections are of length dy and dx, as shown in Fig. A8-1. Each surface is assumed to be 1 unit long in a direction perpendicular to the paper. The flux crossing the vertical surface is $D_x \, dy$ while that crossing the horizontal surface is $D_y \, dx$.

Consider now the simple closed contour C shown in Fig. A8-2. C is the cross section of a cylinder of unit length perpendicular to the page. We seek an expression for the total electric flux emanating outward from the interior of the cylinder. Let $dz = dx + i \, dy$ be an incremental distance along C, as shown. The flux df across the

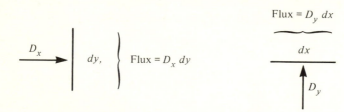

Figure A8-1

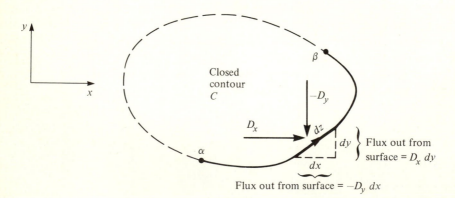

Figure A8-2

surface whose cross section is dz is the sum of the fluxes crossing the projections of this surface on the x- and y-axes. These projections have areas dx and dy. Thus

$$df = D_x \, dy - D_y \, dx.$$

The minus sign occurs in front of D_y because we are computing the outward flux, that is, flux passing from the inside to the outside of the cylinder. The flux passing outward along the surface whose cross section is the solid line connecting α with β in Fig. A8–2 is

$$f_{\alpha\beta} = \int_\alpha^\beta D_x \, dy - D_y \, dx, \tag{A8–2}$$

where we integrate in the positive sense along the solid line, that is, we keep the interior of the cylinder on our left. With the aid of Eq. (2.6–22) we rewrite Eq. (A8–2) in terms of the electrostatic potential $\phi(x, y)$. Thus

$$f_{\alpha\beta} = \int_\alpha^\beta - \varepsilon \frac{\partial \phi}{\partial x} \, dy + \varepsilon \frac{\partial \phi}{\partial y} \, dx, \tag{A8–3}$$

where ε is the permittivity of the surrounding material. Since the stream function $\psi(x, y)$ is the harmonic conjugate of $\phi(x, y)$, we have from the Cauchy–Riemann equations

$$\frac{\partial \phi}{\partial x} = \frac{\partial \psi}{\partial y},$$

$$\frac{\partial \phi}{\partial y} = - \frac{\partial \psi}{\partial x}.$$

This enables us to rewrite $f_{\alpha\beta}$ in Eq. (A8–3) in terms of ψ as follows:

$$f_{\alpha\beta} = - \varepsilon \int_\alpha^\beta \frac{\partial \psi}{\partial y} \, dy + \frac{\partial \psi}{\partial x} \, dx.$$

The integrand here is the exact differential $d\psi$, and the integration is immediately performed with the result that

$$f_{\alpha\beta} = - \varepsilon \int_\alpha^\beta d\psi = \varepsilon [\psi(\alpha) - \psi(\beta)]. \tag{A8–4}$$

Thus the product of ε and the decrease in ψ encountered as we move along the contour connecting α to β is the electric flux crossing the surface whose cross section is this contour (see the solid line representation in Fig. A8–2). Let us now move completely around the closed contour C in Fig. A8–2 and measure the net decrease in ψ (initial value minus final value). Because ψ is typically a branch of a multivalued function, defined by means of a branch cut, the decrease in ψ, which we will call $\Delta\psi$, can be nonzero.

> The total electric flux leaving the closed surface whose cross section is C is $\varepsilon\Delta\psi$, which, according to Gauss's law,[†] is exactly equal to the charge on or enclosed by the surface.

[†] See P. Lorrain, and D. Corson, *Electromagnetism: Principles and Applications* (San Francisco: W. H. Freeman, 1979), ch. 3.

When two conductors are maintained at different electrical potentials, a knowledge of the stream function $\psi(x, y)$ will establish the charge on either conductor and, with the use of Eq. (8.5–23), the capacitance of the system.

Suppose we have a two-dimensional system of conductors whose capacitance per unit length we wish to determine. Let these conductors (actually their cross sections) be mapped from the xy-plane into the uv-plane by means of a conformal transformation. Then we can prove the following:

> The capacitance that now exists between the two image conductors in the uv-plane is precisely the same as existed between the two original conductors in the xy-plane.

To establish this equality, consider the pair of conductors A and B shown in the xy-plane in Fig. A8–3. They are at electrostatic potentials V_0 and zero, respectively. Under a conformal mapping these conductors become the conductors A' and B' in the uv-plane. The new conductors are assigned the potentials V_0 and zero, respectively. The complex potentials in the two planes are $\Phi(x, y) = \phi(x, y) + i\psi(x, y)$ and $\Phi_1(u, v) = \phi_1(u, v) + i\psi_1(u, v)$.

Refer now to Fig. A8–3(a), and consider the curve Γ (shown by the solid line) that goes from (x_1, y_1) to (x_2, y_2) along conductor A. The image of Γ in the w-plane is the curve Γ' that goes from (u_1, v_1) to (u_2, v_2) along conductor A' (see Fig. A8–3b). The amount by which ψ decreases as we move along Γ is $\psi(x_1, y_1) - \psi(x_2, y_2)$, whereas the corresponding change along the image Γ' is $\psi_1(u_1, v_1) - \psi_1(u_2, v_2)$. Since the stream function $\psi(x, y)$ and its transformed version $\psi_1(u, v)$ assume identical values at corresponding image points in the two planes, the expressions for the change in ψ and ψ_1 are equal. Because (x_1, y_1) and (x_2, y_2) were chosen arbitrarily, it follows that the change in ψ occurring if we go completely around the boundary of A must equal the change in ψ_1 if we go completely around the boundary of A'. Thus the amount of charge on A and A' must be identical. The potential difference V_0 between conductors A and B is identical to the potential difference between conductors A' and B'. Since capacitance is the ratio of charge to potential difference, the capacitance between conductors A and B is the same as appears between conductors A' and B'.

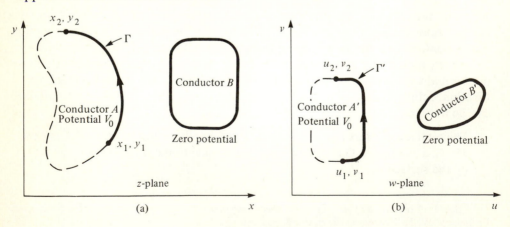

Figure A8–3

SELECTED ANSWERS†

CHAPTER 1

Section 1.1

1. (a) rational, (c) complex, (e) complex, (g) integers; **3.** (c) $3 - 4i$, (e) $-6 + 25i$,

(f) 25; **4.** (b) yes, (d) no; **7.** $2^{187} - i2^{187}$; **8.** (b) $y = 0$, $x = 0$ or -2, (d) $x = 0$, $y = 0$ or -1,

(f) $x = \left[\dfrac{1}{\sqrt{2}} \sqrt{1 + \sqrt{2}} \right]$, $y = \dfrac{1}{(2x)}$ or $x = - \left[\dfrac{1}{\sqrt{2}} \sqrt{1 + \sqrt{2}} \right]$, $y = \dfrac{1}{(2x)}$,

(g) $x = 1$, $y = 0$ or $x = -\dfrac{1}{2}$, $y = \pm \dfrac{\sqrt{3}}{2}$.

Section 1.2

1. (a) $1/2 - i/2$, (c) $5/2 + i$, (g) 1; **4.** (a) no, (c) yes, (e) no, (g) no; **5.** (b) no.

Section 1.3

1. (a) 10, (c) $\dfrac{1}{10} \sqrt{410}$, (f) $\left(\dfrac{1}{\sqrt{2}} \right)^{11}$, (g) $\left(\dfrac{1}{\sqrt{2}} \right)^{17}$; **2.** (b) $-12 - i$; **4.** (b) $x = -3$, $y = -1$;

8. (a) $-1.98 + i0.282$; **9.** (a) $r = \sqrt{2}$, $\theta = -\dfrac{\pi}{4} + 2k\pi$, principal value when $k = 0$,

(c) $r = \sqrt{5}$, $\theta = -2.68 + 2k\pi$, (e) $r = \dfrac{2}{25}$, $\theta = -0.283 + 2k\pi$; **10.** (a) -0.425;

11. (b) $r = \sqrt{\dfrac{65}{34}}$, $\theta = -0.2274$, (c) $\dfrac{3}{(4\sqrt{2})}$, $\theta = \dfrac{-7\pi}{12}$.

Section 1.4

1. (a) $-2^7 - i2^7\sqrt{3} = 2^8 \left/ \dfrac{8\pi}{6} \right.$, (c) $(\sqrt{2})^{13} \left/ \dfrac{11\pi}{12} \right. = -87.4 + i23.4$,

---

†In some cases approximate decimal answers are given.

(f) $\dfrac{1}{16}\Big/\dfrac{-2\pi}{3} = -\dfrac{1}{32} - \dfrac{i\sqrt{3}}{32}$; **3.** (a) $\pm\left(\dfrac{1}{\sqrt{2}} + \dfrac{i}{\sqrt{2}}\right)$, (d) $\pm\left(\dfrac{1}{\sqrt{2}} \pm \dfrac{i}{\sqrt{2}}\right)$,

(f) $\sqrt[4]{2}\left[\cos\left(\dfrac{-\pi}{12} + \dfrac{\pi}{2}k\right) + i\sin\left(\dfrac{-\pi}{12} + \dfrac{\pi}{2}k\right)\right]$, $k = 0, 1, 2, 3$,

(h) $\dfrac{1}{\sqrt[5]{2}}\left[\cos\left(\dfrac{\pi}{15} + \dfrac{2k\pi}{5}\right) + i\sin\left(\dfrac{\pi}{15} + \dfrac{2k\pi}{5}\right)\right]$, $k = 0, 1, 2, 3, 4$;

6. (a) $w = \sqrt[6]{2}\ \text{cis}\left(-\dfrac{\pi}{12} + \dfrac{2k\pi}{3}\right)$, $k = 0, 1, 2$, (d) $w = i[-1 \pm \sqrt{2}\,]$,

(f) $\dfrac{1}{\sqrt[3]{2}}\ \text{cis}\left(-\dfrac{\pi}{6} + \dfrac{2k\pi}{3}\right)$, $k = 0, 1, 2$; **8.** (b) $\sqrt[3]{2}\ \text{cis}\left(\dfrac{\pi}{6} + \dfrac{2k\pi}{3}\right)$, $k = 0, 1, 2$,

(d) $\sqrt[3]{4}\ \text{cis}\left(-\dfrac{\pi}{3} + \dfrac{2k\pi}{3}\right)$, $k = 0, 1, 2.$, (f) $\text{cis}\left(\dfrac{\pi}{3} + \dfrac{2k\pi}{3}\right)$, $k = 0, 1, 2$,

(i) $\text{cis}\left(\dfrac{k\pi}{3}\right)$, $k = 0, 1, 2, \ldots, 5$, (j) $\text{cis}\left(\dfrac{2k\pi}{3}\right)$, $k = 0, 1, 2$.

Section 1.5

1. (e) the points outside a circle, radius 2, center $x = -1, y = 2$, (i) the points on both branches of the hyperbola $(x - 1/2)^2 - y^2 = 1/4$ and all other points of the z-plane except those lying between the branches, (j) the point $x \doteq -0.567, y = 0$, (k) the points inside a circle, center $x = -1, y = 1/2$, radius $\sqrt{5}/2$; **2.** (b) $2 < |z - (3 + i)| \le 4$; **4.** consider points on $y = 2$; **5.** (a) try connecting $(1, 0)$ with $(-1, 0)$, (b) yes connected, not domain; **7.** (a) $z = 3i$ and points on circle $|z - 3i| = 1$ are boundary points; none belong to the set, (e) all the points on the circles $|z| = n\pi/2, n = 1, 5, 9, \ldots$, these not in set, (f) $x = 1, y = 0$ boundary point, not in set; all points in set are boundary points; **8.** (a) yes, (b) no, (g) yes, (i) no; **9.** (c) on the upper hemisphere.

CHAPTER 2

Section 2.1

1. (a) no, (d) $z = i$; **2.** (a) $-0.1 - 0.2i$, (c) $2 + 4i$; **3.** (a) $x^2 - y^2 + 2x + 1 + i(2xy + 2y)$, (c) $x + x^2 - y^2 + i(y - 2xy)$; **4.** (a) $(1 + i)(z^2 - \bar{z}^2)/4i$, (c) $i\bar{z}$.

Section 2.2

5. -4; **6.** (b) 9, (c) $1/(\sqrt{2} - 1)$.

Section 2.3

2. (c) continuous but $f'(0)$ does not exist; **4.** (a) all z, (b) $z = -i$ only, (c) $\text{Im}\,z = 1/2$, (e) all z, (h) $\text{Im}\,z = 0$; **5.** (a) nowhere, (c) $z = 0$, (e) nowhere.

Section 2.4

1. (a) everywhere, (c) nowhere, (e) everywhere, (f) nowhere, (i) all $z \ne 0$;
2. (a) $f'(z) = 3z^2 + 4z + 1$, all z, (c) $f'(z) = 1 - 1/(z - 1)^2$, all $z \ne 1$, (e) $1 - z^{-2}, z \ne 0$;
3. (a) $3[x^2 - y^2 + i2xy]$, (c) $2e^{x^2 - y^2}(x + iy)\,\text{cis}\,(2xy)$; **9.** (a) $u = r^{10}\cos(10\theta)$,
$v = r^{10}\sin(10\theta)$.

Section 2.5

1. Re z = Im z; **2.** (a) no, (c) yes, (d) yes; **4.** (b) $v = 6x^2y^2 - y^4 + y - x^4 +$ const., (c) not identical; **5.** $v = -e^{-x}\sin y - 2x - \frac{1}{2}(x^2 - y^2) +$ const.; **8.** (b) $\frac{1}{2}\log(x^2 + y^2) +$ const.; **9.** $u = -e^{x^3 - 3xy^2}\sin(3x^2y - y^3) + C$; **10.** (a) yes, (d) yes; **12.** (a) $u = 1 = x^2 - y^2$,

(b) $x = \sqrt{\dfrac{1 + \sqrt{2}}{2}}$, $y = \dfrac{1}{2x}$, (c) slopes $2x^2$ and $-\dfrac{1}{2x^2}$, x from part (a); **13.** (c) $x = 0.805$, $y = 1.11$, (d) slopes $\frac{1}{2}$ and -2.

Section 2.6

1. (a) $\phi = -30x - 40y$, (b) $\psi = -30y + 40x$; **2.** (b) $V_x = 0$, $V_y = -\frac{1}{2}$, (c) $(x - 1)^2 + y^2 = 1$; **3.** (a) $-e[\text{cis}(-\frac{1}{2})]$, (c) $D_x = -\varepsilon_0 e\cos\frac{1}{2}$, $D_y = \varepsilon_0 e\sin\frac{1}{2}$, $\varepsilon_0 = 8.85 \times 10^{-12}$ in m.k.s. units; **4.** (b) $\phi = -\dfrac{xy}{\varepsilon}$,

(c) $\psi = \dfrac{x^2 - y^2}{2\varepsilon}$, (d) $\Phi = \dfrac{iz^2}{2\varepsilon}$, (e) $E_x = \dfrac{1}{\varepsilon}$, $E_y = \dfrac{1}{\varepsilon}$;

5. (a) $\phi = x\cos\alpha + y\sin\alpha$, $\phi = c_1, c_2, \ldots$, (b) $\psi = y\cos\alpha - x\sin\alpha$, $\psi = b_1, b_2, \ldots$, (c) $v_x = \cos\alpha$, $v_y = \sin\alpha$, angle α.

CHAPTER 3

Section 3.1

1. (a) $3.99 + i6.22$, (c) $0.073 - i0.114$, (e) 0.073, (h) $1.44 - i0.790$;

2. (a) $u = e^{x^2 - y^2}\cos 2xy$, $v = e^{x^2 - y^2}\sin 2xy$, (c) $u = e^{(x/x^2 + y^2)}\cos\dfrac{y}{x^2 + y^2}$,

$v = -e^{(x/x^2 + y^2)}\sin\dfrac{y}{x^2 + y^2}$, (e) $u = e^{(e^x\cos y)}\cos(e^x\sin y)$, $v = e^{(e^x\cos y)}\sin(e^x\sin y)$;

3. (a) no, (b) yes, (f) yes; **6.** -7.39 and $-7.02 - i2.28$;

7. (a) ± 1, (d) $\pm\left[\pm\dfrac{1}{2} + \dfrac{i\sqrt{3}}{2}\right]$, ± 1; **8.** max e, min e^{-1};

9. (a) $f'(z) = 2e^{2z}$, all z, (c) not analytic, (e) $f'(z) = -z^{-2}e^{1/z}$, all $z \neq 0$;

10. (b) $f'(z) = e^{(e^z + z)}$; **11.** (a) $\phi = e^x\cos y$, (c) $v_x = \dfrac{e}{\sqrt{2}}$, $v_y = \dfrac{-e}{\sqrt{2}}$.

Section 3.2

1. (a) $-3.72 + i0.51$, (c) $0.53 + i3.59$, (e) $-0.010 - i0.965$, (f) $1.18 - i0.485$, (g) 4.68, (j) 1.77; **5.** $z = (2k + 1)\pi/2$, $k = 0, \pm 1, \ldots$; **7.** all z; **9.** (a) cosec z analytic except $z = k\pi$, $k = 0, \pm 1$, etc., (c) $\cos[\sec z]\tan z\sec z$, $-i0.393$.

Section 3.3

2. (a) $1.96 + i3.17$, (b) $0.83 + i0.99$; **7.** (a) all z, (b) $f'(i) = 1.49i$; **8.** (a) 1, (b) 1, (c) 0.

Section 3.4

1. k is any integer, principal values when $k = 0$, (a) $i2k\pi$, (c) $i(\pi + 2k\pi)$, (d) $1.35 - i(3\pi/4 + 2k\pi)$, (f) $2 + i(\pi + 2k\pi)$, (h) $1.61 + i(0.93 + 2k\pi)$, (i) $i(\pi + 2k\pi)$,

(j) $0.161 + i(\pi/2 + 2k\pi)$, (m) $\cos 1 + i(\sin 1 + (2k\pi))$; **2.** (a) $i\sinh(\pi/2 + 2k\pi)$, (b) $-i$, (c) $\text{Log}\,\pi/2 + i\pi/2$; **3.** (a) yes; **4.** (b) $2 + i(\pi/2 + 2k\pi)$, (c) $z = \text{Log}[\pi/2 + 2k\pi] + i(\pi/2 + 2m\pi)$, k nonnegative integer, m any integer, $z = \text{Log}[2k\pi - \pi/2] + i(-\pi/2 + 2m\pi)$, k positive integer, m any integer; **6.** (a) $z = \text{Log}\,2 + i2k\pi$; **7.** (a) $\log z_1 = 1 - i\pi/2$, $\log z_2 = \text{Log}\,2 + i\pi$, $\log(z_1 z_2) = \text{Log}\,2 + 1 + i\pi/2$; **8.** (a) $\log(1 + i) = \text{Log}\,\sqrt{2} + i\pi/4$ and $\log(1 + i)^5 = 5\,\text{Log}\,\sqrt{2} + i5\pi/4$, (b) no.

Section 3.5

2. (b) $2 + i\pi$; **3.** (b) the z-plane with line $y = 0$, $x \le 0$ removed and point $z = 1$ removed;

4. (b) $\phi = 1$ is circle $|z| = \dfrac{1}{e}$, (c) $E_x = \dfrac{x}{x^2 + y^2}$, $E_y = \dfrac{y}{x^2 + y^2}$;

5. (a) $\phi(x, 0) = \text{Log}\left|\dfrac{x + 1}{x - 1}\right|$, (b) if $k = 9$, $\phi = \tfrac{1}{2}\text{Log}\,9$; **7.** (b) $\dfrac{1}{z\,\text{Log}\,z}$.

Section 3.6

1. (a) $e^{-2k\pi}$, $k = 0$ principal value, (c) $e^{-(\pi/2 + 2k\pi)}$, $k = 0$, p.v.,

(e) $e^{-(\pi/2 + 2k\pi)}\,\text{cis}\left(\text{Log}\left(\dfrac{\pi}{2}\right)\right)$ $k = 0$, p.v.,

(g) $-ie^{-2\pi - 8k\pi}$, p.v., $k = 0$, (j) $e^{-2k\pi}\,\text{cis}\,(0.434)$, $k = 0$, p.v.;

4. (a) $f'(z) = z^z[1 + \text{Log}\,z]$, (b) $e^{-\pi/2}(1 + i\,\pi/2)$.

Section 3.7

4. (a) $w = \pi + 2k\pi - i\,\text{Log}\left(1 + \sqrt{2}\right)$ and $w = 2k\pi - i\,\text{Log}\left(\sqrt{2} - 1\right)$,

(c) $\pm 1.023 + i\left(\dfrac{\pi}{2} + 2k\pi\right)$, (e) $-\tfrac{1}{2}(\pi + 2k\pi) - \dfrac{i}{2}\,\text{Log}\,3$,

(g) $\dfrac{\pi}{2} + 2m\pi - i\,\text{Log}\left(\dfrac{\pi}{2} + 2k\pi \pm \sqrt{\left(\dfrac{\pi}{2} + 2k\pi\right)^2 - 1}\right)$,

$k = 0, 1, 2, \ldots, \ m = 0, \pm 1, \pm 2, \ldots,$

or $\left(-\dfrac{\pi}{2} + 2m\pi\right) - i\,\text{Log}\left(-\dfrac{\pi}{2} - 2k\pi \pm \sqrt{\left(\dfrac{\pi}{2} + 2k\pi\right)^2 - 1}\right)$,

$k = -1, -2, \ldots, \ m = 0, \pm 1, \pm 2, \ldots,$ (i) $-i\tanh 2$; **5.** (a) no, (b) yes.

Section 3.8

1. (a) $3i$, (b) $\dfrac{-1}{\sqrt{2}} + \dfrac{i}{\sqrt{2}}$, (c) $\dfrac{1}{\sqrt{2}} + \dfrac{i}{\sqrt{2}}$, (d) $f'(-9) = \dfrac{1}{6i}$, $f'(i) = \dfrac{1}{2}\,\text{cis}\left(\dfrac{-\pi}{4}\right)$;

2. (a) $\sqrt[4]{3}\left[\dfrac{1}{\sqrt{2}} - \dfrac{i}{\sqrt{2}}\right]$; **4.** (a) no; **5.** (a) $\sqrt[6]{2}\,\sqrt[4]{2}\,\text{cis}\left(\dfrac{\pi}{24}\right)$, (b) $\sqrt[6]{2}\,\sqrt[4]{2}\,\text{cis}\left(\dfrac{-\pi}{24}\right)$;

8. (a) $2e^{i2\pi/3}$, (b) $\dfrac{1 + 3i}{\left(\sqrt[3]{2}\right)^2}$; **9.** (a) $f(0) = 0$, $f(1) = 0.8814$, $f(1 + i) = 1.061 + i0.666$,

(b) $f'(1) = \dfrac{1}{\sqrt{2}}$, $f'(1 + i) = \dfrac{1}{\sqrt[4]{5}}\,\text{cis}\,(-0.55357)$.

Appendix

1. (a) $2e^{-t}\cos t$, (b) $-2e^{-t}\sin t$, (c) $-2e^{-t}\sin\left(t + \dfrac{\pi}{3}\right)$,

(e) $e^{-3t}\cos 4t - 2e^{-3t}\sin\left(4t + \dfrac{\pi}{3}\right)$; **2.** (a) phasor 1, $s = -3$, (b) phasor $5e^{i\pi/3}$,

$s = -4 + 6i$, (c) phasor $-i$, $s = 3i$, (e) phasor $-3ie^{i\pi/3}$, $s = -1 + 4i$, (h) no phasor;

7. $\iota(t) = \dfrac{V_0}{L\sigma + R}e^{\sigma t}$; **8.** $\iota(t) = \dfrac{V_0}{R^2 + \dfrac{1}{\omega^2 C^2}}\left[R\sin \omega t + \dfrac{1}{\omega C}\cos \omega t\right]$;

9. $x(t) = \dfrac{F_0}{\left(k - \omega^2 m\right)^2 + \omega^2 \alpha^2}[(k - \omega^2 m)\cos(\omega t) + \omega \alpha \sin(\omega t)]$.

CHAPTER 4

Section 4.1

2. $1 + 2i$; **3.** approximately $0.386 + i0.488$, exactly $\dfrac{1 + i}{3}$; **4.** (a) $\dfrac{e^2}{2} + i(2 - e)$,

(b) $\dfrac{e^2}{2} + i(e - 2)$; **5.** (a) $e - 1$, (c) $1 - e^{1+i}$; **6.** $-\dfrac{1}{4} + \dfrac{i}{4}\left(\dfrac{\pi}{2} - 1\right)$; **9.** (a) elliptic arc,

(b) $\dfrac{7}{2} - 2i$; **10.** πi.

Section 4.2

2. (a) zero; **4.** (a) no, (b) no, (e) no, (g) yes; **7.** (a) $2\pi i$, (b) zero, (d) $-2\pi i$, (e) zero;
8. $6\pi i$.

Section 4.3

2. (a) $\cosh 1 - \cos 1 \cosh e + i\sin 1 \sinh e$,

(b) $\dfrac{e^{1+ie}}{2}[\sin(1 + ie) + \cos(1 + ie)] - \dfrac{e^i}{2}[\sin i + \cos i]$; **3.** (b) i and 1, respectively;

4. (a) $1 - \dfrac{\pi}{2} - i$; **5.** correct value $-\pi i$; **6.** (a) πi; **8.** $\dfrac{2}{3}\left[1 - \dfrac{1}{\sqrt{2}} - \dfrac{i}{\sqrt{2}}\right]$.

Section 4.4

2. (a) $-2\pi i \cos 2 \sin 2$, (b) zero, (c) $\dfrac{2\pi i}{7}\cos 2 \sin 2$; **3.** (a) $\dfrac{-2\pi}{\sqrt{3}}\cosh\left[\dfrac{1 + i\sqrt{3}}{2}\right]$;

5. (a) $\dfrac{\pi i}{2}[1 - \text{Log } 2]$, (b) $\frac{8}{15}\pi i$; **6.** zero; **8.** (a) $\frac{1}{3}[e\cos 1 - e^{-2}\cos 2]$, (d) $\pi i \cosh 1$;

10. (a) 1, (b) 2π, (c) 2π, (d) zero; **12.** (a) min at $(\sqrt{2} - 1)\text{cis}\left(\dfrac{\pi}{4}\right)$ (on boundary),

(b) max at $\dfrac{i}{2}$; **14.** $\min|f(z)| = 0$, $\max|f(z)| = 2 + \sqrt{2}$ occurs at $x = 1 + \sqrt{2}, y = -\sqrt{2}$;

17. $-\dfrac{2}{3\sqrt{3}} < T < \dfrac{2}{3\sqrt{3}}$.

Section 4.5

6. (c) $E_x = \dfrac{-V_0}{\pi}$, $E_y = \dfrac{V_0}{\pi}$; **9.** $\phi(x, y) = \dfrac{-y}{\pi} \displaystyle\int_{-\infty}^{\infty} \dfrac{\phi(u, 0)\, du}{(u - x)^2 + y^2}$.

CHAPTER 5

Section 5.1

1. (a) $x - \dfrac{x^3}{3!} + \dfrac{x^5}{5!} - \cdots$, (b) $\displaystyle\sum_{n=0}^{\infty} c_n x^n$, $c_n = \dfrac{(-1)^{n/2}}{n!}$, n even, $c_n = 0$, n odd;

2. (a) all x, (b) all x, (c) all x, (d) $-1 < x < 1$, (e) $\frac{1}{2} < x < \frac{3}{2}$.

Section 5.3

3. (a) $\displaystyle\sum_{n=1}^{\infty} \dfrac{(-1)^{n-1}}{n}(z - 1)^n$, inside $|z - 1| = 1$, (b) $\displaystyle\sum_{n=0}^{\infty} (-1)^n (n + 1)(z - 1)^n$,

inside $|z - 1| = 1$, (e) $\displaystyle\sum_{n=0}^{\infty} c_n (z + i)^n$, $c_n = \dfrac{\cosh 1}{n!}(-1)^{((n-1)/2)}$, n odd;

$c_n = \dfrac{i \sinh 1}{n!}(-1)^{(n/2)+1}$, n even, valid all z;

4. (b) $1 + \frac{1}{2}(z - 1) - \frac{1}{8}(z - 1)^2 + \cdots$; **5.** (b) inside $\left| z - \dfrac{\pi}{8} \right| = \dfrac{3\pi}{8}$,

(d) inside $|z| = \dfrac{\pi}{2}$, (f) inside $|z - 1| = \sqrt{3}$; **6.** (a) $-3 < x < 1$, (b) $0 < x < 2$;

8. (b) radius $\dfrac{1}{\sqrt{2}}$.

Section 5.4

3. (a) $\dfrac{z + z^2}{(1 - z)^3}$; **6.** $0.2151 + i0.1206$; **7.** (a) $- \displaystyle\sum_{n=0}^{\infty} (z)^{2n+1}$, $|z| < 1$,

(b) $(\frac{1}{8} + \frac{1}{4}) + (-\frac{1}{32} - \frac{1}{8})(z - 3) + (\frac{1}{128} + \frac{1}{16})(z - 3)^2 + \cdots$, $|z - 3| < 2$,

(c) $\frac{1}{4}\left[1 + (z + 1) + \frac{3}{4}(z + 1)^2 + \frac{1}{2}(z + 1)^3 + \cdots\right]$, $|z + 1| < 2$,

(d) $-\dfrac{1}{4}\left[\dfrac{2 \cdot 1}{4} + \dfrac{3 \cdot 2(z + 1)}{8} + \dfrac{4 \cdot 3(z + 1)^2}{16} + \dfrac{5 \cdot 4(z + 1)^3}{32} + \cdots\right]$, $|z + 1| < 2$,

(g) $\displaystyle\sum_{n=0}^{\infty}\left[-1 + \dfrac{1}{3^n} - \dfrac{1}{(2)4^n}\right]\dfrac{(z + 1)^n}{6}$, $|z + 1| < 1$; **8.** $\dfrac{10!}{6}\left[-1 + \dfrac{1}{3^{10}} - \dfrac{1}{(2)4^{10}}\right]$;

9. $1 + 2w^2 + 3w^4 + 4w^6 + 5w^8, \ldots$, $|w| < 1$;

13. $\dfrac{-5}{2} + \dfrac{9}{4}(z + 1) + \displaystyle\sum_{n=2}^{\infty}\left[\dfrac{3}{2}\left(\dfrac{-1}{2}\right)^n - 2(-1)^n\right](z + 1)^n$, $|z + 1| < 1$;

15. (a) $\dfrac{1}{1 + z}$, (c) $|z - \frac{1}{2}| = \frac{3}{2}$, (d) $\dfrac{1}{1 + z}$.

Section 5.5

1. (a) $\cdots + \dfrac{16}{(z - 1)^3} - \dfrac{4}{(z - 1)^2} + \dfrac{1}{(z - 1)}$, $|z - 1| > 4$, (b) $\cdots - \dfrac{(i + 2)^2}{(z - 2)^3}$

$\dfrac{(i+2)}{(z+2)^2} + \dfrac{1}{(z-2)}$, $|z-2| > \sqrt{5}$; **2.** (a) Three series: Taylor: $\frac{1}{3} - \frac{4}{9}(z-2)$

$+ \frac{13}{27}(z-2)^2 + \cdots$, $|z-2| < 1$; Laurent: $\dfrac{1}{2}\left[\cdots + (z-2)^{-3} - (z-2)^{-2}\right.$

$\left[+ (z-2)^{-1} - \dfrac{1}{3} + \dfrac{(z-2)}{9} - \dfrac{(z-2)^2}{27} + \cdots\right]$,

$1 < |z-2| < 3$; Laurent: $\cdots + 13(z-2)^{-4} - 4(z-2)^{-3} + (z-2)^{-2}$, $|z-2| > 3$;

3. (a) $\left[\cdots + \dfrac{2}{3}(z-3)^{-3} - \dfrac{2}{3}(z-3)^{-2} + \dfrac{2}{3}(z-3)^{-1}\right.$

$+ \dfrac{1}{12} - \dfrac{(z-3)}{48} + \dfrac{(z-3)^2}{192} - \cdots\Big]$ for $1 < |z-3| < 4$;

4. (a) $\dfrac{1}{3z} - \dfrac{1}{9} + \dfrac{z}{27} - \dfrac{z^2}{81} + \cdots$, $0 < |z| < 3$,

(b) $-\dfrac{(z+3)^{-1}}{3} - \dfrac{1}{9} - \dfrac{(z+3)}{27} - \dfrac{(z+3)^2}{81} - \cdots$, $0 < |z+3| < 3$,

(e) $\dfrac{e^2}{z-2} + e^2 + \dfrac{e^2}{2}(z-2) + \dfrac{e^2}{3!}(z-2)^2 + \cdots$, $0 < |z-2|$,

(f) $\dfrac{\cos 2}{(z-2)^2} - \dfrac{\sin 2}{z-2} - \dfrac{\cos 2}{2} + \dfrac{\sin 2}{3!}(z-2) + \cdots$, $0 < |z-2|$.

CHAPTER 6

Section 6.1

1. (a) $\dfrac{-2}{3}\pi i$, (b) 0; **2.** (a) $\pi i/12$, (b) πi, (d) 1, (e) zero, (f) $\dfrac{7\pi i}{180}$, (g) $2\pi i$; **5.** (b) $-\pi i$.

Section 6.2

2. (b) $f(0) = 1$, (c) $f\left(\dfrac{\pi}{2}\right) = -1$; **4.** (a) zero, (b) $\dfrac{-1}{2}$; **5.** (a) poles, 2nd order $-\dfrac{1}{2} \pm \dfrac{i\sqrt{3}}{2}$,

(b) 2nd order, $\pm i$, (c) simple poles, $z = k\dfrac{\pi}{2}$, k integer $\neq 0$, (e) simple poles at $i2k\pi$,

$k = 0, \pm 1, \pm 2, \ldots$, (g) poles, 2nd order $\pm i$, (i) pole, 2nd order, $z = 0$.

Section 6.3

1. (a) simple poles at $\pm \dfrac{(1-i)}{\sqrt{2}}$, residues both $\frac{1}{2}$, (c) simple pole $z = 2$, residue $\cos 2$; pole

order 2 at $z = 3$, residue $-\sin 3 - \cos 3$, (d) simple pole $z = 1$, residue 1; **2.** (a) at cis $\frac{2}{3}\pi$,

residue $\dfrac{i}{3\sqrt{3}}$; at cis $\left(\dfrac{-2}{3}\pi\right)$, residue $\dfrac{-i}{3\sqrt{3}}$, (c) at $z = \frac{1}{2}k\pi$, residue $= (-1)^k \frac{1}{2}k\pi$;

k any integer $\neq 0$, (e) residue at $i2k\pi$ is 1,

(g) residue at i is $\dfrac{e^{-i}(1-i)}{4}$, residue at $-i$ is $\dfrac{e^i(1+i)}{4}$, (i) residue $z = 0$ is 1;

3. (a) $\frac{1}{6}$, (c) $1 - \dfrac{1}{2!3!} + \dfrac{1}{4!5!} - \dfrac{1}{6!7!} + \cdots$; **4.** (a) $-\left(k\pi + \dfrac{\pi}{2}\right)$, (c) $\frac{1}{2} - e^\pi$, (e) zero,

(g) -1, (h) e; **6.** (a) $\pi \cos 1$, (c) $2\pi i(e-2)$, (e) $2\pi i\,[2\cosh \pi + 1]$, (g) $\dfrac{\pi i}{\sqrt{2}}$; **8.** (b) 1.

Section 6.4

1. (a) $\dfrac{2\pi}{\sqrt{3}}$, (c) $\dfrac{2\pi}{b}\left[1 - \dfrac{a}{\sqrt{a^2-b^2}}\right]$, (e) $\dfrac{2\pi a}{\left(\sqrt{a^2-b^2}\,\right)^3}$, (f) $\dfrac{2\pi}{3\sqrt{10}}$, (g) $\dfrac{3\pi}{4}$,

(h) $\dfrac{2\pi(-1)^n e^{-|n||a|}}{\sinh|a|}$, $a \neq 0$; **2.** (a) $-\dfrac{\pi}{6}$, (c) $\dfrac{9\pi}{4(\sqrt{20}\,)^3}$, (e) $\dfrac{\pi}{\sqrt{2}}$, (f) $\dfrac{\pi a^2}{1-a^2}$; **3.** (b) V_0;

4. (b) $r\cos\theta$.

Section 6.5

1. (b) no; **2.** (d) no; **4.** (a) $\pi\sqrt{2}$, (b) $\dfrac{\pi}{3}$, (d) $\dfrac{2\pi}{\sqrt{3}}$, (e) $\dfrac{\pi}{6}\sqrt{3}$, (g) $\dfrac{\pi}{4}$; **8.** π.

Section 6.6

1. (a) $\dfrac{\pi}{3}e^{-6}$, (b) $\dfrac{-5\pi}{12}e^{-6}$, (d) zero, (e) $\dfrac{\pi}{5e}[1 + 3\sin 1 - \cos 1]$, (f) $-\dfrac{\pi}{2}e^{-1}\sin 1$

(i) $\dfrac{\pi}{2|d|}\left[e^{-|a+b||d|}\sin[(a+b)c] + e^{-|a-b||d|}\sin[(a-b)c]\right]$;

3. (a) $\pi\left(e^{-3} - \dfrac{e^{-3\sqrt{2}}}{\sqrt{2}}\right)$, (c) $\pi e^{-1}[e^{-i}]$; **4.** (b) πe^{-2};

5. (a) $\dfrac{\pi}{\sqrt{2}}e^{-1/\sqrt{2}}\left[(1-i)\cos\dfrac{1}{\sqrt{2}} - (1+i)\sin\dfrac{1}{\sqrt{2}}\right]$,

(b) $\pi e^{-|\omega|}\cosh 1$ if $|\omega| \geq 1$, $\pi e^{-1}\cosh\omega$ if $|\omega| \leq 1$.

Section 6.7

1. (a) $2\pi i$, (b) $\pi i - 2$; **4.** (a) $\pi\sin 3$, (c) $-(\pi/2)[e^{-2} + \sin 2]$, (d) -1; **5.** (a) π, (b) π.

Section 6.8

2. (a) $\dfrac{\pi\sqrt{2}}{32}\left[\operatorname{Log} 2 - \dfrac{\pi}{4}\right]$.

Section 6.9

1. (a) $\pi e^{-|t|}$ all t, (c) $-2\pi t e^{-t}$, $t \geq 0$; $f(t) = 0$, $t \leq 0$, (e) $\dfrac{\pi}{2}(1+|t|)e^{-|t|}$ all t,

(g) $\dfrac{\pi}{2}e^{-2}\cosh 2t$, $|t| \leq 1$; $\dfrac{\pi}{2}e^{-2|t|}\cosh 2$, $|t| \geq 1$, (i) $f(t) = 0$, $t \geq 2\pi$;
$f(t) = -2\pi\sin t$, $0 \leq t \leq 2\pi$; $f(t) = 0$, $t \leq 0$; **3.** $e^{-i\omega\tau}F(\omega)$;

4. $F(\omega) = \dfrac{e^{-i\omega T/2}}{\pi\omega}\sin\left(\dfrac{\omega T}{2}\right)$;

6. $e^{-i5\omega}F(\omega)$, use $F(\omega)$ from Exercise 4 with $T = 1$; **7.** $v(t) = r(1 - e^{-t/rc})$, $0 \leq t \leq T$;

$v(t) = re^{-t/rc}[e^{T/rc} - 1]$, $T \leq t$; $v(t) = 0$, $t \leq 0$; **8.** $F(\omega) = \frac{1}{2}$ if $|\omega| < \dfrac{2\pi}{T}$, $F(\omega) = 0$ if

$|\omega| > \dfrac{2\pi}{T}$; **11.** $y(x, t) = e^{-|x|}\cosh ct$ if $ct \leq |x|$, $y(x, t) = \Delta e^{-ct}\cosh x$ if $ct \geq |x|$,

assuming $t \geq 0$ all cases.

CHAPTER 7

Section 7.1

1. (a) $\dfrac{e^{-t}}{3} + \dfrac{e^{2t}}{6} - \dfrac{1}{2}$, **(b)** $\dfrac{1}{2}\cos t + \dfrac{3}{2}\sin t - \dfrac{1}{2}\cos\left(\sqrt{3}\,t\right) - \dfrac{\sqrt{3}}{2}\sin\left(\sqrt{3}\,t\right)$,

(c) $e^{-t/2}\left[-\dfrac{1}{\sqrt{3}}\sin\left(\dfrac{\sqrt{3}}{2}t\right) + \cos\left(\dfrac{\sqrt{3}}{2}t\right)\right]$, **(g)** $\frac{1}{2}[t^2 - 2t + 2] - e^{-t}$; **3. (b)** part (ii)

$$\dfrac{-q_0 e^{-(R/2L)t}}{LC\sqrt{\dfrac{1}{LC} - \dfrac{R^2}{4L^2}}}\sin\left(\sqrt{\dfrac{1}{LC} - \dfrac{R^2}{4L^2}}\right)t, \text{ part (iii) } \dfrac{-q_0 t}{LC}e^{-t/\sqrt{LC}};$$

4. (b) $X_1(s) = \dfrac{(s)(s^2 + 1)}{s^4 + 3s^2 + \left(\dfrac{3}{2}\right)}$,

(c) $x_1(t) = \dfrac{1}{2\sqrt{3}}\left[(-1 + \sqrt{3})\cos\left(\sqrt{\dfrac{3 - \sqrt{3}}{2}}\,t\right) + (1 + \sqrt{3})\cos\left(\sqrt{\dfrac{3 + \sqrt{3}}{2}}\,t\right)\right]$;

5. (b) $I_1(s) = \dfrac{\frac{4}{3}}{s^2 + \frac{8}{3}s + \frac{4}{3}}$, **(c)** $i_1(t) = e^{-(2/3)t} - e^{-2t}$; **10. (a)** $f(t) = (t - 1)$ for $t > 1$,

(b) $f(t) = 0$ for $0 < t < 1$.

Section 7.2

1. (a) yes (bounded), **(c)** yes, **(e)** no, **(g)** no; **2. (b)** decay $\beta > 0$, grow $\beta < 0$;
3. (a) unstable (marginally), **(b)** unstable, **(c)** stable, **(g)** unstable; **4. (a)** $\cos(t)$ is possible;
5. $(1/2)e^{-s}$, stable; **7.** stable.

Section 7.3

2. (a) stable, **(b)** unstable (marginally), **(c)** stable, **(d)** unstable, **(e)** unstable; **6. (c)** stable,
(d) unstable, **(e)** unstable.

CHAPTER 8

Section 8.2

2. $v = \dfrac{2au}{1 - a^2}$ is image of $y = ax$; **3. (b)** 1, **(c)** $i2k\pi$; **4. (c)** angle not preserved;

5. (b) $\left|z - \dfrac{1}{2} + \dfrac{i}{2}\right| = \dfrac{1}{\sqrt{2}}$; **6. (a)** 0.05, **(b)** 0.0488; **7. (a)** 0.0739, **(b)** 0.0778.

Section 8.3

1. (a) one boundary is $x^2 - y^2 = u_1$; another is $xy = \dfrac{v_1}{2}$; there are two others; mapping

is one to one; **3. (a)** region lying on and between appropriate branches of the hyperbolas

$$\dfrac{u^2}{\cos^2 a} - \dfrac{v^2}{\sin^2 a} = 1 \text{ and } \dfrac{u^2}{\cos^2 b} - \dfrac{v^2}{\sin^2 b} = 1; \text{ mapping is one to one, (b) mapping not}$$

one to one, consider boundaries, **(c)** mapping one to one onto the half space $\text{Im}\, w \leq 0$;

5. (b) $z = -i \text{Log}[iw + (1 - w^2)^{1/2}]$ where $(1 - w^2)^{1/2} = 1$ for $w = 0$, branch cuts extend from $w = \pm 1$ into lower half plane; **7.** (a) half space $\text{Im} \, w < 0$, (b) $\dfrac{1 - w^{1/2}}{1 + w^{1/2}} = z$ use $w^{1/2} = 1$ when $w = 1$ and take branch cut as positive imaginary axis.

Section 8.4

5. (a) $w = \dfrac{az + b}{bz + a}$; **6.** (a) $\left(u - \frac{1}{2}\right)^2 + \left(v + \frac{1}{2}\right)^2 = \frac{1}{2}$, (c) $v = u - 1$; **7.** (a) $\left(u - \frac{1}{2}\right)^2$ $+ \left(v + \frac{1}{2}\right)^2 = \frac{1}{2}$, (c) $\left(u + \frac{1}{2}\right)^2 + \left(v + \frac{1}{2}\right)^2 = \frac{1}{2}$; **8.** (a) mapped onto $\text{Re} \, w > \frac{1}{2}$, (c) onto $\left|w - \frac{1}{3}\right| < \frac{1}{3}$; **9.** (a) domain between circles $\left|w - \frac{1}{6}\right| = \frac{1}{6}$, $\left|w - \frac{1}{4}\right| = \frac{1}{4}$;

13. $a = a_1 a_2 + c_1 b_2$, etc.; **14.** (b) locus is line $\text{Re} \, A = \frac{1}{2}$; **16.** (a) $\sqrt{2}\left(\dfrac{1}{z} + \dfrac{i - 1}{2}\right) = w$;

17. (b) $w = \dfrac{z^{\pi/\alpha} - i}{z^{\pi/\alpha} + i}$; **18.** (c) $w = \left[\dfrac{-3 - i\sqrt{3} + z(i\sqrt{3} - 1)}{(i\sqrt{3} - z)(1 + i\sqrt{3})}\right]^{3/2}$;

19. (b) $w = e^{-i\pi(z + 2i)/(z - 2i)}$.

Section 8.5

2. (b) $B = T_1$; **3.** $\dfrac{-20i}{\pi} \text{Log}\left(\dfrac{z - 1}{z + 1}\right)$; **4.** (a) $\dfrac{100x}{x^2 + y^2}$; **8.** (d) $\dfrac{2\pi\varepsilon}{\cosh^{-1}\left(\dfrac{A}{k}\right)}$;

10. (a) $\dfrac{100}{\pi} \tan^{-1} \dfrac{4xy}{\left(x^2 + y^2\right)^2 - 1}$ where $0 \le \tan^{-1}(\cdots) \le \pi$;

11. (a) $100\left[\dfrac{1}{2} - \dfrac{1}{\pi} \tan^{-1} \dfrac{\tan\left(\dfrac{\pi}{2}\dfrac{x}{a}\right)}{\tanh\left(\dfrac{\pi}{2}\dfrac{y}{a}\right)}\right]$ where $\dfrac{-\pi}{2} \le \tan^{-1}(\cdots) \le \dfrac{\pi}{2}$.

Section 8.7

2. $w = \dfrac{z^2}{2i}$; **3.** $w = (z^2 + 1)^{1/2}$, connect $z = -i$ and $z = i$ by straight line branch cut;

4. $z = \dfrac{2}{\pi} \text{Log}[iw^{1/2} + (1 - w)^{1/2}] + \dfrac{2}{\pi} w^{1/2}(w - 1)^{1/2}$;

6. (a) $z = \dfrac{-1}{\pi}[\text{Log}(w - 1) - \text{Log}(w + 1)] + i$, (b) $\Phi(w) = -iz$, z given in part (a), $\phi = 1 - (1/\pi) \tan^{-1}\left[2v/(u^2 + v^2 - 1)\right]$;

7. (a) $w = k(1 - e^{\pi z})^{1/2}$, branch cut along negative real z-axis, take $w(z)$ as positive imaginary when z positive real;

(b) $\Phi(z) = \dfrac{i}{\pi}[\text{Log}[(1 - e^{\pi z})^{1/2} - 1] - \text{Log}[(1 - e^{\pi z})^{1/2} + 1]] + 1$.

INDEX